MOLECULAR BIOLOGY
INTELLIGENCE
UNIT

Rise and Fall of Epithelial Phenotype: Concepts of Epithelial-Mesenchymal Transition

Pierre Savagner, Ph.D., D.V.M.
Génotypes et Phénotypes Tumoraux
INSERM
Bâtiment de Recherche en Cancerologie
CRLC Val d'Aurelle-Paul Lamarque
Montpellier Cedex, France

LANDES BIOSCIENCE / EUREKAH.COM
GEORGETOWN, TEXAS
U.S.A.

KLUWER ACADEMIC / PLENUM PUBLISHERS
NEW YORK, NEW YORK
U.S.A.

RISE AND FALL OF EPITHELIAL PHENOTYPE: CONCEPTS OF EPITHELIAL-MESENCHYMAL TRANSITION

Molecular Biology Intelligence Unit

Eurekah.com / Landes Bioscience
Kluwer Academic / Plenum Publishers

Printed in the U.S.A.

Kluwer Academic / Plenum Publishers, 233 Spring Street, New York, New York, U.S.A. 10013
http://www.wkap.nl

Please address all inquiries to the Publishers:
Eurekah.com / Landes Bioscience, 810 South Church Street, Georgetown, Texas, U.S.A. 78626
Phone: 512/ 863 7762; FAX: 512/ 863 0081
http://www.eurekah.com
http://www.landesbioscience.com

ISBN: 0-306-48239-8

Rise and Fall of Epithelial Phenotype: Concepts of Epithelial-Mesenchymal Transition, edited by Pierre Savagner, Landes / Kluwer dual imprint / Landes series: Molecular Biology Intelligence Unit

Library of Congress Cataloging-in-Publication Data

Rise and fall of epithelial phenotype : concepts of epithelial-mesenchymal transition / [edited by] Pierre Savagner.

p. ; cm. -- (Molecular biology intelligence unit)

Includes index.

ISBN 0-306-48239-8

1. Vertebrates--Embryology. 2. Epithelium. 3. Mesenchyme. I. Savagner, Pierre. II. Series: Molecular biology intelligence unit (Unnumbered) [DNLM: 1. Mesoderm--physiology. 2. Cell Differentiation--physiology. 3. Epithelial Cells--physiology. 4. Phenotype. WQ 205 R5945 2005]
QL959.R57 2005
571.8'636--dc22

2005005382

To Mary, who holds my hand

CONTENTS

EDITOR

Pierre Savagner
Génotypes et Phénotypes Tumoraux
INSERM
Bâtiment de Recherche en Cancerologie
CRLC Val d'Aurelle-Paul Lamarque
Montpellier Cedex, France
Email: psavagner@valdorel.fnclcc.fr
Chapters 2, 8

CONTRIBUTORS

Valérie Arnoux
EMI 0229 INSERM: Génotypes et Phénotypes Tumoraux
CRLC Val d'Aurelle-Paul Lamarque
Montpellier, France
Chapter 8

Alfonso Bellacosa
Developmental Genetics of Melanocytes
UMR 146 CNRS-Institut Curie
Orsay Cedex, France
Chapter 14

Avri Ben-Ze'ev
Department of Molecular Cell Biology
The Weizmann Institute of Science
Rehovot, Israel
Email: Avri.ben-zeev@weizmann.ac.il
Chapter 12

Geert Berx
Unit of Molecular and Cellular Oncology
VIB-Ghent University
Ghent, Belgium
Email: Geert.Berx@dmb.rug.ac.be
Chapter 11

Donald P. Bottaro
Urologic Oncology Branch
Center for Cancer Research, NIH
Bethesda, Maryland, U.S.A.
Email: dbottaro@helix.nih.gov
Chapter 13

Brigitte Boyer
Laboratoire Oncogénése et Régulations Cellulaires
UMR CNRS 146
Institut Curie Section de Recherche Bâtiment
Orsay, France
Email: Brigitte.Boyer@curie.u-psud.fr
Chapter 16

Marc E. Bracke
Department of Experimental Cancerology
University Hospital Ghent
Ghent, Belgium
Email: brackemarc@hotmail.com
Chapter 9

Todd D. Camenisch
Departments of Pharmacology and Toxicology
University of Arizona
Tucson, Arizona, U.S.A.
Chapter 4

Marie Causeret
Centre de Recherches de Biochimie Macromoléculaire
Centre de la Recherche Scientifique
Montpellier Cedex, France
Chapter 19

Sophie Charrasse
Centre de Recherches de Biochimie Macromoléculaire
Centre de la Recherche Scientifique
Montpellier Cedex, France
Chapter 19

Christophe Côme
EMI 0229 INSERM: Génotypes et Phénotypes Tumoraux
CRLC Val d'Aurelle-Paul Lamarque
Montpellier, France
Chapter 8

Franck Comunale
Centre de Recherches de Biochimie Macromoléculaire
Centre de la Recherche Scientifique
Montpellier Cedex, France
Chapter 19

Maralice Conacci-Sorrell
Department of Molecular Cell Biology
The Weizmann Institute of Science
Rehovot, Israel
Chapter 12

Regina M. Day
Pulmonary and Critical Care Division
Tupper Research Institute
Tufts University/New England Medical Center
Boston, Massachusetts, U.S.A.
Chapter 13

Véronique Delmas
Developmental Genetics of Melanocytes
UMR 146 CNRS-Institut Curie
Orsay Cedex, France
Chapter 14

Laurence Denat
Developmental Genetics of Melanocytes
UMR 146 CNRS-Institut Curie
Orsay Cedex, France
Chapter 14

Patrick Fafet
Institut de Genetique Moleculaire de Montpellier
CNRS
Montpellier Cedex, France
Chapter 15

Angelina Felici
Laboratory of Cell Regulation and Carcinogenesis
National Cancer Institute
National Institutes of Health
Bethesda, Maryland, U.S.A.
Chapter 13

Atish Ganguly
Department of Molecular Genetics and Microbiology
University of Massachusetts Medical School
Worcester, Massachusetts, U.S.A.
Chapter 7

Cécile Gauthier-Rouvière
Centre de Recherches de Biochimie Macromoléculaire
Centre de la Recherche Scientifique
Montpellier Cedex, France
Email: Gauthier@crbm.cnrs-mop.fr
Chapter 19

Spiro Getsios
Departments of Pathology and Dermatology and the Robert H. Lurie Cancer Center
Northwestern University Feinberg School of Medicine
Chicago, Illinois, U.S.A.
Chapter 10

Christine Gilles
Laboratory of Tumor
and Developmental Biology
University of Liege
CHU-Sart-Tilman
Liege, Belgium
Chapter 20

Lisa M. Godsel
Departments of Pathology
and Dermatology and the Robert H.
Lurie Cancer Center
Northwestern University Feinberg
School of Medicine
Chicago, Illinois, U.S.A.
Chapter 10

Kathleen J. Green
Department of Pathology
Northwestern University Medical School
Chicago, Illinois, U.S.A.
Email: kgreen@northwestern.edu
Chapter 10

Elizabeth D. Hay
Harvard Medical School
Department of Cell Biology
Boston, Massachusetts, U.S.A.
Chapter 1

Ronald L. Heimark
Department of Surgery
University of Arizona
Tucson, Arizona, U.S.A.
Chapter 4

Laurie G. Hudson
UNM Health Sciences Center
Albuquerque, New Mexico, U.S.A.
Chapter 8

Y. Tony Ip
Program in Molecular Medicine
University of Massachusetts School
of Medicine
Worcester, Massachusetts, U.S.A.
Chapter 7

Nanae Izumi
Department of Cell Pharmacology
Nagoya University, Graduate School
of Medicine
Showa, Nagoya, Japan
Chapter 17

Sylvia Julien-Grille
Developmental Genetics of Melanocytes
UMR 146 CNRS-Institut Curie
Orsay Cedex, France
Chapter 14

Kozo Kaibuchi
Department of Cell Pharmacology
Nagoya University
Graduate School of Medicine
Showa, Nagoya, Japan
Email: kaibuchi@med.nagoya-u.ac.jp
Chapter 17

Hideki Katow
Marine Biological Station
Graduate School of Science
University of Tohoku
Asamushi, Aomori, Japan
Email: hkatow@mail.tains.tohoku.ac.jp
Chapter 6

Scott E. Klewer
Department of Pediatrics
University of Arizona
Tucson, Arizona, U.S.A.
Chapter 4

Michael W. Klymkowsky
Molecular, Cellular and Developmental
Biology
University of Colorado, Boulder
Boulder, Colorado, U.S.A.
Email: michael.klymkowsky@colorado.edu
Chapter 18

Donna F. Kusewitt
Department of Veterinary Biosciences
The Ohio State University
Columbus, Ohio, U.S.A.
Chapter 8

Lionel Larue
Developmental Genetics of Melanocytes
UMR 146 CNRS-Institut Curie
Orsay Cedex, France
Chapters 2, 14

Sonja J. McKeown
Murdoch Children's Research Institute
Department of Paediatrics
Royal Children's Hospital
Parkville, Victoria, Australia
Chapter 3

Robert Moore
Developmental Genetics of Melanocytes
UMR 146 CNRS-Institut Curie
Orsay Cedex, France
Chapter 14

Olivier G. Morali
Developmental Genetics of Melanocytes
UMR 146 CNRS-Institut Curie
Orsay Cedex, France
Chapters 2, 14

Masato Nakagawa
Department of Cell Pharmacology
Nagoya University, Graduate School of Medicine
Showa, Nagoya, Japan
Chapter 17

Donald F. Newgreen
Embryology Laboratory
Murdoch Children's Research Institute
Royal Children's Hospital
Melbourne, Australia
Chapters 3, 20

Raymond B. Runyan
University of Arizona
Departments of Cell Biology and Anatomy
Tucson, Arizona, U.S.A.
Email: rrunyan@u.arizona.edu
Chapter 4

Michael P. Sarras, Jr.
Department of Anatomy and Cell Biology
University of Kansas Medical School
Kansas City, Kansas, U.S.A.
Email: msarras@kumc.edu
and
Department of Cell Biology and Anatomy
Chicago Medical School
North Chicago, Illinois, U.S.A.
Email: msarras@finchcms.edu
Chapter 5

Hiroshi Sato
Department of Molecular Virology and Oncology
Cancer Research Institute
Kanazawa University
Kanazawa, Japan
Chapter 20

Erik W. Thompson
University of Melbourne
Department of Surgery
Melbourne, Australia
Email: ewt@unimelb.edu.au
Chapter 20

Veerle L. Van Marck
Department of Experimental Cancerology
University Hospital Ghent
Ghent, Belgium
Email: Veerle.vanmarck@rug.ac.be
Chapter 9

Frans Van Roy
Unit of Molecular Cell Biology
Department for Molecular Biomedical Research
VIB-Ghent, Belgium
Chapter 11

Marie-Luce Vignais
Institut de Genetique Moleculaire de Montpellier
CNRS
Montpellier Cedex, France
Email: vignais@igm.cnrs-mop.fr
Chapter 15

Gary M. Wessel
Department of Molecular and Cellular Biology
Department of Biochemistry
Brown University
Providence, Rhode Island, U.S.A.
Email: rhet@brown.edu
Chapter 6

PREFACE

Cell phenotype is a comprehensive term describing the general appearance and behavior of the cell. It reflects a dynamic stage of differentiation, proliferation, or apoptosis. It also reflects cell group organization, cell motility and cell interaction status with the local cellular and extracellular environment. The two "classic" cell phenotypes, epithelial and mesenchymal, are distinguished by a number of characteristics. Essentially, epithelial phenotype is defined at the level of a group of cells. Epithelial cells form cohesive groups usually ordered along a bidimensional layer and organized in mono- or pluristratified structures. Epithelial phenotype can be modulated under a variety of physiological and pathological conditions. The rapid and sometimes reversible conversion to a mesenchymal-like phenotype is called epithelial-mesenchymal transition (EMT). This term does not imply that the epithelial cell fully transdifferentiates into a fibroblast, a tissue-specific and differentiated cell type. Rather, it means the cell adopts a fibroblast-like phenotype by going through an activation process presenting common features in a wide range of apparently unrelated situations.

The original concept of EMT arose from in vivo studies characterizing developmental stages involving dramatic phenotype remodeling. The original definition of EMT was later extended, based on in vitro studies of epithelial cells that became individualized motile cells, as part of an activation process stimulated by growth factors or extracellular matrix components. Early examples of in vitro EMT include lens epithelial cell transformation described by Prof. E. Hay (see Chapter 1), FGF-treated NBT-II carcinoma cells,[1] and HGF-treated MDCK cells.[2-4] Many more models have been described since. More recently, the concept of EMT has been extended to describe "dedifferentiation" occurring during pathological events such as chronic fibrosis pathologies affecting the kidney and other organs. It may be more appropriate to use the term epithelial-mesenchymal transformation to describe such pathological situations involving a dysregulation of cell phenotype. These processes probably reflect hyperactivation of signaling pathways, for example marked activation of TGFβ-stimulated pathways in the case of kidney fibrosis.[5] EMT-related transformation taking place during chronic fibrosis will be reviewed in the next edition of this book.

It is attractive to hypothesize that EMT takes place during carcinoma progression, as EMT in vitro typically generates motile and invasive cells that are apparently well suited for cancer progression. Nevertheless, the occurrence of EMT during cancer progression remains controversial (see Chapter 9, Van Marck et al). It has been suggested that EMT might take place during basement membrane invasion and intravasation stages by carcinoma cells.[6-8] Beyond the clinically rare example of carcinosarcoma in which there is good evidence for EMT, observations in clinical carcinoma samples do not provide clear indications of EMT, but rather emphasize the complexity of tumor structure and organization in vivo. Carcinoma cells typically express marked phenotypic heterogeneity related to tumor type and stage. Generally, tissue architecture and cell organization are significantly disrupted, in association with a significant remodeling of cytoskeleton, cell-cell and cell-matrix adhesion structures in tumor cells. Some tumors, such as those of invasive lobular carcinoma of the breast, no longer express cell-cell adhesion structures, a clear EMT-like transformation. However, there is no evidence these tumors are more aggressive than invasive ductal carcinoma, the dominant breast carcinoma type, which still expresses E-cadherin at the protein level.

In fact, it is possible that only limited numbers of tumor cells undergo EMT-like events to generate individualized cells responsible for invasion and metastasis. However, this has been difficult to demonstrate. More typically, tumor cells show partial downregulation of cell-cell adhesion structures while migrating as sheets, cords, tubules or isolated cells. Maintenance of partial but decreased cohesiveness is also found in motile cell populations during physiological events such as wound healing (see Chapter 8) or branching and tubulogenesis during organogenesis. Such a mechanism could be involved in the "mass invasion" process seen in solid tumors. It appears that maintenance of some level of cell-cell adhesion in invasive tumors can actually serve as a strategy for tumor progression and metastasis.[9,10]

In this book, the concept of EMT is first described by the pioneer in the field, Prof. E. Hay, in Chapter 1. EMT concept rised from a combination of in vitro and in vivo observations. Accordingly, Part I reviews early morphogenetic events involving EMT in various animal models. The first developmental stage to involve EMT is gastrulation, when the mesoderm first emerges. This very intricate process is reviewed in three different chapters, covering mouse (Morali et al, Chapter 2), Drosophila (Ganguly et al, Chapter 7) and sea urchin (Wessel, Chapter 6). The second classic and well documented EMT example occurs during emergence and migration of neural crest cells from the neural tube. Neural crest cells individualize and emigrate from a cohesive epithelial sheet, the neuroepithelium, to undergo wide-ranging migrations before undergoing further differentiation into a variety of cell types (Newgreen, Chapter 3). A third classic example of EMT is provided by heart morphogenesis in vertebrate embryos. The multistep differentiation of the mitral and tricuspid valves and the interventricular septum includes an EMT step that has been extensively studied in chick models (Runyan et al, Chapter 4). In these three examples of EMT during development, extracellular matrix components are crucial in providing an appropriate substrate to differentiating cells.

The role of extracellular matrix in cell plasticity is also described during epithelial regeneration in hydra, a diblastic animal devoid of mesenchyme (Chapter 5, Sarras). It may be surprising to find a chapter on epithelial regeneration in diblastic organisms in this book, but the whole process shows interesting similarities to developmental EMT and to cutaneous wound healing in vertebrates, another process involving EMT-like stages. Reepithelialization is initiated by a cell activation stage, including partial and transient cell-cell dissociation and cell migration (see Chapter 8, Arnoux et al). It is striking that this physiological process involves molecular pathways that are known for their involvement in other instances of EMT.

Characterizing the pathways driving EMT is key to understanding the whole process and is therefore the focus of Part II. Because of the diversity of physiological and pathological situations in which EMT occurs, it is still problematic to define "EMT-specific" genes or pathways. Initiation of EMT is typically linked to downregulation of cell-cell adhesion structures, particularly desmosomes and adherens junctions. Regulation of desmosomes is reviewed in Chapter 10 (Getsios et al). Desmosomes are sturdy cell-cell adhesion structures that show unexpected plasticity. One of the best known cell-cell adhesion molecules is E-cadherin, a critical component of the adherens junction. Its role in maintaining cell-cell cohesiveness has been demonstrated both in vitro and in vivo. Its integrity depends on associated proteins, the catenins, that link E-cadherin to the cytoskeleton. Regulation of the main molecular components of adherens junctions, E-cadherin and catenins, is reviewed in Chapters 11 (G. Berx et al), and 12 (M. Conacci-Sorrell et al). Destabilization of cell-cell adhesion structures is a general feature of EMT and commonly involves downregulation of these components at protein or RNA level. Induction of EMT typically involves growth factors, including members of HGF (Chapter 13, R.M. Day et al), FGF, IGF (Chapter 14, S. Julien-Grille et al) and TGFβ (Chapter 15, M.L. Vignais et al) families. Mesenchymal cells are often responsible for secreting these factors. The specificity of the induction signal depends on the ability of the target cell to recognize these growth factors through specific receptors, usually tyrosine-kinase receptors. The growth factor signal typically activates ras and sarc pathways, as described in Chapter 16 (B. Boyer). In response, the cytoskeleton undergoes dramatic remodeling mediated by the Rho family members and associated proteins (Chapter 17, M. Nakagawa). Expression of cytokeratin intermediate filaments is altered at the transcriptional level, and the actin cytoskeleton undergoes drastic reorganization, reflecting the induction of motility. Accordingly, the microtubule network is also modulated, as reviewed in Chapter 19 (C. Gauthier-Riviere). A distinct EMT pathway described both in vitro and in vivo implicates a large family of extracellular activators, known as Wnt, as signals for EMT. Their complex biological functions, mediated by the Frizzled family of receptors, include developmental EMT, reviewed in Chapter 18 by M. Klymkowsky. Activation of matrix metalloproteases is usually associated with EMT. Matrix metalloprotease substrates include extracellular matrix molecules as well as membrane proteins involved in cell-cell adhesion, as reviewed in Chapter 20 by C. Gilles et al. Eventually, induction of EMT is controlled at the transcriptional level and the EMT response, in most cases, depends on members of the snail family of transcription factors, described in several chapters. However, none of the pathways described above is functionally specific for EMT, since these pathways can also be involved in unrelated functions such as cell proliferation or apoptosis, in other circumstances. The specificity of these pathways for EMT induction is apparently provided by the cellular environment of the target cells and the timing of the signaling.

In conclusion, this book attempts, for the first time, to bring together multiple perspectives on EMT. It is now very clear that the concept of EMT links processes involving common molecular pathways and can bring valuable conceptual tools to the study of dynamic processes ranging from early embryo development to wound healing and carcinoma progression. The field is expanding very rapidly with the institution of an international conference in 2003 and the recent establishment of TEMTIA (The EMT International Association: http://www.magicdatabases.com/TEMTIA/temtia.html). Some newer aspects of EMT, including molecular pathways and potential inducers have been omitted in this book and will be included in an updated version. I am personally very thankful to the numerous authors who contributed to this multidisciplinary effort.

Acknowledgments

I thank D.F. Kusewitt and M. Savagner for helpful suggestions. Financial support is provided by the Fondation de France and Ligue Nationale contre le Cancer.

References

1. Boyer B, Tucker GC, Valles AM et al. Rearrangements of desmosomal and cytoskeletal proteins during the transition from epithelial to fibroblastoid organization in cultured rat bladder carcinoma cells. J Cell Biol 1989; 109:1495-1509.
2. Li Y et al. Effect of scatter factor and hepatocyte growth factor on motility and morphology of MDCK cells. In Vitro Cell Dev Biol 1992; 28A:364-8.
3. Weidner KM, Sachs M, Birchmeier W. The met receptor tyrosine kinase transduces motility, proliferation, and morphogenic signals of scatter factor/hepatocyte growth factor in epithelial cells. J Cell Biol 1993; 121:145-54.
4. Uehara Y, Kitamura N. Expression of a human hepatocyte growth factor/scatter factor cDNA in MDCK epithelial cells influences cell morphology, motility, and anchorage-independent growth. J Cell Biol 1992; 117:889-94.
5. Zeisberg M et al. Renal fibrosis: Collagen composition and assembly regulates epithelial-mesenchymal transdifferentiation. Am J Pathol 2001; 159:1313-21.

6. Savagner P, Boyer B, Valles A et al. Modulations of the epithelial phenotype during embryogenesis and cancer progression. Cancer Treat Res 1994; 71:229-249.
7. Birchmeier W, Birchmeier C. Epithelial-mesenchymal transitions in development and tumor progression. Exs 1995; 74:1-15.
8. Thiery J. Epithelial–mesenchymal transitions in tumour progression. Nature Rev 2002; 2:442-454.
9. Ikeguchi M, Makino M, Kaibara N. Clinical significance of E-cadherin-catenin complex expression in metastatic foci of colorectal carcinoma. J Surg Oncol 2001; 77:201-7.
10. Graff JR, Gabrielson E, Fujii H et al. Methylation patterns of the E-cadherin 5' CpG island are unstable and reflect the dynamic, heterogeneous loss of E-cadherin expression during metastatic progression. J Biol Chem 2000; 275:2727-32.

Chapter 1

EMT Concept and Examples from the Vertebrate Embryo

Elizabeth D. Hay

Abstract

Epithelial-mesenchymal transformation (EMT) creates a family of invasive cell types from the relatively sedentary epithelial cells that line the surfaces of the body. The mesenchymal cell's primary trait is that, unlike the epithelium of origin, it can invade extracellular matrix and migrate great distances in the embryo. It is a bipolar cell with a very active front end rich in filopodia that apparently drive motility by providing new actin cortex for the myosin endoplasm of the fibroblast to slide forward on. In this chapter, we describe the progression of EMTs as they actually occur in the vertebrate embryo from primitive streak stages to craniofacial remodeling stages in older embryos. We propose a mechanism of TGFβ driven LEF-1 transcription that may be responsible for most, if not all, embryonic EMTs that result in formation of fully active, invasive mesenchymal cells, and we emphasize the importance of studying physiologically relevant signal transduction pathways that lead to the acquisition of invasive motility in vivo, rather than pathways that give rise to nonmotile, stress-fiber rich cells in vitro.

Introduction

The importance of epithelial mesenchymal transformation to the development and evolution of the embryos of the higher vertebrates is enormous. EMT is the process enabling the chordate phylum to evolve beyond its ancestral epithelial architecture. For example, all tissues of our adult chordate ancestor, Amphioxus, are epithelial. Even the axial skeleton is an epithelial rod with the characteristics of true epithelium discussed by Morali et al in Chapter 2 of this book. What the creation of mesenchyme did first for the evolving chordate was to provide a significant source of inwandering cells, derived from epithelium, that invades the extracellular matrices (ECM) separating the epithelial sheets and produces local connective tissue, as in the heart (see Chapter 4, Runyan et al), or that migrates some distance within the embryo, before differentiating into cartilage, bone, and other connective tissues suitable to support the enlargement of the new animals.

As these enlarging animals came out of water onto land, the amphibians continued to lay their eggs in ponds, but the higher vertebrates called amniotes developed substitute ponds, called amniotic sacs, enclosing their embryos and allowing their eggs to develop on land or inside the mother. A mechanism had to be invented, however, to replace the mesoderm, ectoderm and endoderm germ layers that formerly originated from the outside of the egg and now were part of the avian and mammalian extraembryonic membranes. This second major function of mesenchyme in higher vertebrates is accomplished by using mesenchymal cells from the primitive streak to recreate by mesenchymal to epithelial transformation (MET) the

Rise and Fall of Epithelial Phenotype: Concepts of Epithelial-Mesenchymal Transition, edited by Pierre Savagner.

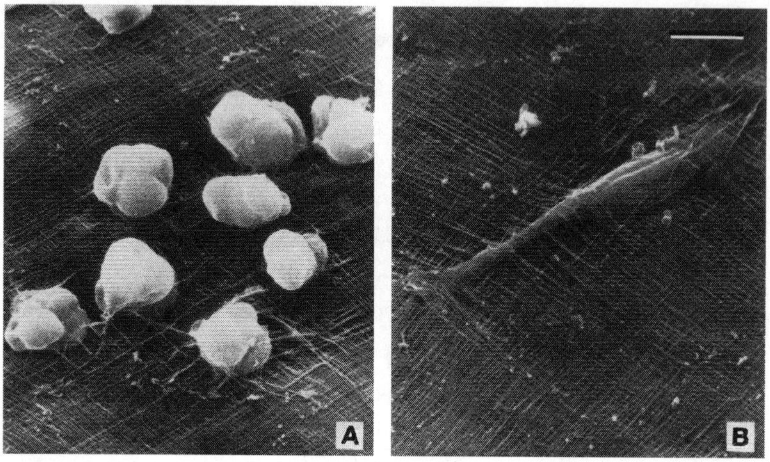

Figure 1. The response of the mesenchymal cell to ECM is different than that of the epithelial cell. In the experiment shown here, epithelial cells (A) and fibroblasts (B) were isolated as single cells from larval frog skin. They were then placed separately on the acellular dermis of the tadpole, an ECM consisting of many layers of orthogonally arranged collagen fibrils. The epithelial cells assumed a cuboical shape and a flat basal surface on top of the matrix (A). The mesenchymal cell, however, elongated and moved into the layers of collagen fibrils (B). Reprinted from Overton, 1977.[1]

mesodermal, ectodermal, and endodermal epithelia in the right places (enabling ontogeny to repeat phylogeny).

Thirdly, as a result of the evolution of neural crest mesenchyme, the peripheral nervous system, the pigment system, and many endocrine glands now formed, followed by other desirable structures such as crest-derived facial connective tissue (Amphioxus has no head). Separately from neural crest, a fourth EMT mechanism occurs much later to remodel vertebrate facial features following fusion of various embryonic primordia to form the lip, nose, and palate.

In this chapter, we shall describe these embryonic EMT's in more detail and discuss a few of the molecular mechanisms that have actually been shown to operate in embryonic tissue transformations such as these.

Formation of the Mesenchymal Cell in the Embryo

Epithelial cells are programmed to attach to the ECM, usually on top of it. Unlike mesenchyme, they do not invade ECM as single cells. If they are isolated by trypsin or E.D.T.A. as single cells and set down on top of ECM (Fig. 1A), they round up[1] and flatten[2] their basal surfaces, and gradually move together to form contiguous sheets on top of ECM. On the contrary, when an isolated mesenchymal cell is placed on top of ECM, it elongates and develops a front end (Fig. 1B, upper right) that invades the ECM.[1]

The epithelium from which newly forming embryonic mesenchyme arises in vivo has a flat basal surface resting on basement membrane (Fig. 2A, bm). When these normally quiescent cells receive the signal to transform into mesenchymal cells, they extend filopodia from their basal surface (Fig. 2A, cell labeled x) and may or may not digest the basement membrane. The cell surface with actin-rich filopodia is the new front end of the mesenchymal cell. It has been shown[3] that newly synthesized actin polymerizes in this front end where actin mRNA localizes (Fig. 2B).

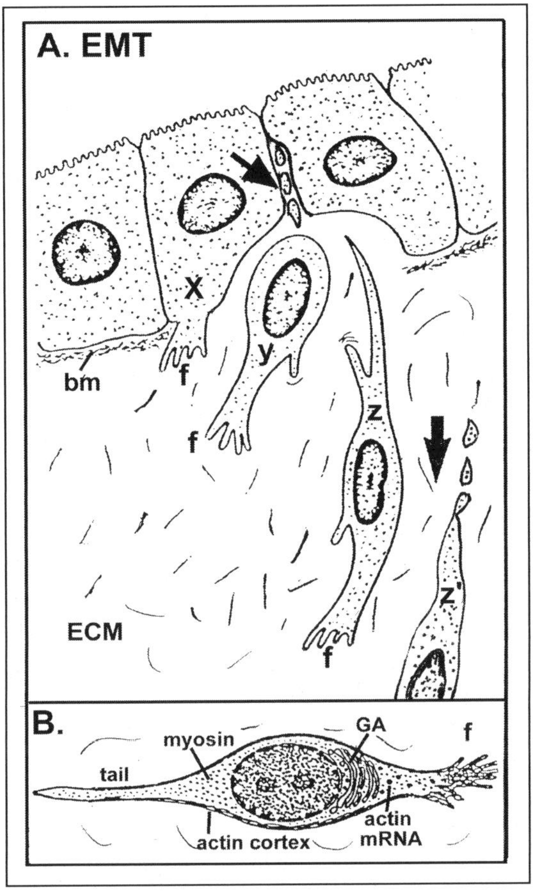

Figure 2. Acquisition of the ability to invade ECM is essential to EMT (A), as well as to subsequent movement through the matrix (B). The activated epithelial cell extends filopodia (f) from its basal surface, as at cell x (A). These become part of the new front end (cell y), where new actin is polymerizing and attaching via ECM receptors (integrins) to the underlying collagen matrix. To get out of the epithelium, cell y has moved its endoplasm into the continuously forming front end, and left its tail end (arrow) behind in pieces in the epithelium (see Bilozur and Hay, 1989). As cell z' moves out, a new tail forms that is also left behind (arrow). B) It has been shown that the Golgi apparatus (GA) and mRNA for actin are in the front end of the migrating fibroblast, where actin polymerizes in filopodia (f) and attaches to the cell membrane via integrins located in the membrane. The Golgi apparatus sends these membrane proteins to the front end (see text).

The actin cortex forms all around the surface of the mesenchymal cell where it attaches via integrins (ECM receptors) to the surrounding collagen fibrils, probably using endogenous fibronectin. $\alpha5\beta1$ has been shown to be the integrin of choice during EMT of lens suspended in 3D ECM.[4] It seems likely that myosin moves the endoplasm forward into the front end by sliding on actin cortex fixed to integrin-ECM complexes.[5,6] Mesenchymal cells contain vimentin intermediate filaments and well developed microtubules and it is possible that by attaching to them, myosin can easily pull whole endoplasm of the cell forward, leaving behind actin cortex in membrane blebs (arrows, Fig. 2) Newly synthesized membrane proteins (Fig. 2B), such as integrins, are sent to the front end of the mesenchymal cell[7,8] from its anterior positioned Golgi complex (GA, Fig. 2B).

Thus, the basal pole of the epithelial cell that is preparing to undergo EMT totally alters its morphology from a sedentary flat surface to a new filopodia-rich front end (cell x, Fig. 2) that is synthesizing a new actin cortex that the mobilized myosin cytoskeleton apparently slides into. What happens to the tail end? It has been shown that the tail end of the transforming mesenchymal cell attached to the adjacent cell at the terminal bar pinches off the cell in blebs.[9] The mesenchymal cell leaving the epithelium becomes pear-shaped (Fig. 2A, cell y) as a result of leaving the tail behind in the lateral compartment (Fig. 2A, upper arrow) of the epithelium of origin. The cell is now ready to invade the underlying ECM using the new front end mechanism and it elongates as its endoplasm slides forward (Fig. 2A, cell z). Moving cells continue to leave cytoplasm behind in blebs (Fig. 2A, cell z') and become pear-shaped.[10] They apparently use the same mechanism[6] they invented to depart from the sticky epithelium to invade sticky matrix.

Definition of the Mesenchymal Cell

This mechanism of movement of the mesenchymal cell in vivo is compatible with its fixed cortex morphology[5] and is the one most likely to explain how a mesenchymal cell leaves the epithelium during EMT.[6] As the transforming mesenchymal cell prepares to leave the epithelium, it extends filopodia into the underlying ECM. Once it leaves the epithelium, and while it still has migratory activity, it is often called a fibroblast. As it moves out into the matrix, it breaks off its rear end and leaves it behind in the epithelium (Fig. 2A). It is important to emphasize that the function of the mesenchymal cell is to invade ECM and that an active mesenchymal cell will always assume an elongated bipolar profile when placed on top of ECM (Fig. 1) or inside it (Fig. 3, lower two cells).

Thus, the best definition of a mesenchymal cell is that it is a bipolar cell with a filopodia-rich front end capable of invading ECM. Vimentin is often used as a chemical marker for mesenchyme, but vimentin also occurs in some epithelial cells.[10] Most of the earlier studies of mesenchymal cells used invasion of ECM as the definition of mesenchyme,[11,12] and we believe this should be continued. At the very least, a mesenchymal cell should be defined as a bipolar cell with motile morphology (including filopodia on the front end). Grown on glass or other planar 2D substrata, fibroblasts flatten, develop stress fibers and ruffled membranes (Fig. 3).

Pollard and associates studied the effects of this morphology on motility. They documentated the fact that cells with stress fibers are unable to migrate.[13] The practice of using the stress fiber to define a mesenchymal cell thus is undesirable. The stress fiber is an in vitro artifact that is never seen in vivo in a fibroblast.[14,15] Signaling pathways (e.g., rhoA) that produce stress fibers in vitro can hardly be expected to operate in vivo or to produce an invasive cell.

Examples of EMT in Higher Vertebrate Embryos

Formation of the Germ Layers in Amniotes

The evolution of the extraembryonic membranes in amniotes enabled their embryos to develop on land in artificial "ponds", but left the egg with only an internal flat disc from which to develop ectoderm, mesoderm, and endoderm. All three germ layers of the bird and mammal are coded in the epiblast (top layer of the disc), which is the equivalent of the presumptive ectoderm, mesoderm, and endoderm lining the outside of the amphibian eggs that relocate during gastrulation to the inside or outside as epithelia.

To relocate these germ layers, the epiblast of the early avian[16] and mammalian embryo develops a linear "primitive" streak of EMT activity (Fig. 4A) that relocates endoderm and mesoderm under the presumptive ectoderm. These mesenchymal cells form epithelia (MET). Axial mesoderm consists of paired epithelial somites that recapitulate the morphology (Fig. 4A) of the mesodermal epithelium of lower chordates. The ventromedial wall of the epithelial somite (Scler., Fig. 5B) gives rise to the sclerotome mesenchyme that differentiates into vertebrata. The dorsolateral wall (Derm., Fig. 5A) forms dermis just as in the frog.

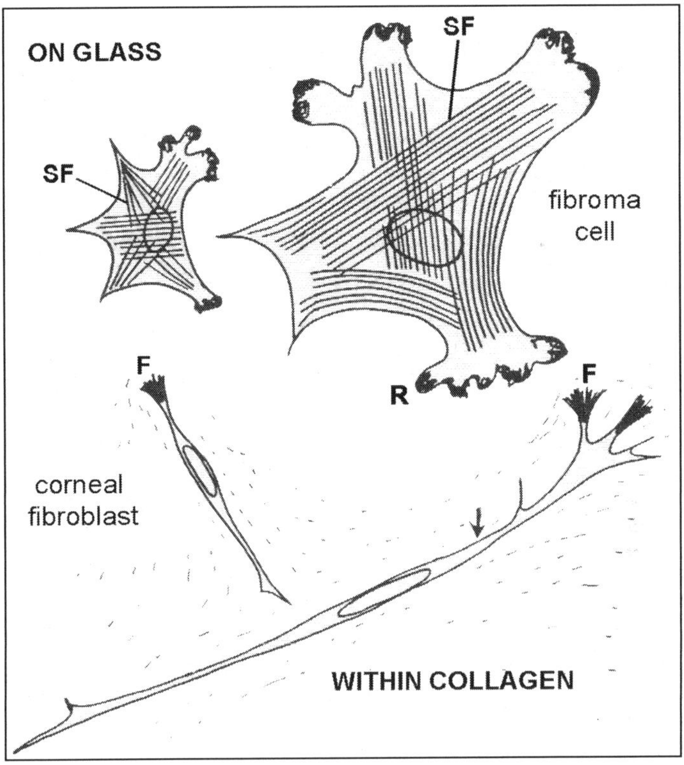

Figure 3. The mesenchymal cell does not develop a migratory morphology when placed on a glass surfaces or other inert planar substrata (top). In this experiment, corneal fibroblasts and a very big mesenchymal cell, the gerbil fibroma cell, were grown for one day on glass, where they developed abundant stress fibers throughout the cytoplasm and multiple ruffling psendopodia. These cells were unable to migrate. When their counterparts were placed within ECM, however, they became bipolar, developed filopodia, and migrated extensively through the ECM (bottom). Reprinted from Hay, 1985.[14]

Not all of the mesodermal epithelia revert to mesenchyme. The myotome (Myo., Fig. 5A) differentiates directly into trunk musculature. The nephric pattern of the lower vertebrate is repeated in the intermediate mesoderm: pronephros first formed, replaced by a mesonephros (emptying into W. Duct, Fig. 5C), which is replaced by the metanephros. All of the kidney epithelia, including that of the late-formed metanephros, appear to derive from primitive streak mesenchyme by MET. This embryonic MET has received intensive study.

Formation of the Neural Crest

The neural crest is a primitive mesenchyme having evolved, as we noted earlier, in the lower chordates. In the vertebrate embryo, it appears in the trunk at almost exactly the same time as the sclerotome mesenchyme (Fig. 5B). As soon as the neural tube closes, its crest begins to express TGFβ3 (Fig. 6E), and the cells send out filopodia to move into the adjacent ECM by the mechanisms described at the beginning of this chapter.[17] Signaling mechanisms for crest EMT have been studied (see Chapter 3, Newgreen and McKeown), and a great deal is known about the specification of crest cell fates.[18] A role for the transcription factors, Slug and Snail, in neural crest formation has recently received considerable attention.[19]

In a study underway currently in our laboratory, Jia Shi and Toni Burbidge have analysed the relation of Slug and TGFβ3 signaling in avian trunk neural crest at 2 days of development.

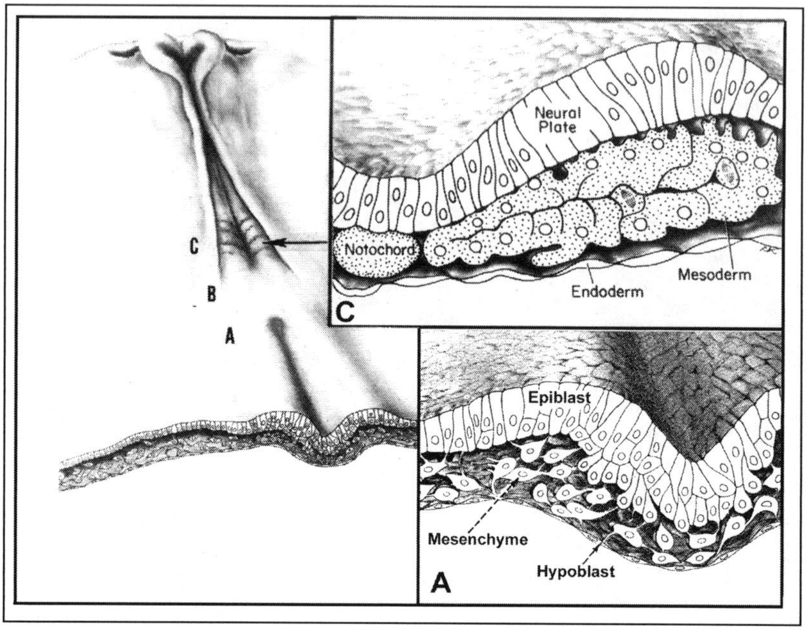

Figure 4. Drawing of a four somite chick embryo (stage 8) sectioned through the primitive streak (inset). The primitive streak (A) gives off mesenchymal cells that create epithelial somites (C) and other mesoderm derivatives after it regresses posteriorly. Thus, MET follows this EMT activity of the epiblast. The reformation of epithelium from mesenchyme involves aggregation (level B) of the cells via E-cadherin, followed by formation (C) of a free surface (inner somite cavity), and laying down of basement membrane on the basal somite surface facing ECM. Reprinted from Hay, 1968.[16]

The experimental approach is to dissect anterior and posterior neural tube (Fig. 6A) separately, as anterior is ahead of posterior in its development. Grown on ECM in vitro, the neural tube sheds neural crest mesenchyme that migrates onto the substratum (Fig. 6B). Administration of exogenous Slug upregulates TGFβ3, a growth factor commonly involved in embryonic EMT (see Chapter 4, Runyan et al and Chapter 15, Vignais and Fafet). In addition to TGFβ3 (Fig. 6D, E), β-catenin and LEF-1 are produced by neural crest during EMT (Fig. 6F, G). LEF-1 is a member of the TCF transcription factor family that has been shown to induce EMT in vitro without the addition of any other factors, except β-catenin and/or Smads (to activate its transcription) (see Chapter 12, Conacci-Sorrell and Ben-Ze'ev and Chapter 18, Klymkowsky).[20]

In a decisive experiment, we treated neural tubes with blocking antibodies to TGFβ3 during the administration of Slug and completely inhibited crest EMT (Shi et al, submitted). We then added exogenous LEF-1, which induced crest EMT, using endogenous β-catenin (Fig. 6F) for its activation. Exogenous LEF-1 induced all aspects of crest EMT, without the need of TGFβ3 or Slug. In the next section, we describe additional experiments on embryos that confirm this important role of LEF-1 in embryonic EMT.

Role of EMT in Remodeling of the Face

In the second half of the embryogenesis of most vertebrates, facial morphogenesis gets underway. Neural crest cells and remnants of the branchial arches are used to create facial features not present in fishes. The maxillary process of the first branchial arch grows across the front of the putative face to fuse with the lateral nasal process and differentiates into the maxilla on each side. The ventral portion of each maxillary process fuses with the intermaxillary segment of the medial nasal process to form the lip. Inside the developing mouth, maxillary processes form the

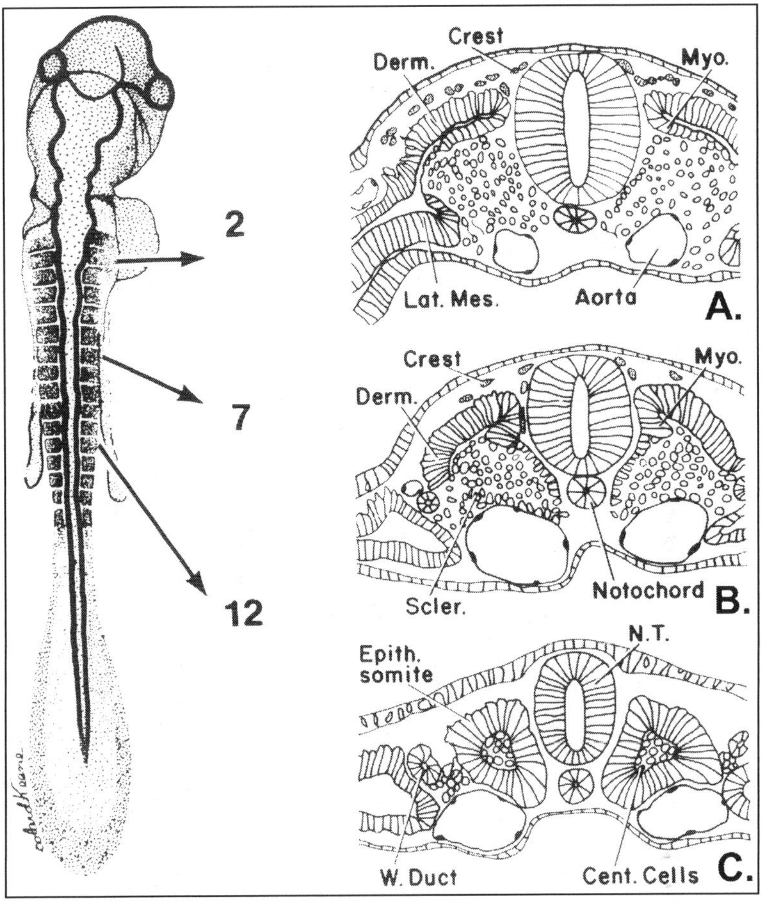

Figure 5. Camera lucida drawings of sections of the 16 somite (stage 12) chick embryo at 2 days. The levels illustrated show (C) the 100% epithelial (Epith.) somite (level of somite 12); B) beginning of EMT from somite sclerotome (Scler.) and the neural tube crest (level of somite 7); and (A) a more advanced stage of mesenchyme formation (level of somite 2). The dermatome will soon become mesenchymal and the only epithelial somite layer left is the myotome. Derm.= dermatome; Myo.= myotome; Lat. Mes.= lateral mesoderm; N.T.= neural tube; W.= Wolffian; Cent.= Central. Reprinted from McCarthy and Hay, 1991.[17]

palatal shelves that subsequently fuse in the midline. Both the lip[21] and the palate (Fig. 7) form epithelial seams along the planes of fusion of these embryonic processes. Initially, it was argued that apoptosis removed such seams from the embryo, but we were able to show in both the lip and the palate that the epithelial seams are removed by EMT.[21-24] The studies to be described now are limited to the palate, because the lip cannot be cultured long enough for meaningful experiments.

The process of EMT in the mouse palatal seam begins about 12 hr after adherence of the two shelves (Fig. 7B). Fitchett and Hay described by TEM,[22] steps in the process of EMT (arrowheads, Fig. 7B). Griffith and Hay[23] developed a carboxyfluorescein cell marker that could be seen at both the light and electron microscope levels to identify the mesenchymal cells and their migratory paths (Fig. 7C, arrowheads). This was extended to avian palates, where it was shown unequivocally that TGFβ3 causes seam EMT.[24] Other studies used DiI to follow palatal EMT in frozen sections.[25] Griffith was able to describe the labeled progeny by TEM.[23]

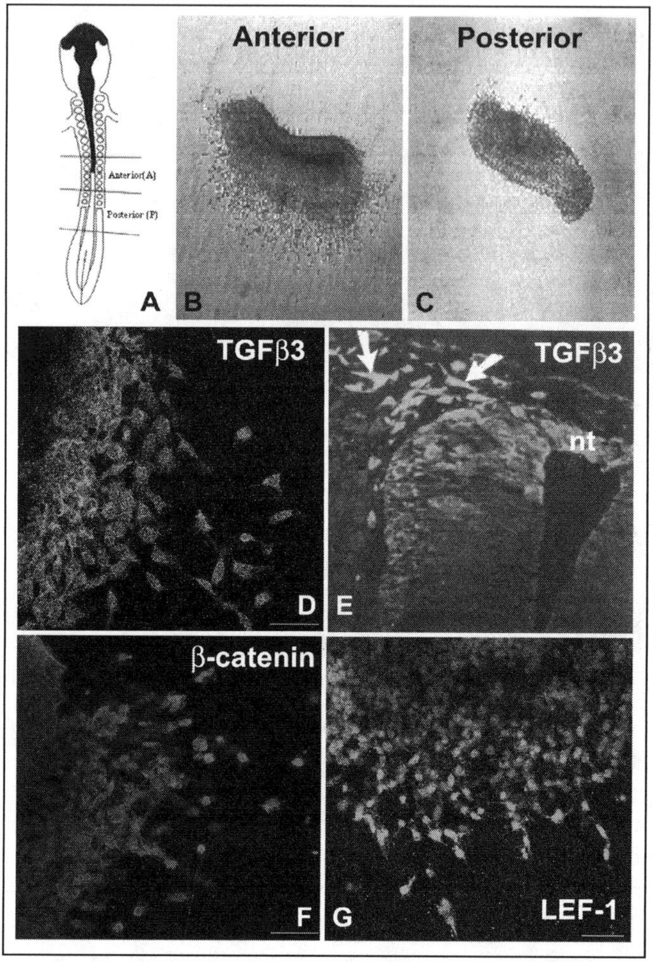

Figure 6. For experimental studies of neural crest in vitro, the neural tube is enzymatically isolated from the stage 12 chick embryo and anterior and posterior pieces (A) set up on collagen substrates (B, C) on which the crest cells will emigrate. This figure shows staining for TGFβ3 in the emigrating crest population on collagen (D) and in a section of the neural tube (E) at the same stage in vivo. β-catenin and LEF-1 are also present in the neural crest at this time (F, G). From Shi et al, 2003. E) is courtesy of Dr. Charles Vanderburg.

A recent study (Nawshad and Hay, submitted) has confirmed the essential role of TGFβ3[23-26] in EMT of the palatal seam and shown for the first time that LEF-1 is the transcription factor that is directly involved in transformation of the palate seam epithelial cell to an actively migrating mesenchymal cell. However, β-catenin does not appear to be the factor activating the transcription of LEF-1 in this case. Immunohistochemical stains for β-catenin show it is not present in palatal epithelial nuclei (Fig. 8C, D). Possible involvement of TGFβ3 in activation of palatal LEF-1 was indicated by an experiment showing that LEF-1 by itself was insufficient to rescue palatal EMT in the presence of antibodies blocking TGFβ3 (Nawshad, unpublished). Active seams expressed abundant phosphoSmad2 (Fig. 8A), but no Smad was present in nuclei of palatal seams treated with TGFβ3 blocking antibody (Fig. 8B). This mechanism of Smad/TCF interaction was anticipated by Labbe et al[27] who recently showed in elegant experiments

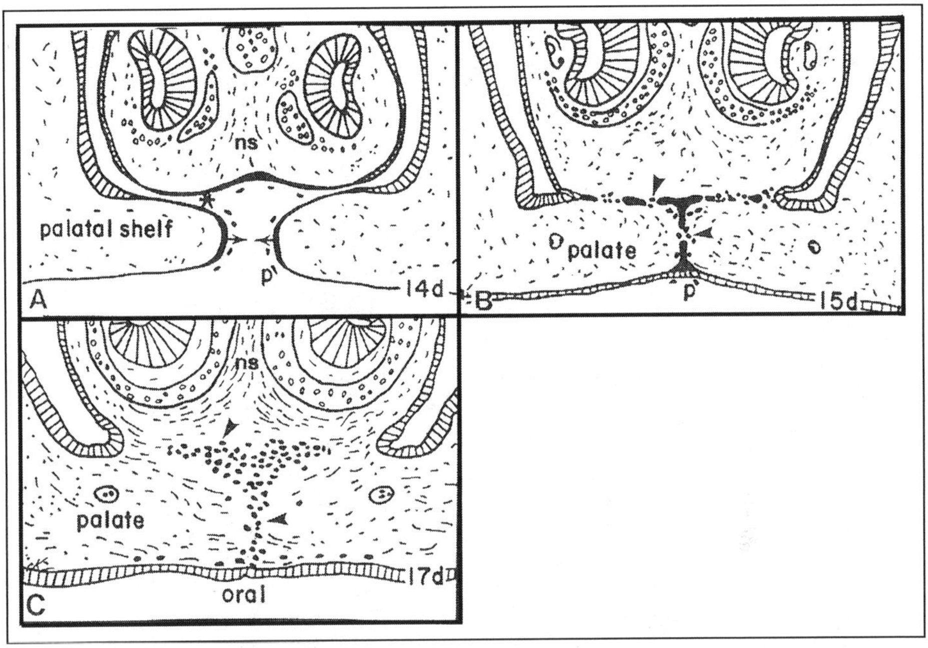

Figure 7. Camer-lucida drawings showing the successive stages in the fusion of the secondary palatal shelves in the mouse. Based on our previous observations (Fitchett and Hay, 1989; Griffith and Hay, 1992) the epithelia contributing mesenchymal cells during palatal fusions are shown in black. A) By 14 days, the palatal shelves have rotated to a horizontal position and are approaching each other. At the time of apposition and fusion of the palatal shelves with each other and with the nasal septum, the outer epithelial layer, the periderm (p), dies and sloughs off prior to contact of the basal epithelial cells. B) The fused basal layer form midline palatal and horizontal palate nasal seams , which break up into epithelial islands that transform into mesenchymal cells (arrowheads, 15d). C) The result is mesenchymal confluence across the palate midline and between the nasal septum (ns, 17d) and dorsal palate, The final location of the epithelium-derived mesenchymal cells (arrowheads, 17d) is based on data presented by Griffith and Hay, 1992. Reprinted from Griffith and Hay, 1992.[23]

that LEF-1 transcription can be activated by either Smads or β-catenin, as well as by a combination of Smads and β-catenin in Xenopus. We also showed that TGFβ3 upregulated LEF1 mRNA, as well as stimulated LEF-1 transcriptional activity in developing palates (Nawshad and Hay, submitted).

While many other pathways have been implicated in EMT in in vitro studies using various cell lines, our study of the palate makes us doubt that any of these are operating in vivo. It has, for example, been reported in vitro that TGFβ stimulates EMT by the RhoA pathway (see also Chapter 15, Vignais and Fafet).[28] The endpoint for EMT in this case was stress fiber formation, not a good indication of physiologically relevant EMT. We found that inhibitors of RhoA and MEK have no significant effect at all on palatal EMT. We confirmed the inhibitory effect of PI3K,[29] but since PI3K is part of the TGFβ pathway, this is to be expected.

It is our hypothesis that the TCF/LEF-1 pathway activated by either β-catenin or Smads or both will prove to be the only pathway that activates all of the physiologically relevant components of EMT (including invasive motility). On the basis of recent experimental studies,[30] we suggest that the Smad pathway is more closely regulated by the cell than is the β-catenin pathway, is less likely to initiate neoplasms, and is the one preferred by embryonic EMT in vivo.[31]

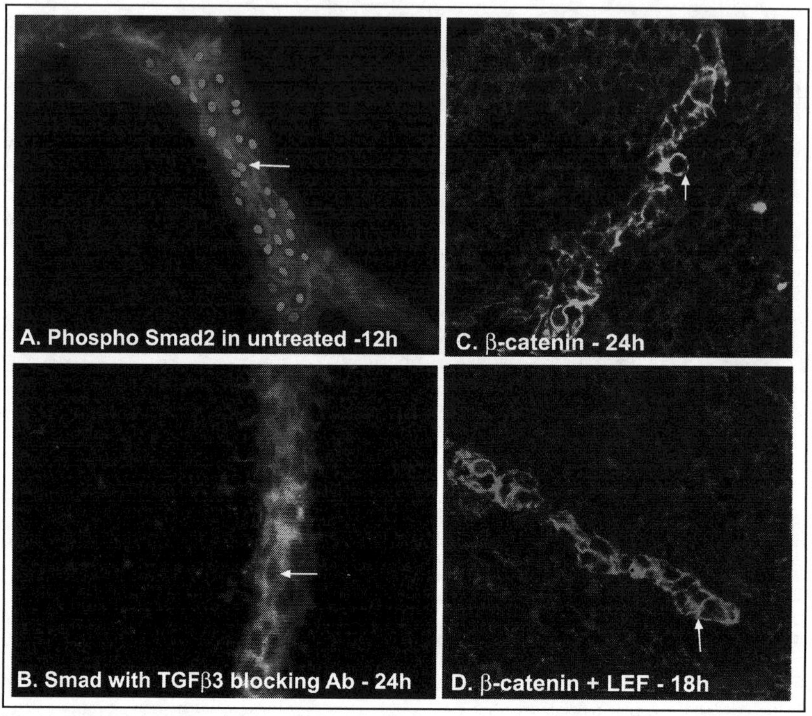

Figure 8. Immunohistochemical localization of Smads and β-catenin in the developing mouse palate. A) The transforming midline epithelial seam at the stage shown in Figure 7B has abundant phospho-Smad2 in all nuclei of epithelial cells about to undergo EMT. B) In palates treated with blocking antibodies to TGFβ3, Smads are not phosphorylated and do not enter the midline seam nuclei, where they would be able to activate LEF-1 and EMT (Labbe et al, 2000). C,D) The palate epithelium contains enough β-catenin to support EMT, but does it does not enter the nucleus where it could activate LEF-1 transcription leading to EMT.

References

1. Overton J. Response of epithelial and mesenchymal cells to culture on basement lamella observed by scanning microscopy. Exp Cell Res 1977; 105:313-323.
2. Sugrue SP, Hay ED. Response of basal epithelial cell surface and cytoskeleton to solubilized extracellular matrix molecules. J Cell Biol 1981; 91:45-54.
3. Lawrence JB, Singer RH. Intracellular localization of messenger RNAs for cytoskeletal proteins. Cell May 9 1986; 45:407-415.
4. Zuk A, Hay ED. Expression of β1 integrins changes during transformation of avian lens epithelium to mesenchyme in collagen gels. Dev Dyn 1994; 201:378-393.
5. Hay ED. Interaction of migrating embryonic cells with extracellular matrix. Expl Biol Med 1985; 10:174-193.
6. Hay ED. Theory for epithelial-mesenchymal transformation based on the "fixed cortex" cell motility model. Cell Motil Cytoskeleton 1989; 14:455-457.
7. Kupfer A, Kronebusch PJ, Rose JK et al. A critical role for the polarization of membrane recycling in cell motility. Cell Motil Cytoskeleton 1987; 8:182-189.
8. Zuk A, Matlin KS, Hay ED. Type I collagen gel induces Madin-Darby canine kidney cells to become fusiform in shape and lose apical-basal polarity. J Cell Biol 1989; 108:903-919.
9. Bilozur ME, Hay ED. Cell migration into neural tube lumen provides evidence for the "fixed cortex" theory of cell motility. Cell Motil Cytoskeleton 1989; 14:469-484.
10. Bard JB, Hay ED. The behavior of fibroblasts from the developing avian cornea. Morphology and movement in situ and in vitro. J Cell Biol 1975; 67:400-418.

11. Behrens J, Vakaet L, Friis R et al. Loss of epithelial differentiation and gain of invasiveness correlates with tyrosine phosphorylation of the E-cadherin/β-catenin complex in cells transformed with a temperaturesensitive v-SRC gene. J Cell Biol 1993; 120:757-766.
12. Greenburg G, Hay ED. Epithelia suspended in collagen gels can lose polarity and express characteristics of migrating mesenchymal cells. J Cell Biol 1982; 95:333-339.
13. Herman IM, Crisona NJ, Pollard TD. Relation between cell activity and the distribution of cytoplasmic actin and myosin. J Cell Biol 1981; 90:84-91.
14. Hay ED. Extracellular matrix, cell polarity and epithelial- mesenchymal transformation. In: GM Edelman, ed. Molecular Determinants of Animal Form (UCLA Symposia on Molecular and Cellular Biology). New York: Alan R. Liss, Inc., 1985:31:293-318.
15. Walpita D, Hay ED. Studying actin-dependent processes in tissue culture. Nat Rev Mol Cell Biol 2002; 3:137-141.
16. Hay ED. Organization and fine structure of epithelium and mesenchyme in the developing chick embryo. In: Fleischmajer R, Billingham RE, eds. Epithelial-mesenchymal interactions. Baltimore: Williams and Wilkins Co., 1968:31-55.
17. McCarthy RA, Hay ED. Collagen I, laminin, and tenascin: Ultrastructure and correlation with avian neural crest formation. Int J Dev Biol 1991; 35:437-452.
18. LeDouarin N. The Neural Crest. Cambridge: Cambridge Univ Press, 1982:259.
19. Nieto MA. The snail superfamily of zinc-finger transcription factors. Nat Rev Mol Cell Biol 2002; 3:155-166.
20. Kim K, Lu Z, Hay ED. Direct evidence for a role of β-catenin/LEF-1 signaling pathway in induction of EMT. Cell Biol Int 2002; 26:463-476.
21. Sun D, Baur S, Hay ED. Epithelial-mesenchymal transformation is the mechanism for fusion of the craniofacial primordia involved in morphogenesis of the chicken lip. Dev Biol 2000; 228:337-349.
22. Fitchett JE, Hay ED. Medial edge epithelium transforms to mesenchyme after embryonic palatal shelves fuse. Dev Biol 1989; 131:455-474.
23. Griffith CM, Hay ED. Epithelial-mesenchymal transformation during palatal fusion: Carboxyfluorescein traces cells at light and electron microscopic levels. Development 1992; 116:1087-1099.
24. Sun D, Vanderburg CR, Odierna GS et al. TGFβ3 promotes transformation of chicken palate medial edge epithelium to mesenchyme in vitro. Development 1998; 125:95-105.
25. Shuler CF, Halpern DE, Guo Y et al. Medial edge epithelium fate traced by cell lineage analysis during epithelial-mesenchymal transformation in vivo. Dev Biol 1992; 154:318-330.
26. Proetzel G, Pawlowski SA, Wiles MV et al. Transforming growth factor-beta 3 is required for secondary palate fusion. Nat Genet 1995; 11:409-414.
27. Labbe E, Letamendia A, Attisano L. Association of Smads with lymphoid enhancer binding factor 1/T cell-specific factor mediates cooperative signaling by the transforming growth factor-β and wnt pathways. Proc Natl Acad Sci USA 2000; 97:8358-8363.
28. Bhowmick NA, Ghiassi M, Bakin A et al. Transforming growth factor-β1 mediates epithelial to mesenchymal transdifferentiation through a RhoA-dependent mechanism. Mol Biol Cell 2001; 12:27-36.
29. Kang P, Svoboda KK. PI-3 kinase activity is required for epithelial-mesenchymal transformation during palate fusion. Dev Dyn 2002; 225:316-321.
30. Di Guglielmo GM, Le Roy C, Goodfellow AF et al. Distinct endocytic pathways regulate TGFβreceptor signalling and turnover. Nat Cell Biol 2003; 5:410-421.
31. Nawshad A, LaGamba D, Hay ED. Transforming growth factor beta signalling in palatal growth, apoptosis and epithelial mesenchymal transformation (EMT). Review. Arch Oral Biol 2004; 49:675-689.

CHAPTER 2

Epithelium-Mesenchyme Transitions Are Crucial Morphogenetic Events Occurring during Early Development

Olivier G. Morali, Pierre Savagner and Lionel Larue

Abstract

Developmental biology constitutes a unique field to study cell dynamics within an organism. Transitions from epithelial to mesenchymal architectures represent major morphogenetic events during development. In this chapter, trophectoderm and mesoderm formation in the mouse is analyzed in detail to exemplify general features of epithelial to mesenchyme transition at the level of the organism, tissues, cells and molecules. As a conclusion, importance of regulation of these processes in embryos and adults is stressed, dysregulations leading to cancer formation and progression.

Introduction

The term "morphogenesis" originates from the combination of two Greek words, "Morphê" (= form, shape) and "Génésis" (= principle, origin). Consequently, it means "origin of forms". Because of this etymological definition, the word has numerous domains of application in Biology. Life sciences, by definition, deal with a dynamic environment, and "morphogenesis" designates the formation and organization, and also the deformation, movement and disappearance of biological objects. The biological objects at the sub-cellular level for example are the different organelles. At the cellular level, morphogenesis is at least involved in two major cellular phenomena, cell division and the general shape of the cell. For tissues, morphogenesis refers to the correct arrangement of the cells forming the tissue. Finally, for whole organisms, morphogenesis designates all the morphological transformations of the organism. In embryology in particular, morphogenesis designates the harmonious transformations leading from the zygote to an adult organism. Developmental biology is consequently a field involving the integration of a series of morphogenetic events that depend on molecular events and that can be analyzed at the sub cellular, cellular and tissue levels.

The study of morphogenesis has been fundamental to the classification of animal species and hence to considerations of evolution. Indeed, phylogenetic divisions are based on the appearance of relative morphogenetic characteristics. In particular, the appearance of intercellular adhesion and the subsequent apparition of different types of supracellular architectures were essential. The first division is based on the appearance of multicellular structures (= colonial protists) from unicellular organisms (= protozoans) as shown in Figure 1. Seventeen known types of colonial protists present development phases during which they form multicellular assemblies. An example of this type of protist is *Dictyostelium discoideum* that forms a multicellular assembly, a pseudo-plasmodium, in absence of nutrients. The molecular mechanism of

Rise and Fall of Epithelial Phenotype: Concepts of Epithelial-Mesenchymal Transition,
edited by Pierre Savagner.

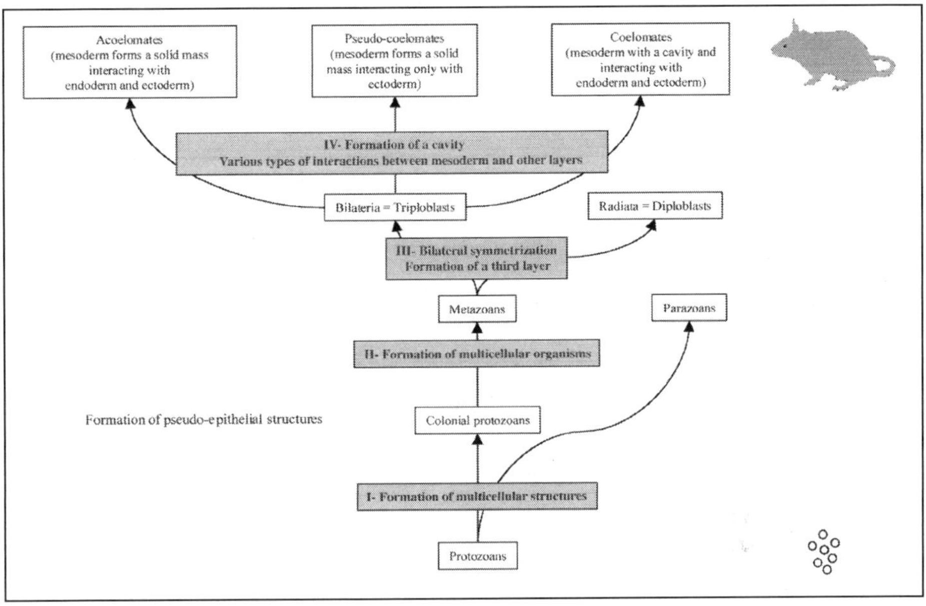

Figure 1. Simplified phylogenetic tree. This simplified phylogenetic tree present the major evolutionary steps from Protozoans to Mammals. The first step consists in the apparition of multicellular structures followed by multicellular organisms. The following steps correspond to the simultaneous apparition of bilateral symmetry and of the first mesenchymal layer, mesoderm. The interaction of mesoderm with the two other layers defines the further subclassification.

intercellular adhesion involves the sequential expression of transmembrane glycoproteins. The initial adhesion is mediated by the expression at the membrane of one glycoprotein.[1] This adhesion is stabilized by the expression of a second membrane glycoprotein.[2] In a third phase, a third membrane glycoprotein is expressed, and mediates the cohesion of particular cells during the migration of the pseudo-plasmodium. Loomis proposed the hypothesis that this third intercellular adhesion system allowed the separation of two cell types: the future spore cells and the future stem cells. Thus, very early during evolution, intercellular adhesion, morphogenesis and cell differentiation are closely related.

The major second step that occurred during evolution was the formation of multicellular organisms, the Metazoans. A supracellular and organized architecture, the epithelium, had appeared (see Chapter 5, Sarras). It is only from after this step was made that the term "embryo" can be used to describe the first developmental stages of an animal. By definition, an embryo is a multicellular assembly in which distinct cell types become individualized (cell differentiation).

The third accepted criterion appearing during evolution, used for classification, is the appearance of bilateral symmetry. This is associated with the appearance of a third basic layer, the mesoderm, initially a loose tissue. This evolutionary step corresponds to the acquisition of a particular supracellular organization in the embryo, and the appearance of the mesenchyme. Thus, there was a transition from the epithelial architecture to the mesenchymal architecture, called the Epithelial-Mesenchymal Transition (or EMT). The presence of the mesoderm and bilateral symmetry distinguishes triblastics from diblastics. Triblastics can be classified into two groups (coelomates and pseudo-coelomates) according to the presence of a cavity within the embryo and specific interactions of the mesoderm layer with the two other basic epithelial layers (ectoderm and endoderm). The coelomates are defined by the presence of a cavity called the coelom limited by epithelial cells. These epithelial cells are mesoderm cell derivatives, corresponding to a Mesenchymal-Epithelial Transition or MET. Coelom development is induced by ectoderm and endoderm cells. The pseudo-coelomates have no such cavity in the mesoderm, which forms a solid mass of cells in contact with ectoderm and endoderm.

Thus, the phylogenetic classification of animals is based on morphogenetic events that are closely linked to the apparition and interconversion of two types of cell architectures: epithelium, a tight aggregation of polarized cells, and the mesenchyme, a loose association of poorly polarized cells. Here, we apply a strict dichotomy between epithelium and mesenchyme. Although this is as arbitrary as any binary representation of complex events, it is legitimized by the general outline of evolution, presented above, and is widely accepted. We will first define epithelium and mesenchyme, and then describe MET and EMT as occurs during early mammalian development. We will consider trophectoderm formation and gastrulation at all levels from the organism to molecular, using the mouse as a reference animal model. Finally, regulatory processes for EMT in embryos will be compared to those in adults.

Supracellular Architecture: The Dichotomy between Epithelium and Mesenchyme

Epithelial Organization: A Cohesive Assembly of Cells

The word "epithelium" has for etymology "épi" (on) and 'thêlê" (nipple). Consequently, "epithelium" designated, at the origin, the aggregation of cell recovering the surface of the nipple. The epithelial organization was invented in the animal reign with the diblastics and allowed the production of tissues and organs. Epithelial organization has two fundamental features: (i) separating two distinct biological compartments, the "interior" and the "exterior" and (ii) conferring to an assembly of cells a transportation function vector. The vectorial character results from the structural and functional polarity of the epithelial cells (Fig. 2).

Epithelial characteristics at the cellular level can be defined by the five following criteria in various in vivo or in vitro study systems: (i) the cohesive interaction between cells allowing the formation of continuous cell layers, (ii) the existence of three types of membrane domains: apical, lateral and basal, (iii) the existence of tight junctions between apical and lateral domains, (iv) the polarized distribution of the different organelles and components of the cytoskeleton and, (v) the quasi immobility of the group of epithelial cells relative to the local environment.

These structural features allow three important types of function: (a) the formation of vast surfaces for exchange (for example, microvilli) and also of cavities by the overall folding of epithelial layers (for example, intestinal tube or nervous tube), (b) the formation of biological compartments of different ionic compositions (low ionic strength and serous), due to selective permeability of the cells, (c) the absorption, the transcytosis and the vectorial secretion of macromolecules.

The ontogeny, the maintenance and the dynamics of epithelial structure require appropriate functioning of the cells individually and the epithelial tissue as a whole. Thus, the functioning of the epithelial cells, and the implementation of their genetic programs, has to be coherent and coordinated both in time and space.

The adhesive systems of the epithelial cells can be classified into cell-cell adhesion involving mainly the lateral sides of the cells, and cell-matrix adhesion involving mainly the basal surfaces of the cells. Cell-cell adhesion confers the cohesiveness of the epithelium, and is a characteristic of epithelia, whereas cell-matrix adhesion is also found for mesenchymal cells. There are several epithelial cell-cell adhesion systems including gap junction, adherens junctions, desmosomes, and tight junctions. Different families of proteins are involved in these different systems.

Mesenchymal Organization

The term "mesenchyme" originates from the Greek words Mésos (environment), In (in) and Chymos (juice). A supracellular mesenchymal architecture can be defined by contrast to a

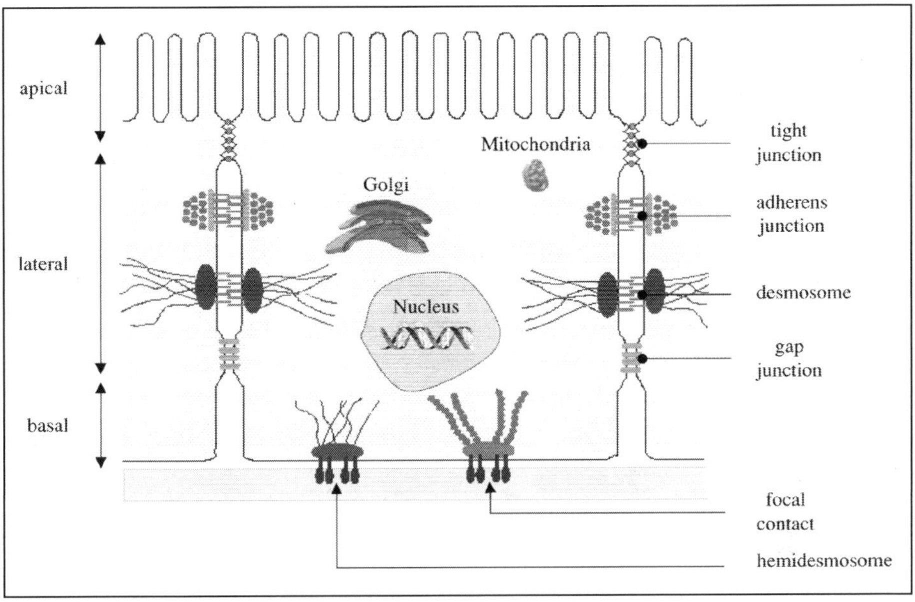

Figure 2. This scheme presents the main characteristics of an epithelial cell: the intercellular adhesion mediated by different types of junctions (tight, adherens, desmosomal, gap), the definition of distinct membrane regions (apical, lateral, basal) and the polarised distribution of organelles. It should be noted that cell-matrix adhesion complexes are located exclusively on the basal region and involves different types of junctions (focal contacts and hemi-desmosomes). In addition, an epithelial assembly is characterised by its immobility and its exchanges of molecules between compartments of different composition.

supracellular epithelial organization as: (i) loose or no interaction between the cells, such that a continuous cell layer is not formed. (ii) The absence of clear apical and lateral membranes. (iii) The nonpolarized distribution of the different organelles and components of the cytoskeleton. (iv) The motility and even invasiveness of the cells.

During development and cancer progression, mesenchymes may be temporary intermediates in the formation of an epithelial structure from another epithelial structure. However, the mesenchymal architecture can be a lasting organization. Mesenchymal functions include support and nutrient supply.

In summary, the epithelial organization is mainly dependent on the tightness of the cell-cell junctions. Cell-cell adhesion is dependent on transmembrane glycoproteins, including the epithelial marker E-cadherin. The mesenchymal status can be considered to be the only, and default, alternative of the epithelial status.

During the genesis of epithelium from individualized cells, four phases can be distinguished (i) an intercellular aggregation step, (ii) a polarization cell step, (iii) a cell differentiation step, and (iv) an integration of the differentiated cells into a functional organization. Thus, a parallel can be drawn between embryonic development and evolution. This general scheme applies to embryonic development as well as to physiological and pathological processes in adult; the gain or loss of the epithelial organization associated with the loss or gain of the mesenchymal status.

Embryonic Morphogenesis: A Harmonious Series of Transitions from One Cellular Architectural Type to Another

Here, we will describe the interconversions between epithelium and mesenchyme during early mouse embryonic development, until gastrulation. In mammals, the first MET occurs

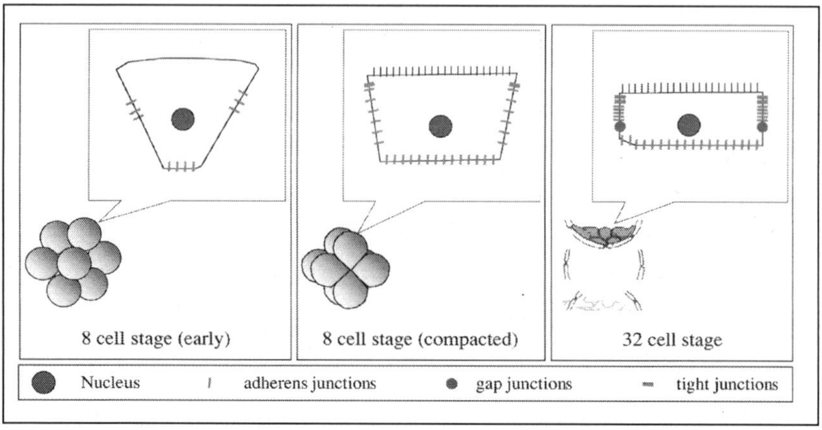

Figure 3. At the early 8 cell stage, blastomeres are characterised by a intercellular adhesion mediated by E-cadherin. At the compacted 8 cell stage, in addition to the formation of additional adherens junctions, some tight junctions appear on the apical side. At the 32 cell stage, gap junctions appear in addition to the existing adherens and tight junctions.

during the preimplantation period with the formation of the trophectoderm and the first EMT during the gastrulation period with the formation of the mesoderm.

The First MET in Embryonic Epithelium Leads to the Formation of the Trophectoderm

The first stages of Metazoans embryonic development involve a series of rapid cell divisions, called the segmentation phase. The cells at this stage are called blastomeres. The morphogenetic characteristics of these first stages are limited, although the cell fates are already determined after the first cell divisions in organisms such as the *Tunicates*, *Caenorhabditis elegans* or sea urchin (see Chapter 6, Wessel). For Mammals, the determination seems to occur at a later stage during development. However, the first real morphogenetic event is the formation of the trophectoderm.

After segmentation, the embryo looks like a small blackberry, mora in Latin; thus, the 8-cell stage is termed the "morula". The shape and morphology of these 8 cells, the blastomeres are highly similar. The morula then becomes a compacted morula with 16 cells; the blastomeres at the periphery flatten progressively, become polarized and all types of junctions appear. These cells become the first embryonic epithelium, the trophectoderm. The cells inside the compacted morula become the inner cell mass (ICM), and are totipotent. The trophectoderm is an active epithelium and gives rise to the blastocoele.[3,4] There are three axes of symmetry: antero-posterior, dorso-ventral and left-right axis. The dorso-ventral axis is defined at the time of the blastocyst. The future dorsal part of the embryo, contained in the ICM, is located at the contact of the blastocoele. A series of cellular and molecular modifications during early development lead to the formation of the trophectoderm and the ICM.

Modulation of Cell-Cell Adhesion

E-cadherin (or Cdh1) is a cell adhesion molecule important during compaction and the formation of the trophectoderm. E-cadherin is present in the zygote as a maternal protein. At the end of the two-cell stage, zygotic E-cadherin production starts. In noncompacted morulas (Fig. 3B), the blastomeres are linked. The maternal and zygotic E-cadherin is progressively redistributed and concentrated on the future baso-lateral side of the blastomeres at the periphery of the future compacted morula.[5,6]

By the 16-cell stage (Fig. 3C), the amount of E-cadherin on the baso-lateral side of the blastomere is greater, and cell-cell adhesion tighter. Noncompacted morulas incubated with anti-E-cadherin antibodies are not able to compact or to form a blastocyst.[5,7-9] E-cadherin knockout mice, Cdh1 -/-, have been produced. These embryos cannot form a functional epithelium and die at the peri-implantation stage. Interestingly, the compaction of Cdh1-/- embryos is not perturbed, indicating that the maternal E-cadherin is sufficient. The zygotic E-cadherin is however required to allow the embryo to form a blastocyst.[10]

Cell Polarization

Scanning electron microscopy analysis of compacted and uncompacted embryos shows that compaction is associated with morphological modifications, such as formation of microvilli scattered in a polarized manner.[11] It appears that the cells polarize after cell-cell contact. When compaction is artificially reproduced by pairing two noncompacted embryo blastomeres, the membranes have two distinct fates. The free membranes form microvilli and the cell-cell interacting membranes stay smooth.[6] The main proteins involved in adhesion at 2-cell stage are the proteins of the cadherin-catenin complex. At the noncompacted, 8-cell stage (Fig. 3B), the membranes of the blastomeres are associated by functional gap junctions, allowing the exchange of ions and small molecules between the cells of the forming epithelial layer. These gap junctions are formed by connexin octamers. In parallel, E-cadherin and ZO-1 are recruited at the cell-cell contact site leading to adherens and tight junctions on the baso-lateral sides of the future trophectoderm epithelial cells. Maternal E-cadherin is redistributed at the cell surface and zygotic E-cadherin and ZO-1 are addressed to the future adherens and tight junctions. On the future apical membrane, endosomes accumulate. Reorganization of cortical and cytoplasmic actin and of microtubules is observed at microvilli.

Cingulin is recruited at the tight junctions at the 16-cell stage (Fig. 3C), before the junctions are functional. Functional adhesive junctions are scattered along the baso-lateral sides, reinforcing regional cell-cell adhesion. Large amounts of protein migrate in the plasma membrane; the apical proteins, such as the Na^+/glucose carrier, and the baso-lateral proteins, such as EGFR (epidermal growth factor receptor) relocate.[12,13] The polarized distribution of the organelles is continues within the cells; the mitochondria and lipid vesicles concentrate on the baso-lateral side, and the lysosomes, the Golgi apparatus and the nucleus at the central basal part of the cells.

Function Acquisition by the Active Epithelium

The tight junctions established in the earliest stages become functional at the 32-cell stage (Fig. 3D). Numerous desmosomes form on the baso-lateral sides by desmogleins and desmoplakins which are synthesized during the morula stage.[14] The size of the blastocoele, and therefore the blastocyst, increases continuously due to the activity of the trophectoderm cells. The resulting pressure on the zona pellucida is responsible, at least in part, for hatching. The hatched embryo then interacts with and implants into the uterus.

The formation of the first embryonic epithelium is associated with the development of both complex cell-cell adhesion machinery and also cellular and the sub-cellular polarization. This first epithelial supracellular architecture subsequently leads to the first mesenchymal supracellular architecture: the mesoderm.

During post-implantation, the trophectoderm regionalizes into distinct epithelia with various morphologies and different rates of proliferation.[15] The parietal trophectoderm is the part of the trophectoderm that is not in contact with the ICM. This cell population stops dividing, although the DNA continues to replicate. The chromosomes of these cells become polytenic, and the cells themselves become giant cells.[16] The visceral trophectoderm is in contact with the ICM. This population of cells contributes to the formation of the extraembryonic ectoderm during gastrulation.

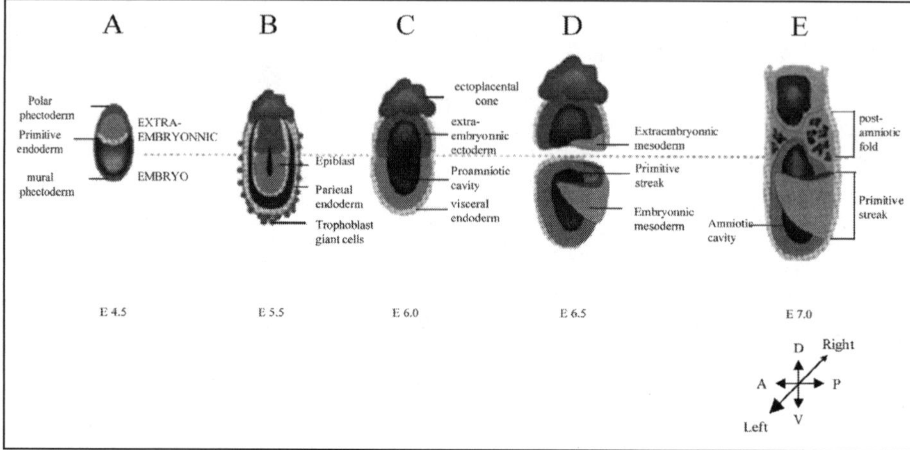

Figure 4. Like in all coelomates, gastrulation in the mouse is characterised by cellular movements occurring between embryonic days E4.5 and E7.5 which enable the reorganisation of the embryonic cells and the formation of the definitive plan of the animal. A to E: the major steps of this reorganisation between E4.5 and E7.0 are shown on schematic sections along the dorso-ventral axis (D-V). In particular, the migration of mesoderm progenitor cells (in red) from the posterior to the anterior side of the embryo between the ectoderm (in blue) and endoderm (in yellow) is presented on sections D and E. Reprinted with permission from Hogan B, Beddington R, Costantini F et al. Manipulating the mouse embryo: a laboratory manual, 2nd ed. 1994:58. ©1994 Cold Spring Harbor Laboratory Press.

Formation of the First Embryonic Mesenchyme: The Mesoderm Layer

Following the formation of trophectoderm, at E4.0, a second epithelial layer is formed at the interface between the cells of the ICM and of the blastocoele. This second epithelium is called the primitive endoderm and contributes to the formation of the visceral endoderm (which remains associated with the epiblast) and of the parietal endoderm, which covers the surface of the blastocoele and the parietal trophectoderm. The parietal endoderm and the visceral endoderm do not contribute to the embryonic endoderm.[17] The blastocyst implants into the uterine wall between E4.5 and E5.5 (Fig. 4). Spatial constraints and the attachment of the blastocyst to the uterine wall cause the cells of the ICM and of the trophectoderm to grow towards the interior of the blastocoele cavity.[18] The cells of the ICM then form a third epithelium: the epiblast. This is accompanied by the formation of a central cavity called the amniotic pro-cavity. This cavity is not the result of the formation of epithelium, but rather of localized cell death.[19] At this stage the embryo, called the "egg cylinder", has the shape of a cup made of two layers: the epiblast, surrounded by the visceral endoderm. The embryonic and extraembryonic regions are clearly distinct, and the dorso-ventral and proximo-distal axes are apparent.

Mouse gastrulation displays highly coordinated epithelium to mesenchyme conversions, cell migration, cell proliferation and differentiation. The general organization of the embryo emerges, with the formation of a new layer, the mesoderm, between the ectoderm and the endoderm. During gastrulation cell populations which were previously separated come together to form tissues and organs, and cells are transiently or permanently changed to allow permissive or instructive induction. The formation of the mesoderm layer involves a transition from the epithelial organization of the epiblast (tight and polarized) to the mesenchymal organization of the mesoderm (loose and apolar).

The mesoderm cells are recruited from the cells of the epiblast, into a structure called the primitive streak. The location of the primitive streak defining the posterior pole and this is the first assignment of an antero-posterior axis to the embryo. The cells recruited into the primitive streak undergo an epithelium-mesenchyme transition, then ingress between the primitive ectoderm and the endoderm form the mesoderm and of part of the endoderm.[20] The newly formed mesoderm cells form a new tissue, which spreads between the endoderm and the ectoderm, symmetrically from the primitive streak like "wings".

Triggering of Gastrulation

Gastrulation does not start before the epiblast contains a certain number of cells.[21-23] The triggering of gastrulation also depends upon a chronological checkpoint.[24-26] Cell proliferation is the main driving force leading to the development of the primary layers.[27-30] Additional morphogenetic forces are involved in the formation of the mesoderm layer, for example ingression of epiblast cells into the primitive streak and their subsequent tendency to move away from it.

Formation of the Primitive Streak

The formation of the primitive streak occurs in the posterior and ventral epiblast, at the junction with the extraembryonic ectoderm.[31] The streak progresses to the extremity of the cylinder. Initiation of the formation of the primitive streak is poorly understood. The streak may lengthen due to division, recruitment and intercalation of cells of the epiblast between the extremities of the developing primitive streak. Thus, the cells at the distal end of the streak are of posterior origin, having been the first recruited when the streak formed (Lawson et al, 1991).

Epithelium to Mesenchyme Transition

There is little available data about EMT in the mouse. Nevertheless, findings for rabbit and rat appear to apply to the mouse. An epithelium-mesenchyme transition occurs in epiblast cells before they penetrate the primitive streak. In the mouse, the epiblast is a pseudo-stratified epithelium constituted of high and columnar cells.[28,32,33] The epiblast presents various characteristics of epithelia, including polarization, specialization of the apical surface with short and separated microvilli.[34,35] junctions (principally adhesive junctions, but also tight junctions, communicating junctions and desmosomes) in the apical portion of the lateral faces and a basal lamina.[14,33]

Loss of Polarization

When EMT begins, organelles in certain cell populations of the epiblast are relocated to the apical face[34,36] This step is called "cytoplasmic hyperpolarization". Concomitantly, an apical constriction appears, with an enlargement distorting and destroying the basal lamina at the basal pole (Fig. 5). As in amphibians, these cells are named "bottle cells".[36] After EMT, cells which separated from the epiblast show mesenchymal features: irregular outlines and apolar distribution of organelles. The cells that penetrate the primitive streak show a characteristic distribution of the Na^+-K^+ ATPase pump. This enzyme is basal in epiblast cells outside of the primitive streak, but is apical in cells in the primitive streak, and is present throughout the cytoplasm of mesoderm cells after they have gone through the primitive streak.[37]

Modification of the Cytoskeleton

During EMT the intermediary filaments of the cytokeratin cytoskeleton of the epiblast cells are replaced by vimentin.[38] If the gene coding for vimentin or cytokeratin 8, an epithelial cytokeratin, is knocked out, the mice do not present any abnormality during gastrulation.[39,40] In cultures of primitive streak stage mouse embryos in vitro, immunoreactivity for desmoplakins and for E-cadherin is lost by epiblast cells going through the primitive streak.[41] Desmoplakins are found in mesoderm cells during migration, suggesting that desmosomal junctions contribute to the cohesion of the migrating mesoderm cell layer.[42]

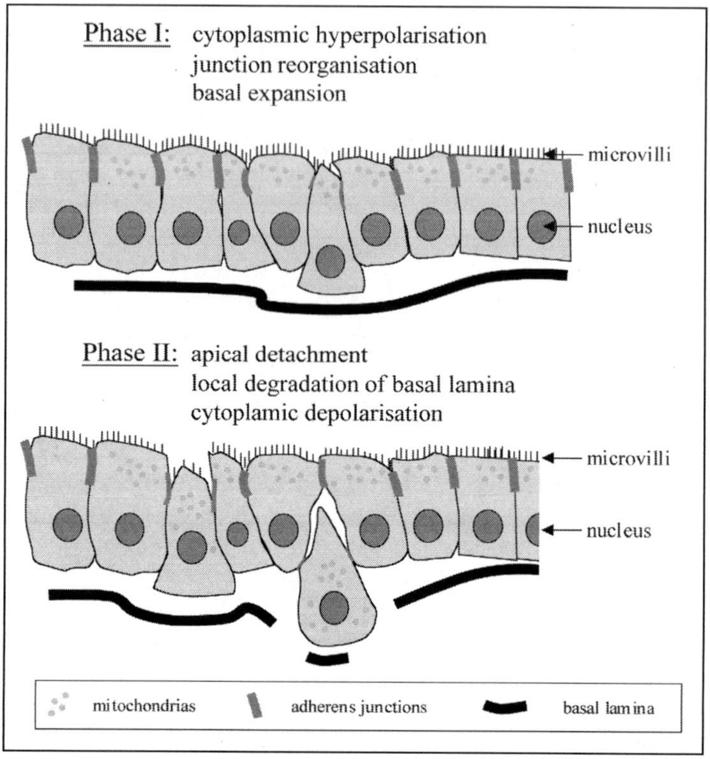

Figure 5. This model for epiblast to mesoderm transition in Mammals was proposed by Viebahn in 1995 from observations using transmission electron microscopy.[36] In phase I, organelles are relocalised to the apical side of the cells and a basal enlargement appears. In phase II, cytoplasmic polarisation is gradually lost and apical detachment of the cells undergoing epiblast to mesoderm transition occurs. This second phase comprises also a local degradation of the extracellular matrix.

Modification of Cell-Cell Adhesion

The reorganization of adhesive junctions during the formation of "bottle cells" has been studied by electron microscopy.[36] The resulting model explains the delamination of mesoderm cells from the epithelium and their ingression into the primitive streak.

The molecular mechanisms of this EMT in vivo are poorly understood because functional studies in the embryo in utero are particularly difficult. In the mouse, the transformation of the epiblast epithelium into mesoderm mesenchyme involves the loss of E-cadherin immunoreactivity.[43] In the chicken, N-cadherin is first expressed by cells penetrating the primitive streak.[44] In the mouse, it is not clear whether there is similar replacement of E-cadherin by another cadherin. For example, cadherin-11 is expressed by mesoderm cells only after they start to migrate.[45,46] β-catenin is a protein linking the conventional cadherins to the actin cytoskeleton. Its knockout results in embryonic death at E7.0, associated with the absence of mesoderm formation.[47] γ-catenin appears to be unable to compensate for β-catenin deficiency at this developmental step. However, in γ-catenin gene knockout mice, functional compensation by β-catenin is possible in certain tissues at E10.5.[48,49]

Epiblast explants have been cultured in vitro.[50] Under basal conditions, they form epithelial clusters of cells, show characteristic adhesive and migratory behavior and express the characteristic markers of the epiblast. Antibodies blocking E-cadherin activity at the surface of explants of the posterior epiblast destroy the epithelial organization and induce a mesenchymal organization. This involves radical changes in the cell-cell and cell-extracellular matrix adhesive properties, the acquisition of far more developed migratory properties and the corresponding changes in the expression of molecular markers for mesoderm and epiblast cells. This EMT is not conventional insofar as it leads to the formation of a "mesenchymal layer", in which mesenchyme cells are not individualized. This mesenchymal layer state is preserved during migration after ingression into the primitive streak. This state represents a feature of mesoderm formation in Vertebrates, not found in Invertebrates.

Modification of the Rate of Cell Proliferation

BrdU staining and a stathmokinetic analysis demonstrated that the mean length of the cell cycle in the rat primitive streak is 3 hours, whereas that of an epiblast cell before penetrating into the primitive streak is 7.5 hours.[30] The difference is due to shortened G1 and G2 phases.

Modification of the Extracellular Matrix

Initially, there is a thin layer of extracellular matrix in the form of a basal lamina between the epiblast and the primitive endoderm. It is constituted of hyaluronic acid, fibronectin, laminin, type IV collagen and thrombospondin.[51-55] The basal lamina is discontinuous, even absent, under bottle cells which go on to form the mesoderm, whereas it is intact at the more posterior part of the epiblast (Fig. 5).[33,35,42]

Aggregation experiments with mutant ES cells with wild-type blastocyst cells demonstrated that the knock-out of the genes coding for the components of the extracellular matrix or for their receptors at the cell membrane such as fibronectin and a5 or b1 integrins does not perturb the behavior of the mutant cells during gastrulation.[56-58]

Modification of Gene Expression

Molecules of the signaling pathways for growth factors such as TGFβ (transforming growth factor β) or FGF (fibroblast growth factor) are involved in mesoderm formation in mouse.[59] T (or brachyury) gene expression is induced and E-cadherin gene expression decreases during mesoderm formation. Knockout mutant mice for T, msd, eed, snail show defects in mesoderm formation and, consequently, defective conversion of the epithelial epiblast into mesenchymal mesoderm see also Chapter 11, Berx and Van Roy.[60-62] The brachyury mutant is particularly interesting because T expression appears to require FGF-2 activity for mesoderm formation, as in amphibians.[63] In mouse, anterior epiblast explants treated with 10 to 50 ng/ml of FGF-2 behaved in the same way as anterior epiblast explants treated with anti-E-cadherin-blocking antibodies.[64] Functional compensation by other growth factors of the FGF family or of families resulting in a functionally identical signaling could explain why the fgf2 knockout does not impair mesoderm formation.[65,66] Mouse mutants for the major gastrulation FGF-2 receptor, fgfr1, show defects in mesoderm formation, due to defective mesoderm cell migration after ingression. Conversely, defects in epithelium to mesenchyme transition could not be demonstrated directly.[67]

Model for the EMT from Epiblast to Mesoderm

The generally accepted model distinguishes two phases (Fig. 5). First, cytoplasmic hyperpolarization and junction remodeling (apical constrictions and basal enlargement) occur. Secondly, there is apical detachment and local destruction of the basal lamina. Two types of molecule directly controlling the EMT leading to mesoderm cell formation have been identified: a growth factor (FGF-2) and a cell-cell adhesion molecule (E-cadherin).

Ingression into the Primitive Streak

Once the primitive streak has been established, cells in contact with it ingress. The first cells to go through the primitive streak are those that will form the extraembryonic mesoderm.[68] As the streak elongates towards the anterior pole of the epiblast, the cells forming the embryonic mesoderm and endoderm migrate through it; the further from the primitive streak and closer to the anterior pole of the epiblast they are, the later they start.[68-71]

Cell fate determination at gastrulation has been extensively studied in invertebrates such as sea urchin (see Chapter 6, Wessel) or drosophila (see Chapter 7, Ganguly), and vertebrates including amphibians, fish and birds. In mammals and in particular in the mouse, in vivo studies are complicated because embryonic development is internal. Cell fate maps have been established in the mouse by the technique of orthotopic transplantation of stained epiblast cells. Like "organizers" in other Vertebrates, the "node" at the posterior end of the primitive streak announces the organization properties of the embryo. Heterotopic transplantation of a node removed from a primitive streak stage gastrula into a gastrula at the same stage induces the formation of a second axis containing neural and somitic structures originating from the host, but induced by the transplant.[72] In contrast to amphibians, fish and birds, the induced axis in the mouse does not contain any anterior structures. This suggests that there is a structure in the mouse, in addition to the node, able to induce and to organize the anterior structures.[73] The mouse node structure only appears during gastrulation, whereas the equivalent structures in other Vertebrates are present before gastrulation.

Regionalization of mesoderm populations induced during the course of the gastrulation in the mouse has been correlated with sites of ingression into the primitive streak.[74-77] The dorso-ventral axis of the embryo is the antero-posterior axis of the primitive streak: the axial mesoderm originates from cells crossing the streak primitive in the vicinity of the node region, the paraxial mesoderm originates from cells crossing the primitive streak in the perinodal region, i.e., anterior region of the streak, the lateral mesoderm originates from cells crossing the primitive streak in its middle part, and the extraembryonic mesoderm originates from cells crossing the streak in its distal, and therefore posterior end.

Numerous transcription factors, membrane receptors and growth factors act as mesoderm inducers and regionalizers in lower Vertebrates. However, the molecular mechanisms of the steps between induction of these various mesoderm lineages to their ingression into the streak are mostly unknown.[78,79] In vivo gene knockouts and in vitro assays indicate that growth factors control epiblast cell migration through the primitive streak and the consequent cell determination. Growth factors and receptors of the FGF, TGF β and Wnt families have been implicated in certain cell population allocation defects at the time of ingression. FGF-2 elicits the conversion of epiblast explants into cells displaying cellular and molecular features similar to those of mesoderm cells in vivo.[64] Knocking-out the gene encoding the main FGF receptor at this stage, fgfr1, results in cells having completed ingression being retained in the neighborhood of the primitive streak, leading to a defect in paraxial mesoderm formation. This has been interpreted as a mesoderm dorsalizing defect.[67,80,81] Cell migration abnormalities after ingression have also been described in embryos carrying T mutations; T is a transcription factor the expression of which in other vertebrates is specific to developing mesoderm cells and is modulated by FGFs.[82,83]

Likewise, homozygote mutants for wnt3a show a paraxial mesoderm formation defect in the trunk: epiblast cells can cross the primitive streak, but cannot migrate laterally or differentiate into neural tissue.[84,85] Conversely, homozygote mutants for bmp4 (bone morphogenetic protein 4) show a defect in the formation of ventral mesoderm.[86]

Mesoderm induction and regionalization involve restriction of differentiation potential, corresponding to phases of cell differentiation. Epiblast cells are multipotent.[68,75,87,88] Epiblast cells of any location grafted heterotopically differentiate with the cells at the site of the graft. Epiblast cell multipotency is such as they can form either somatic tissues, or germ line cells. These are the cells that are used to establish embryonic stem cell lines (ES cells).

Epiblast cells corresponding to the embryonic cells at the late primitive streak stage can differentiate into any mesoderm lineage.[89] Nevertheless, cells from the most anterior region of the primitive streak contribute only to neural tissues.[90] Therefore, the transformation of the epiblast into primitive ectoderm during gastrulation is associated with a restriction of the development potential of the cells. Anterior primitive ectoderm cells cannot differentiate into hematopoietic lineage cells in vitro.[91] Conversely, ES cells treated with activin or BMP-4, regain the potential to differentiate in vitro into hematopoietic lineage cells.[91,92] the loss of differentiation potential can also be interpreted as an indicator of the availability of an instructive or permissive inducer in the environment of a cell population.

The cell differentiation potential is restricted when epiblast cells ingress into the primitive streak. Mesoderm cells grafted back into the epiblast recover some of the potential of epiblast cells, including both that to ingress, again, into the primitive streak and that to form mesoderm populations. Nevertheless, they cannot colonize lateral mesoderm.[93] Epiblast cells grafted into the mesoderm contribute to all mesoderm populations and also to lineages to which mesoderm cells at the grafting site cannot contribute.[93] This supports Beddington's suggestion that ingression of epiblast cells into the primitive streak results in a loss of differentiation potential.[94] Furthermore, the differentiation of epiblast cells into a given type of tissue is not dependent on the initial location of these cells, but on the route they follow and the molecular and cellular environments they meet during gastrulation.

Migration Associated with Gastrulation

Epiblast cells migrate as a cluster towards the primitive streak, during gastrulation and after ingression. Clonal analysis shows that cells derived from the same parental cells are very rarely close to one another during this migration: the migration of the primitive ectoderm cells is not clonal.[75] At gastrulation, epiblast cells that will contribute to the future neurectoderm migrate from the distal and anterior part of the epiblast towards the proximal and posterior part of the embryo. These cells migrate as a cluster of cells and they strictly follow a very precise route.[95] Once ingressed into the primitive streak, the cells contributing to the embryonic mesoderm and endoderm, and those contributing to the extraembryonic mesoderm migrate together in the opposite direction to the presumptive ectoderm cells, i.e., from the postero-distal to the antero-proximal part of the embryo.

After ingression, the migration of the precursors of the cranial and cardiac mesoderm pushes back the precursors of extraembryonic mesoderm towards the extraembryonic region. These are among the first ingressed cells to cross the primitive streak and are localized at the antero-proximal pole during the early phases of gastrulation. Under the pressure of cardiac and cranial mesoderm precursors, this population of cells is pushed back to the extraembryonic region.[68] The same type of migratory movements by propulsion have been observed for precursor cells of the embryonic endoderm, which move preexisting visceral endoderm cells towards the antero-proximal pole, in the yolk sac.[70,71,96] The physical basis of this propulsion has not been clearly elucidated. It is currently believed, in the case of the propulsion of the visceral endoderm towards the yolk sac, that the intercalation of newly ingressed endoderm cells into the posterior and distal region of the embryo pushes the preexisting visceral endoderm in the antero-proximal direction; in addition the expansion of the yolk sac in this same direction creates a traction on the visceral endoderm. All mouse mutants presenting a phenotypic defect in the formation of the endoderm, extraembryonic cavity or constrictions at the interface of the embryonic and extra-embryonic regions also present defects in the formation of the primitive layers.[97-100]

Little is known about the molecular mechanisms involved in these cell migrations. Cell-cell adhesion molecules and matrix-cell adhesion molecules may well be involved in this process.[44,50,101] During the EMT associated with ingression in the primitive streak, the range of cell adhesion molecules expressed at the surface of the cells is modified, and in particular E-cadherin disappears.[9] After ingression, cohesion between the mesodermal cells is reestablished via N-cadherin and cadherin-11.[45,46] However, N-cadherin and cadherin-11 knockouts do not present any obvious phenotype at gastrulation.[102,103] E-cadherin knockout embryos, in contrast die at the time of perimplantation, and this early lethality prevents analysis of the effect during gastrulation.[10] Nevertheless, Cdh1-/- ES cells strongly express T, a mesodermal marker.[104] Finally, in the absence of FAK, a protein kinase involved in matrix-cell adhesion, mesodermal cells present clear migratory defects.[105]

Table 1. EMT examples during embryonic development

Example	Stage (Mouse)	Transition	Specific Related Factors
Gastrulation	E6.5	Epiblast -> Mesoderm	TGFβ /BMP, FGF, Wnt
Prevalvular mesenchyme (Heart formation)	E8	Endothelium -> Atrial and ventricular septum	TGFβ, Slug
Neural crest cells	E8	Neural plate -> Numerous derivatives	Notch/delta, shh, BMP Slug
Somitogenesis and sclerotome differentiation	E9	Somite walls -> Sclerotome	Notch/delta, shh, slug?
Palate formation	E13.5	Oral epithelium -> Mesenchymal and epithelial cells + apoptosis	TGFβ
Mullerian tract regression	E15	Mullerian tract -> Mesenchymal cells combined with apoptosis	TGFβ, anti-Mullerian hormone, β catenin

In conclusion, the formation of the first embryonic mesenchyme is associated with the conversion of some epithelial cells of the primitive ectoderm. One of the main features of this conversion is the loss of the (mostly E-cadherin-dependent) cell-cell adhesion. The neo mesenchymal cells then lose subcellular polarization and their gene expression is modified. The cells present novel properties including higher rates of proliferation and motility. This EMT results in the genesis of novel cell types, and tissues and a new general organization in the embryo. Some of the cells derived from the epiblast do not undergo EMT, and will form the ectoderm. This developmental step is therefore associated with a restriction of multipotentiality.

EMT Later during Embryonic Development and Adult Stages

Several epithelium-to-mesenchyme conversions occur later during embryonic development (Table 1). Examples are the formation of neural crest cells from the neural tube on embryonic day 8 (= E8) (see Chapter 3, Newgreen and McKeown), of the atrial and ventricular septum from the endothelium during the formation of the heart on E8 (see Chapter 4, Runyan et al), of the sclerotome from the somite on E9, of mesenchymal cells of the palate from oral epithelium at E13.5 (see Chapter 1, Hay), and of mesenchymal cells during the regression of the Mullerian tract on E15. The main molecular events associated with these transitions are similar but the regulation is different. These EMT will not be described here.

Regulation of EMT

Soluble factors were thought to be important in EMT for many years, but they are not the only factors involved. Indeed, morphogen gradients are not the sole molecular mechanism of EMT induction. Transmembrane proteins such Notch/Delta or cadherin/catenin are essential in the induction and control of EMT (see Chapter 11, Berx and Van Roy).

During embryonic development, EMT are regulated by the combination of (i) the execution of a genetic program within cells that are undergoing EMT, and (ii) the signals emanating from the environment, such as growth factors, and physical constraints.

Regulation of gastrulation-related EMT arises from the migration of mesenchymal cells to environments in which EMT triggering factors are not produced. In addition, execution of genetic programs is responsible for producing pro-epithelial molecules. EMT is a normal process in the maintenance and repair of tissues (for example maintenance of the intestinal epithelium, and wound healing heightened on Chapter 8, Arnoux et al) in the adult. The regulation of adult EMT is similar to that for gastrulation-related EMT. Some abnormal processes in the adult, such as primary tumor or metastasis formation, are also associated with EMT. These EMT are not regulated because: (i) EMT triggering signals are produced continuously after mutations in their genes; (ii) cells undergoing EMT have become insensitive to their environment and therefore, the process never stops.

Conclusion

In conclusion, it appears that the embryonic development involves a series of conversions of epithelial architecture to mesenchymal architecture and of mesenchymal architecture to epithelial architecture. These successive MET <-> EMT interconversions can affect the same original cell. Consequently, the acquisition of the epithelial or mesenchymal status by a cell is not final. In other words, there is a single mechanism which is defined as MET or EMT according to the direction in which it works.

References

1. Knecht DA, Fuller DL, Loomis WF. Surface glycoprotein, gp24, involved in early adhesion of Dictyostelium discoideum. Dev Biol 1987; 121:277-83.
2. Loomis WF. Cell-cell adhesion in Dictyostelium discoideum. Dev Genet 1988; 9:549-59.
3. Chisholm JC, Johnson MH, Warren PD et al. Developmental variability within and between mouse expanding blastocysts and their ICMs. J Embryol Exp Morphol 1985; 86:311-36.
4. Collins JE, Fleming TP. Epithelial differentiation in the mouse preimplantation embryo: Making adhesive cell contacts for the first time. Trends Biochem Sci 1995; 20:307-12.
5. Hyafil F, Morello D, Babinet C et al. cell surface glycoprotein involved in the compaction of embryonal carcinoma cells and cleavage stage embryos. Cell 1980; 21:927-34.
6. Ziomek CA, Johnson MH. Cell surface interaction induces polarization of mouse 8-cell blastomeres at compaction. Cell 1980; 21:935-42.
7. Hyafil F, Babinet C, Jacob F. Cell-cell interactions in early embryogenesis: A molecular approach to the role of calcium. Cell 1981; 26:447-54.
8. Peyrieras N, Hyafil F, Louvard D et al. nonintegral membrane protein of early mouse embryo. Proc Natl Acad Sci USA 1983; 80:6274-7.
9. Vestweber D, Kemler R. Rabbit antiserum against a purified surface glycoprotein decompacts mouse preimplantation embryos and reacts with specific adult tissues. Exp Cell Res 1984; 152:169-78.
10. Larue L, Ohsugi M, Hirchenhain J et al. E-cadherin null mutant embryos fail to form a trophectoderm epithelium. Proc Natl Acad Sci USA 1994; 91:8263-7.
11. Reeve WJ, Ziomek CA. Distribution of microvilli on dissociated blastomeres from mouse embryos: Evidence for surface polarization at compaction. J Embryol Exp Morphol 1981; 62:339-50.
12. Wiley LM, Lever JE, Pape C et al. Antibodies to a renal Na+/glucose cotransport system localize to the apical plasma membrane domain of polar mouse embryo blastomeres. Dev Biol 1991; 143:149-61.
13. Dardik A, Smith RM, Schultz RM. Colocalization of transforming growth factor-alpha and a functional epidermal growth factor receptor (EGFR) to the inner cell mass and preferential localization of the EGFR on the basolateral surface of the trophectoderm in the mouse blastocyst. Dev Biol 1992; 154:396-409.
14. Jackson BW, Grund C, Schmid E et al. Formation of cytoskeletal elements during mouse embryogenesis. Intermediate filaments of the cytokeratin type and desmosomes in preimplantation embryos. Differentiation 1980; 17:161-79.
15. Nichols J, Gardner RL. Heterogeneous differentiation of external cells in individual isolated early mouse inner cell masses in culture. J Embryol Exp Morphol 1984; 80:225-40.
16. Varmuza S, Prideaux V, Kothary R et al. Polytene chromosomes in mouse trophoblast giant cells. Development 1988; 102:127-34.
17. Gardner RL, Beddington RS. Multi-lineage 'stem' cells in the mammalian embryo. J Cell Sci Suppl 1988; 10:11-27.
18. Copp AJ. Interaction between inner cell mass and trophectoderm of the mouse blastocyst. II. The fate of the polar trophectoderm. J Embryol Exp Morphol 1979; 51:109-20.
19. Coucouvanis E, Martin GR. Signals for death and survival: A two-step mechanism for cavitation in the vertebrate embryo. Cell 1995; 83:279-87.
20. Tam PP, Williams EA, Chan WY. Gastrulation in the mouse embryo: Ultrastructural and molecular aspects of germ layer morphogenesis. Microsc Res Tech 1993; 26:301-28.
21. Snow MHL, Tam PPL. Is compensatory growth a complicating factor in mouse teratology? Nature 1979; 279:557-559.

22. Tam PP. Postimplantation development of mitomycin C-treated mouse blastocysts. Teratology 1988; 37:205-12.
23. Power MA, Tam PP. Onset of gastrulation, morphogenesis and somitogenesis in mouse embryos displaying compensatory growth. Anat Embryol (Berl) 1993; 187:493-504.
24. Buehr M, McLaren A. Size regulation in chimaeric mouse embryos. J Embryol Exp Morphol 1974; 31:229-34.
25. Lewis NE, Rossant J. Mechanism of size regulation in mouse embryo aggregates. J Embryol Exp Morphol 1982; 72:169-81.
26. Rands GF. Size regulation in the mouse embryo. II. The development of half embryos. J Embryol Exp Morphol 1986; 98:209-17.
27. Snow MH, Aitken J, Ansell JD. Role of the inner cell mass in controlling implantation in the mouse. J Reprod Fertil 1976; 48:403-4.
28. Poelmann RE. The formation of the embryonic mesoderm in the early post-implantation mouse embryo. Anat Embryol (Berl) 1981; 162:29-40.
29. Snow MHL. Gastrulation in the mouse: Growth and regionalisation of the epiblast. J Embryol Exp Morph 1977; 42:293-303.
30. Mac Auley A, Werb Z, Mirkes PE. Characterization of the unusually rapid cell cycles during rat gastrulation. Development 1993; 117:873-883.
31. Hashimoto K, Fujimoto H, Nakatsuji N. An ECM substratum allows mouse mesodermal cells isolated from the primitive streak to exhibit motility similar to that inside the embryo and reveals a deficiency in the T/T mutant cells. Development 1987; 100:587-98.
32. Spiegelman M, Bennett D. Fine structural study of cell migration in the early mesoderm of normal and mutant mouse embryos (T-locus: t-9/t-9). J Embryol Exp Morphol 1974; 32:723-8.
33. Batten BE, Haar JL. Fine structural differentiation of germ layers in the mouse at the time of mesoderm formation. Anat Rec 1979; 194:125-41.
34. Enders AC, Schlafke S, Hubbard NE et al. Morphological changes in the blastocyst of the western spotted skunk during activation from delayed implantation. Biol Reprod 1986; 34:423-37.
35. Viebahn C, Mayer B, Miething A. Morphology of incipient mesoderm formation in the rabbit embryo: A light- and retrospective electron-microscopic study. Acta Anat (Basel) 1995; 154:99-110.
36. Viebahn C. Epithelio-mesenchymal transformation during formation of the mesoderm in the mammalian embryo. Acta Anat (Basel) 1995; 154:79-97.
37. Stern CD, MacKenzie DO. Sodium transport and the control of epiblast polarity in the early chick embryo. J Embryol Exp Morphol 1983; 77:73-98.
38. Jackson BW, Grund C, Winter S et al. Formation of cytoskeletal elements during mouse embryogenesis. II. Epithelial differentiation and intermediate-sized filaments in early postimplantation embryos. Differentiation 1981; 20:203-16.
39. Colucci-Guyon E, Portier MM, Dunia I et al. Mice lacking vimentin develop and reproduce without an obvious phenotype. Cell 1994; 79:679-94.
40. Baribault H, Price J, Miyai K et al. Mid-gestational lethality in mice lacking keratin 8. Genes Dev 1993; 7:1191-202.
41. Klinowska TC, Ireland GW, Kimber SJ. A new in vitro model of murine mesoderm migration: The role of fibronectin and laminin. Differentiation 1994; 57:7-19.
42. Franke WW, Grund C, Jackson BW et al. Formation of cytoskeletal elements during mouse embryogenesis. IV. Ultrastructure of primary mesenchymal cells and their cell-cell interactions. Differentiation 1983; 25:121-41.
43. Damjanov I, Damjanov A, Damsky CH. Developmentally regulated expression of the cell-cell adhesion glycoprotein cell-CAM 120/80 in peri-implantation mouse embryos and extraembryonic membranes. Dev Biol 1986; 116:194-202.
44. Hatta K, Takeichi M. Expression of N-cadherin adhesion molecules associated with early morphogenetic events in chick development. Nature 1986; 320:447-9.
45. Hoffmann I, Balling R. Cloning and expression analysis of a novel mesodermally expressed cadherin. Dev Biol 1995; 169:337-46.
46. Kimura Y, Matsunami H, Inoue T et al. Cadherin-11 expressed in association with mesenchymal morphogenesis in the head, somite, and limb bud of early mouse embryos. Dev Biol 1995; 169:347-58.
47. Haegel H, Larue L, Ohsugi M et al. Lack of beta-catenin affects mouse development at gastrulation. Development 1995; 121:3529-37.
48. Bierkamp C, Schwarz H, Huber O et al. Desmosomal localization of beta-catenin in the skin of plakoglobin null-mutant mice. Development 1999; 126:371-81.
49. Bierkamp C, McLaughlin KJ, Schwarz H et al. Embryonic heart and skin defects in mice lacking plakoglobin. Dev Biol 1996; 180:780-5.

50. Burdsal CA, Damsky CH, Pedersen RA. The role of E-cadherin and integrins in mesoderm differentiation and migration at the mammalian primitive streak. Development 1993; 118:829-44.
51. Solursh M, Morriss GM. Glycosaminoglycan synthesis in rat embryos during the formation of the primary mesenchyme and neural folds. Dev Biol 1977; 57:75-86.
52. Wartiovaara J, Leivo I, Vaheri A. Expression of the cell surface-associated glycoprotein, fibronectin, in the early mouse embryo. Dev Biol 1979; 69:247-57.
53. Leivo I, Vaheri A, Timpl R et al. Appearance and distribution of collagens and laminin in the early mouse embryo. Dev Biol 1980; 76:100-14.
54. Wan YJ, Wu TC, Chung AE, Damjanov I. Monoclonal antibodies to laminin reveal the heterogeneity of basement membranes in the developing and adult mouse tissues. J Cell Biol 1984; 98:971-9.
55. Miosge N, Gunther E, Becker-Rabbenstein V et al. Ultrastructural localization of laminin subunits during the onset of mesoderm formation in the mouse embryo. Anat Embryol (Berl) 1993; 187:601-5.
56. George EL, Georges-Labouesse EN, Patel-King RS et al. Defects in mesoderm, neural tube and vascular development in mouse embryos lacking fibronectin. Development 1993;119:1079-91.
57. Yang JT, Rayburn H, Hynes RO. Embryonic mesodermal defects in alpha 5 integrin-deficient mice. Development 1993; 119:1093-105.
58. Fassler R, Meyer M. Consequences of lack of beta 1 integrin gene expression in mice. Genes Dev 1995;9:1896-908.
59. Tam PP, Behringer RR. Mouse gastrulation: The formation of a mammalian body plan. Mech Dev 1997; 68:3-25.
60. Herrmann BG. Expression pattern of the Brachyury gene in whole-mount TWis/TWis mutant embryos. Development 1991; 113:913-7.
61. Holdener BC, Faust C, Rosenthal NS et al. Msd is required for mesoderm induction in mice. Development 1994; 120:1335-46.
62. Faust C, Schumacher A, Holdener B et al. The eed mutation disrupts anterior mesoderm production in mice. Development 1995; 121:273-85.
63. Schulte-Merker S, Smith JC. Mesoderm formation in response to Brachyury requires FGF signalling. Curr Biol 1995; 5:62-7.
64. Burdsal CA, Flannery ML, Pedersen RA. FGF-2 alters the fate of mouse epiblast from ectoderm to mesoderm in vitro. Dev Biol 1998; 198:231-44.
65. Dono R, Texido G, Dussel R et al. Impaired cerebral cortex development and blood pressure regulation in FGF-2-deficient mice. Embo J 1998; 17:4213-25.
66. Ortega S, Ittmann M, Tsang SH et al. Neuronal defects and delayed wound healing in mice lacking fibroblast growth factor 2. Proc Natl Acad Sci USA 1998; 95:5672-7.
67. Yamaguchi TP, Harpal K, Henkemeyer M et al. fgfr-1 is required for embryonic growth and mesodermal patterning during mouse gastrulation. Genes Dev 1994; 8:3032-44.
68. Parameswaran M, Tam PP. Regionalisation of cell fate and morphogenetic movement of the mesoderm during mouse gastrulation. Dev Genet 1995; 17:16-28.
69. Frohman MA, Boyle M, Martin GR. Isolation of the mouse Hox-2.9 gene; analysis of embryonic expression suggests that positional information along the anterior-posterior axis is specified by mesoderm. Development 1990; 110:589-607.
70. Lawson KA, Meneses JJ, Pedersen RA. Cell fate and cell lineage in the endoderm of the presomite mouse embryo, studied with an intracellular tracer. Dev Biol 1986; 115:325-39.
71. Lawson KA, Pedersen RA. Cell fate, morphogenetic movement and population kinetics of embryonic endoderm at the time of germ layer formation in the mouse. Development 1987; 101:627-52.
72. Beddington RS. Induction of a second neural axis by the mouse node. Development 1994; 120:613-20.
73. Kodjabachian L, Lemaire P. Embryonic induction: Is the Nieuwkoop centre a useful concept? Curr Biol 1998; 8:R918-21.
74. Tam PP, Beddington RS. The formation of mesodermal tissues in the mouse embryo during gastrulation and early organogenesis. Development 1987; 99:109-26.
75. Lawson KA, Meneses JJ, Pedersen RA. Clonal analysis of epiblast fate during germ layer formation in the mouse embryo. Development 1991; 113:891-911.
76. Smith JL, Gesteland KM, Schoenwolf GC. Prospective fate map of the mouse primitive streak at 7.5 days of gestation. Dev Dyn 1994; 201:279-89.
77. Wilson V, Beddington RS. Cell fate and morphogenetic movement in the late mouse primitive streak. Mech Dev 1996; 55:79-89.
78. Sasaki H, Hogan BL. Differential expression of multiple fork head related genes during gastrulation and axial pattern formation in the mouse embryo. Development 1993; 118:47-59.

79. Tam PP, Trainor PA. Specification and segmentation of the paraxial mesoderm. Anat Embryol (Berl) 1994; 189:275-305.
80. Deng CX, Wynshaw-Boris A, Shen MM et al. Murine FGFR-1 is required for early postimplantation growth and axial organization. Genes Dev 1994; 8:3045-57.
81. Ciruna BG, Schwartz L, Harpal K et al. Chimeric analysis of fibroblast growth factor receptor-1 (Fgfr1) function: A role for FGFR1 in morphogenetic movement through the primitive streak. Development 1997; 124: 2829-41.
82. Wilson PA, Melton DA. Mesodermal patterning by an inducer gradient depends on secondary cell-cell communication. Curr Biol 1994; 4:676-86.
83. Isaacs HV, Pownall ME, Slack JM. eFGF regulates Xbra expression during Xenopus gastrulation. Embo J 1994; 13:4469-81.
84. Takada S, Stark KL, Shea MJ et al. Wnt-3a regulates somite and tailbud formation in the mouse embryo. Genes Dev 1994; 8:174-89.
85. Yoshikawa Y, Fujimori T, McMahon AP et al. Evidence that absence of Wnt-3a signaling promotes neuralization instead of paraxial mesoderm development in the mouse. Dev Biol 1997; 183:234-42.
86. Winnier G, Blessing M, Labosky PA et al. Bone morphogenetic protein-4 is required for mesoderm formation and patterning in the mouse. Genes Dev 1995; 9:2105-16.
87. Lawson KA, Hage WJ. Clonal analysis of the origin of primordial germ cells in the mouse. Ciba Found Symp 1994; 182:68-84; discussion 84-91.
88. Tam PP, Zhou SX. The allocation of epiblast cells to ectodermal and germ-line lineages is influenced by the position of the cells in the gastrulating mouse embryo. Dev Biol 1996; 178:124-32.
89. Beddington RS. An autoradiographic analysis of tissue potency in different regions of the embryonic ectoderm during gastrulation in the mouse. J Embryol Exp Morphol 1982; 69:265-85.
90. Carey FJ, Linney EA, Pedersen RA. Allocation of epiblast cells to germ layer derivatives during mouse gastrulation as studied with a retroviral vector. Dev Genet 1995; 17:29-37.
91. Kanatsu M, Nishikawa SI. In vitro analysis of epiblast tissue potency for hematopoietic cell differentiation. Development 1996; 122:823-30.
92. Johansson BM, Wiles MV. Evidence for involvement of activin A and bone morphogenetic protein 4 in mammalian mesoderm and hematopoietic development. Mol Cell Biol 1995; 15:141-51.
93. Tam PP, Parameswaran M, Kinder SJ et al. The allocation of epiblast cells to the embryonic heart and other mesodermal lineages: The role of ingression and tissue movement during gastrulation. Development 1997; 124:1631-42.
94. Beddington RS, Rashbass P, Wilson V. Brachyury—a gene affecting mouse gastrulation and early organogenesis. Dev Suppl 1992:157-65.
95. Quinlan GA, Williams EA, Tan SS et al. Neuroectodermal fate of epiblast cells in the distal region of the mouse egg cylinder: Implication for body plan organization during early embryogenesis. Development 1995; 121:87-98.
96. Tam PP, Beddington RS. Establishment and organization of germ layers in the gastrulating mouse embryo. Ciba Found Symp 1992; 165:27-41; discussion 42-9.
97. Farrington SM, Belaoussoff M, Baron MH. Winged-helix, Hedgehog and Bmp genes are differentially expressed in distinct cell layers of the murine yolk sac. Mech Dev 1997; 62:197-211.
98. Ang SL, Rossant J. HNF-3 beta is essential for node and notochord formation in mouse development. Cell 1994; 78:561-74.
99. Shawlot W, Behringer RR. Requirement for Lim1 in head-organizer function. Nature 1995; 374:425-30.
100. Ang SL, Jin O, Rhinn M et al. A targeted mouse Otx2 mutation leads to severe defects in gastrulation and formation of axial mesoderm and to deletion of rostral brain. Development 1996; 122:243-52.
101. Takeichi M. The cadherins: Cell-cell adhesion molecules controlling animal morphogenesis. Development 1988; 102:639-55.
102. Manabe T, Togashi H, Uchida N et al. Loss of Cadherin-11 Adhesion Receptor Enhances Plastic Changes in Hippocampal Synapses and Modifies Behavioral Responses. Mol Cell Neurosci 2000; 15:534-546.
103. Radice GL, Rayburn H, Matsunami H et al. Developmental defects in mouse embryos lacking N-cadherin. Dev Biol 1997; 181:64-78.
104. Larue L, Antos C, Butz S et al. A role for cadherins in tissue formation. Development 1996; 122:3185-94.
105. Furuta Y, Ilic D, Kanazawa S et al. Mesodermal defect in late phase of gastrulation by a targeted mutation of focal adhesion kinase, FAK. Oncogene 1995; 11:1989-95.

CHAPTER 3

The Neural Crest:
A Model Developmental EMT

Donald F. Newgreen and Sonja J. McKeown

Introduction

The NC Is a Valuable Model for EMT

The concept of the epithelial-mesenchymal transition (EMT) arose from developmental biology, (see ref. 1) where EMT occurs in many situations, each being predictable, stereotyped and with the outcomes often dramatic. The EMT of the neural crest (NC) is an example of this event in development. In its own right; the NC has been ranked as a "fourth germ layer" by Hall.[2] The tissues and structures to which the NC gives rise is widespread in the body and diverse in cell types, ranging from craniofacial connective tissues to peripheral nerve and glial cells to skin pigment cells.[3] In addition, abnormalities involving NC development seem to be disproportionately represented in human birth defects. The NC is important for evolutionary research too, because it is the only organ system unique to vertebrates. Its appearance in evolution is suggested to have enabled the massive adaptive radiation of these chordates. Technically, the experimental approaches for developmental biology, described as "cutting, labeling and pasting",[4] have shown the NC to be a particularly accessible, manipulatable and robust subject, with relatively straight-forward evaluation of the results in terms of altered developmental patterns. Thus because of its importance and technical advantages, the NC is probably the most studied developmental EMT.

Defining the NC

The NC consists of those cells that arise at the border of the neural and epidermal ectoderm epithelia in early development of vertebrate embryos, and which undergo EMT and migrate away from the ectoderm into adjacent tissues (Fig. 1). Thus the EMT is central to the definition of this important cell lineage. Definitions based on cell behaviours in vivo may be difficult to apply in experimental systems where normal events are perturbed. Thus molecular markers of the NC and its EMT have been sought.

The EMT of the NC Is Stereotyped

The NC occurs at almost all axial levels, and is most broadly classified as cranial NC rostrally and trunk NC more caudally: there are some molecular and behavioural differences in the NC levels. Broadly speaking the NC undergoes EMT as a timetabled wave that sweeps from rostral to caudal (Figs. 1, 2). At most axial levels in most vertebrates, the NC can be recognized towards the end of neurulation at the dorsal margin of the newly formed neural tube. The presumptive NC cells participate in the cellular bridge between the epidermal and the neural epithelia. At some axial levels the epidermis and neural tube separate then the NC cells undergo EMT and migration from the latter, as in the trunk levels of

Rise and Fall of Epithelial Phenotype: Concepts of Epithelial-Mesenchymal Transition, edited by Pierre Savagner.

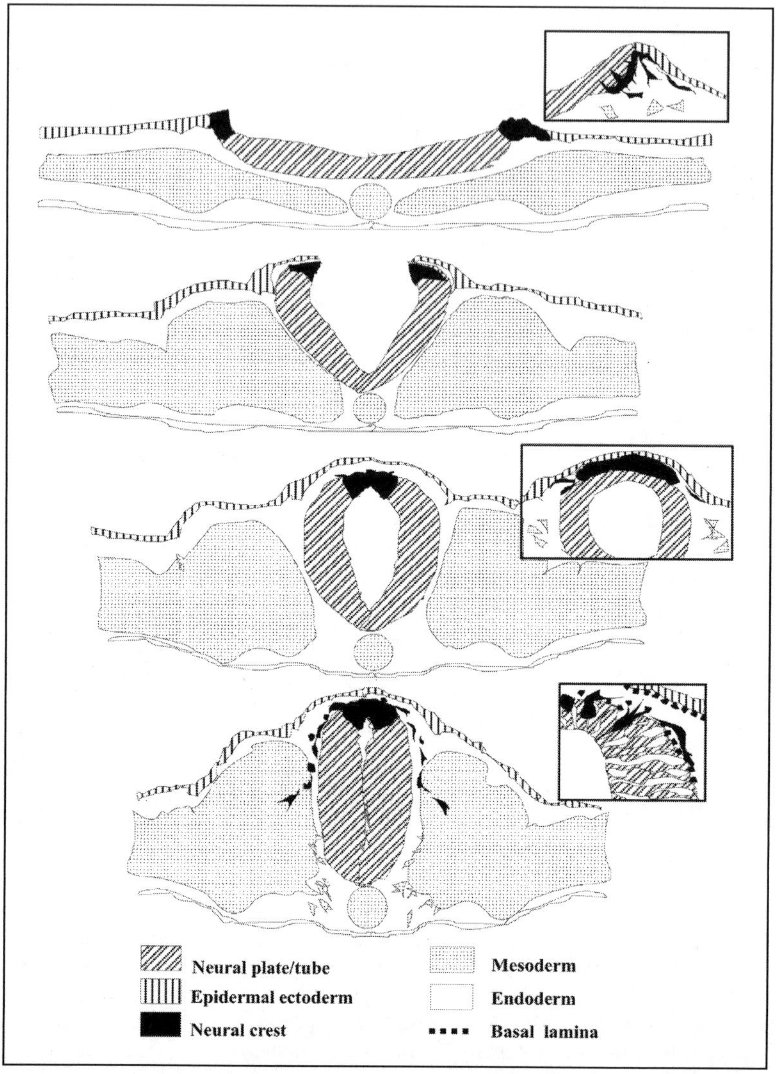

Figure 1. Diagrammatic transverse sections through an early vertebrate during neurulation (modeled on trunk levels of the chick embryo) showing the region of ectoderm that gives rise to NC cells, in relation to the epidermal ectoderm and the neural plate/tube. The NC cells typically undergo EMT and detach from the neural tube after the neural tube plus NC has separated from the epidermis. However, at cranial level in mammals, NC cells undergo EMT and migration at the neural plate stage, before the neural plate has separated from the epidermis (upper inset). At cranial levels of birds, the NC forms the junction between epidermis and neural tube; its EMT and migration separates the neural tube and the epidermis (middle inset). The changes in shape and position of NC cells during EMT, and the localized absence of basal lamina is shown at higher magnification in the lower inset.

most vertebrates (Fig. 1).[5,6] However, sometimes the EMT and emigration of the NC is brought forward relatively earlier in the neurulation process. In the most extreme case seen in cranial levels of mammals, the first cells to commence EMT do so well before the neural plate has rolled up to form a tube. This requires a process termed delamination in which the

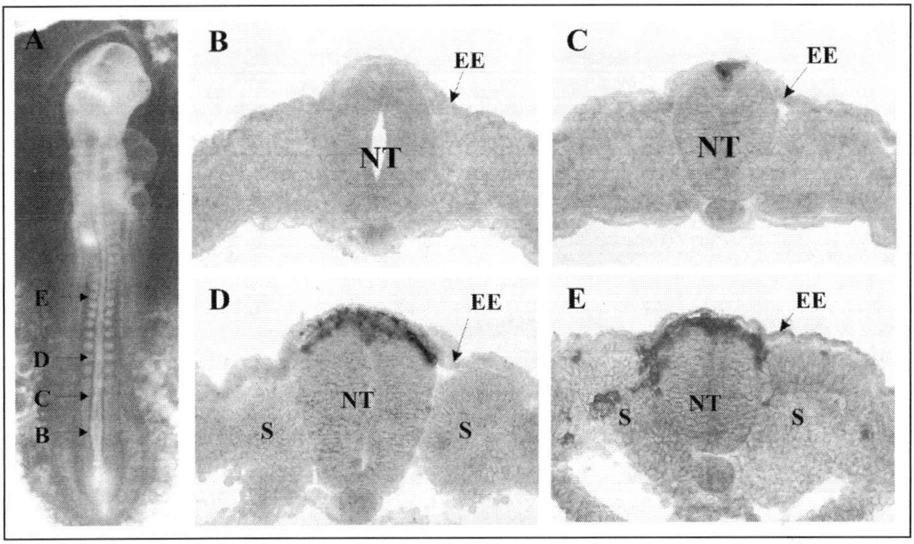

Figure 2. A 2-day incubation quail embryo is shown in A, with the levels of the transverse sections B-E shown by arrows. NC development proceeds in a rostral to caudal wave, so that most caudal section (B) will be developmentally youngest and the most rostral section (E) will be developmentally oldest. This is shown with in situ hybridization by the increasing expression of *Sox10* in NC cells in the dorsal neural tube (NT), and the eventual migration of these *Sox10*+ve mesenchymal cells out of the neural tube epithelium into adjacent cell-free space then into the mesodermal somites (S).

neural and epidermal epithelia remain conjoined despite the loss by emigration of NC cells from the basal surface of the neuro-epidermal border (upper inset, Fig. 1).[7] In avian cranial NC (middle inset in Fig. 1) EMT is also brought forward. The separation of neural and epidermal moieties in this case seems to be actually accomplished by NC cell EMT and emigration, which dissolves the cell mass that links these two epithelial layers.[8] In Urodele amphibians the NC population at trunk levels seems to be extruded from the dorsal neural tube, upon which it forms a separate coherent string whose cells only later migrate away.[9] Thus, NC EMT is a part of neurulation but is not tied to any particular stage in the morphogenesis of the neural tube. Moreover, NC emigration occurs in neural tubes formed by rolling up of a plate (primary neurulation) and by tubes formed by cavitation of a solid rod (secondary neurulation).[10]

Despite these variations, at the cell level NC undergo a typical EMT (lower inset in Fig. 1). Neural epithelial cells stretch from the luminal surface to the external surface of the neural tube. The apical or luminal ends exhibit numerous cell-cell junctional specializations including adherens junctions, while the external or basal surfaces are in contact via cell matrix adhesions to the basal lamina extracellular matrix. These cells are markedly polarized apicobasally not only in these features but also in the microtubular and microfilament (actin) cytoskeletal systems. In the prospective NC region the apicobasal polarity alters as the cells initially become more rounded or complex in shape, have fewer more irregularly placed cell-cell junctions, and the basal lamina is lacking or discontinuous.[5,8,11] The cells then project fine processes which contact interstitial extracellular matrix fibrils, and the cells extend out of the epithelium, showing front-back polarity.[12] Eventually these definitive NC cells lose contact with the epithelium of origin. The migrating cells often show extensive areas of membrane apposition to their neighbours, although junctional specializations as seen at the earlier epithelial stage are not visible.[13]

The Induction, EMT and Migration of NC Are Linked

In principle, the early development of the NC can be divided into three stages: a lineage decision that marks the NC as different from both the neural tube and the epidermal epithelial populations; the morphogenetic realization of this difference in the EMT of these cells; and then subsequent migration that separates them from their epithelial tissues of origin. But in practice, separation into these three categories is difficult.

The NC Is Induced by Cell Interactions

The ectodermal epithelial layer develops a neural and epidermal character, the former being marked by a thickened form called the neural plate, while the latter remains a flattened layer termed the epidermis. This neural induction is in detail controversial but is likely to involve the activation and, equally important, suppression of intercellular signaling molecules including BMP-4, FGFs and Wnts.[14-19] On the basis of this general neural induction, the NC is generated at the border of the neural plate and epidermis, a site which soon elevates as a neural fold. Both the neural (thick) and epidermal (thin) region of the border can produce migratory NC cells.[7] Influences on the ectoderm from underlying paraxial mesoderm plays a part in NC induction, as demonstrated by recombination experiments with tissues from the frog, *Xenopus*.[20] But inductive interactions within the plane of the ectoderm are also important. NC can be generated by any part of the neural epithelium juxtaposed to epidermal ectoderm. This was shown in amphibian and avian embryos by transplanting neural plate which is not NC-fated into distant (not juxta-neural) epidermis.[21,22] At the site of contact with the epidermis, mesenchymal cells emigrated from this neural epithelial transplant and expressed NC-specific markers. This showed that signaling between neural and epidermal sheets could induce a third lineage, the NC, at the interface, which could then undergo EMT and migration. These tissue combination experiments also showed that the mode of signaling and the response repertoire for NC induction are not restricted to the cells at the normally fated NC origin. However, the competence to generate NC was restricted in time in both the neural plate[23] and epidermal ectoderm.[24]

NC Induction and EMT Decisions Are Incomplete and Provisional

Even within the neural folds which contain nascent NC cells, there is some heterogeneity: when single cells were labelled to trace their fate, some fold cells later exhibited NC properties of migration and differentiation, but some were found in the epidermis and the neural tube.[25] In addition, when tested by insertion back into the epithelial neural tube, even post-EMT, migrating NC cells could revert to a neural epithelial type.[26] Thus, although the EMT and migration is normally a watershed in NC identity in that they do not go back either physically or in identity, this morphogenetic event is not necessarily a commitment decision. Instead at least for some cells it is conditional on the microenvironmental change achieved by EMT and migration. Interestingly, the role of the microenvironment (including cells, matrix growth factors) in augmenting or preventing EMT-like events is also observed in pathological examples.[27]

Growth Factors Induce NC and EMT from Neural Epithelium

It is clear that the inductive environment for NC is complex in spatial origins of the signals and the temporal windows of responsiveness. These experiments set the scene for searches for the inductive molecules.

Fibroblast growth factor 2 (FGF2) have been implicated in *Xenopus* as early paraxial mesoderm-derived influences to induce NC, as shown by perturbation of FGF and FGF Receptor functions in vivo and in vitro.[28] However, this effect may be indirect, via Wnt mediation.[29] FGFs have also been implicated in the triggering of pathological EMT.[30]

Wnt family members and their receptors Frizzled[31] have been shown to be present during the early stages at and prior to induction of the NC in embryos of the Zebrafish (Wnt8), *Xenopus* (Xwnt1) and chick (Wnt6), and in vitro assays have shown certain Wnt proteins to be capable of instigating the mesenchymal transition and NC marker expression of neural epithelial cells not fated to become NC.[32-35] Wnt family members have also been implicated in the generation of invasive cell phenotypes in various carcinoma models, consistent with an inductive role for Wnts in pathological EMT.[36]

In vitro and in vivo experiments have shown that the TGF-β family members BMP-4 and 7 are important for NC induction and, perhaps as a separable event, for the sequential EMT.[37] A persuasive case, at least for *Xenopus*, has been made that NC induction depends on an intermediate level of BMP-4 and 7 signaling, with higher levels in epidermal layers and lower levels in central neural epithelium destined for the neural tube.[38] BMP-4 action as a gradient appears to generate NC identity and, also dorsal neural tube characteristics in the neural epithelial cells slightly more distant. In avian embryos, the case for BMP-4 as an early NC inducer is more problematic. Part of the response to epidermal BMP-4 is the homogenetic induction of BMP-4 gene expression in the presumptive NC cells themselves, and this later source of BMP-4 may be crucial for the EMT of cells that already have a NC identity.[39] However the timing of the EMT of the NC, which spreads as a rostro-caudal wave, is largely due to the spatiotemporal availability of the BMP-4 inhibitor Noggin.[40] The involvement of TGF-β family members is a recurrent theme of EMT, both controlled in development and uncontrolled as in carcinoma[41] and fibrosis.[42]

Genes and Molecules in the NC during Induction, EMT and Migration

Many genes are now known to be up- or down-regulated in early NC cells, or otherwise play a role in the process. Many are known from previous studies elsewhere, but increasing numbers arise from directed searches in the embryonic NC. Most successful of these so far are based on exposure of neural epithelial cells to the growth factor BMP-4 or to inducing epidermis, and comparing these to untreated cells by subtractive[43] and array[44] techniques.

Induction of NC cells, exemplified in avian embryos, leads to expression of numerous down-stream targets[44] including transcription factors such as the zinc finger transcription factor *Slug*. Also expressed are *Sox9*, then *Sox10* (Fig. 2) and *Sox8* of the SoxE subgroup of HMG-box transcription factors.[45] *Sox9* is the first to be expressed, at a similar time to *FoxD3* (a winged helix/forkhead transcription factor),[46] overlapping with expression of *BMP4* (a growth factor; see above), *cadherin 6b* (cell adhesion molecule), and *RhoB*, (a GTPase possibly involved in cytoskeletal dynamics).[43,47] *Sox10* is expressed shortly after *Slug*, and *Sox8* expression begins soon after *Sox10*, prior to migration from the neural tube.[47] As migration commences and proceeds at trunk levels, *Slug*, *RhoB*, *N-cadherin* and *cadherin 6b* are downregulated,[43,48-49] as are *Sox8* and *Sox9* (SJM, unpublished). In contrast *Sox10* and *FoxD3* continue to be expressed in migratory NC cells.[46,50,51] At cranial but not trunk levels *Slug* also continues expression after emigration onset. Some migrating neural crest cells then upregulate *cadherin-7*[49] and, in chick embryos, practically all exhibit the carbohydrate epitope, HNK-1,[52] which appears distinctly after EMT and after the start of migration.[13]

Specific Transcription Factors Are Expressed by Nascent NC Cells

Currently the marker most frequently used to mark nascent NC cells is the expression of the genes for zinc-finger transcription factors of the *Slug/Snail* family. These genes are also expressed in many normal and pathological EMTs.[53] Interestingly they predate NC evolution[54] and in the NC there has been an evolutionary change in vertebrate classes in the usage of *Slug* or *Snail* at similar times of NC formation.[55] In the EMT of the avian NC, *Slug* is expressed, but experimentally *Slug* and *Snail* are functionally interchangeable.[56] The *Slug* gene seems to be a target of Wnt signaling, via the Lef/β-catenin transcription complex.[35] It is thought that the Snail-family proteins act as a transcriptional repressor.[57] Interestingly, in a breast cancer line, Snail acts specifically to repress the *E-cadherin* gene, and this repression promotes cell invasion.[58] Experimental expression of *Slug* in cranial levels of avian embryos increases the expression of RhoB, a marker for the NC region prior to migration, and the number of HNK-1 immunoreactive NC cells that migrate away. Interestingly, this NC amplification induced by this one gene is not seen in *Xenopus*[29] or in avian embryos at trunk levels, indicating that different or additional factors are at play in this region.[56]

The SoxE group of genes drive interesting changes in NC induction and EMT when they are over-expressed in avian embryos.[47,59] A characteristic result of over-expression of all SoxE genes in trunk levels of avian embryos is increase in the amount of the neural tube that undergoes EMT, and the more disorganized EMT and local migration of these cells. The HNK-1 marker, normally expressed after the start of NC migration, is induced by SoxE over-expression prior to emigration, while *RhoB* expression is reduced. *FoxD3* expression is slightly reduced with *Sox9* and also *Sox10* over-expression, but increases later with *Sox9*.[47] *Slug* was reportedly briefly upregulated by *Sox9* expression, but overall *Slug* changes little especially when compared to the amplification of the area that undergoes EMT. *Sox8* and *Sox10* over-expression belatedly reduced *Sox9* expression, but *Sox9* over-expression could increase *Sox10* and *Sox8* levels (SJM, unpublished).

Over-expression of *FoxD3*[46,51] in chick embryos stimulates NC formation and EMT much like the SoxE genes, including wide–spread premigratory expression of HNK-1 and *cadherin-7*. However, neither *Slug* nor *RhoB* are induced, and there seems to be no synergies for NC induction when *FoxD3* and *Slug* are up-regulated together. There was a lack of up-regulation of *Slug* and *Sox9* but up-regulation of *Sox10* and, briefly, *Sox8* was seen when *FoxD3* was over-expressed (SJM, unpublished). In contrast, over-expression of FoxD3 in *Xenopus*[60] has the opposite effect on NC generation.

A great deal of work has been made to arrange these genes in cascades, although the results are not yet convincing, perhaps due to the confounding effects of redundancy and the possibility that the genes and molecules in these events are linked as networks rather than as cascades.

Motor Molecules and Genes in the NC that Contribute to EMT

Newgreen and Gibbins[5] defined four general conditions that must be or become permissive of EMT and migration onset in the NC system (i) low NC cell cohesion with the neural epithelium, (ii) adhesion to migration substrate molecules, (iii) cytoskeletal motility mechanisms, and (iv) absence of barriers to emigration. Although proposed with scant molecular and genetic knowledge, this is still a useful framework for discussing the mechanisms of EMT in the NC.

Low NC Cell Cohesion with the Neural Epithelium

The EMT and onset of migration NC cells is associated with a loss or reduction of adherens junctions from the apical ends of the cells by electron microscopy and a reduction of cell-cell adhesions by in vitro adhesion assays.[5] In addition, transient reduction of calcium-dependent cell adhesion (that is, cadherin-mediated) can trigger a coordinated EMT in avian neural epithelial cells,[61] and the initial response is too rapid to involve gene regulation. This is due to the down-regulation of the adherens junction-associated adhesion molecule N-cadherin. The *N-cadherin* gene is down-regulated during and after EMT, and over-expression impairs the ability of some NC cells to leave the neural epithelium.[62] But at the protein level, the most important rapid change may be not the amount of N-cadherin but its de-localization from the junctional apparatus.[63] The cadherin family of homophilic cell-cell adhesion molecules has a long history of involvement in EMTs (see ref. 1) where they are typically de-localized and down-regulated (see above for Snail regulation of *E-cadherin*). The nascent and early migrating NC cells are however, still adherent to some degree, judging by their extensive yet unspecialized areas of membrane apposition,[13] and contributing to this may be residual N-cadherin. However, other cadherins are actually expressed at this time, such as *cadherin-7*, which appears after migration commences in a sub-group of comigrating NC cells.[49]

A consequence of reduction of cadherin is likely to be a liberation of the molecules that link organized cadherin junctions to the actin cytoskeleton, such as catenins. In particular, β-catenin is of interest since it has a role as a transcription cofactor, binding to Lef/Tcf transcription repressors, the complex transforming to transcriptional activation. This effect is prominent in Wnt signaling, where β-catenin availability is increased by inhibition of GSK3β in the proteosomal pathway for its degradation.[64] The activity of Wnt signaling in NC induction has been noted above. Genes activated by the β-catenin/Lef complex include many genes associated with the mesenchymal state, cell migration and invasion.[65] In addition, nuclear localization of β-catenin[66] is associated with EMT/invasion in some carcinomas.[67] Mutations and silencing of cadherin and β-catenin that reduce association of cadherins and actin are associated with EMT and invasion in carcinoma models.[68]

Adhesion to Migration Substrate Molecules

For mouse and avian NC cells, the cells give the morphological impression of hauling themselves out of the epithelium of origin via contacts to surrounding extracellular matrix, which seems to be the initial adhesion substrate for NC cell migration. Changes in the presence and activity of integrin matrix receptors occur about the time of EMT, and are of functional importance, as shown when they are perturbed.[69] However, the usage of particular integrins by NC cells may show species variability,[70] and multiple integrins are coexpressed in NC cells.[71] In particular, the α4β1 integrin receptor for fibronectin seems to be important for migratory mesenchymal NC.[72] Integrin receptors for laminin are also of interest,[73] and there seems to be differences in avian NC cells at cranial and trunk levels.[74] Changes in matrix molecules themselves are also important, and for the NC string in axolotl, exposure to ECM from older embryos will stimulate premature migration.[75]

Cytoskeletal Motility Mechanisms

Dissociation of the N-cadherin junctions would have direct consequences on the adjacent circumferential F-actin cytoskeletal ring. This would be expected to reduce the pronounced apicobasal cell polarity typical of the epithelial state, and perhaps allow actin to participate in other arrangements, for example, association with matrix adhesions and cell motility processes. The Rho family of GTPases is important in actin-based dynamics of cell motility.[76] Consistent with this, *RhoB* was identified as a gene induced in chick neural epithelium by BMP-4, and was normally expressed transiently in the NC region just before EMT.[43] Also, broad spectrum Rho inhibition with the C3 exotransferase peptide, albeit at rather high concentrations, inhibited emigration of HNK-1 immunoreactive NC cells in in vitro assays. However, the early marker *Slug* was still detected, leading to the notion that RhoB was required for the EMT and migration of NC cells but not for the earlier induction stage.[43] RhoB is expressed before but not during migration, and over-expression of *Slug* induces RhoB expression. However, *SoxE* and *FoxD3* over-expression (which also increase NC migration; see above) suppress *RhoB* expression (see above). Thus the expression of RhoB does not seem to be crucial for EMT, and its down-regulation suggests it is not involved in migration. It is at present difficult to say whether RhoB specifically effects actin in the NC cells, and in addition, relating Rho activity to *Rho* expression is not direct. Therefore a possible effect on actin dynamics might lead to stimulation of EMT in the NC or, conversely, the timing of EMT might be regulated by prevention of premature EMT. In this latter regard, activity of the Rho family is important in maintaining the integrity of cadherin-based cell-cell adhesions[77] as seen prior to EMT. In addition by analogy with effects of RhoA on stem cell lineage choice,[78] RhoB may be involved in NC induction. In this regard, cross-talk has been observed between Rho and Wnt pathways,[79] the latter being prominent in NC induction.

Perturbation of actin organization in avian neural epithelia in vitro by pharmacological inhibition of atypical PKC[80] leads to EMT. This involves immediate actin dissociation from cell-cell junctions, cadherin delocalization and functional adhesive and motile responses typical of NC cells, and later expression of the NC marker, HNK1.[81] Interestingly, the initial EMT changes are so rapid that they must preceed gene expression changes. In addition, the final NC phenotype in this pharmacologically induced EMT is indistinguishable from spontaneously generated NC cells, but the order of change of molecular systems is different.[82]

Absence of Barriers to Emigration

In many cases the NC is not faced with a complete basal lamina, as in the avian and mammalian trunk,[5,83] but in the cranial level of mammals, NC cells clearly fragment the basal lamina to commence migration.[84] This strongly suggests proteolytic activity, and NC cells express Ets-family transcription factors[85,86] whose genes are FGF-regulated,[87] and which are known to be regulators of proteases in development, wound healing and cancer.[88] NC cells show unusually strong plasminogen activator expression[89] and activity,[90] and this activity is positively associated with migratory ability in cell culture assays.[91,92] ADAM13, a multidomain protease, is important for cranial NC migration in *Xenopus*.[93] Interestingly, the MMPs are important for NC cells even in regions where there is no basal lamina. MMP-8 is expressed in NC cells during EMT in mouse embryos.[94] MMP-2 is expressed transiently in avian trunk NC cells at the time of EMT, but not during later migration. MMP-2-specific knockdown with morpholino antisense oligonucleotides, and MMP inhibition using drugs confirm that that MMP-2 is required for neural crest EMT, but these do not inhibit migration if applied after the EMT.[95,96] Since there is no basal lamina to breach in the avian trunk, it is moot what the plasminogen activator and MMP function is, but one possibility is that it is generating ECM fragments that could act as motility signals.[97]

Conclusions

The EMT of the vertebrate NC is probably the best described and investigated EMT, and already numerous genes and molecules are known to be involved in the process. The rapid application of new functional and analysis tools such as electroporation and arrays is allowing finer and finer degrees of knowledge of the spatial and temporal relationships of molecules, and their functional relationships. Although this is revealing how complex this process is, it also holds the promise that EMT can be understood.

Acknowledgements

This work was supported by the MCRI and NHMRC Project Grant 209118. Tanya Hatzistavrou provided the in situ hybridisations shown in Figure 2.

References

1. Hay ED. An overview of epithelio-mesenchymal transformation. Acta Anat (Basel) 1995; 154(1):8-20.
2. Hall BK. The neural crest as a fourth germ layer and vertebrates as quadroblastic not triploblastic. Evol Dev 2000; 2(1):3-5.
3. Le Douarin NM, Kalcheim C. The Neural Crest. Cambridge: Cambridge University Press, 1999:445.
4. Schoenwolf GC. Cutting, pasting and painting: Experimental embryology and neural development. Nat Rev Neurosci 2001; 2(11):763-71.
5. Newgreen D, Gibbins I. Factors controlling the time of onset of the migration of neural crest cells in the fowl embryo. Cell Tissue Res 1982; 224(1):145-60.
6. Martins-Green M, Erickson CA. Development of neural tube basal lamina during neurulation and neural crest cell emigration in the trunk of the mouse embryo. J Embryol Exp Morphol 1986; 98:219-36.
7. Nichols DH. Neural crest formation in the head of the mouse embryo as observed using a new histological technique. J Embryol Exp Morphol 1981; 64:105-20.
8. Tosney KW. The segregation and early migration of cranial neural crest cells in the avian embryo. Dev Biol 1982; 89(1):13-24.
9. Epperlein HH, Lofberg J. The development of the neural crest in amphibians. Anat Anz 1993; 175(6):483-99.
10. Schoenwolf GC, Nichols DH. Histological and ultrastructural studies on the origin of caudal neural crest cells in mouse embryos. J Comp Neurol 1984; 222(4):496-505.
11. Tosney KW. The early migration of neural crest cells in the trunk region of the avian embryo: An electron microscopic study. Dev Biol 1978; 62(2):317-33.

12. Newgreen DF et al. Ultrastructural and tissue-culture studies on the role of fibronectin, collagen and glycosaminoglycans in the migration of neural crest cells in the fowl embryo. Cell Tissue Res 1982; 221(3):521-49.
13. Newgreen DF, Powell ME, Moser B. Spatiotemporal changes in HNK-1/L2 glycoconjugates on avian embryo somite and neural crest cells. Dev Biol 1990; 139(1):100-20.
14. Wilson PA, Hemmati-Brivanlou A. Vertebrate neural induction: Inducers, inhibitors, and a new synthesis. Neuron 1997; 18(5):699-710.
15. Streit A, Stern CD. Neural induction. A bird's eye view. Trends Genet 1999; 15(1):20-4.
16. Harland R. Neural induction. Curr Opin Genet Dev 2000; 10(4):357-62.
17. Streit A et al. Initiation of neural induction by FGF signalling before gastrulation. Nature 2000; 406(6791):74-8.
18. Sasai Y. Regulation of neural determination by evolutionarily conserved signals: Anti-BMP factors and what next? Curr Opin Neurobiol 2001; 11(1):22-6.
19. Wilson SI, Edlund T. Neural induction: Toward a unifying mechanism. Nat Neurosci 2001; 4(Suppl):1161-8.
20. Bonstein L, Elias S, Frank D. Paraxial-fated mesoderm is required for neural crest induction in Xenopus embryos. Dev Biol 1998; 193(2):156-68.
21. Moury JD, Jacobson AG. Neural fold formation at newly created boundaries between neural plate and epidermis in the axolotl. Dev Biol 1989; 133(1):44-57.
22. Dickinson ME et al. Dorsalization of the neural tube by the nonneural ectoderm. Development 1995; 121(7):2099-106.
23. Basch ML, Selleck MA, Bronner-Fraser M. Timing and competence of neural crest formation. Dev Neurosci 2000; 22(3):217-27.
24. Ruffins S, Bronner-Fraser M. A critical period for conversion of ectodermal cells to a neural crest fate. Dev Biol 2000; 218(1):13-20.
25. Selleck MA, Bronner-Fraser M. Origins of the avian neural crest: The role of neural plate-epidermal interactions. Development 1995; 121(2):525-38.
26. Ruffins S, Artinger KB, Bronner-Fraser M. Early migrating neural crest cells can form ventral neural tube derivatives when challenged by transplantation. Dev Biol 1998; 203(2):295-304.
27. Kenny PA, Bissell MJ. Tumor reversion: Correction of malignant behavior by microenvironmental cues. Int J Cancer 2003; 107(5):688-95.
28. Monsoro-Burq AH, Fletcher RB, Harland RM. Neural crest induction by paraxial mesoderm in Xenopus embryos requires FGF signals. Development 2003; 130(14):3111-24.
29. LaBonne C, Bronner-Fraser M. Neural crest induction in Xenopus: Evidence for a two-signal model. Development 1998; 125(13):2403-14.
30. Strutz F et al. Role of basic fibroblast growth factor-2 in epithelial-mesenchymal transformation. Kidney Int 2002; 61(5):1714-28.
31. Wu J, Saint-Jeannet JP, Klein PS. Wnt-frizzled signaling in neural crest formation. Trends Neurosci 2003; 26(1):40-5.
32. Deardorff MA et al. A role for frizzled 3 in neural crest development. Development 2001; 128(19):3655-63.
33. Garcia-Castro MI, Marcelle C, Bronner-Fraser M. Ectodermal Wnt function as a neural crest inducer. Science 2002; 297(5582):848-51.
34. Yanfeng W, Saint-Jeannet JP, Klein PS. Wnt-frizzled signaling in the induction and differentiation of the neural crest. Bioessays 2003; 25(4):317-25.
35. Vallin J et al. Cloning and characterization of three Xenopus slug promoters reveal direct regulation by Lef/beta-catenin signaling. J Biol Chem 2001; 276(32):30350-8.
36. Vincan E. Frizzled/WNT signalling: The insidious promoter of tumour growth and progression. Front Biosci 2004; 9:1023-34.
37. Liem Jr KF et al. Dorsal differentiation of neural plate cells induced by BMP-mediated signals from epidermal ectoderm. Cell 1995; 82(6):969-79.
38. Mayor R, Aybar MJ. Induction and development of neural crest in Xenopus laevis. Cell Tissue Res 2001; 305(2):203-9.
39. Selleck MA et al. Effects of Shh and Noggin on neural crest formation demonstrate that BMP is required in the neural tube but not ectoderm. Development 1998; 125(24):4919-30.
40. Sela-Donenfeld D, Kalcheim C. Regulation of the onset of neural crest migration by coordinated activity of BMP4 and Noggin in the dorsal neural tube. Development 1999; 126(21):4749-62.
41. Pasche B. Role of transforming growth factor beta in cancer. J Cell Physiol 2001; 186(2):153-68.
42. Zeisberg M, Kalluri R. The role of epithelial-to-mesenchymal transition in renal fibrosis. J Mol Med 2004; 82(3):175-81.

43. Liu J, Jessell T. A role for rhoB in the delamination of neural crest cells from the dorsal neural tube. Development 1998; 125(24):5055-5067.
44. Gammill L, Bronner-Fraser M. Neural crest specification: Migrating into genomics. Nature Reviews Neuroscience 2003; 4:795-805.
45. Bowles J, Schepers G, Koopman P. Phylogeny of the SOX family of developmental transcription factors based on sequence and structural indicators. Dev Biol 2000; 227(2):239-55.
46. Kos R et al. The winged-helix transcription factor FoxD3 is important for establishing the neural crest lineage and repressing melanogenesis in avian embryos. Development 2001; 128(8):1467-79.
47. Cheung M, Briscoe J. Neural crest development is regulated by the transcription factor Sox9. Development 2003; dev 00808.
48. Akitaya T, Bronner-Fraser M. Expression of cell adhesion molecules during initiation and cessation of neural crest cell migration. Dev Dyn 1992; 194(1):12-20.
49. Nakagawa S, Takeichi M. Neural crest cell-cell adhesion controlled by sequential and subpopulation-specific expression of novel cadherins. Development 1995; 121(5):1321-32.
50. Cheng Y et al. Chick sox10, a transcription factor expressed in both early neural crest cells and central nervous system. Brain Res Dev Brain Res 2000; 121(2):233-41.
51. Dottori M et al. The winged-helix transcription factor Foxd3 suppresses interneuron differentiation and promotes neural crest cell fate. Development 2001; 128(21):4127-38.
52. Tucker GC et al. Identical reactivity of monoclonal antibodies HNK-1 and NC-1: Conservation in vertebrates on cells derived from the neural primordium and on some leukocytes. Cell Differ 1984; 14(3):223-30.
53. Blanco MJ et al. Correlation of Snail expression with histological grade and lymph node status in breast carcinomas. Oncogene 2002; 21(20):3241-6.
54. Manzanares M, Locascio A, Nieto MA. The increasing complexity of the Snail gene superfamily in metazoan evolution. Trends Genet 2001; 17(4):178-81.
55. Sefton M, Sanchez S, Nieto MA. Conserved and divergent roles for members of the Snail family of transcription factors in the chick and mouse embryo. Development 1998; 125(16):3111-21.
56. del Barrio MG, Nieto MA. Overexpression of Snail family members highlights their ability to promote chick neural crest formation. Development 2002; 129(7):1583-1593.
57. LaBonne C, Bronner-Fraser M. Snail-related transcriptional repressors are required in Xenopus for both the induction of the neural crest and its subsequent migration. Dev Biol 2000; 221(1):195-205.
58. Cano A et al. The transcription factor snail controls epithelial-mesenchymal transitions by repressing E-cadherin expression. Nat Cell Biol 2000; 2(2):76-83.
59. Honore SM et al. Sox10 is required for the early development of the prospective neural crest in Xenopus embryos. Developmental Biology 2003; 260(1):79-96.
60. Pohl BS, Knochel W. Overexpression of the transcriptional repressor FoxD3 prevents neural crest formation in Xenopus embryos. Mech Dev 2001; 103(1-2):93-106.
61. Newgreen DF, Gooday D. Control of the onset of migration of neural crest cells in avian embryos. Role of Ca++-dependent cell adhesions. Cell Tissue Res 1985; 239(2):329-36.
62. Nakagawa S, Takeichi M. Neural crest emigration from the neural tube depends on regulated cadherin expression. Development 1998; 125(15):2963-71.
63. Monier-Gavelle F, Duband JL. Control of N-cadherin-mediated intercellular adhesion in migrating neural crest cells in vitro. J Cell Sci 1995; 108(Pt 12):3839-53.
64. Nelson WJ, Nusse R. Convergence of Wnt, beta-catenin, and cadherin pathways. Science 2004; 303(5663):1483-7.
65. Schwartz DR et al. Novel candidate targets of beta-catenin/T-cell factor signaling identified by gene expression profiling of ovarian endometrioid adenocarcinomas. Cancer Res 2003; 63(11):2913-22.
66. Sharpe C, Lawrence N, Martinez Arias A. Wnt signalling: A theme with nuclear variations. Bioessays 2001; 23(4):311-8.
67. Wong NA, Pignatelli M. Beta-catenin—a linchpin in colorectal carcinogenesis? Am J Pathol 2002; 160(2):389-401.
68. Conacci-Sorrell M, Zhurinsky J, Ben-Ze'ev A. The cadherin-catenin adhesion system in signaling and cancer. J Clin Invest 2002; 109(8):987-991.
69. Lallier T, Bronner-Fraser M. Inhibition of neural crest cell attachment by integrin antisense oligonucleotides. Science 1993; 259(5095):692-5.
70. Pietri T, Thiery JP, Dufour S. Differential expression of beta3 integrin gene in chick and mouse cranial neural crest cells. Dev Dyn 2003; 227(2):309-13.
71. Testaz S, Delannet M, Duband J. Adhesion and migration of avian neural crest cells on fibronectin require the cooperating activities of multiple integrins of the (beta)1 and (beta)3 families. J Cell Sci 1999; 112(Pt 24):4715-28.

72. Testaz S, Duband JL. Central role of the alpha4beta1 integrin in the coordination of avian truncal neural crest cell adhesion, migration, and survival. Developmental Dynamics 2001; 222(2):127-40.
73. Lallier T, Bronner-Fraser M. Alpha 1 beta 1 integrin on neural crest cells recognizes some laminin substrata in a Ca(2+)-independent manner. J Cell Biol 1992; 119(5):1335-45.
74. Lallier T et al. Cranial and trunk neural crest cells use different mechanisms for attachment to extracellular matrices. Development 1992; 116(3):531-41.
75. Lofberg J et al. Stimulation of initial neural crest cell migration in the axolotl embryo by tissue grafts and extracellular matrix transplanted on microcarriers. Dev Biol 1985; 107(2):442-59.
76. Raftopoulou M, Hall A. Cell migration: Rho GTPases lead the way. Dev Biol 2004; 265(1):23-32.
77. Fukata M, Kaibuchi K. Rho-family GTPases in cadherin-mediated cell-cell adhesion. Nat Rev Mol Cell Biol 2001; 2(12):887-97.
78. McBeath R et al. Cell shape, cytoskeletal tension, and RhoA regulate stem cell lineage commitment. Dev Cell 2004; 6(4):483-95.
79. Malliri A, Collard JG. Role of Rho-family proteins in cell adhesion and cancer. Curr Opin Cell Biol 2003; 15(5):583-9.
80. Minichiello J et al. Induction of epithelio-mesenchymal transformation of quail embryonic neural cells by inhibition of atypical protein kinase-C. Cell Tissue Res 1999; 295(2):195-206.
81. Newgreen DF, Minichiello J. Control of epitheliomesenchymal transformation. I. Events in the onset of neural crest cell migration are separable and inducible by protein kinase inhibitors. Dev Biol 1995; 170(1):91-101.
82. Newgreen DF, Minichiello J. Control of epitheliomesenchymal transformation. II. Cross-modulation of cell adhesion and cytoskeletal systems in embryonic neural cells. Dev Biol 1996; 176(2):300-12.
83. Martins-Green M, Erickson CA. Basal lamina is not a barrier to neural crest cell emigration: Documentation by TEM and by immunofluorescent and immunogold labelling. Development 1987; 101(3):517-33.
84. Nichols DH. Ultrastructure of neural crest formation in the midbrain/rostral hindbrain and preotic hindbrain regions of the mouse embryo. Am J Anat 1987; 179(2):143-54.
85. Maroulakou IG, Papas TS, Green JE. Differential expression of ets-1 and ets-2 proto-oncogenes during murine embryogenesis. Oncogene 1994; 9(6):1551-65.
86. Fafeur V et al. The ETS1 transcription factor is expressed during epithelial-mesenchymal transitions in the chick embryo and is activated in scatter factor-stimulated MDCK epithelial cells. Cell Growth Differ 1997; 8(6):655-65.
87. Munchberg SR, Steinbeisser H. The Xenopus Ets transcription factor XER81 is a target of the FGF signaling pathway. Mech Dev 1999; 80(1):53-65.
88. Remy P, Baltzinger M. The Ets-transcription factor family in embryonic development: Lessons from the amphibian and bird. Oncogene 2000; 19(55):6417-31.
89. Pietri T et al. The human tissue plasminogen activator-Cre mouse: A new tool for targeting specifically neural crest cells and their derivatives in vivo. Dev Biol 2003; 259(1):176-87.
90. Valinsky JE, Le Douarin NM. Production of plasminogen activator by migrating cephalic neural crest cells. Embo J 1985; 4(6):1403-6.
91. Agrawal M, Brauer PR. Urokinase-type plasminogen activator regulates cranial neural crest cell migration in vitro. Dev Dyn 1996; 207(3):281-90.
92. Erickson CA, Isseroff RR. Plasminogen activator activity is associated with neural crest cell motility in tissue culture. J Exp Zool 1989; 251(2):123-33.
93. Alfandari D et al. Xenopus ADAM 13 is a metalloprotease required for cranial neural crest- cell migration. Curr Biol 2001; 11(12):918-30.
94. Giambernardi TA et al. Neutrophil collagenase (MMP-8) is expressed during early development in neural crest cells as well as in adult melanoma cells. Matrix Biol 2001; 20(8):577-87.
95. Duong TD, Erickson CA. MMP-2 plays an essential role in producing epithelial-mesenchymal transformations in the avian embryo. Dev Dyn 2004; 229(1):42-53.
96. Cai DH, Brauer PR. Synthetic matrix metalloproteinase inhibitor decreases early cardiac neural crest migration in chicken embryos. Dev Dyn 2002; 224(4):441-9.
97. Knapp DM, Helou EF, Tranquillo RT. A fibrin or collagen gel assay for tissue cell chemotaxis: Assessment of fibroblast chemotaxis to GRGDSP. Exp Cell Res 1999; 247(2):543-53.

Chapter 4

Epithelial-Mesenchymal Transformation in the Embryonic Heart

Raymond B. Runyan, Ronald L. Heimark, Todd D. Camenisch and Scott E. Klewer

Abstract

The progenitors of the mitral and tricuspid valves and the membranous interventricular septum in the heart arise by an epithelial-mesenchymal cell transformation (EMT) from embryonic endothelial cells. Experiments using collagen gel cultures to mimic the three dimensional environment in the embryonic heart cushions showed that EMT was triggered by an inductive stimulus produced by the adjacent myocardium and that myocardium from a nontransforming region was insufficient to induce transformation. Studies using chick embryos have demonstrated distinct and sequential roles for two isoforms of TGFβ, Bone Morphogenetic Proteins, and roles for several TGFβ-family receptors in this EMT. Additional studies have identified at least two separate transcription factors required for EMT, Slug and Mox-1. The complexity of the inductive signal is not yet fully understood, but recent studies have shown roles for several Wnt proteins and a requirement for signaling by extracellular hyaluronan. The embryonic heart provides an experimental model of relevance to congenital heart diseases that are likely to continue to provide novel insight into the regulation of EMT as a normal process of development.

Introduction

While not the first instance of phenotypic change by an epithelium in the embryo (see Chapter 2, Morali et al), the formation of mesenchymal cells in the embryonic heart provides an instructive model of epithelial mesenchymal transformation (EMT). Though there are several EMT processes that take place in the heart this chapter will focus on the formation of mesenchymal cells from endothelial cells lining the atrioventricular (AV) canal. These mesenchymal cells form the earliest progenitors of the fibroblasts of the septum intermedium and the mitral and tricuspid valves. Two other observed EMT processes in the heart include a similar but not identical process in the outflow tract of the heart that forms the precursors of the aortic and pulmonary valves[1] and an EMT of the epicardium that produces both the fibroblasts of the myocardium and the vascular precursors of the coronary circulation. Though some mention of complementary or conflicting information may be mentioned, there is no effort to be comprehensive concerning these other cardiac EMTs.

Study of the embryonic AV canal has several advantages for investigators attempting to understand the process of phenotypic shape change by an epithelium. EMT in the embryonic heart is an induced process that occurs with very defined timing in the embryo. Unlike several other EMTs in the embryo described in the present volume, the valvular progenitors do not appear to participate in widely divergent differentiation programs and there is no evidence to

Rise and Fall of Epithelial Phenotype: Concepts of Epithelial-Mesenchymal Transition, edited by Pierre Savagner.

suggest that AV canal endothelial cells are divided into subsets of committed and uncommitted cell populations. The ability to identify and collect heart tissues prior to EMT enables considerable advantage in obtaining cells and tissues for examination. Studies over the last 20 years, particularly in culture, have provided significant insight into EMT and the present review will attempt to summarize the progress made in this system.

Description of EMT in Vivo

Description of embryonic heart development dates from the time of Aristotle. The obvious external location and motion of the heart in the chick embryo made it a focal point of observation. Using the light microscope, cellularity of the cardiac cushions of the AV canal was obvious although the role of these cells was not well understood.[2] The modern focus on the development of the cardiac cushions and the origins of their cellular constituents dates from the investigations of Markwald and Manasek and their colleagues in the late 1970s. Using electron microscopy, light microscopy and histochemical procedures, these investigators described the formation of cushion mesenchymal cells in the AV canal and explored their interaction with the extracellular matrix that is the substance of the cardiac cushion.[3-6] These investigators focused on the origin of mesenchyme from the endothelium of the atrioventricular canal and explored the interactions with the extracellular matrix. Manasek[6] speculated that the formation of mesenchyme in the AV canal was due to a tissue interaction with the myocardium, but there was no experimental evidence to support this conjecture at that time. There remained some controversy as to the origin of the cardiac mesenchyme. Some argued that the myocardium of the heart was the source of mesenchyme in the AV canal. This controversy was resolved by the work of Kinsella and Fitzharris.[7] These investigators utilized a microscope and a movie camera to view a cross-section of the AV canal in a culture chamber and photographed epithelial mesenchymal cell transformation in situ strictly from the endothelium. These studies combined to provide the basic picture of EMT in vivo summarized below. There has been some more recent work concerning "myocardialization" of the heart valves by cells migrating from the myocardial layer at much later stages, but it is not clear that this process represents an EMT.[8]

As shown in Figure 1, the looped heart of the stage 17 chick[9] embryo shows an expanded extracellular matrix in the atrioventricular canal that forms two opposing cushions. These cushions consist of an acellular extracellular matrix covered by an endothelium. Radioactive studies with labeled glucosamine suggest that the vast majority of the extracellular matrix is produced by the myocardium.[10] By this stage, the endothelium shows morphological signs of EMT. Endothelial cells are hypertrophied,[3] have polarized golgi apparatus,[10] show loss of cell-cell adhesion[11] and have extended filopodia.[3,7] Mesenchymal cells have invaded the underlying extracellular matrix.[3,12] Movies by Kinsella and Fitzharris enabled the appreciation of the remarkable length of these filopodia as they appeared to probe more than 50 μm into the ECM.[7] Scanning electron microscope pictures[11,13] showed that the hypertrophy of the endothelia was accompanied by a loss of cell-cell adhesion as openings became visible in the endothelial sheet and their appeared to be lateral migration in the endothelial layer. After probing the matrix, a subset of endothelial cells left the endothelial sheet and entered the ECM of the cushion tissue. The ECM of the cushion was shown to by a hyaluronan-rich ECM while the passage of mesenchymal cells through the cushion produced a matrix that was rich in chondroitin sulfate proteoglycans.[14] EMT by the AV canal endothelium continued at least through stage 19 and produced highly cellular cushions that would be remodeled into valvular structures with subsequent development. Figure 2 shows that the mitral and tricuspid valves and the membranous septum of the ventricle are developed by remodeling of the progenitor cells that invaded the extracellular matrix of the cardiac cushions. It is worth noting that although there are micrographs that suggest a broad outline of cushions remodeling into valves,[15,16] there is very little knowledge concerning these mechanisms and the development of connections to the papillary muscles. For example, during mouse heart development immunolocalization results indicate

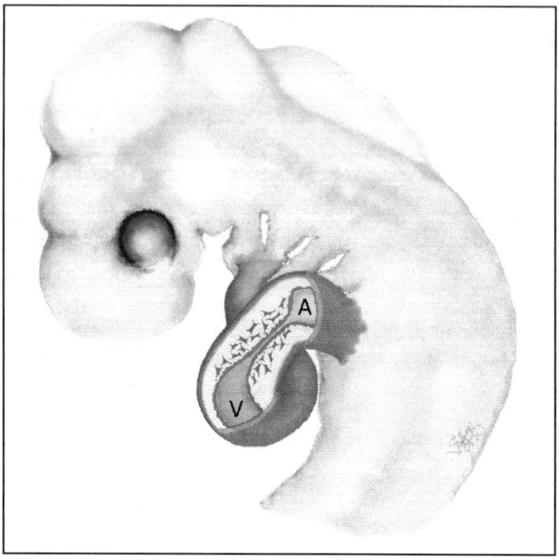

Figure 1. EMT in the looped heart is shown by a cutaway view of the atrioventricular canal in a stage 17 chick embryo. Cardiac cushions are shown projecting into the lumen of the heart between the atrium (A) and the ventricle (V). Mesenchymal cells are shown within the extracellular matrix of the each cushion. These mesenchymal cells arise from the endothelium lining the lumen and covering the cardiac cushions.

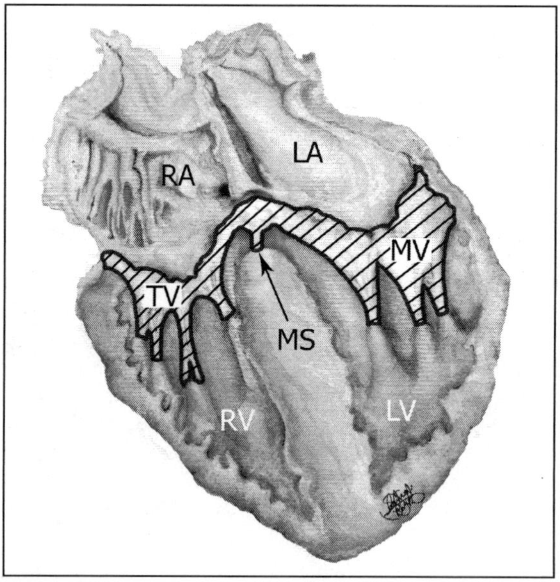

Figure 2. Structures in the adult heart formed by an EMT in the atrioventricular canal. The mitral (MV) and tricuspid (TV) valves and the membranous ventricular septum (MS) are populated by fibroblastic cells produced by the embryonic EMT. RA=right atrium, LA= left atrium, RV=right ventricle and LV=left ventricle.

expression of type VI collagen during the period when the AV endocardial cushions differentiate into valve leaflets and the membranous septa. These findings support the hypothesis of a role of collagen VI in normal AV endocardial cushion differentiation. In the more mature AV valves of the mouse, high expression of collagen VI is confined to a thin layer on the ventricular side of the AV valve leaflets.[15]

Development of Collagen Gel Culture

Our understanding of the mechanisms underlying EMT in the heart arose out of the development of a three-dimensional collagen gel culture system. Though the utilization of collagen gels as substrates for three-dimensional migration arose out of the work of Elsdale and Bard,[17] it is not clear that cells were actually invasive into their dense substrates. However, using gelling solutions of collagen enabled several investigators of the early 1980s to explore the formation and migration of mesenchyme or fibroblasts in three dimensions.[18-21] Differences in behavior by epithelial cells and mesenchyme were observed in natural ECM at approximately the same time by Overton.[22] Collagen gels were first used in the heart by Bernanke and Markwald[21] to explore the interactions between cushion mesenchyme and extracellular glycosaminoglycans.

In 1983, collagen gel cultures were used to demonstrate that, as predicted by Manasek,[6] that EMT in the heart was the product of a tissue interaction between the endothelium and the myocardium.[23] These experiments consisted of the timed culture and removal of the myocardium with staged AV canal explants. These data showed that the myocardium produced an inductive stimulus that was required to initiate the EMT and that ventricular explants had little potential for EMT. Subsequent experiments[24-26] showed that the stimulus for EMT could be provided by either an extract of the ECM or by myocardium-conditioned medium. Comparison of ventricular and AV canal explants supported the regional specificity of EMT in vivo. Ventricular myocardium was incapable of inducing EMT in either ventricular endothelium or AV canal endothelium.[26,27] The timing and appearance of the cultures suggested a faithful replication of the EMT process in vivo. These cultures have been subsequently used by a number of investigators to explore the molecular and cellular aspects of EMT in this system. Figure 3 depicts the EMT process in vitro and summarizes the regional nature of the inductive stimulus.

Identification of Components of the EMT Process

The initial exploration of the EMT process focused upon the role of the extracellular matrix. As EMT was shown to be a response to a myocardial stimulus, the first approach was to collect and test the extracellular matrix intervening between the cell layers. Krug et al[10] showed that a hyaluronidase-treated fraction of heart ECM produced an apparent activation by endothelial cells but it was insufficient to induce invasion of collagen cultures by mesenchymal cells. Subsequently it was shown that either an ECM extract of the heart collected without hyaluronidase or myocardium-conditioned medium was sufficient to induce EMT in a competent monolayer of stage 14 endothelial cells.[24,25] Though, at the time, experiments combining hyaluronidase-treated ECM with exogenous hyaluronan did not produce EMT,[10] it should be noted that hyaluronidase disrupts AV endocardial cushion EMT in vitro, and an active role for hyaluronan in EMT was recently demonstrated in heart explant tissues from genetically engineered mouse models (Fig. 4).[28] A similar phenotype to the Has2$^{-/-}$ mouse embryos occurs in Zebrafish containing a mutated UDP-glucose dehydrogenase gene, encoding an upstream enzyme of Has2 (*jekyll* mutant).[29]

Because of its ubiquitous distribution in the cardiac jelly and the embryonic ECM, it is unlikely that hyaluronan alone stimulates AV canal EMT. Based on their role as upstream activators of Ras, restricted patterns of expression during cardiac morphogenesis, and cardiac phenotypes in null mutants, members of the ErbB family of receptor tyrosine kinases may act coordinately with hyaluronan to mediate this process. ErbB2, ErbB3 and ErbB4 are upstream activators of Ras and are important in heart development.[30,31] We recently demonstrated that the induction of AV endothelial cushion EMT by hyaluronan involves ErbB2 and ErbB3. The

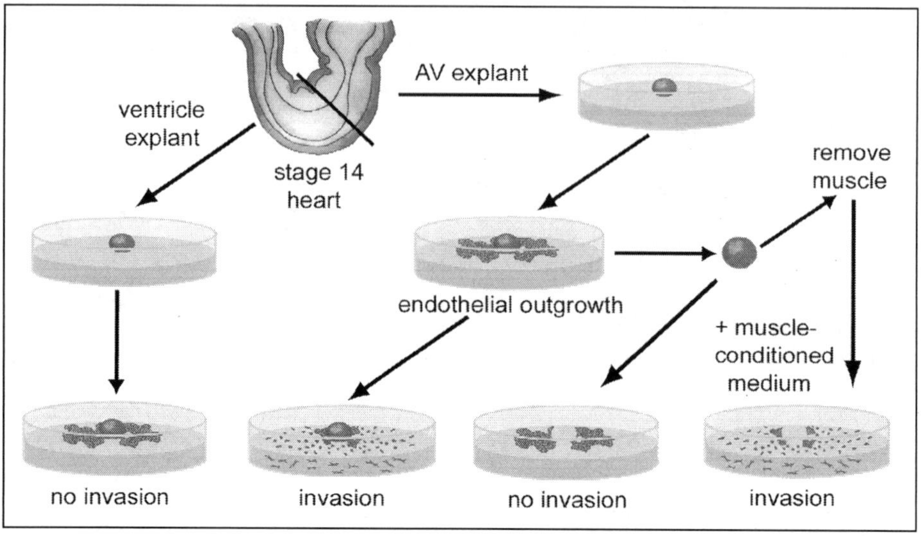

Figure 3. Utilization of the collagen gel assay system to explore EMT. An atrioventricular explant is collected from a stage 14 (preEMT) heart and placed on the surface of a collagen gel. Endothelial cells grow out from the explant over the surface of the gel. Continued culture of the explant produces a population of mesenchymal cells derived from the endothelium. Removal of the muscular portion of the explant prevents the EMT. Addition of muscle-conditioned medium restores EMT by the endothelium. An equivalent explant of ventricular tissue produces little, if any, EMT.

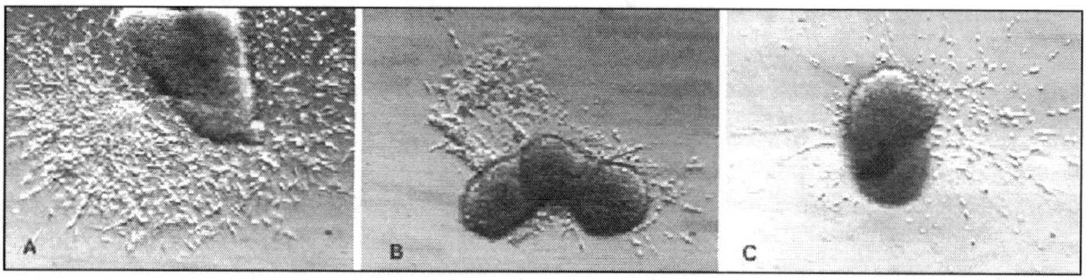

Figure 4. Demonstration of requirement for hyaluronan for EMT. Panel A shows a normal chick heart explant with a large outgrowth of activated endothelial cells and invaded mesenchyme. Panel B is an equivalent chick explant grown in the presence of streptomyces hyaluronidase to remove hyaluronan. Panel C shows an equivalently staged mouse heart atrioventricular canal explant (E9.5) from a $Has2^{-/-}$ animal. Outgrowth and EMT is inhibited by both genetic and enzymatic loss of hyaluronan.

ErbB3 receptor is activated in response to exogenous hyaluronan rescue in $Has2^{-/-}$ AV canals explants, and EMT is rescued in $Has2^{-/-}$ AV canals explants by the ErbB receptor ligand heregulin. In addition, soluble ErbB3 and ErbB2 inhibitors block these signaling events. These studies demonstrate a costimulatory pathway involving hyaluronan and a tyrosine kinase receptor-ligand family are required for AV endocardial cushion EMT.[28]

One approach to explore the process of EMT, was the production of a polyclonal antiserum against the inductive ECM fraction (ES antibodies). This antiserum recognized a number of ECM components and was capable of blocking EMT in the collagen gel cultures.[32] Due to the success of this approach, the ECM extract was used as an antigen for monoclonal antibody production. Clones were tested for activity against EMT cultures. One of these antibodies was particularly effective and as it recognized a 130 KD protein called ES/130.[33] By a variety of criteria, the antigen recognized by ES/130 appeared to be a critical mediator of EMT.[34] However analysis of the ES/130 sequence suggests that it is most likely an intracellular protein that participates in regulated secretion (Krug, personal communication). The presence of the ES/130 antigen in the ECM of the heart may have to do with the extensive secretory activity of the myocardium during the development of the cardiac cushions. Under the conditions of specimen preparation utilized by these investigators, packets of secreted material may remain associated with the ES/130 antigen. The basis for EMT inhibition produced by ES/130 is unclear. Perhaps the antibody perturbs interaction between the antigen and one or more components of the ECM. This immune approach towards the identification of critical components of the extracellular matrix also resulted in the identification of transferrin as a mediator of cardiac cushion development.[35] An additional molecule identified by this approach, hLAMP1, remains under investigation.[36]

An alternative approach, to identification of mediators of EMT was to test a series of growth factors against competent endothelium (AV endothelial cell cultures with the myocardium removed prior to induction of EMT). There were few differences observed between the ECM of the ventricle and the AV canal in two dimensional protein gels but those that were observed were in a low molecular weight area consistent with growth factors.[10] A series of growth factors, including EGF, FGF2, and TGFβ, were tested.[27] Alone, none of these growth factors produced EMT. However, when cultured in combination with a ventricular explant, TGFβ produced EMT.[27] (For a review of TGFβ see Chapter 15, Vignais and Fafet, and also Roberts et al[37]). EGF and FGF2 had no effect on ventricular cultures. The original experiment utilized the TGFβ1 isoform but it is clear that the exogenous delivery of any of the three TGFβ isoforms was sufficient to produce EMT.[27,38] Confirmation that TGFβ was a mediator of EMT was provided by the observation that a pan-specific antibody against all of the TGFβ isoforms blocked EMT in intact AV canal explant cultures.[27]

TGFβ as a Mediator of EMT

The demonstration of TGFβ as a mediator of EMT resulted in a series of experiments to resolve the parameters of the interaction. RNase protection assays were undertaken to define the TGFβ isoforms present in the chick heart. These experiments showed a greater concentration of TGFβ3 in the AV canal than in the adjacent ventricle but approximately equal amounts of TGFβ2 mRNA in the AV canal and the ventricle. TGFβ4 (the avian equivalent of mammalian TGFβ1[39] was not found in the heart during early valve development.[40] In contrast, studies by Akhurst and colleagues[41,42] in the mouse heart showed that TGFβ1 was present largely in the endothelium and that TGFβ2 was present in the myocardium and in the mesenchyme of the cardiac cushions after EMT. Though these investigators found little TGFβ3 in the mouse heart, a recent study shows this isoform appears in the mouse heart cushions after EMT.[43] In situ hybridization studies by Barnett et al[44] and Boyer et al[45] have produced the following pattern of TGFβ transcription in the chick heart. Prior to EMT, TGFβ2 mRNA is present throughout the myocardium and endothelium of the heart. In contrast, TGFβ3 transcripts are only present in the myocardium. Subsequent to endothelial activation, TGFβ3 is detected in the endothelium and mesenchyme of the AV canal. Migratory mesenchymal cells continue to produce TGFβ2 as they migrate in the cushion. Considering the preservation of specific TGFβ isoform homology between species, the differences between chick and mouse hearts in terms of isoform distribution and utilization (below) are not completely explained.

Functional utilization of TGFβ isoforms in the AV canal was first explored by the use of antisense oligonucleotides designed against each of the published TGFβ sequences. Modified antisense oligonucleotides were delivered to explant cultures and the effect on EMT was observed. Oligonucleotides directed against TGFβ3, alone, were able to perturb EMT in the explant cultures.[46] Though it was noted that the antisense oligonucleotides against TGFβ3 did not completely recapitulate the phenotype produced by the pan-specific anti-TGFβ antibody it was accepted that TGFβ3 was the critical member of this family involved in cardiac EMT.

Studies by Doetschman and colleagues[47-49] were focused on the production of null mice for each of the TGFβ isoforms. The surprising observation in the mouse was that only the TGFβ2 null mouse had heart defects. Defects in this mouse include structures derived from the cardiac cushions and had a similarity to defects seen in the human as the Tetrology of Fallot as well as an atrial septal defect. The TGFβ3 null mouse had defects in EMT in the palate but its heart was normal.

In light of these findings, the role of TGFβ2 in the chick was reexamined. As only one antisense oligonucleotide was tested against TGFβ2 in the previous study, it was possible that the lack of inhibition seen in the earlier study was due to an ineffective oligonucleotide sequence. Studies were undertaken with isoform-specific, blocking antibodies in collagen gel cultures. This approach uncovered separate and sequential activities for TGFβ2 and TGFβ3 during EMT in the chick heart. As seen before with antisense TGFβ3 oligonucleotides, anti-TGFβ3 antibodies blocked EMT after cell separation by preventing invasion of the collagen gel matrix. However, anti-TGFβ2 antibodies blocked the initial step of cell-cell separation similar to the effect seen with the pan-specific anti-TGFβ antibody used in a previous study.[27,45] This apparent difference between mice and chicks was tested using mouse heart explants in collagen gel cultures. As seen in the TGFβ null mice, inhibition of TGFβ2, not TGFβ3 blocked EMT in vitro. TGFβ3 levels in the AV canal of the mouse rise only after EMT.[43]

TGFβ Receptors

There are three principal TGFβ receptors in most biological systems (reviewed by ref. 50). A signaling complex is formed between the TGFβ Type II receptor and the Type I receptor after binding TGFβ. The Type II receptor has higher affinity for TGFβ1 and TGFβ3 than for TGFβ2.[51] The Type I receptor does not appear to bind TGFβ in the absence of the Type II receptor but provides the serine-threonine kinase activity that initiates signal transduction into the cell.[52] A third receptor, betaglycan or the TGFβ Type III receptor, has a large extracellular domain and a small cytoplasmic domain. This receptor has a somewhat greater affinity for TGFβ2 compared to the other isoforms and has been proposed to present this isoform to the Type II receptor.[53]

Studies examining the TGFβ receptors in the heart were made possible by the efforts of Barnett and colleagues who cloned the avian homologues of the TGFβ receptors.[54-56] Initial studies with antibodies towards the TGFβ Type II receptor showed that this receptor was expressed throughout the endothelium lining the developing heart and the blood vessels of the embryo. The antibodies were shown to block receptor function and EMT in collagen gel culture was inhibited.[55] Similar studies were performed after the cloning and production of antibodies to the Type III receptor.[56] Unlike the Type II receptor, immunostaining for the Type III receptor in the endothelium was limited to the developing cardiac cushions. This antibody proved to block EMT in collagen gel cultures as well.[56] Examination of antibody-treated cultures revealed some differences between the two receptors.[57] Cultures treated with antibody towards TGFβ Type III receptor had a close similarity with cultures treated with a blocking antibody towards TGFβ2.[45] In both cases, the endothelial outgrowth of the cardiac explants remained cohesive and "unactivated". In contrast, cultures blocked with antibody towards the Type II receptor demonstrated a separated, "activated" endothelium where there were numerous fusiform cells on the cell surface. This was similar to the inhibition previously seen with either antisense oligonucleotides or antibodies against TGFβ3.[45,46] These data suggest that although ligand and receptor affinity differences do not appear to be significant in vitro, there is specificity between ligand isoforms and receptor types that can be distinguished in situ.

TGFβ Activation

One aspect of TGFβ biology in the heart that is difficult to assess is the role of ligand activation. Each of the TGFβ isoforms is normally secreted in an inactive form with its amino terminal sequence attached and functional as a block to receptor binding.[58] Both pH and proteolytic treatment can be performed in vitro to activate the ligand. In vivo, it appears that that TGFβ isoforms associate with a latent TGFβ binding protein and that this association is critical for either activation or presentation of the active ligand to its receptor.[59] Brauer et al[60] demonstrated that ninety five percent of the TGFβ3 found in the extracellular matrix of the heart outflow tract is un-activated. Similar levels of unactivated TGFβs were found by McCormick.[61] Nakajima and colleagues showed that inhibition of latent TGFβ binding protein-1 (LTBP-1) by a blocking antibody could prevent EMT in mouse heart explants in culture.[62] Utilization of both antisense oligonucleotides and antibodies against LTBP-1 in chick heart cultures confirms this finding and shows that the TGFβ2-mediated endothelial separation is particularly sensitive to these reagents (Berkompas and Runyan, in preparation). These data show that regulation of the TGFβ-mediated embryonic induction could be accomplished by secretion or activation of proteases into the ECM between the endothelium and the myocardium. Studies by McGuire and colleagues[63-65] have identified several proteases, including urokinase and matrix metalloproteinases 2 and 9, in the heart that could mediate TGFβ activation. While the work of these authors has focused largely on the role of these molecules in cell migration, inhibitory antibodies towards urokinase will block EMT in vitro (Romano and Runyan, unpublished).

Extracellular Matrix in EMT

The structure of a cardiac cushion prior to EMT is essentially a large area of extracellular matrix (ECM) bounded by the myocardium and the endothelium. This ECM is rich in hyaluronan (HA), chondroitin sulfate proteoglycan, type VI collagen, fibulins, and fibrillins.[66,67] In addition to serving as primitive valves, the cushion matrix is remodeled to form the mature valves and septum. Molecular genetic studies in the mouse demonstrate that both hyaluronan (HA) and versican, a HA binding proteoglycan, are required for formation of the cardiac cushions. Animals lacking these matrix molecules die at E9.5 to 10.5, shortly after the time of normal development of the AV and outflow tract cushions. Evidence for a critical role of versican in heart development was provided by the Heart Defect (*hdf*) transgenic mouse line.[68,69] This line resulted from the transgenic insertion of a putative enhancer-lacZ reporter gene construct into the region encoding the glycosaminoglycan attachment domain of versican. *Hdf* mice do not survive past E10.5, and the AV and outflow tract cushions are absent.

The data support both space filling and ligand type roles for HA. While deficits in vivo correspond to a loss of space-filling HA, many of the in vitro studies suggest that a ligand type of role is equally important. For example, the removal of HA by hyaluronidase digestion of chick heart ECM produced an incomplete EMT in culture.[10] ECM extracts were fully inductive when extracted without hyaluronidase.[25] In whole rat embryo cultures, hyaluronidase blocked cardiac looping, cushion development and resulted in collapsed cushions.[70] The addition of streptococcus hyaluronidase to normal E9.5 AV canal cushion mouse explants blocked endothelial transformation into mesenchyme where, presumably, space-filling is less important.[71] Gene targeting of the principle hyaluronan synthase, Has2, also results in embryonic lethality ~E9.5 in part due to endocardial cushion defects both of the AV canal and OFT.[72] Embryonic cushion cultures established from $Has2^{-/-}$ embryos in the collagen gel invasion assay lack mesenchyme formation which can be rescued by adding HA or reintroducing the Has2 gene into the explanted cultures.[72] Finally, it was recently demonstrated that HA cooperates with growth factors to signal specified endocardium to transform into cushion mesenchyme.[28] Together, these data suggest that both HA and versican are essential for formation of the acellular premigratory endocardial cushion cardiac jelly. It is not yet clear if other matrix components in this matrix also play essential roles, although collagen VI has been implicated in the cardiac defects common in trisomy 21 (Down Syndrome).[73]

In addition to space-filling and ligand roles, components of the ECM provide a motility substrate for the cell migration component of EMT. The effect of adhesion signals on cellular behavior is complex. Cell migration shows a biphasic dependence on adhesion, with maximal cell movement occurring at intermediate adhesion levels.[74] This balance requires exquisite control of cellular activities with the ECM. In this regard, integrins are a well-characterized family of cell surface receptors capable of mediating these activities. Integrins are heterodimers composed of α and β subunits that function as bidirectional signaling molecules.[75] Variation in levels of specific matrix molecules has been demonstrated to govern integrin-mediated cell migration.[76] Modulation of integrin-ECM binding affinity,[77-79] and changes in levels of integrins expressed on the cell surface can alter cell migration rates.[80] Integrin binding of the ECM is well known to initiate signals that are transmitted into the cell altering cell adhesion, migration, proliferation, differentiation, and cell survival.[81] Loeber et al[82] used the collagen gel assay to directly examine the effects of disrupting integrin-ECM binding during AV canal EMT. Function blocking $\beta 1$ integrin antibodies, RGD peptides against FN, and YIGSR peptides against laminin blocked EMT and cell migration. These observations demonstrate that $\beta 1$ integrin binding to the ECM is essential for AV canal EMT. Studies are now investigating distinct integrin heterodimer modulation of AV canal EMT.

Other Growth Factors in EMT

As described in Chapter 18, Klymkowsky, in this volume, Wnt proteins are associated with several EMTs during embryonic development. A recent report of the expression of the soluble Wnt inhibitor, FrzB, showed expression in the cardiac cushions.[83] While such expression may indicate a role in EMT for Wnts, it is conceivable that Wnt activity mediates other developmental events in the heart. To test the activity of Wnts in cardiac EMT, several antisense oligonucleotides were prepared against FrzB. Collagen gel experiments with these oligonucleotides showed a two-fold increase in the number of mesenchymal cells compared to controls. Conversely, the application of mouse FRP-3, a homologue of FrzB, to collagen gel cultures blocked EMT (Person, Klewer and Runyan, in preparation). Preliminary experiments have identified 6 different Wnt proteins in the heart at the time of EMT. Experiments are underway to identify which of these are involved in EMT (unpublished data).

The TGFβ isoforms are not the only members of the TGFβ superfamily that appear to play a role in cardiac EMT. Lyons et al[84] observed that BMP4 (formerly BMP2A) was expressed in a collar of myocardium around the AV canal in the mouse embryo at the time of EMT. While it was tempting to suggest a functional role in EMT in the mouse, null animals for BMP4 died around the time of EMT and it was difficult to determine whether normal EMT had taken place. More recently, expression in the AV canal of the mouse heart and the chick heart that suggested both BMP2 and BMP4 might be involved. Nakajima and coworkers[85,86] found that BMP2 was synergistic with TGFβ3 in promoting EMT in chick AV canal tissue cultures. Further, BMPs 5, 6 and 7 have all been found in the heart.[87] However, experiments utilizing misexpression of the BMP inhibitor, Noggin, and mutations in either BMP receptors or BMPs 6 and 7 have consistently shown stronger effects in the outflow tract than the AV canal.[87,88] These data suggest that EMT in the outflow tract of the heart may be somewhat differently regulated.

Signal Transduction during EMT

Studies of signal transduction during EMT began with the observation that pertussis toxin was a potent blocker of EMT. Pertussis toxin ADP-ribosylates the alpha subunit of G proteins in the Gi or Gq class of G proteins. This inactivates and inhibits the functions of these molecules.[89] This study also showed that EMT was also inhibited by inhibitors of protein kinase C and serine-threonine kinase (and that activation of endothelial cells was accompanied by a flux in intracellular calcium.[89] Subsequent studies of signal transduction in the heart focused on the TGFβ receptors described above. Evidence for ErbB and Ras function during EMT suggests that these signal transduction mechanisms are also involved see also Chapter 16, Boyer.[28] Though not yet integrated into a adequate picture of EMT in the heart, a modulator of Ras signaling, NF-1, is known to be present and critical in this system for normal valve development.[90]

Cell-Cell Regulation of EMT

Early in cardiac development distinct cadherin subtypes mediate adhesion and segregation of endocardial and myogenic precursors into the heart forming fields.[91,92] Following fusion of the paired precardiac mesoderm, the primary heart tube consists of an endocardium lining the lumen, and an outer myocardium. Adhesion between neighboring cells of the endocardium is mediated by homophilic interaction of VE-cadherin and PECAM-1D and in the myocardium by N-cadherin.[93] VE-cadherin/catenin complex is found in adherens junctions of vascular endothelium and has been show to inhibit migration and proliferation.[94] Disruption of the cadherin/catenin complex is a critical step in the transformation of epithelium to mesenchyme.

During the transformation of endothelium to mesenchyme in the cardiac cushions, endothelial cells repress endothelial genes and activate expression of mesenchymal genes. The majority of endocardial cells will remain as VE-cadherin positive endothelial, and go on to express adult endothelial markers, such as factor vWF.[95] A subset of cells will display a loss of cell-cell contact and will undergo an EMT to invade the underlying extracellular matrix. VE-cadherin mRNA expression is lost by endocardial cells that have undergone cell transformation in the endocardial cushion (Heimark, unpublished). PECAM-1 mRNA down regulation has been described in endocardial cells that have migrated to form mesenchymal cells in the AV canal (E11.5) during cell transformation.[70] Regulation of the balance of multiple cell-cell adhesion molecules is likely to play a role in control of cell invasion during cell transformation.

Transcriptional Regulation of EMT

It is clear that EMT in the heart is a product of multiple inductive and permissive influences. One approach undertaken in this system has been to explore the transcriptional regulators required for EMT. The zinc finger transcription factor slug is required for EMT during neural crest migration and gastrulation in the chick embryo.[96] Studies exploring a potential role for slug during endocardial cushion formation were undertaken to determine whether it is functional in this EMT as well. Antibody and antisense oligonucleotide studies showed that slug was located in the atrioventricular canal and that it was required for cell separation. Loss of slug expression produced a polygonal endothelial morphology consistent with that previously obtained by treatment with antiTGFβ2.[97] Subsequent experiments showed that expression of slug in collagen gel cultures could overcome inhibition by anti-TGFβ2 antibody.[98] These data suggest that regulation of slug expression is the major function of TGFβ2 in the atrioventricular canal. While slug is likely to transcriptionally regulate several molecules, it has been implicated in the loss of cadherins during cell transformation. Slug and several related molecules, snail, zeb1 and zeb2, bind to the Ebox elements to repress E-cadherin (see also Chapter 11, Berx and Van Roy).[99] It has recently been shown that snail is also a positive regulator of matrix metalloprotease-2 (MMP-2) expression during snail-induced EMT in carcinoma cells.[100] This enzyme is required for mesenchymal cell migration in the AV canal and may be concomitantly regulated by Slug.[64]

A second transcription factor, Mox-1 (Meox-1) was first identified in the mouse as a marker of mesenchymal cells throughout the embryo (including in the outflow tract cushions.[101] The avian version of this molecule was cloned and its distribution and function was explored in the chick heart (Huang and Runyan, submitted; Wendler, Klewer and Runyan, submitted). Unlike Mox-1 in the mouse, we found this molecule distributed in the endothelium and the myocardium of the chick. Antisense oligonucleotides towards Mox-1 specifically blocked EMT in the collagen gel assays after cell separation but prior to invasion. As this pattern was similar to the effects of anti-TGFβ3 we determined that Mox-1 expression could be blocked by anti-TGFβ3. However, overexpression of Mox-1 in endothelial cells was not sufficient to cause EMT.

Several additional transcription factors critical to EMT are less well characterized but likely involved. It was observed that large T mouse mutants had defects in AV canal development.[102] The T gene was subsequently identified as brachyury, the prototype of the Tbx family of transcription factors.[103] In situ hybridization shows that cells undergoing EMT in the heart express brachyury (Huang and Runyan, unpublished) but experiments with antisense techniques have not yet identified the specific role played by brachyury in the AV canal. NFATc1 was identified as an expressed transcription factor in the AV canal and Outflow tract cushions of the mouse.[104,105] However, there is some divergence of opinion as to whether NFATc1 plays a functional role in AV canal development. This transcription factor interacts with calcineurin in gene regulation[106] but experiments to date with cyclosporine as an inhibitor of this pathway have shown no effect on AV canal EMT (Runyan, unpublished data). Additional transcription factors expressed in the AV canal but without a clear understanding of a role in EMT include the bHLH repressor, Id2, and GATA4.[107,108]

Clinical Significance of EMT in the Heart

The formation of valves in the heart requires precise interactions among multiple cell types. The high frequency of congenital heart defects (CHD) reflects the complexity of these developmental events. One of the most common, significant forms of CHD are atrioventricular septal defects (AVSD). AVSD consist of a deficiency of the inlet ventricular septum, a common AV valve, and an incomplete atrial septum primum. This anatomy is reminiscent of a 3-4 month human embryonic heart, and suggests that a developmental perturbation of normal morphogenetic pathways might be responsible for this CHD. AVSD communications between the right and left sides of the heart lead to symptoms of heart failure shortly after birth; surgery is usually required by 6 months of age. AVSD account for approximately 10% of all CHD and are the second most common CHD repaired in the first year of life.[109,110] Post-operatively, these patients require close follow-up with a likelihood of additional surgeries for AV valve insufficiency or pulmonary hypertension.

Nearly 70 percent of AVSD are diagnosed in infants with trisomy 21 (Down syndrome).[111] Likewise, the majority of CHD diagnosed in trisomy 21 infants are AVSD.[110,112] This association has led to speculation that chromosome 21 genes are important in AV valve development. While most trisomies include an entire extra chromosome 21, partial trisomies have been useful in identifying a region of this chromosome that corresponds specifically to CHD in these patients.[111] What is striking is that the candidate genes in this region include both cell adhesion molecules (CAMs) and ECM molecules (Collagen VI chains). This observation fits with the observations of Kurnit and colleagues[113] that fibroblasts from trisomy 21 patients are more adhesive and that increased adhesiveness could account for altered morphology in the AV canal.[114]

There are likely additional forms of CHD attributable to perturbation of EMT in the heart. Sheffield and colleagues[115] identified a large family with inherited endocardial cushion defects. While identification of the specific basis for this problem has eluded detection, the defect has been mapped to a region of chromosome 1 in the vicinity of the TGFβ Type III receptor. The power of genetics to identify candidate molecules and our ability to test for function in AV canal cultures should be productive in identifying additional mediators of CHD.

Questions and Future Directions

There are several questions raised by these studies and those of our colleagues in other chapters of this work. Among these, what are the common elements of EMT and are there specific mechanisms or molecules that are common to all or many of the EMTs observed in the embryo. As perusal of this volume suggests, there appear to be a variety of different regulators and it is clear that not all EMTs are identical. We have begun to systematically test a number of molecules found in other embryonic EMTs as found in a survey of embryonic gene expression (http://geisha.biosci.arizona.edu). To date, it appears that some of these molecules will be functional in the heart and others will not be. The characterization of these molecules may enable the clustering of EMTs into classes and shed light on pathologic EMTs in postnatal life.

While we have a basic understanding of TGFβ-mediated EMT in the chick heart, there are some significant differences between the mouse and the chick. We are very interested in understanding the basis for this difference and whether the human uses TGFβ isoforms more like the chick or the mouse. Confounding analysis of EMT in the AV canal is evidence in both species that there are a great number of ligands, receptors and signal transduction mechanisms functional in the heart. The complexity appears to increase on a daily basis. We have embarked on a microarray analysis of transcriptional regulation for many of the identified signal transduction processes in the heart. It is our hope that a bioinformatics approach can help identify the intersection or independence of the various pathways.

Together, these approaches should be relevant to the entire field of EMT as represented in this volume and the usefulness of the heart in resolving aspects of this field will be validated.

Acknowledgements

The authors thank Shannon Shoemaker for drawing the illustrations used in Figures 1-3. Original research in each of the author's laboratories was supported by a program project grant from the National Heart, Lung and Blood Institute of the National Institutes of Health (P01 HL63926).

References

1. Eisenberg LM, Markwald RR. Molecular regulation of atrioventricular valvuloseptal morphogenesis. Circulation Research Jul 1995; 77(1):1-6.
2. Barry A. The functional significance of the cardiac jelly in the tubular heart of the chick embryo. Anatomical Record 1948; 102:289-298.
3. Markwald RR, Fitzharris TP, Manasek FJ. Structural development of endocardial cushions. American J Anatomy 1977; 148(1):85-120.
4. Markwald RR, Fitzharris TP, Smith WN. Structural analysis of endocardial cytodifferentiation. Dev Biol 1975; 42(1):160-180.
5. Manasek F. Glycoprotein synthesis and tissue interactions during establishment of the functional embryonic chick heart. J Mol Cell Cardiol 1976; 8:389-402.
6. Manasek F. Heart development: Interactions involved in cardiac morphogenesis. In: Poste G, Nicholson G, eds. The Cell Surface in Animal Embryogenesis and Development. Elsevier/North-Holland Biomedical Press, 1976:545-598.
7. Kinsella M, Fitzharris T. Origin of cushion tissue in the developing chick heart: Cinematographic recordings of in situ formation. Science 1980; 207:1359-1360.
8. van den Hoff MJB, Moorman AFM, Ruijter JM et al. Myocardialization of the cardiac outflow tract. Dev Biol Aug 15 1999; 212(2):477-490.
9. Hamburger V, Hamilton HL. A series of normal stages in the development of the chick embryo. J Morphol 1951; 88:49-92.
10. Krug EL, Runyan RB, Markwald RR. Protein extracts from early embryonic hearts initiate cardiac endothelial cytodifferentiation. Dev Biol 1985; 112(2):414-426.
11. Markwald RR, Fitzharris T, Bolender DL et al. Structural analysis of cell: Matrix association during the morphogenesis of atrioventricular cushion tissue. Dev Biol 1979; 69:634-654.
12. Bolender DL, Markwald RR. Epithelial-mesenchymal transformation in chick atrioventricular cushion morphogenesis. SEM 1979; (3):313-321.
13. Icardo JM, Ojeda JL, Hurle JM. Endocardial cell polarity During the looping of the heart in the chick-embryo. Dev Biol 1982; 90(1):203-209.
14. Markwald RR, Fitzharris T, Bank H et al. Structural analysis on the matrical organization of glycosaminoglycans in developing endocardial cushions. Dev Biol 1978; 62(2):292-316.
15. Kitten GT, Kolker SJ, Krob SL et al. Type VI collagen in the cardiac valves and connective tissue septa during heart development. Braz J Med Biol Res 1996; 29:1189-1193.
16. De La Cruz M, Gimenez-ribotta M, Saravalli O. The contribution of the inferior endocardial cushion of the atrioventricular canal to cardiac septation and to the development of the atrioventricular valve: Study in the chick embryo. Am J Physiol Regul Integr Comp Physiol 1983; 166:63-72.
17. Elsdale T, Bard J. Collagen substrata for studies on cell behavior. J Cell Biol 1972; 54:626-637.
18. Schor S. Cell proliferation and migration on collagen substrata in vitro. J Cell Sci 1980; 41:159-175.
19. Greenburg G, Hay ED. Epithelia suspended in collagen gels can lose polarity and express characteristics of migrating mesenchymal cells. J Cell Biol 1982; 95:333-339.
20. Bernanke DH, Markwald RR. Migratory behavior of cardiac cushion tissue-cells in a collagen-lattice culture system. Dev Biol 1982; 91(2):235-245.

21. Bernanke DH, Markwald RR. Effects of hyaluronic acid on cardiac cushion tissue cells in collagen matrix cultures. Texas Reports in Biology and Medicine 1979; 39:271-285.
22. Overton J. Differential response of embryonic-cells to culture on tissue matrices. Tissue & Cell 1979; 11(1):89-98.
23. Runyan RB, Markwald RR. Invasion of mesenchyme into three-dimensional collagen gels: A regional and temporal analysis of interaction in embryonic heart tissue. Dev Biol 1983; 95(1):108-114.
24. Mjaatvedt CH, Lepera RC, Markwald RR. Myocardial specificity for initiating endothelial-mesenchymal cell transition in embryonic chick heart correlates with a particulate distribution of fibronectin. Dev Biol 1987; 119(1):59-67.
25. Krug EL, Mjaatvedt CH, Markwald RR. Extracellular matrix from embryonic myocardium elicits an early morphogenetic event in cardiac endothelial differentiation. Dev Biol 1987; 120(2):348-355.
26. Mjaatvedt CH, Markwald RR. Induction of an epithelial-mesenchymal transition by an in vivo adheron-like complex. Dev Biol 1989; 136(1):118-128.
27. Potts JD, Runyan RB. Epithelial-mesenchymal cell transformation in the embryonic heart can be mediated, in part, by transforming growth factor beta. Dev Biol 1989; 134(2):392-401.
28. Camenisch TD, Schroeder JA, Bradley J et al. Heart-valve mesenchyme formation is dependent on hyaluronan-augmented activation of ErbB2-ErbB3 receptors. Nat Med 2002; 8(8):850-855.
29. Walsh E, Stainier DYR. Novel role for glycosaminoglycans in cell signaling events during heart valve initiation: Cloning of the zebrafish jekyll mutation. Dev Biol 2001; 235(1):229-229.
30. Gassmann M, Casagranda F, Orioli D et al. Aberrant neural and cardiac development in mice lacking the Erbb4 neuregulin receptor. Nature 1995; 378(6555):390-394.
31. Meyer D, Birchmeier C. Multiple essential functions of neuregulin in development. Nature 1995; 378(6555):386-390.
32. Mjaatvedt CH, Krug EL, Markwald RR. An antiserum (ES1) against a particulate form of extracellular matrix blocks the transition of cardiac endothelium into mesenchyme in culture. Dev Biol 1991; 145(2):219-230.
33. Sinning AR, Krug EL, Markwald RR. Multiple glycoproteins localize to a particulate form of extracellular matrix in regions of the embryonic heart where endothelial cells transform into mesenchyme. Anatomical Record 1992; 232(2):285-292.
34. Rezaee M, Isokawa K, Halligan N et al. Identification of an extracellular 130-kDa protein involved in early cardiac morphogenesis. J Biol Chem 1993; 268(19):14404-14411.
35. Isokawa K, Rezaee M, Wunsch A et al. Identification of transferrin as one of multiple EDTA-extractable extracellular proteins involved in early chick heart morphogenesis. J Cell Biochem 1994; 54:207-218.
36. Sinning AR. Partial purification of HLAMP-1 provides direct evidence for the multicomponent nature of the particulate matrix associated with cardiac mesenchyme formation. J Cell Biochem 1997; 66(1):112-122.
37. Roberts A, Flanders K, Kondaiah P et al. Transforming growth factor β: Biochemistry and roles in embryogenesis, tissue repair and remodeling, and carcinogenesis. Recent Prog Horm Res 1988; 44:157-197.
38. Nakajima Y, Krug EL, Markwald RR. Myocardial regulation of transforming growth factor-β expression by outflow tract endothelium in the early embryonic chick heart. Dev Biol 1994; 165:615-626.
39. Burt DW. Evolutionary grouping of the transforming growth factor-β superfamily. Biochem Biophys Res Commun 1992; 184(2):590-595.
40. Potts JD, Vincent EB, Runyan RB et al. Sense and antisense TGF-Beta 3 messenger-RNA levels correlate with cardiac-valve induction. Dev Dyn 1992; 193(4):340-345.
41. Akhurst RJ, Lehnert S, Faissner A et al. TGF beta in murine morphogenetic processes: The early embryo and cardiogenesis. Development 1990; 108(4):645-656.
42. Dickson MC, Slager HG, Duffie E et al. RNA and protein localisations of TFG beta 2 in the early mouse embryo. Development 1993; 117:625-639.
43. Camenisch TD, Molin DGM, Person A et al. Temporal and distinct TGF beta ligand requirements during mouse and avian endocardial cushion morphogenesis. Dev Biol 2002; 248(1):170-181.
44. Barnett JV, Moustakas A, Lin W et al. Cloning and developmental expression of the chick type II and type III TGF beta receptors. Dev Dyn 1994; 199(1):12-27.
45. Boyer AS, Ayerinskas II, Vincent EB et al. TGF beta 2 and TGF beta 3 have separate and sequential activities during epithelial-mesenchymal cell transformation in the embryonic heart. Dev Biol 1999; 208(2):530-545.
46. Potts JD, Dagle JM, Walder JA et al. Epithelial mesenchymal transformation of embryonic cardiac endothelial-cells is inhibited by a modified antisense oligodeoxynucleotide to transforming growth factor beta 3. Proc Nat Acad Sci USA 1991; 88(4):1516-1520.

47. Diebold RJ, Eis MJ, Yin M et al. Early-onset multifocal inflammation in the transforming growth factor beta 1-null mouse is lymphocyte mediated. Proc Nat Acad Sci USA 1995; 92(26):12215-12219.
48. Proetzel G, Pawlowski SA, Wiles MV et al. Transforming growth factor-beta 3 is required for secondary palate fusion. Nat Genet 1995; 11(4):409-414.
49. Sanford LP, Ormsby I, Gittenberger-de Groot AC et al. TGFβ2 knockout mice have multiple development defects that are non overlapping with other TGFβ knockout phenotypes. Development 1997; 124:2659-2670.
50. Massague J. TGFβ signaling: Receptors, transducers, and Mad proteins. Cell 1996; 85:947-950.
51. Wrana JL, Attisano L, Wieser R et al. Mechanism of activation of the TGF-β receptor. Nature 1994; 370:341-347.
52. Wrana JL, Attisano L, Carcamo J et al. TGF beta signals through a heteromeric protein kinase receptor complex. Cell 1992; 71:1003-1014.
53. Lopez-Casillas F, Wrana JL, Massague J. Betaglycan presents ligand to the TGF beta signaling receptor. Cell 1993; 73(7):1435-1444.
54. Barnett JV, Moustakas A, Lin W et al. Cloning and developmental expression of the chick type II and type III TGF⇓ receptors. Dev Dyn 1994; 199:12-27.
55. Brown CB, Boyer AS, Runyan RB et al. Antibodies to the type II TGF beta receptor block cell activation and migration during atrioventricular cushion transformation in the heart. Dev Biol 1996; 174(2):248-257.
56. Brown CB, Boyer AS, Runyan RB et al. Requirement of type III TGF-beta receptor for endocardial cell transformation in the heart. Science 1999; 283(5410):2080-2082.
57. Boyer AS, Runyan RB. TGFβ Type III and TGFβ Type II receptors have distinct activities during epithelial-mesenchymal cell transformation in the embryonic heart. Dev Dyn 2001; 221(4):454-459.
58. Massague J. The transforming growth factor-beta family. Ann Rev Cell Biol 1990; 6:597-641.
59. Annes JP, Munger JS, Rifkin DB. Making sense of latent TGF beta activation. J Cell Sci 2003; 116(2):217-224.
60. Ghosh S, Brauer PR. Latent transforming growth factor-β is present in the extracellular matrix of embryonic hearts in situ. Dev Dyn 1996; 205(2):126-134.
61. McCormick KM. TGF beta 2 activation status during cardiac morphogenesis. Dev Dyn 2001; 222(1):17-25.
62. Nakajima Y, Miyazono K, Kato M et al. Extracellular fibrillar structure of latent TGF beta binding protein-1: Role in TGF beta-dependent endothelial-mesenchymal transformation during endocardial cushion tissue formation in mouse embryonic heart. J Cell Biol 1997; 136(1):193-204.
63. McGuire PG, Orkin RW. Urokinase activity in the developing avian heart: A spatial and temporal analysis. Dev Dyn 1992; 193:24-33.
64. Alexander SM, Jackson KJ, Bushnell KM et al. Spatial and temporal expression of the 72-kDa type IV collagenase (MMP-2) correlates with development and differentiation of valves in the embryonic avian heart. Dev Dyn 1997; 209(3):261-268.
65. McGuire PG, Alexander SM. Urokinase production by embryonic endocardial-derived cells: Regulation by substrate composition. Dev Biol (Orlando) 1993; 155(2):442-451.
66. Bouchey D, Argraves WS, Little CD. Fibulin-1, vitronectin, and fibronectin expression during avian cardiac valve and septa development. Anat Rec 1996; 244(4):540-551.
67. Kitten GT, Markwald RR, Bolender DL. Distribution of basement membrane antigens in cryopreserved early embryonic hearts. Anat Rec 1987; 217(4):379-390.
68. Mjaatvedt CH, Yamamura H, Capehart AA et al. The Cspg2 gene, disrupted in the hdf mutant, is required for right cardiac chamber and endocardial cushion formation. Dev Biol (Orlando) 1998; 202(1):56-66.
69. Yamamura H, Zhang M, Markwald RR et al. A heart segmental defect in the anterior-posterior axis of a transgenic mutant mouse. Dev Biol (Orlando) 1997; 186(1):58-72.
70. Baldwin HS, Lloyd TR, Solursh M. Hyaluronate degradation affects ventricular function of the early postlooped embryonic rat heart in situ. Circ Res 1994; 74(2):244-252.
71. Camenisch TD, Biesterfeldt J, Brehm-Gibson T et al. Regulation of cardiac cushion development by hyaluronan. Exp Clin Cardiol 2001; 6:4-10.
72. Camenisch TD, Spicer AP, Brehm-Gibson T et al. Disruption of hyaluronan synthase-2 abrogates normal cardiac morphogenesis and hyaluronan-mediated transformation of epithelium to mesenchyme. J Clin Invest 2000; 106(3):349-360.
73. Klewer SE, Krob SL, Kolker SJ et al. Expression of type VI collagen in the developing mouse heart. Dev Dyn 1998; 211(3):248-255.

74. DiMilla PA, Stone JA, Quinn JA et al. Maximal migration of human smooth muscle cells on fibronectin and type IV collagen occurs at an intermediate attachment strength. J Cell Biol 1993; 122(3):729-737.
75. McDonald JA. Integrins minireview series. J Biol Chem 2000; 275(29):21783.
76. Palecek SP, Loftus JC, Ginsberg MH et al. Integrin-ligand binding properties govern cell migration speed through cell-substratum adhesiveness. [erratum appears in Nature 1997; 388(6638):210.]. Nature 1997; 385(6616):537-540.
77. Akiyama SK, Nagata K, Yamada K. Cell surface receptors for extracellular matrix components. Biochimica et Biophysica Acta 1990; 1031:91-110.
78. Duband JL, Dufour S, Yamada SS et al. Neural crest cell locomotion induced by antibodies to beta 1 integrins. A tool for studying the roles of substratum molecular avidity and density in migration. J Cell Sci 1991; 98(Pt 4):517-532.
79. Kuijpers TW, Mul EP, Blom M et al. Freezing adhesion molecules in a state of high-avidity binding blocks eosinophil migration. Journal of Experimental Medicine 1993; 178(1):279-284.
80. Keely PJ, Fong AM, Zutter MM et al. Alteration of collagen-dependent adhesion, motility, and morphogenesis by the expression of antisense alpha 2 integrin mRNA in mammary cells. J Cell Sci 1995; 108(Pt 2):595-607.
81. Giancotti FG, Ruoslahti E. Integrin signaling. Science 1999; 285(5430):1028-1032.
82. Loeber CP, Runyan RB. A comparison of fibronectin, laminin, and galactosyltransferase adhesion mechanisms during embryonic cardiac mesenchymal cell-migration in vitro. Dev Biol Aug 1990; 140(2):401-412.
83. Ladher RK, Church VL, Allen S et al. Cloning and expression of the Wnt antagonists Sfrp-2 and Frzb during chick development. Dev Biol 2000; 218(2):183-198.
84. Lyons KM, Pelton RW, Hogan BL. Organogenesis and pattern formation in the mouse: RNA distribution patterns suggest a role for bone morphogenetic protein-2A (BMP-2A). Development 1990; 109(4):833-844.
85. Yamagishi T, Nakajima Y, Miyazono K et al. Bone morphogenetic protein-2 acts synergistically with transforming growth factor-beta 3 during endothelial- mesenchymal transformation in the developing chick heart. J Cell Physiol 1999; 180(1):35-45.
86. Nakajima Y, Yamagishi T, Hokari S et al. Mechanisms involved in valvuloseptal endocardial cushion formation in early cardiogenesis: Roles of transforming growth factor (TGF)-beta and bone morphogenetic protein (BMP). Anat Rec 2000; 258(2):119-127.
87. Kim RY, Robertson EJ, Solloway MJ. Bmp6 and Bmp7 are required for cushion formation and septation in the developing mouse heart. Dev Biol 2001; 235(2):449-466.
88. Allen SP, Bogardi JP, Barlow AJ et al. Misexpression of noggin leads to septal defects in the outflow tract of the chick heart. Dev Biol 2001; 235(1):98-109.
89. Runyan RB, Potts JD, Sharma RV et al. Signal transduction of a tissue interaction during embryonic heart development. Cell Regul 1990; 1(3):301-313.
90. Lakkis MM, Epstein JA. Neurofibromin modulation of ras activity is required for normal endocardial-mesenchymal transformation in the developing heart. Development - Supplement 1998; 125(22):4359-4367.
91. Garcia-Castro MI, Vielmetter E, Bronner-Fraser M. N-cadherin, a cell adhesion molecule involved in establishment of embryonic left-right asymmetry. Science 2000; 288(5468):1047-1051.
92. Breier G, Albrecht U, Sterrer S et al. Expression of vascular endothelial growth-factor during embryonic angiogenesis and endothelial-cell differentiation. Development 1992; 114(2):521-532.
93. Linask KK, ImanakaYoshida K, Knudsen KA. N-cadherin/beta-catenin interaction is necessary for chick cardiomyocyte compartmentalization and myofibrillogenesis. Dev Biol 1997; 186(2):B65-B65.
94. Caveda L, Martin-Padura I, Navarro P et al. Inhibition of cultured cell growth by vascular endothelial cadherin (cadherin-5/VE-cadherin). J Clin Invest 1996; 98(4):886-893.
95. Coffin JD, Harrison J, Schwartz S et al. Angioblast differentiation and morphogenesis of the vascular endothelium in the mouse embryo. Dev Biol 1991; 148(1):51-62.
96. Nieto MA, Sargent MG, Wilkinson DG et al. Control of cell behavior during vertebrate development by slug, a zinc finger gene. Science 1994; 264:835-839.
97. Romano LA, Runyan RB. Slug is a mediator of epithelial-mesenchymal cell transformation in the developing chicken heart. Dev Biol 1999; 212(1):243-254.
98. Romano LA, Runyan RB. Slug is an essential target of TGFbeta2 signaling in the developing chicken heart. Dev Biol 2000; 223(1):91-102.
99. Cano A, Perez-Moreno MA, Rodrigo I et al. The transcription factor Snail controls epithelial-mesenchymal transitions by repressing E-cadherin expression. Nat Cell Biol FEB 2000; 2(2):76-83.

100. Yokoyama K, Kamata N, Fujimoto R et al. Increased invasion and matrix metalloproteinase-2 expression by Snail-induced mesenchymal transition in squamous cell carcinomas. Int J Oncol 2003; 22(4):891-898.
101. Candia AF, Hu J, Crosby J et al. Mox-1 and Mox-2 define a novel homeobox gene subfamily and are differentially expressed during early mesodermal patterning in mouse embryos. Development 1992; 116(4):1123-1136.
102. Bayna EM, Runyan RB, Scully NF et al. Cell-surface galactosyltransferase as a recognition molecule during development. Mol Cell Biochem 1986; 72(1-2):141-151.
103. Herrmann BG, Lehrach H. From phenotype to gene: Molecular cloning in the Brachyury (T) locus region. Curr Top Microbiol Immunol 1988; 137:77-81.
104. de la Pompa JL, Timmerman LA, Takimoto H et al. Role of the NF-ATc transcription factor in morphogenesis of cardiac valves and septum [see comments]. Nature 1998; 392(6672):182-186.
105. Ranger AM, Grusby MJ, Hodge MR et al. The transcription factor NF-ATc is essential for cardiac valve formation. Nature 1998; 392(6672):186-190.
106. Molkentin JD, Lu JR, Antos CL et al. A calcineurin-dependent transcriptional pathway for cardiac hypertrophy. Cell 1998; 93(2):215-228.
107. Evans SM, Obrien TX. Expression of the helix-loop-helix factor id during mouse embryonic-development. Dev Biol 1993; 159(2):485-499.
108. Kelley C, Blumberg H, Zon LI et al. GATA-4 is a novel transcription factor expressed in endocardium of the developing heart. Development 1993; 118(3):817-827.
109. Ferencz C, Rubin JD, Mccarter RJ et al. Congenital heart disease: Prevalence at livebirth; the BaltimoreWashington infant study. Am J Epidem 1990; 121:31-36.
110. Carmi R, Boughman JA, Ferencz C. Endocardial cushion defect: Further studies of "isolated" versus "syndromic" occurrence. Am J Med Genet 1992; 43(3):569-575.
111. Korenberg JR, Bradley C, Disteche CM. Down Syndrome: Molecular mapping of the congenital heart disease and duodenal stenosis. Am J Hum Genet 1992; 50:294-302.
112. Adams MM, Erickson JD, Layde PM et al. Down's syndrome. Recent trends in the United States. JAMA 1981; 246(7):758-760.
113. Kurnit D, Aldridge J, Matsuoka R et al. Increased adhesiveness of trisomy 21 cells and atrioventricular canal malformations in down syndrome: A stochastic model. Am J Med Genet 1985; 20:385-399.
114. Jongewaard IN, Lauer RM, Behrendt DA et al. Beta 1 integrin activation mediates adhesive differences between trisomy 21 and nontrisomic fibroblasts on type VI collagen. Am J Med Genet 2002; 109(4):298-305.
115. Sheffield VC, Pierpont ME, Nishimura D et al. Identification of a complex congenital heart defect susceptibility locus by using DNA pooling and shared segment analysis. Hum Mol Genet 1997; 6(1):117-121.

Chapter 5

Epithelial-Extracellular Matrix (Cell-ECM) Interactions in Hydra

Michael P. Sarras, Jr.

Abstract

As a member of the Phylum, Cnidaria, hydra is organized a simple gastric tube with a head and foot pole. The entire body wall of hydra is organized as a epithelial bilayer with an intervening extracellular matrix (ECM). The major components of hydra ECM are highly conserved and reflect those seen in vertebrate systems. These components include laminin, collagen type IV, and a fibrillar collagen that is similar to a vertebrate type I/II class. The supramolecular organization of hydra ECM is seen as two basal lamina containing laminin and collagen type IV (one associated with the basal plasma membrane of the ectoderm and endoderm) with a central interstitial-like matrix containing fibrillar collagens. Because of the unique biophysical properties of hydra ECM, decapitation of hydra (or any wound to the bilayer) results in retraction of the matrix from the wound site. While the epithelial bilayer will seal within one hour, the matrix remains retracted resulting in a bilayer lacking an intervening ECM. This triggers an upregulation of matrix component mRNA within 3 hours of wounding and de novo biogenesis of hydra ECM that is completed within 24-72 hours from the initial time of decapitation or wounding. While the ECM of hydra is symmetrical (two basal laminin to the periphery of the central interstitial matrix), the synthesis of matrix components from the epithelium is asymmetrical. For example, laminin is secreted from the endoderm while collagen type IV and hydra fibrillar collagen (Hcol-I) are both secreted from the ectoderm. The timing of matrix component secretion is also irregular in that laminin and collagen type IV are secreted and integrated within the newly forming basal lamina within 6-12 hours of decapitation or wounding while fibrillar collagen is not secreted until at least 24 hours. Antisense studies in which hyda laminin translation was blocked, indicates that secretion of hydra fibrillar collagen is depending on secretion and polymerization of hydra laminin. Lastly, biogenesis of hydra ECM is accompanied by the upregulation of hydra matrix metalloproteinase. Functional studies of shown that blockage in the translation of any of these ECM components results in a blockage of hydra regeneration (morphogenesis). Other studies have also shown that cell-ECM interactions in hydra is also coupled to cell differentiation processes. In summary, hydra morphogenesis is dependent on ECM biogenesis as are processes related to cell differentiation.

General Introduction to Hydra Structure and Tissue Dynamics

As a member of the Phylum Cnidaria, hydra arose early during metazoan evolution prior to the divergence of the protosome and deuterostome branches. The body plan of hydra is organized as a gastric tube with a hypostome and ring of tentacles at the head pole (apical pole) and a peduncle and basal disk at the foot pole (basal pole). The head pole functions in feeding while

Rise and Fall of Epithelial Phenotype: Concepts of Epithelial-Mesenchymal Transition, edited by Pierre Savagner.

the foot pole functions for attachment to the substratum. The entire body wall of hydra is organized as a simple epithelial bilayer with an intervening extracellular matrix (ECM). Previous studies have established that hydra ECM has a similar molecular composition to that of vertebrate species[1-5] and functional studies have established that cell-ECM interactions are critical to developmental processes in hydra.[2-12] In addition, pattern formation in hydra is regulated by epitheliopeptides[13] and classical growth factor signaling pathways.[14] The organism has about 20 different cell types that arise from three cell lineage pathways to include: (1) ectodermal cells, (2) endodermal cells, and (3) interstitial cells.[13,14] Cells of these lineages have particular distribution patterns along the longitudinal axis. For example, battery cells are restricted to the tentacle ectoderm and basal disk cells are restricted to the base of the foot process. Despite this restricted distribution pattern, the cells of hydra are in constant division and turnover. This division occurs by stem cells in the body column that lead to differentiated body column cells that are constantly displaced toward the poles.[14] In the case of the ectoderm, for example, as cells are displaced into the tentacles or the basal disk, they trans-differentiate into battery cells and basal disk cells, respectively.[14] Battery cells and basal disk cells are eventually shed from the body column when they reach the tip of the tentacles or central portion of the base of the foot pole. The division rate of interstitial stem cells varies from that of epithelial cells as does the patterns of transdifferentiation. As a consequence of this extensive cell turnover, hydra are highly regenerative.[14,15] This high regenerative capacity allows one to analyze developmental processes using a variety of molecular and cell biological approaches. It is important to keep in mind, however, that previous studies have established that morphogenesis in hydra only requires the presence of epithelial cells[14] given the fact that epithelial hydra that have no interstitial cells can regenerate both a head and foot pole or form complete hydra from pellets of only epithelial cells.[14] With this in mind, the focus of this chapter is to highlight the role of epithelial-ECM (cell-ECM) interactions in pattern formation and regeneration in hydra. The chapter has been organized to first give the reader a general introduction to hydra structure and tissue dynamics and then discuss the structure and biogenesis of hydra ECM as related to pattern formation and regeneration. The chapter ends with a section on the regulation of cell-ECM interactions by hydra metalloproteinases.

Composition and Supramolecular Organization of Hydra Extracellular Matrix

Two decades ago, experiments were begun to utilize hydra as a developmental model to study the role of cell-ECM interactions during pattern formation in epithelial systems. The specific questions being asked related to the role of hydra ECM in cell differentiation and morphogenetic processes. As a first step, it was essential to clarify the molecular composition and structure of hydra ECM. The strategy implemented involved: (1) isolation of hydra ECM and use of biochemical and immunological approaches to analyze the purified matrix preparation, (2) use of the purified hydra ECM preparation to generate a battery of hydra specific polyclonal and monoclonal antibodies, (3) use of these hydra specific antibodies as reagents to screen expression cDNA libraries and as probes to characterize the distribution of matrix components in hydra ECM using morphological techniques, and (4) use of hydra-specific antibodies and isolated matrix component domains as blocking reagents to study the role of cell-ECM interactions in hydra using a number of regeneration bioassays.

Initial biochemical and immunological studies using antibodies generated to vertebrate ECM components indicated that hydra ECM had a spectrum of matrix components similar to those observed in higher invertebrates and vertebrates. Specifically, evidence for the presence of collagen type IV, laminin, heparan sulfate proteoglycan and fibronectin-like molecules was presented.[16] Pulse-chase autoradiographic studies in conjunction with translational and post-translational processing inhibitor studies supported the presence of collagen and proteoglycan components Sarras et al.[17] Use of hydra-specific monoclonal antibodies[15] in combination with special ultrastructural staining techniques[18] clarified that hydra ECM contained

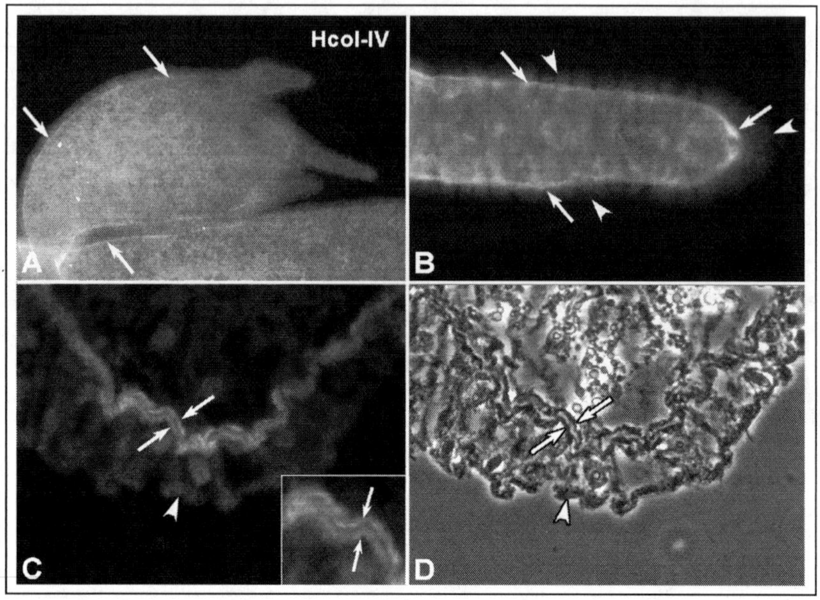

Figure 1. Localization of hydra collagen type IV to the basal lamina of hydra ECM. Whole mount immunofluorescent images are shown in A and B while frozen sections are shown in C and D. Arrows in A and B indicate the signal from ECM along the longitudinal axis while the arrowhead indicates the apical border of the ectodermal cells. Arrows in C indicate the signal associated with the basal lamina. Because the ectodermal and endodermal cell layer both have their own basal lamina in hydra, a "railroad track" pattern is seen (C). The inset in C is an enlargement of the image shown in C. A phase image of fluorescent image of C is shown in D. In D, arrows indicate the basal lamina regions while the arrowhead indicates the apical border of the ectoderm.

distinct structural regions. Adjacent to the basal plasma membrane border of each epithelial layer was a defined basal lamina-like region named the subepithelial zone and intervening between these two basal lamina-like regions was a central fibrous zone that appeared similar to interstitial matrix. Subsequent cloning studies resolved this issue by showing that laminin chains were confined to the subepithelial zones[18] (i.e., basal lamina) and type I-like collagen was confined to the central fibrous zone[2] (i.e., interstitial matrix). Recent unpublished studies (Sarras, et al) using collagen type IV NC1-derived anti-peptide antibodies have established that hydra collagen type IV[19] is also localized to the subepithelial zone (basal laminin). As shown in Figure 1, immunostaining using this collagen type IV NC1 antibody shows the typical railroad track pattern reflective of the two basal lamina to the periphery of the central interstitial matrix of hydra ECM. Based on these studies, an overall structure of hydra ECM is shown in Figure 2 and a discussion of each of the major components (laminin, collagen type IV and collagen type I/II) of hydra ECM follows. A list of the major components of hydra ECM with their general properties is shown in Table 1.

Laminins

The Laminin Family of Matrix Proteins

Laminins represent a family of glycoproteins that are a major component of basement membranes (basal lamina). To date, twelve different laminin heterotrimers have been identified in mammals and a number of laminins have been identified in invertebrates such as *Drosophila, C. elegans*, and sea urchin.[20-27] Laminins are involved in basement membrane assembly and are

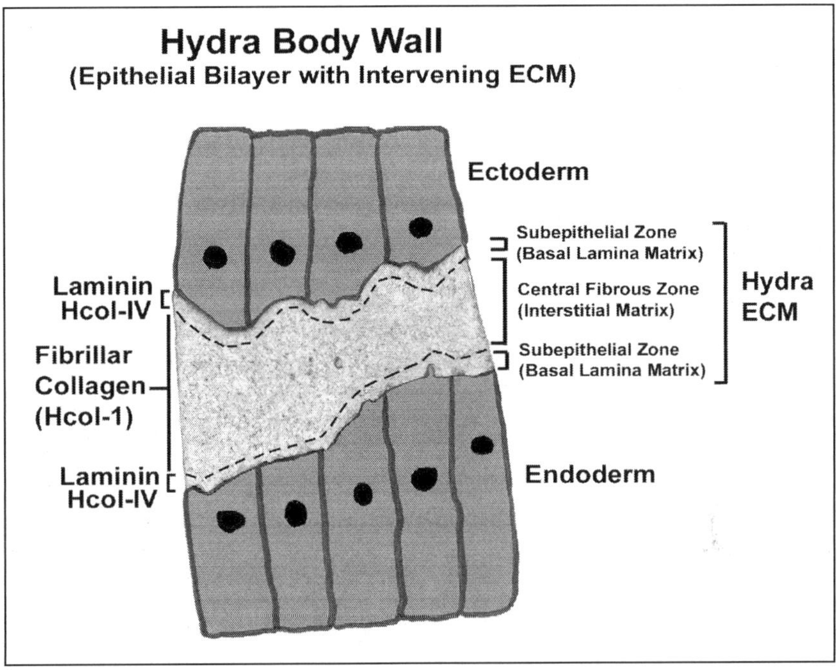

Figure 2. Diagrammatic representation of the supramolecular structure of hydra ECM. Only the major components are shown (hydra laminin, collagen type IV, and fibrillar collagen).

an important component in the supramolecular architecture of matrix.[28-30] The heterotrimer isoforms are generated from three types of subunits, namely, the α, β, and γ chains. In vertebrates, we find five different α chains and three different β and γ chains. Various combinations of these chains leads to the twelve heterotrimers referred to above and at least three chains may exist as alternatively spliced forms.[31-33] ECM suprastructure is based on assembly of a self-binding laminin mesh-like polymer, formation of a self-binding type IV collagen network polymer, and subsequent binding of various other ECM components to these networks.[34] Interaction of laminin domains to cell surface ECM-binding proteins such as integrins and dystroglycan is involved with matrix assembly processes[35] and with cell signaling events[36] associated with cell differentiation and morphogenesis.

Hydra Laminin

Hydra laminin chains have been cloned and functionally analyzed.[6,15,18] As stated, hydra laminin is localized to the two subepithelial zones (basal lamina) of hydra ECM. The complete sequence for a hydra β1-like chain[6,18] and a partial sequence of an α-1-like chain[6] has been reported. No γ chain has thus far been identified, although the typical trimeric form for isolated hydra laminin has been observed by rotary shadow TEM analysis.[6] While hydra laminin is localized to the two subepithelial zones (basal lamina) of hydra ECM, it is synthesized exclusively by the endoderm, which means that the molecules have to diffuse through the mesoglea to reach the ectodermal layer Shimizu et al.[7] The location at the basal region of both cell layers suggests that laminin is required for proper cell function and differentiation. Consistent with this proposal, laminin secretion from the endoderm precedes secretion of hydra collagen-I/II that arises from the ectoderm and inhibition of laminin secretion will block collagen secretion Shimizu et al.[7] Earlier studies have already established that antibodies to hydra laminin will block hydra morphogenesis Sarras et al[15] and other ECM-related processes such as cell

Table 1. Major components of hydra extracellular matrix

ECM Component	General Properties
Laminin	Contains α, β, and γ subunits in a trimeric cruciate (HLM) structure. Cloning studies indicate the alpha subunit is a vertebrate α1-like or *Drosophila*-like chain while the beta subunit is a vertebrate β1-like chain containing at least one defined cell binding domain (FTGTQ).
Collagen type IV (Hcol-IV)	A homotrimeric glycoprotein formed from three collagen IV α1-like chains. Each subunit contains a collagen domain at its N-terminus and a smaller non-collagenous domain (NC1) at its C-terminus. The collagen domain contains RGD cell binding motifs. TEM rotary shadow studies indicate that polymerization of Hcol-IV molecules within the ECM involves interaction of the NC1 domains. A typical 7S domain at the N-terminus appears not to be present in the mature ECM.
Fibrillar Collagen type I (Hcol-I)	A homotrimeric glycoprotein formed from three collagen type I α1-like chains. Each chain contains a N-terminal propeptide and C-terminal propeptide. During processing the C-terminal pro-peptide is removed, but the N-terminal propeptide is retained in the mature molecule. This results in the formation of flexible fibrils but not thickened banded fibrils as typically seen in vertebrate type I collagens. Hcol-I's flexibility is enhanced by a reduction in its proline content and a loss of critical lysines involved in lysyl-cross bridging.

migration.[37] While the mechanism(s) of signal transduction in hydra is not fully known, some published data suggest the involvement of integrins, the primary class of ECM-receptors in higher animals,[36] as being important to cell-ECM interactions in hydra.[6,38] This hydra data mainly pertains to a region in the short arm of the hydra β1-like chain. In this regard, sequence analysis of the β1-like chain indicates the substitution of a FTGTQ sequence for the YIGSR receptor-binding sequence observed in vertebrates. Although the role of the YIGSR sequence in signal transduction-mediated processes has been questioned, recent studies indicate (1) its potential use as an inhibitor of human preB leukaemic cell growth and metastasis using SCID mice models[39] and (2) its role in the guidance of axon growth cones.[40] Such studies and others support its involvement in cell signaling processes. The substituted FTGTQ sequence in the hydra β1-like chain has also been shown to interact with the cell surface under both in-vitro and in-vivo conditions[18] and affinity purification studies indicate that the FTGTQ sequence can interact with a hydra integrin-like protein.[6] Further analysis of (1) laminin-mediated cell signaling processes and (2) the role of laminin in the biogenesis and assembly of hydra ECM is required to fully understand the relationship of pattern formation to ECM structure and assembly in this invertebrate organism.

Collagens

The Collagen Family of Matrix Proteins

Collagens are found in all animals and are the most abundant protein of the extracellular matrix. The basic structure of this large protein family consists of multiple Gly-X-Y repeats. Extensive structural and functional diversity among collagens is accomplished by introducing interruptions in the triple helical domains and inclusion of various globular domains.[41,42] Some collagens, like the basement membrane collagen type IV are found in all animals, whereas others are limited to particular groups. The fibrillar collagens for instance have previously only been found in vertebrates and have not been identified in invertebrates. In contrast, specialized collagens such as the cuticle collagens *of C elegans*[43] and the mini-collagens of hydra nematocyte capsules[43] have only been found in invertebrates. Surprisingly, Kramer[44] has reported that the cuticle collagens of *C. elegans* are encoded by more than a 100 genes. These observations indicate how diversity in collagen structure is utilized to meet the needs of a broad spectrum of specialized extracellular matrices. Accordingly, analysis of the structure of specialized collagens provides us with insight into the organization and function of ECM in both vertebrates and invertebrates.

While indirect evidence had suggested the existence of collagens in the ECM of hydra,[10,45] only recent structural and functional analysis has provided a clear understanding of the types of collagens that exist in this invertebrate.[2,19] These collagens include a basement membrane-type (hydra collagen type IV, Hcol-IV) and an interstitial-type (hydra fibrillar collagen, Hcol-I). As will be discussed, these collagens have been characterized at both the cDNA and protein level.

Hydra Collagen Type IV (Hcol-IV)

Collagen type IV is the second most prominent constituent of basement membranes after laminin. These collagens are glycoproteins composed of three subunits that form a polymerized network in conjunction with laminin.[34,46,47] Currently, six different types of collagen type IV subunits are known to exist in vertebrates.[29,40,47] Invertebrate collagen type IV molecules have previously been identified in such organisms as *Drosophila,*[48] *C. elegans*[49] and a number of other species.[50] Recent studies by Fowler et al[19] report that hydra ECM also contains a collagen type IV. Analysis of the cDNA clone revealed a protein of 1723 amino acids, including an interrupted 1455 residue collagenous domain and a 228 residue carboxyl-terminal noncollagenous domain. Hcol-IV is similar to all known α(IV) chains, but again, most closely resembles vertebrate and invertebrate $\alpha 1$(IV) chains. Like hydra fibrillar collagen, Hcol-IV also forms homotrimeric molecules. Electron microscopy reveals an irregular network of rod-like structures interrupted by globular domains. This network can be depolymerized by reducing agents to dimeric collagen molecules, joined via their C-terminal noncollagenous domains. Under extensive denaturing conditions, depolymerization can only be taken to the dimeric; but not monomeric stage; indicating that the individual polypeptide chains, are quantitatively held together by nonreducible cross-links in addition to disulfide bonds. This behavior is quite different from the vertebrate collagen type IV that needs pepsin digestion for solubilization. For vertebrate collagen type IV, a model has been proposed in which four molecules aggregate via their N-terminal domains to form a spider-like structure. The interactions are stabilized via disulfide bonds and lysine derived cross-links, resulting in a highly protease resistant 7S domain. In addition, the C-terminal globular domain, NC1, binds to itself, mainly via disulfide bridges, to form a linear dimer. Both interactions at the N-terminal and C-terminal ends lead to the proposal of an open network structure that can further polymerize via lateral aggregation of the triple helical domains.[29,50] In contrast, in hydra while C-terminal interactions and lateral aggregation occurs, a stable 7S domain is not formed. A similar collagen type IV has also been reported in the worm, *Ascaris suum.*[51]

Hydra Fibrillar Collagen (Hcol-I)

Fibrillar collagens make up the majority of matrix components within the interstitial matrix. Likewise, hydra fibrillar collagen is the major component of hydra ECM.[2] The cDNA for the hydra fibrillar collagen, Hcol-I, encodes a protein of 1412 amino acids. The polypeptide isolated from hydra ECM has an apparent molecular weight of 155 kDa. The subunit chains of Hcol-I form homotrimeric molecules that constitute the majority of the fibrils within the central fibrous zone (interstitial matrix). Sequence comparisons clearly define Hcol-I as a fibrillar collagen. The highest similarity is to the α chains of vertebrate collagens type I and II. A similar degree of similarity is found between Hcol-I and invertebrate sea urchin collagen[52] and a sponge fragment.[53] Corresponding to the similarity at the sequence level, Hcol-I also exhibits the characteristic domain structure of fibrillar collagens, consisting of a central triple helical domain flanked by an N-terminal propeptide-like domain and a C-terminal propeptide. It is note worthy that the triple helical domain with 340 uninterrupted GLY-X-Y repeats has exactly the same length as the fibrillar collagens of vertebrates, suggesting similar fibril forming possibilities.

Despite marked similarities in the primary structure, there are distinct differences in the supramolecular organization of vertebrate fibrillar collagen networks as compared to that seen in hydra ECM. Hcol-I forms a network of fine fibrils[2] rather than thicker banded fibrils as seen by electron microscopy of vertebrate interstitial matrices. In contrast to vertebrate collagens that require pepsin digestion for solubilization, large polymeric structures of Hcol-I can be isolated from the ECM of hydra under native conditions.

Several factors are responsible for the special structure of hydra Hcol-I. These factors include: (1) a low content of proline in the triple helical domain that is only about 40% that of vertebrate collagens: (2) a reduced degree of inter-chain cross-linking due to the lack of classical consensus sequences for lysine/lysine-aldehyde derived covalent bonds; and (3) most importantly, altered post-translational processing that results in retention of the N-terminal propeptide-like domain in the mature molecule. Combined, these factors result in a more flexible collagen that can bend on itself as suggested by the early ultrastructural studies of Davis and Haynes.[9]

In addition to Hcol-I, recent studies by Boot-Handford[54] have shown that hydra has at least two additional fibrillar collagens (Hcol-2 and Hcol-3). While it is known that these fibrillar collagens contain NC2 domains, their localization to ECM has not been determined.

In summary, the basic components of hydra ECM are organized into two basal lamina containing at least laminin and collagen type IV and an interstitial collagen containing at one three fibrillar collagen. The hydra EST project (Bode and Steele, Univ. of California, Irvine) has identified a number of additional matrix components such as fibrillin; however, their localization to hydra ECM has not been determined.

Biogenesis of Hydra ECM during Regeneration, Epithelial Repair Along the Longitudinal Axis, and during Normal Cell Turnover in the Adult Polyp

Based on previous studies, the sequential phases of ECM biogenesis during head regeneration in hydra are summarized in Figure 3. These phases involve, (1) ECM retraction at the time of decapitation, (2) early epithelial wound healing, (3) biogenesis of the two basal lamina, and (4) subsequent biogenesis of the interstitial matrix. As will be discussed in this section, these phases, along with functional studies, indicate that ECM biogenesis in hydra involves cross-talk between the two epithelial layers as well as signaling between the ECM and each epithelial cell layer.[7]

Hydra is unique among metazoans in that surgical excision of the head pole (or foot pole) gives rise to a significant area of tissue not underlain by the ECM even after the cut edges of the bilayer has fused and the wound site is sealed. In this process, the mechanisms underlying loss of the ECM likely involve a combination of factors related to (1) the intrinsic flexibility of hydra matrix as discussed by Sarras and Deutzmann[55] and (2) alterations in ECM-epithelial adhesions that result in subsequent changes in the relative position of epithelial cells to the underlying ECM.[56] Loss of the ECM at the regenerating pole triggers rapid morphological changes in cells of the head pole that is most evident in the ectoderm cell layer.

The rapid changes in the structure of cells that lose an ECM association (from a high to a low cuboidal morphology in the case of the ectoderm) may be related to signaling processes involving tensegrity mechanisms as previously discussed by Ingber and associates.[57-60] Tensegrity mechanisms involve structures that are mechanically stable not because of the strength of individual members, but because of the way the entire structure distributes and balances mechanical stresses. In the case of cell-ECM interactions, these structural balances involve the inter-relation of the ECM, plasma membrane ECM receptors, and cytoskeletal systems within the cell that are linked to the ECM via cell surface receptor systems.[57] Such alterations in cell morphology resulting from the presence or absence of an ECM have commonly been reported

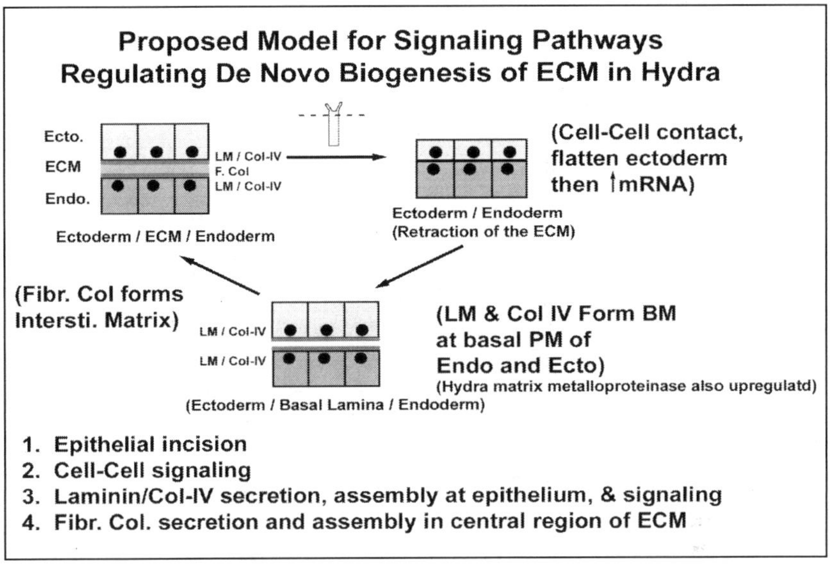

Figure 3. Steps in the biogenesis of hydra ECM during head regeneration. The steps deplicted are described in detail in the section dealing with ECM biogenesis.

for cells under in vitro conditions (e.g., cell culture studies),[61] but have been less frequently reported for cells under in vivo conditions. One notable exception to this is the process of wound healing in mammals. As recently reviewed by Nedelec et al[62] an incision to the skin results in a migration of fibroblasts through the ECM at the site of the wound. Subsequent wound healing involves reepithelialization and reformation of an intact basement membrane associated with the reformed epidermis. In regard to the epithelial components, this mimics what is observed in hydra and suggests that these processes have been highly conserved during evolution.

Within 3 hours of decapitation, when the ECM is no longer in contact with the head pole bilayer, epithelial cells of the ectoderm and endoderm layer up-regulate genes for ECM components of both the basement membrane (HLM-β1, Hcol-IV) and interstitial matrix (Hcol-I) (Fig. 4). The spatial and temporal pattern for expression of ECM components in hydra is more complicated than originally envisioned. While the symmetrical organization of hydra ECM [two peripheral subepithelial-associated basement membranes and one central interstitial matrix][55] might suggest that both epithelial layers would be involved with secretion of all basement membrane components, it is now clear from that at least one basement membrane components, namely hydra laminin (chains HLM-β1 and HLM-α1), is produced solely by the endoderm while at least one interstitial matrix component, namely Hcol-I, is produced solely by the ectoderm. Like Hcol-I, hydra collagen type IV (Hcol-IV) is also produced by the ectoderm[19] and becomes associated with the basal lamina during ECM biogenesis (Figs. 1-5). As was seen with laminin (Fig. 4) hydra collagen type IV is observed associated with the reforming basal lamina by about 7 hours following decapitation of hydra and its immunofluorescent signal reaches a maximum at about 12 hours following decapitation (Figs. 4 and 5). Although not anticipated, hydra matrix metalloproteinase (HMMP) is also expressed during ECM biogenesis and as with laminin, is produced by the endoderm.[7] The temporal pattern observed for the secretion of matrix components implies that a coordinated cross-epithelial signaling process is occurring during the biogenesis of hydra ECM (Fig. 3). This cross signaling would occur in two phases. The first phase would occur immediately following decapitation when the ECM retracts and the ectoderm and endoderm directly contact one another along their basal plasma

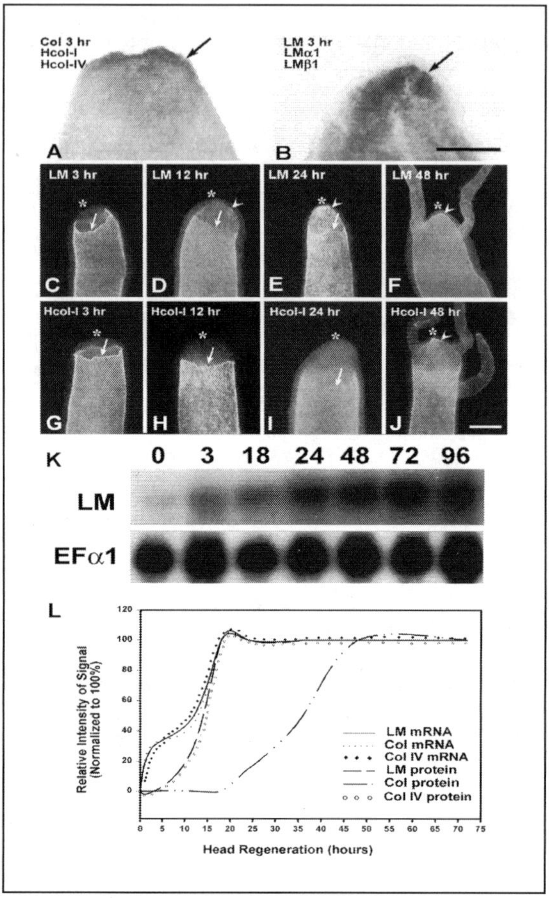

Figure 4. Spatial and temporal pattern of the expression of hydra ECM components as analyzed at the mRNA and protein level. Morphological and biosynthetic events occurring within 3 to 96 hours following decapitation as monitored by whole mount in situ hybridization for hydra collagen I (Col) (A) [same pattern seen for hydra collagen type IV (Hcol-IV)] and hydra laminin β1 chain (LM) [same pattern seen for hydra laminin α1 chaiin (HLM-α1)] (B), whole mount immunofluorescence for LM (C-F) and Col (G-J), and Northern blot analysis (K). As shown in A and B, up-regulation at 3 hours following decapitation of hydra collagen is associated with the ectoderm (A, arrow) while up-regulation of hydra laminin is associated with the endoderm (B, arrow). While the epithelial bilayer has already fused at the apical pole (asterisk in C-J) a hiatus in the ECM still exists at 3 hours following decapitation (C and G). The original cut edge of the ECM can still be detected through 24 hours following decapitation as monitored with antibodies to LM (C-E) and Col (G-I). Reformation of a continuous ECM at the regenerating head pole is first observed with antibodies for LM between 7-12 hours following decapitation (D, arrowhead) and this signal continues through 24 - 48 hours of regeneration (E and F, respectively; arrowhead) (the same pattern is shown for hydra collagen type IV as seen for hydra laminin as shown in Fig. 5). In contrast, an ECM associated signal for hydra Col is only weakly detected by 15-24 hours (not evident at the magnification shown in Fig. 2) while an easily observed signal is seen between 24-48 hours at this same magnification (compare I and J, arrowhead). Up-regulation of mRNA for LM (K) and Col (data not shown) precedes the appearance of immunofluorescent signals for proteins associated with the reforming ECM. Elongation factor α1 (Efα1) is used as a loading control for Northern blot analysis of the mRNA lanes shown in K. The relative fluorescent and Northern blot signals for LM and Col (both fibrillar collagen the hydra collagen type IV) over 72 hours following decapitation are shown in graph form in L. Scale bar in B (A and B), 200 μm. Scale bar in J (C-J), 250 μm.

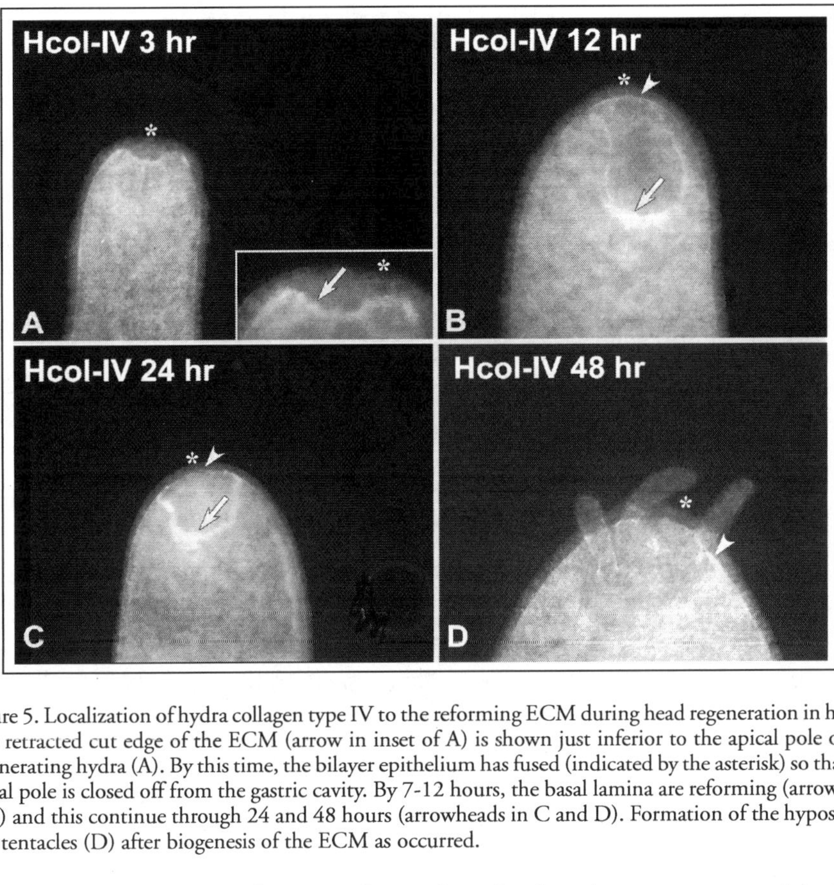

Figure 5. Localization of hydra collagen type IV to the reforming ECM during head regeneration in hydra. The retracted cut edge of the ECM (arrow in inset of A) is shown just inferior to the apical pole of the regenerating hydra (A). By this time, the bilayer epithelium has fused (indicated by the asterisk) so that the apical pole is closed off from the gastric cavity. By 7-12 hours, the basal lamina are reforming (arrowhead in B) and this continue through 24 and 48 hours (arrowheads in C and D). Formation of the hypostome and tentacles (D) after biogenesis of the ECM as occurred.

membrane surfaces. The combination of ECM loss plus the subsequent contact of the basal plasma membrane surfaces of the bilayer results in an up-regulation of mRNA for matrix components and HMMP by 3 hours following decapitation. In the second phase, laminin is secreted from the endoderm and collagen type IV is secreted from the ectoderm into the inter-basal plasma membrane compartment of the bilayer. Laminin and collagen type IV seed to the basal plasma membrane of both the ectoderm and endoderm. This directly supports the work of Colognato and Yurchenco[34] who have proposed that similar laminin seeding processes occur in mammalian systems. Binding of laminin to the bilayer then stimulates ectodermal cells[2] to later begin to secrete Hcol-I (between 15-24 hours following decapitation) that then polymerizes in the central fibrous zone of hydra ECM. This polymerizing interstitial matrix is seen as an easily detectable fluoresescent signal between 24-48 hours. An alternative explanation is that the delay in the appearance of Hcol-I relates to some changes in the normal processing of Hcol-I following its secretion and this altered processing prevents recognition by our monoclonal antibody to Hcol-I. This seems unlikely however, because use of a polyclonal antibody raised to mature Hcol-I also gives the same results (unpublished data). In addition, a causal relationship between laminin binding and the discharge of Hcol-I follows from the antisense studies described in the current study. These studies clearly indicated that blockage of the discharge of laminin resulted in the lack of appearance of Hcol-I at the site of ECM formation. Previous studies by Agbas and Sarras[38] and Sarras et al[18] as well as more recent studies by Zhang et al[6] have shown that hydra has ECM receptors for laminin and some of these receptors appear to be of the integrin class. The role of collagen type IV in this process is less clear and requires further investigation. Taken in total, these studies indicate that while the body wall of hydra is structurally reduced to an epithelial bilayer with an intervening ECM, matrix biogenesis by this bilayer is complicated and involves signaling events between both the ectoderm and endoderm. These signaling events function to coordinate the expression and final polymerization of both basement membrane and interstitial matrix components and also involve the simultaneous expression of matrix metalloproteinases.

The sequence of events following excision of the head pole (i.e., fusion of the epithelium, retraction of the ECM resulting in an epithelial bilayer with no underlying matrix, shape changes of epithelial cells and subsequent biosynthesis of a new ECM), is not unique to regeneration processes in hydra. Epithelial repair experiments in which a transverse incision of the body column was inflicted, resulted in the same ECM-associated events observed during head regeneration; namely retraction of the ECM at the incision site and subsequent up-regulation of basement membrane components, interstitial matrix components and hydra matrix metalloproteinase.[7] This coincidence of cell-ECM events strongly suggests that the de-novo biogenesis of a matrix following injury to the epithelium is a fundamental process of the bilayer that is not unique to head or foot regeneration. Grafting of hydra also induces ECM biogenesis.[7] In some cases grafting results in abnormalities in the cylindrical shape of the body column. Like decapitation, surgical bisection of hydra results in a loss of the ECM at the graft site followed by a de novo biogenesis of a new matrix between the two grafted halves. The shape abnormalities induced in the body wall in some grafts could result from a number of variables such as poor adhesion of the two cut epithelial surfaces[63] in combination with a loss of the ECM at the time of grafting. Because all grafts have a retraction and reformation of the ECM, but not all grafts show deformation of the body column, it is not likely that cell-ECM interactions are the sole factor contributing to body abnormalities following grafting. These studies do indicate however, that ECM biogenesis does occur whenever grafting is performed and therefore cell-ECM interactions are one of a number of factors that must be considered when evaluating such body shape abnormalities.

It should be noted that a certain degree of ECM biogenesis and turnover is always occurring along the body column of hydra. This has been shown for laminin,[64] fibrillar collagen,[2] type IV collagen,[19] and hydra matrix metalloproteinase.[64] Levels of ECM biogenesis and turnover vary along the longitudinal axis. Relatively higher levels are associated with positions of cell transdifferentiation (e.g., base of tentacles and basal disc region). Therefore, when stated that incision of the epithelium induces ECM biogenesis to occur, this is a relative statement and means that incision induces a significant increase in the expression of ECM components over the normal background levels. This increase is clearly related to the fact that incision results in a loss of ECM at the wound site. Loss of an ECM then results in a complete de novo biogenesis of matrix components.

Epithelial Morphogenesis in Hydra Is Dependent on ECM Biogenesis as Monitored during Head Regeneration and Epithelial Repair Following Surgical Incision of the Bilayer

Previous studies with hydra cell aggregates (morphogenesis of intact hydra from a pellet formed from dissociated hydra cells) have shown that reagents that perturb cell-ECM interactions such as antibodies to matrix components or fragments of matrix components can block epithelial morphogenesis.[15,37] The recent studies of Shimizu et al[7] indicate that cell-ECM interactions are inherent to a wide variety of morphogenetic processes in hydra to include simple incision of the body column. In this regard, the term epithelial morphogenesis is being used in a broad sense to include a spectrum of developmental processes such as cell differentiation, cell shape, and the establishment of three-dimensional form, etc. Cell-ECM interactions may affect one or more of these processes. As shown by Shimizu et al[7] the region of the epithelial bilayer undergoing head morphogenesis is always located apical to the original cut edge of the ECM. Therefore head morphogenesis in hydra always involves de novo biogenesis and polymerization of ECM. This is consistent with that reported for other epithelial systems such as the pulmonary system, salivary glands, the mammary gland, and renal system of vertebrates; however, these later studies utilize organ culture conditions[65] while the studies described for hydra represent in vivo conditions. This further supports the contention that ECM biogenesis is closely coupled to epithelial morphogenesis and that this fundamental relationship has been maintained throughout evolution.

In hydra, the importance of ECM biogenesis to head morphogenesis is functionally confirmed by antisense experiments in which blockage in the translation of the hydra basal lamina components such as laminin chains[6,7] and collagen type IV[19] results in a perturbation of head morphogenesis. Recent studies by Deutzmann et al[2] indicate that blockage in the translation of an interstitial matrix component such as Hcol-I can also inhibit head regeneration. Therefore, inhibition in the translation of either basement membrane or interstitial matrix components will lead to a blockage in head regeneration. These cell-ECM interactions are likely multifaceted and involve (1) the role of ECM as an extracellular structural entity whose assembly and presence affects the three-dimensional shape of tissues; (2) the role of ECM as a polymerized network of macromolecules that have endogenous signaling sequences such as RGD or YIGSR (either open or cryptic) that can interact with cell surface matrix receptors; and (3) the role of ECM as a scaffolding for the attachment of signaling molecules such as growth factors, small peptides, or other signaling compounds. There is evidence for each of these processes occurring in hydra.[55] In the studies described above, blockage in the translation of laminin is likely to initially affect head morphogenesis because of perturbations in the normal polymerization of the basal lamina associated with the ectoderm and endoderm. Parallel studies by Zhang et al[6] indicate that blockage of the hydra laminin α chain also results in an inhibition of head regeneration. The fact that these two convergent studies yield the same result strengthens the proposal that laminin biosynthesis is essential for normal head regeneration to occur. As discussed previously, laminin has also been reported to be a seeding molecule that promotes basement membrane assembly.[34] The lack of incorporation of laminin into the polymerizing matrix would be expected to have profound effects on the basic structure of the ECM that would in turn affect the overlying epithelium. Structural changes in hydra ECM have been observed at the ultrastructural level when matrix polymerization has been perturbed.[37] In addition, the lack of incorporation of laminin into the ECM would prevent the presentation of cell binding domains to epithelial cells such as the FTGTQ sequence of the laminin β1 chain. This sequence has been shown to be important for cell-ECM interactions in hydra[18] and recent studies have shown that it can bind to an integrin-class protein in hydra.[6] The absence of this sequence could potentially prevent epithelial signaling cascades that normally occur during head regeneration. As discussed in the section on ECM biogenesis, it should be noted that blockage in the translation of laminin also prevents subsequent incorporation of hydra fibrillar collagen into the ECM.[7] Therefore blockage in the translation of laminin would have the added effect of further perturbing ECM structure by affecting interstitial matrix assembly. It is not known if the incorporation of hydra type IV collagen in ECM is also affected in this case. Besides being an important structural component of the hydra ECM,[37] hydra type IV collagen is known to contain RGD sequences;[19] however it is not known if they are involved in cell signaling events during regeneration.

As a final note, it should be pointed out that biogenesis of ECM is not restricted to regeneration and epithelial repair events in hydra. In situ studies during bud formation indicate that up-regulation of ECM components and HMMP also occurs in this asexual reproductive process (Sarras et al unpublished data). This is of interest, because in the case of budding no loss of ECM occurs prior to the time of bud emergence. Rather, the ECM is continuous at the sites of bud formation and what occurs is simply an increase in the expression of HLM-β1, HLM-α1, Hcol-I, Hcol-IV, and HMMP as evagination of the bud progresses. In situ analysis at the earliest times of bud formation (placode stage) before evagination of the bud occurs, indicates up-regulation of at least HMMP has already occurred. High expression of both basement membrane and interstitial matrix components occurs throughout all stages of bud formation. A

Table 2. Developmental and cellular processes in hydra that have been reported to involve cell-ECM interactions

Developmental Process	Authors with References
1. Cell proliferation during morphogenesis as monitored in hydra cell aggregates.	Zhang et al, 1994[37]
2. Cell migration under in vivo conditions in grafting experiment and under in vitro conditions with isolated hydra nematocytes using ECM coated culture plates.	Zhang and Sarras, 1994; Stidwill and Christen, 1998; Ziegler and Stidwill, 1992; Gonzalez-Agosti and Stidwill, 1991.[87,110-112]
3. Cell differentiation/transdifferentiation of basal disk cells and battery cells and cell differentiation in hydra cell aggregates.	Leontovich et al, 2000; Yan et al, 2000, 1995; Zhang et al, 1994[37,64,66,67]
4. Epithelial morphogenesis of the head or foot as monitored in regeneration experiments, and in the adult polyp as monitored in hydra cell aggregate experiments.	Yan et al, 2000; Deutzmann et al, 2000; Fowler et al, 2000; Yan et al, 1995; Sarras et al, 1994; Sarras et al, 1993; Sarras et al, 1991; Barzansky and Lenhoff, 1974; Shimizu et al, 2002; Zhang et al, 2002; Leontovich et al, 2000[2,6,7,15,16,18,19,64,66,67,113]

summary of developmental and regenerative processes that involve cell-ECM interactions in hydra is shown in Table 2.

Interestingly, ECM biogenesis in hydra is always accompanied by the expression of hydra matrix metalloproteinase (HMMP). Antisense studies indicate that blockage in the translation of this proteinase also results in an inhibition of head regeneration. HMMP has been shown to degrade a broad spectrum of hydra ECM components[64] and therefore its up-regulation during ECM biosynthetic events raises the obvious question of why should a matrix degrading enzyme be expressed at time when formation of an intact ECM is occurring? There is no clear answer to this question, but in line with that reported for vertebrates, it is likely that HMMP may function at multiple levels to include: (1) to assist in the assembly of hydra ECM, (2) in the exposure of cryptic ECM signaling sites (e.g., RGD-like sequences) and (3) in the exposure of ECM-associated latent growth factor-like molecules that are involved in signaling pathways during head morphogenesis and differentiation.[64,66,67] Thus the function of HMMP could be more complex than originally suggested by its name.

Role of Metalloproteinases in the Regulation of Cell-ECM Interactions in Hydra

Besides HMMP, hydra expresses a number of enzymes of the metalloproteinase superfamily. Numerous studies involving both invertebrate and vertebrate organisms have established that members of this superfamily have a critical role in a broad spectrum of developmental processes.[68,69] Through studies designed to analyze the role of cell-ECM interactions in hydra morphogenesis and cell differentiation, a number of metalloproteinases have been identified that have been found to play an important role in a variety of processes related to hydra development. These hydra metalloproteinases fall into three classes that include: (1) astacin-class metalloproteinases (e.g., hydra metalloproteinase-1, HMP-1; and hydra metalloproteinase-2, HMP-2),[66,67,113] (2) matrix metalloproteinases (e.g., Hydra MMP, HMMP),[64] and (3) neprilysin-class metalloproteinases (e.g., Hydra endothelin converting enzyme, HECE).[70] For the purposes of this chapter we will focus on only two of these metalloproteinses, namely,

HMMP and HMP-1 because these two enzymes have been shown to be involved with cell-ECM interactions in hydra.

Hydra Metalloproteinase-1 (HMP-1)

Hydra metalloproteinase-1 (HMP-1) was isolated based on its enzymatic activity in cleaving gelatin as a substrate.[66] HMP-1 was originally detected as a gelatinase activity with a molecular mass of 25-29 x 10^3 that localized to the upper body of the organism. By applying a series of biochemical chromatography procedures, this enzymatic activity was successfully purified from cell extracts of *Hydra vulgaris.* Purified HMP-1 migrated as a single band with a molecular mass of approximately 25.7 x 10^3 that retained its gelatinase activity. N-terminal sequence results suggested that HMP-1 was not a classic matrix metalloproteinase, but rather an astacin metalloproteinase family member.[71] This was further confirmed when the full-length HMP-1 cDNA was cloned (GenBank Accession number U22380) and sequenced.[67]

Based on the amino acid sequence deduced from cDNA, HMP-1 is composed of 285 amino acids with a predicted molecular mass of 32.7 x 10^3. The domain structure of HMP-1 resembles that of other astacin family members. An N-terminal hydrophobic region of 21 residues with the characteristics of a putative signal sequence suggests that HMP-1 is a secreted proteinase. A 30-residue prodomain was identified based on its homology to the same region of *Podocoryne* metalloproteinase-1 (PMP-1).[72] The existence of the prodomain was further supported by the N-terminal sequence of purified HMP-1, which suggested a proteolytic cleavage of secreted HMP-1 between Phe(51) and Lys(52). The processed HMP-1 would have a predicted molecular mass of 27 x 10^3, fitting well with the size of purified HMP-1. Mature HMP-1 has a relatively simple structure consisting of a well-conserved astacin domain followed by a Cys-rich domain that was also identified in PMP-1 as a toxin homology (TH) domain.[72] A zinc-binding motif and a Met-turn, both characteristics of astacin proteinases were also well conserved in HMP-1.[71]

Preliminary to thc study of HMP-1 function, its expression pattern was determined by in situ hybridization and immunohistochemistry.[66,67] Interestingly, HMP-1 mRNA is expressed in a gradient pattern along the longitudinal axis[67] with the highest levels of mRNA at the head pole. At the protein level, HMP-1 is synthesized in endodermal cells in the upper body column as revealed by the cytoplasmic immunostaining of these cells. However, once these cells migrate into the tentacle region, HMP-1 protein is released into the extracellular space, as suggested by the disappearance of the cytoplasmic staining of tentacle cells and the concomitant appearance of HMP-1 staining in the extracellular matrix.[66,67]

The gradient expression pattern of HMP-1 suggested that this enzyme may be involved in pattern formation. This hypothesis was further supported by the finding that other astacin family members such as tolloid[73-76] play a proactive role in morphogenesis.[77] To approach the problem of the role of HMP-1 in morphogenesis, advantage was taken of the high regenerative capacity of hydra to determine the detailed expression pattern of HMP-1 during head regeneration.[66,67] In summary, these studies indicated that at 2 hours after decapitation, HMP-1 mRNA is only expressed by a ring of cells along the cut edge. At 12 hours, the number of HMP-1 positive cells increased and these cells began to migrate to the upper body column of the regenerating animal. At 36 hours, a large population of endodermal cells of the upper body column expressed high levels of HMP-1. This was accompanied by a transient expression of HMP-1 by some ectodermal cells in the very apical pole of the regenerating animal. By 48 hours, the HMP-1 expression gradient was reestablished and was followed by head regeneration with functional tentacles observed by 72 hours. Interestingly, the expression pattern of HMP-1 was restricted within clear boundary zones during this regeneration process. For example, HMP-1 was only expressed by endodermal cells in the upper body column, but not by any cells in the tentacles. These lines of evidence further suggest that HMP-1 is involved in regulating head regeneration.

The role of HMP-1 in controlling morphogenesis was directly studied using neutralizing antibodies and antisense thio-oligonucleotides.[66,67] A neutralizing antibody raised against

HMP-1 was shown to inhibit head regeneration in a reversible fashion.[66] In addition, this antibody was capable of blocking tentacle battery cell differentiation as shown by the progressive loss over time of annexin XII staining in the tentacle of antibody-treated animals.[66] The involvement of HMP-1 in hydra head regeneration was further confirmed by antisense treatment in which HMP-1 protein expression was specifically inhibited by the introduction of antisense thio-oligonucleotides to the head pole of hydra. Subsequently, head regeneration was significantly inhibited as judged by hypostome and tentacle formation.[67] The mechanism(s) underlying the control of morphogenesis by HMP-1 in hydra are unclear. Based on studies of homologous proteins in other species, it has been proposed that HMP-1 executes its regulatory functions through multiple pathways. These regulatory pathways may involve generation of growth factor gradients like their distantly homologous counterparts in Drosophila,[78] Zebrafish,[79] Xenopus,[80] and sea urchin.[81] Alternatively, HMP-1 could function by releasing biologically active peptides via proteolytic processing of precursor proteins.[82,83] Interestingly, the expression of a recently identified third hydra astacin metalloproteinase, foot activator responsive matrix metalloproteinase (Farm1), is regulated by such peptides.[84] This suggests a reciprocal regulatory mechanism between proteinases and peptides. Lastly, it is also possible that HMP-1 acts directly on extracellular matrix (ECM) proteins and effect morphogenesis by cleavage and modification of ECM structure or mechanics.[85,86] Given the capacity of HMP-1 in cleaving gelatin, a denatured collagen molecule, one could envision that these proteinases play an active role in modifying basement membrane structure either alone or together with other proteinases, such as hydra matrix metalloproteinase (HMMP).[64]

Hydra Matrix Metalloproteinase (HMMP)

Studies of extracellular matrix in hydra have established that it has a similar molecular composition to that seen in vertebrate species[16,18,55,87] and functional studies have established that cell-ECM interactions are critical to developmental processes in hydra.[2,6,7,15,16,18,19,55,87] Hydra ECM is in a constant state of turnover (Bode, 2003); thus indicating that matrix-degrading enzyme systems must be in place in hydra to execute these proteolytic processes. These observations coupled with numerous other studies that have tied MMP action to developmental processes (see review by Werb et al[88] led to the logical conclusion that an important aspect of cell-ECM interactions in hydra likely involves MMPs.

The number of MMP families has grown to over 25 with each type containing members from a broad spectrum of vertebrate species.[68,89-91] In contrast, relatively few MMPs have been identified in invertebrates. These invertebrate MMPs include envelysin from sea urchin,[92-94] a Drosophila MMP[95,96] and at least three separate MMPs from *Caenorhabditis elegans*.[97] Additional putative MMPs have been reported in plants such as soybean (*Glycine max*),[98] *Arabidopsis thaliana*,[99,100] and green alga [gamete lytic enzyme].[101]

In studies by Leontovich et al[64] a full-length clone (1642bp) of a hydra MMP (HMMP) was obtained from a cDNA library (GenBank accession number AF162688). Analysis of the deduced amino acid sequence indicated that HMMP had a 34% identity with human MMP-3 and contained a signal peptide, a pro-domain, a catalytic domain, a hinge region, and a C-terminal hemopexin domain as shown in. The calculated mass of the protein was 55.4×10^3. Northern blot analysis revealed that two HMMP mRNA transcripts could be identified with sizes of 1.6 kb and 1.9 kb.

Comparison of the deduced amino acid sequence of HMMP with other MMPs shows that HMMP has an overall 30-35% identity with known MMPs. The least identical region is the C-terminal hemopexin domain (23-38%). A detailed analysis of the HMMP sequence indicates that this enzyme has many similarities to vertebrate MMPs but it also contains a number of unique features. A possible furin cleavage site for activation of the pro-enzyme intracellularly exists at amino acid residues R^{101} R^{102} $-Y.^{103}$ The sequences of two critical regions, the Cys-switch and catalytic Zn^{++}-binding region, are well conserved in HMMP. An unusual feature of HMMP is that the multiple prolines commonly found in the hinge region between the catalytic domain and the hemopexin domain of MMPs are reduced to only two in HMMP. The hinge region of HMMP is somewhat hydrophilic in nature suggesting that this region is exposed as shown by the crystal structure of the full-length porcine MMP-1 Yan et al.[66] This region in vertebrate MMPs typically contains four or more proline residues; although it should be noted that some vertebrate MMPs completely lack a hinge region.[68,89] In addition, two cysteines commonly found in the hemopexin domain of MMPs are substituted with methionine (Met^{288}) and phenylalanine (Phe^{481}) in HMMP. This is of particular interest in regard to the highly conserved cysteines that are normally found at the N-terminus of the hemopexin domain of vertebrate MMPs. Another region of variation in HMMP is the cysteine-switch region where a leucine residue (Leu^{89}) substitutes for the valine residue typically seen in many vertebrate MMPs. A similar substitution has been reported for MMP-19 in humans[102] and MMP-7 in felines. Variations in this region are more common among MMPs as reflected by the divergence of this region in the three MMPs of *C. elegans*,[97] the MMP of soybean leaf, and the MMP of sea urchin (envelysin). In the case of these invertebrate and plant MMPs, a leucine residue replaces the arginine normally seen as the second amino acid residue of the cystine switch region.[103] These cysteines are conserved in the hemopexin domain of all vertebrate and invertebrate MMPs that have been previously analyzed.

Recombinant HMMP expressed in E. coli and folded to an active state[64] was able to cleave not only gelatin, transferrin, and synthetic fluorogenic substrates, but also hydra extracellular matrix. Recombinant HMMP has its maximal activity at about pH 6.5-7.5; a typical optimal pH for known MMPs.[104] The fact that the activity of recombinant HMMP was blocked by a specific MMP inhibitor such as recombinant human TIMP-1[105] and by synthetic MMP inhibitors such as GM6001[106] and Matlistatin provided strong evidence that this enzyme had characteristics reflective of bonafide MMPs. Recombinant HMMP was also able to cleave hydra laminin as monitored by Western blot analysis and analysis of SDS-PAGE profiles indicated that HMMP was able to degrade Hcol-1 (hydra fibrillar collagen 1). While HMMP was unable to cleave mammalian laminin, fibronectin, Type IV collagen, and Type I collagen, it was able to cleave heat-denatured Type I collagen (gelatin). The structural features of HMMP that preclude the digestion of these mammalian ECM components are not obvious, but may be related to the structure of HMMP's C-terminal hemopexin domain. Previous studies have shown that the hemopexin domain of MMPs can be important for substrate interactions. For example, MMP-1, MMP-8, and MMP-13 lacking their hemopexin domain lose the ability to cleave native interstitial collagens.[107-109]

In situ whole mount analysis indicated that HMMP mRNA was expressed in the endoderm along the entire longitudinal axis of the adult polyp, but at relatively higher levels in the tentacles and in the foot process just superior to the basal disk cells. The expression pattern for the hydra ECM component, laminin mirrored what was observed for HMMP suggesting that high levels of ECM turnover occur at regions of cell transdifferentiation. The expression of HMMP was monitored during foot regeneration using both Northern blot and whole mount in situ hybridization techniques.[64] These analyses indicated that the normally high expression level of HMMP was lost between 1-3 hours after excision of the foot process, but then progressively increased in the endoderm by 4 to 10 hours of foot regeneration. Similar patterns were observed during head regeneration.[7]

The effects of GM6001 (an inhibitor of MMP activity) and recombinant human TIMP-1 (an endogenous mammalian inhibitor of MMP activity) were analyzed during foot regeneration. At concentrations known to inhibit recombinant HMMP, both GM6001 and recombinant human TIMP-1 blocked foot regeneration. Blockage of foot regeneration was also observed in antisense experiments performed in parallel with the MMP inhibitor studies. Using a spectrum of antisense thio-oligo nucleotides, it was demonstrated that foot regeneration was significantly blocked by a number of antisense thio-oligonucleotides, but was unaffected by sense or control oligonucelotides.[64] This effect was reversible, in that hydra recovered from blockage when observed 5-7 days from the initial time of inhibition. Similar results were observed during head regeneration.[7] In adult polyps (nonregenerating), basal disk cells of hydra treated with GM6001 or TIMP-1 failed to produce normal amounts of mucous, and had significant reductions in the expression of peroxidase activity, one of their differentiation markers. When the inhibitors were removed, differentiated basal disk cells reformed within 2 days indicating that the effect was reversible as observed with foot regeneration. These studies indicate that HMMP is important to both morphogenesis and to cell differentiation processes in hydra.

In summary, functional studies with HMP-1 and HMMP point to a critical role for these metalloproteinases in hydra morphogenesis and cell differentiation. Biochemical and morphological studies tie these enzymes to the hydra matrix and indicate that they function through the regulation of cell-ECM interactions in this simple metazoan.

Acknowledgements

The studies described in this chapter were supported in part from the National Institutes of Health of the PHS, U.S.A. The author wishes to thank Eileen Roach for her assistance with the figures presented in this chapter.

References

1. Sandblom P. Creativity and Disease. How illness affects literature, art, and music. New York: Marion Boyars, 1995.
2. Deutzmann R, Fowler S, Zhang X et al. Molecular, biochemical and functional analysis of a novel and developmental important fibrillar collagen (Hcol-I) in hydra. Development 2000; 127:4669-4680.
3. Hess A. Observations on the structure of Hydra as seen with the electron and light microscope. Quart J Micr Sci 1957; 98:315-326.
4. Wood RL. The fine structure of intracellular and mesoglea attachments of epithelial cells in Hydra. In: Lenhoff HLW, ed. The biology of Hydra. Coral Gables, FL: Miami Press University, 1961.
5. Gauthier GF. Cytological studies on the gastroderm of Hydra. J Exp Zool 1963; 152:13-40.
6. Zhang X, Fei K, Yan L et al. Structural and functional analysis of an early divergent form of laminin in the Cnidarian, Hydra. Dev Genes E 2002; 212:159-172.
7. Shimizu H, Zhang XM, Zhang JS et al. Epithelial morphogenesis in hydra requires de novo expression of extracellular matrix components and matrix metalloproteinases. Development 2002; 129:1521-1532.
8. Haynes JF, Burnett AL, Davis LE. Histological and ultrastructural study of the muscular and nervous systems in Hydra. I. The muscular system and the mesoglea. J Exp Zool 1968; 167:283-293.
9. Davis LE, Haynes JF. An ultrastructural examination of the mesoglea of Hydra. Z Zellforsch Mikrosk Anat 1968; 92:149-158.
10. Barzansky B, Lenhoff HM. On the chemical composition and developmental role of the mesogelea of Hydra. Amer Zool 1974; 14:575-581.
11. Barzansky B, Lenhoff HM, Bode H. Hydra mesoglea: Similarity of its amino acid and neutral sugar composition to that of vertebrate basal lamina. Comp Biochem Physiol [B] 1975; 50:419-424.
12. Shostak S, Patel NG, Burnett AL. The role of mesoglea in mass cell movement in Hydra. Dev Biol 1965; 12:434-450.
13. Fujisawa T. Hydra regeneration and epitheliopeptides. Dev Dyn 2003; 226:182-189.
14. Bode H. Head regeneration in Hydra. Developmental Dynamics 2003; 226:225-236.
15. Sarras Jr MP, Zhang X, Huff JK et al. Extracellular matrix (mesoglea) of Hydra vulgaris III. Formation and function during morphogenesis of hydra cell aggregates. Dev Biol 1993; 157:383-398.
16. Sarras Jr MP, Madden ME, Zhang XM et al. Extracellular matrix (mesoglea) of Hydra vulgaris. I. Isolation and characterization. Dev Biol 1991; 148:481-494.
17. Sarras Jr MP, Meador D, Zhang XM. Extracellular matrix (mesoglea) of Hydra vulgaris. II. Influence of collagen and proteoglycan components on head regeneration. Dev Biol 1991; 148:495-500.
18. Sarras Jr MP, Yan L, Grens A et al. Cloning and biological function of laminin in Hydra vulgaris. Dev Biol 1994; 164:312-324.
19. Fowler SJ, Jose S, Zhang X et al. Characterization of Hydra Type IV Collagen. Type IV collagen is essential for head regeneration and it expression is up-regulated upon exposure to glucose. J Biol Chem 2000; 275:39589-39599.
20. Burgeson RE, Chiquet M, Deutzmann R et al. A new nomenclature for the laminins. Matrix Biol 1994; 14:209-211.

21. Champliaud MF, Lunstrum GP, Rousselle P et al. Human amnion contains a novel laminin variant, laminin 7, which like laminin 6, covalently associates with laminin 5 to promote stable epithelial-stromal attachment. J Cell Biol 1996; 132:1189-1198.
22. Livanainen A, Sainio K, Sariola H et al. Primary structure and expression of a novel human laminin alpha 4 chain. FEBS Lett 1995; 365(2-3):183-188.
23. Martin PT, Ettinger AJ, Sanes JR. A synaptic localization domain in the synaptic cleft protein laminin beta-2 (s-laminin). Science 1995; 269(5222):413-416.
24. Miner JH, Lewis RM, Sanes JR. Molecular cloning of a novel laminin chain, alpha 5, and widespread expression in adult mouse tissues. J Biol Chem 1995; 270:28523-28526.
25. Richards AJ, al Imara L, Carter NP et al. Localization of the gene (LAMA4) to chromosome 6q21 and isolation of a partial cDNA encoding a variant laminin A chain. Genomics 1994; 22:237-239.
26. Hutter H, Vogel BE, Plenefisch JD et al. Conservation and novelty in the evolution of cell adhesion and extracellular matrix genes. Science 2000; 287:989-994.
27. Benson S, Page L, Ingersoll E et al. Developmental characterization of the gene for laminin alpha-chain in sea urchin embryos. Mech Dev 1999; 81:37-49.
28. Timpl R, Brown JC. Supramolecular assembly of basement membranes. Bioessays 1996; 18:123-132.
29. Yurchenco PD, O'Rear JJ. Basal lamina assembly. Curr Opin Cell Biol 1994; 6:674-681.
30. Yurchenco PD, Cheng YS. Laminin self-assembly: A three-arm interaction hypothesis for the formation of a network in basement membranes. Contrib Nephrol 1994; 107:47-56.
31. Gallinano MF, Aberdam D, Aguzzi A et al. Cloning and complete primary structure of teh mouse laminin alpha 3 chain. Distinct expression pattern of the laminin alpha 3A and alpha 3B chain isoforms. Journal of Biological Chemistry 1995; 270[37]:21820-21826.
32. Kallunki P, Sainio K, Eddy R et al. A truncated laminin chain homologous to the B2 chain: Structure, spatial expression, and chromosomal assignment. J Cell Biol 1992; 119:679-693.
33. Schönherr E, O'Connell BC, Schittny J et al. Paracrine or virus-mediated induction of decorin expression by endothelial cells contributes to tube formation and prevention of apoptosis in collagen lattices. Eur J Cell Biol 1999; 78:44-55.
34. Colognato H, Yurchenco PD. Form and function: The laminin family of heterotrimers. Dev Dyn 2000; 218:213-234.
35. Colognato H, Winkelmann DA, Yurchenco PD. Laminin polymerization induces a receptor-cytoskeleton network. J Cell Biol 1999; 145:619-631.
36. Darribere T, Skalski M, Cousin HL et al. Integrins: Regulators of embryogenesis. Biol Cell 2000; 92:5-25.
37. Zhang X, Hudson BG, Sarras Jr MP. Hydra cell aggregate development is blocked by selective fragments of fibronectin and type IV collagen. Dev Biol 1994; 164:10-23.
38. Agbas A, Sarras Jr MP. Evidence for cell surface extracellular matrix binding proteins in Hydra vulgaris. Cell Adhes Commun 1994; 2:59-73.
39. Yoshida N, Ishii E, Nomizu M et al. The laminin-derived peptide YIGSR (Tyr-Ile-Gly-Ser-Arg) inhibits human preB leukaemic cell growth and dissemination to organs in SCID mice. Br J Cancer 1999; 80:1898-1904.
40. Hopker VH, Shewan D, Tessier-Lavigne M et al. Growth-cone attraction to netrin-1 is converted to repulsion by laminin- 1. Nature 1999; 401:69-73.
41. Prockop DJ, Kivirikko KI. Collagens: Molecular biology, diseases, and potentials for therapy. Annu Rev Biochem 1995; 64:403-434.
42. Olsen BR, Ninomiya Y. Collagens. In: Kreis T, Vale R, eds. Guidebook to the Extracellular Matrix, Anchor, and Adhesion Proteins. Oxford: Oxford University Press, 1999:380-408.
43. Kurz EM, Holstein TW, Petri BM et al. Mini-collagens in hydra nematocytes. J Cell Biol 1991; 115:1159-1169.
44. Kramer JM. Structures and functions of collagens in Caenorhabditis elegans. FASEB J 1994; 8:329-336.
45. Hausman RE, Burnett AL. The mesoglea of Hydra: IV. A quantitative radiographic study of teh protein component. J Exp Zool 1971; 177:435-446.
46. Hudson BG, Kalluri R, Gunwar S et al. Structure and organization of type IV collagen of renal glomerular basement membrane. Contrib Nephrol 1994; 107:163-167.
47. Borza DB, Miner JH, Hudson BG. Basement membrane and cellular components of the nephron. In: Massry SG, Glassock RJ, eds. Massry and Glassock's Textbook of Nephrology. New York: Lippincott Williams and Wilkins Publishers, 2000:37-42.
48. Blumberg B, MacKrell AJ, Fessler JH. Drosophila basement membrane procollagen alpha 1(IV). II. Complete cDNA sequence, genomic structure, and general implications for supramolecular assemblies. J Biol Chem 1988; 263:18328-18337.

49. Guo XD, Johnson JJ, Kramer JM. Embryonic lethality caused by mutations in basement membrane collagen of C. elegans. Nature 1991; 349:707-709.
50. Kuehn K. Basement membrane (type IV) collagen. Matrix Biol 1994; 14:439-445.
51. Noelken ME, Wisdom Jr BJ, Dean DC et al. Intestinal basement membrane of Ascaris suum. Molecular organization and properties of the collagen molecules. J Biol Chem 1986; 261:4706-4714.
52. Exposito JY, Garrone R. Characterization of a fibrillar collagen gene in sponges reveals the early evolutionary appearance of two collagen gene families. Proc Natl Acad Sci USA 1990; 87:6669-6673.
53. Exposito JY, D'Alessio M, Solursh M et al. Sea urchin collagen evolutionarily homologous to vertebrate pro-alpha 2(I) collagen. J Biol Chem 1992; 267:15559-15562.
54. Boot-Handford RP, Tuckwell DS. Fibrillar collagen: The key to vertebrate evolution? Bioessays 2003; 25:142-151.
55. Sarras Jr MP, Deutzmann R. Hydra and Niccolo Paganini (1782-1840) - two peas in a pod? The molecular basis of extracellular matrix structure in the invertebrate, Hydra. Bioessays 2001; 23:716-724.
56. Campbell RD. Tissue dynamics of steady state growth in Hydra littoralis. II. Patterns of tissue movement. J Morphol 1967; 121:19-28.
57. Dike LE, Chen CS, Mrksich M et al. Geometric control of switching between growth, apoptosis, and differentiation during angiogenesis using micropatterned substrates. In Vitro Cell Dev Biol Anim 1999; 35:441-448.
58. Chen CS, Ingber DE. Tensegrity and mechanoregulation: From skeleton to cytoskeleton. Osteoarthritis Cartilage 1999; 7:81-94.
59. Wang N, Butler JP, Ingber DE. Mechanotransduction across the cell surface and through the cytoskeleton. Science 1993; 260:1124-1127.
60. Ingber DE. Tensegrity: The architectural basis of cellular mechanotransduction. Annu Rev Physiol 1997; 59:575-599.
61. Madden ME, Sarras Jr MP. A morphological and biochemical characterization of a human pancreatic ductal cell line (PANC-1). Pancreas 1988; 3:512-528.
62. Nedelec B, Ghahary A, Scott PG et al. Control of wound contraction. Basic and clinical features. Hand Clin 2000; 16:289-302.
63. Shimizu H, Sawada Y, Sugiyama T. Minimum tissue size required for hydra regeneration. Dev Biol 1993; 155:287-296.
64. Leontovich A, Zhang JS, Shimokawa K et al. A novel hydra matrix metalloproteinase (HMMP) functions in extracellular matrix degradation, morphogenesis and the maintenance of differentiated cells in the foot process. Development 2000; 127:907-920.
65. Relan NK, Schuger L. Basement membranes in development. Pediatr Dev Pathol 1999; 2:103-118.
66. Yan L, Pollock GH, Nagase H et al. 25.7 x 10(3) M(r) hydra metalloproteinase (HMP1), a member of the astacin family, localizes to the extracellular matrix of Hydra vulgaris in a head-specific manner and has a developmental function. Development 1995; 121:1591-1602.
67. Yan L, Leontovich A, Fei KY et al. Hydra metalloproteinase 1: A secreted astacin metalloproteinase whose apical axis expression is differentially regulated during head regeneration. Dev Biol 2000; 219:115-128.
68. Woessner JF, Nagase H. Matrix metalloproteinases and TIMPs. Oxford, UK: Oxford University Press, 2000.
69. McCawley LJ, Matrisian LM. Matrix metalloproteinases: They're not just for matrix anymore! Curr Opin Cell Biol 2001; 13:534-540.
70. Zhang J, Leontovich A, Sarras Jr MP. Molecular and functional evidence for early divergence of an endothelin-like system during metazoan evolution: Analysis of teh Cnidarian, Hydra. Development 2001; 1607-1615.
71. Bond JS, Beynon RJ. The astacin family of metalloendopeptidases. Protein Sci 1995; 4:1247-1261.
72. Pan T, Groger H, Schmid V. A toxin homology domain in an astacin-like metalloproteinase of teh jellyfish Podocoryne carnea with a dual role in digestion and development. Dev Genes E 1998; 208:259-266.
73. Shimell MJ, Ferguson EL, Childs SR et al. The Drosophila dorsal-ventral patterning gene tolloid is related to human bone morphogenetic protein 1. Cell 1991; 67:469-481.
74. Ferguson EL, Anderson KV. Dorsal-ventral pattern formation in the Drosophila embryo: The role of zygotically active genes. Curr Top Dev Biol 1991; 25:17-43.
75. Ferguson EL, Anderson KV. Localized enhauncement and repression of the activity of the TGF-beta family member, decapentaplegic, is necessary for dorsal-ventral pattern formation in the Drosophila embryo. Development 1992; 114:583-597.
76. Finelli AL, Bossie CA, Xie T et al. Mutational analysis of the Drosophila tolloid gene, a human BMP-1 homolog. Development 1994; 120:861-870.

77. Sarras Jr MP. BMP-1 and the astacin family of metalloproteinases: A potential link between the extracellular matrix, growth factors and pattern formation. Bioessays 1996; 18:439-442.
78. Marques EJ, Weiss J, Stand M. Molecular characteristrics of a fucosyltransferase encoded by Schistosoma mansoni. Mol Biochem Parasitol 1998; 93:237-250.
79. Blader P, Rastegar S, Fischer N et al. Cleavage of the BMP-4 antagonist chordin by zebrafish tolloid. Science 1997; 278:1937-1940.
80. Piccolo S, Agius E, Lu B et al. Cleavage of Chordin by Xolloid metalloprotease suggests a role for proteolytic processing in the regulation of Spemann organizer activity. Cell 1997; 91:407-416.
81. Ware JL, Angerer LM, Angerer RC et al. Regulation of BMP signaling by the BMP1/TLD-related metalloprotease, SpAN. Developmental Biology 1999; 206:63-72.
82. Grens A, Shimizu H, Hoffmeister SA et al. The novel signal peptides, Pedibin and Hym-346, lower positional value thereby enhancing foot formation in hydra. Development 1999; 126:517-524.
83. Mitgutsch C, Hauser F, Grimmelikhuijzen CJ. Expression and developmental regulation of teh Hydra-RFamide and Hydra-LWamide proprohormone genes in Hydra: Evidence for transient phases of head formation. Developmental Biology 2003; 207:189-203.
84. Dumpfmuller G, Rybakin V, Takahashi T et al. Identification of an astacin matrix metalloprotease as target gene for Hydra foot activator peptides. Dev Genes Evol 1999; 209:601-607.
85. Uzel MI, Scott IC, Babakhaniou-Chase H et al. Multiple bone morphogenetic protein 1-related mammalian metalloproteinases process pro-lysyl oxidase at the correct physiological site and control lysyl oxidase activation in mouse embryo fibroblast cultures. J Biol Chem 2001; 276:22537-22543.
86. Chicurel ME, Chen C, Ingber DE. Cellular control lies in the balance of forces. Curr Opin Cell Biol 1998; 10:232-239.
87. Zhang X, Sarras Jr MP. Cell-extracellular matrix interactions under in vivo conditions during interstitial cell migration in Hydra vulgaris. Development 1994; 120:425-432.
88. Werb Z. ECM and cell surface proteolysis: Regulating cellular ecology. Cell 1997; 91:439-442.
89. Nagase H, Woessner Jr JF. Matrix metalloproteinases. J Biol Chem 1999; 274:21491-21494.
90. Sternlicht MD, Werb Z. How matrix metalloproteinases regulate cell behavior. Annu Rev Cell Dev Biol 2001; 17:463-516.
91. Lohi J, Wilson CL, Roby JD et al. Epilysin, a novel human matrix metalloproteinase (MMP-28) expressed in testis and keratinocytes and iin response to injury. Journal of Biological Chemistry 2003; 276:10134-10144.
92. Lepage T, Gache C. Early expression of a collagenase-like hatching enyzme gene in the sea urchin embryo. Embo J 1990; 9:3003-3012.
93. Nomura K, Tanaka H, Kikkawa Y et al. The specificity of sea urchin hatchiing enzyme (envelysin) places it in the mammalian matrix metalloproteinase family. Biochemistry 1991; 30:6115-6123.
94. Nomura K, Shimizu T, Kinoh H et al. Sea urchiin hatching enzyme (envelysin): cDNA cloning and deprivation of protein substrate specificity by autolytic degradation. Biochemistry 1997; 36:7225-7238.
95. Llano E, Pendás AM, Aza-Blanc P et al. Dm1-MMP, a matrix metalloproteinase from Drosophila with a potential role in extracellular matrix remodeling during neural development. J Biol Chem 2000; 275:35978-35985.
96. Page-McCaw A, Serano J, Sante JM et al. Drosophila matrix metalloproteinases are required for tissue remodeling, but not embryonic development. Dev Cell 2003; 4:95-106.
97. Wada K, Sato H, Kinoh H et al. Cloning of three Caenorhabditis elegans genes potentially encoding novel matrix metalloproteinases. Gene 1998; 211:57-62.
98. Pak JH, Liu CY, Huangpu J et al. Guidebook to the Extracellular Matrix, Anchor, and Adhesion Proteins. FEBS Lett 1997; 404:283-266.
99. Massova I, Kotra LP, Fridman R et al. Matrix metalloproteinases: Structures, evolution, and diversification. FASEB J 1998; 12:1075-1095.
100. Maidment JM, Moore D, Murphy GP et al. Matrix metalloproteinase homologues from Arabidopsis thaliana - Expression and activity. J Biol Chem 1999; 274:34706-34710.
101. Kinoshita T, Fukuzaw H, Shimada T et al. Primary structure and expression of a gamete lytic enzyme in Chiamydomonas reinhardtii: Similarity of functional domains to matrix metalloproteinases. Proc Natl Acad Sci USA 1992; 89:4693-4697.
102. Sedlacek R, Mauch S, Kolb B et al. Matrix metalloproteinase MMP-19 (RASI 1) is expressed on the surface of activated peripheral blood mononuclear cells and is detected as an autoantigen in rheumatoid arthritis. Immunobiology 1998; 198:408-423.
103. Ghiglione C, Lhomond G, Lepage T et al. Structure of teh sea urchin hatchiing enzyme gene. Eur J Biochem 1994; 219:845-854.

104. Nagase H. Human stromelysins 1 and 2. In: Barrett AJ, ed. Methods of Enzymology. London, UK: Academic Press, 1995:449-470.
105. Itoh Y, Ito A, Iwata K et al. Plasma membrane-bound tissue inhibitor of metalloproteinases (TIMP)-2 specifically inhibits matrix metalloproteinase 2 (Gelatinase A) activated on the cell surface. J Biol Chem 1998; 273:24360-24367.
106. Grobelny D, Poncz L, Galardy RE. Inhibition of human skin fibroblasts collagenase, thermolysin, and Pseudomonas aeruginosa elastase by peptide hydroxamic acids. Biochemistry 1992; 31:7152-7154.
107. Clark IM, Cawston TE. Fragments of human fibroblast collagenase. Purification and characterization. Biochem J 2000; 263:210-206.
108. Knauper V, Wilhelm SM, Seperack PK et al. Direct activation of human neutophil procollagenase by recombinant stromelysin. Biochemical Journal 1993; 295(Pt 2):581-586.
109. Knauper V, Lopez-Otin C, Smith B et al. Biochemical characterization of human collagenase-3. Journal of Biological Chemistry 1996; 271:1544-1550.
110. Stidwill RP, Christen M. Alteration of fibronectin affinity during differentiation modulates the in vitro migration velocities of Hydra nematocytes. Cell Motil Cytoskeleton 1998; 41:68-73.
111. Ziegler U, Stidwill RP. The attachment of nematocytes from the primitive invertebrate Hydra to fibronectin is specific and RGD-dependent. Exp Cell Res 1992; 202:281-286.
112. Gonzalez Agosti C, Stidwill R. In vitro migration of Hydra nematocytes: The influence of teh natural extracellular matrix (the mesoglea, of collagen type IV and type, laminin, and fibronectin) on cell attachment, migration parameters and patterns of cytoskeletal proteins. Cell Motil Cytoskeleton 1991; 20:215-227.
113. Yan L, Fei KY, Zhang JS et al. Identification and characterization of hydra metalloproteinase 2 (HMP2): A meprin-like astacin metalloproteinase that functions in foot morphogenesis. Development 2000; 127:129-141.

CHAPTER 6

Regulation of the Epithelial-to-Mesenchymal Transition in Sea Urchin Embryos

Gary M. Wessel and Hideki Katow

Introduction

The sea urchin has been an instructive animal for studies of gene regulation in development, morphogenesis, and cell migration. This is in part due to their convenient external development, their ease in culturing, and their marvelous transparency that makes them optically accessible for many developmental and cellular events. The embryo displays a multitude of complex movements and features, but involves relatively few cells (< 500 during gastrulation). The embryo is resilient to micromanipulation studies whereby cells or portions of tissues may be readily microinjected, transplanted, or photo-ablated to test mechanistically the crafting of the embryo. The whole genome project, its correlated EST identification and arrays (see http://sugp.caltech.edu), combined with the simple technology of introducing exogenous genes, reporters, mRNA, morpholinos or over-expressed proteins prove to be a powerful combination of capabilities with the nearly limitless number of embryos (several million per adult female) make the biochemistry and molecular biology of manipulated embryos readily available.[40] Finally, its status as a basal deuterostome, representing an early branch in the evolution of vertebrates, makes its molecular and cellular changes ripe for comparison in understanding the origin of body plan, embryonic mechanism, gene regulation and function.

This chapter will focus on one aspect of the morphogenesis of the sea urchin embryo – the transition of epithelial cells to a mesenchyme phenotype. This occurs in two major places in the embryo, each of which is accessible to direct observation in vivo, to in vitro culturing, to manipulation by micro-dissection, and to molecular perturbations by microinjection of reagents that alter gene function. We will begin with a brief explanation of the development of this embryo, emphasizing the aspects most relevant for the epithelial-mesenchyme transitions (EMT). We will then focus on a phenotypic analysis of the two major EMTs in this embryo, and follow with an examination of the molecular mechanism for this cellular transition.

Sea Urchin Development: Summary

Eggs are shed from adult females, having completed meiosis, with their chromosomes (mostly) transcriptional inactive in a pronucleus. Fertilization results in a rapid and phenotypically obvious change in the egg by formation of the fertilization envelope – a modified extracellular matrix that protects the embryo during early development, and that blocks subsequent sperm from fusing with the egg. Cleavage divisions (early embryonic cell divisions) proceed rapidly, synchronously, and invariantly such that the polarity and developmental fates of each early blastomere is apparent by the fourth cell division. Depending on the species and the temperature for culturing, cleavage divisions may proceed every 30-90 minutes. The dividing cells form a single cell layer surrounding a blastocoel, and this extracellular space will enlarge with

Rise and Fall of Epithelial Phenotype: Concepts of Epithelial-Mesenchymal Transition, edited by Pierre Savagner.

cellular secretion of extracellular matrix (ECM) molecules. Each cell of the embryo grows a cilium at the blastula stage, and the embryo is capable of swimming once it hatches from the fertilization envelope. Gastrulation, the processes that result in formation of the developmentally distinct primary germ layers (ectoderm – skin and nervous system; mesoderm – skeleton, muscle, pigment cells, coelom; endoderm – digestive system) begins shortly after hatching, and the major morphogenetic changes begin at one end of the embryo, the so called vegetal pole, or pole opposite the site of the original meiotic divisions. The two major morphogenetic complexes leading to gastrulation in this animal are (1) invagination of the epithelial sheet followed by convergent extension rearrangements that propel the future digestive system across the blastocoel to a predictable target site for fusion with the overlying ectoderm to form the mouth opening (the initial site of invagination/gastrulation in this embryo, as in all deuterostomes, become the anus;[27,28,31] for discussion of biomechanical properties of invagination in the sea urchin). The other major morphogenetic mechanism during gastrulation in this embryo is (2) ingression – a term given to describe the characteristic epithelial-to-mesenchymal transition.

The cells that undergo EMT are well defined, and of great interest for the development of this embryo, as their cell interactions lead to inductive events critical for the development of the fundamental body plan of the embryo. The first population of cells undergoing EMT is the primary mesenchyme cells (PMCs). These cells are descendants of the four micromeres that form at the fourth cell division during early cleavage. These cells will give rise to PMC, and only PMC. The sole fate of these cells later in development is in construction of the larval skeletal system. These cells are present in the forming epithelium of the early embryo, and remain at the vegetal pole. Indeed, no noticeable cell rearrangements occur during early development until the EMT of the PMCs. These cells then undergo a characteistic series of migratory events in the blastocoel.[101]

The sequence of events surrounding PMC formation has been described in the sea urchin embryo in detail in the living embryo[32,33,45,46] and at the ultrastructural level (Fig. 1).[2,67] The basal lamina subjacent to the presumptive PMCs is disrupted and the cells exhibit blebbing activity at their basal surfaces. The presumptive PMC detach form the outer hyaline layer and lose apical cell junctions as they elongate into the blastocoel.[67] This behavior is due in part to a decrease in affinity of the mesechyme cells for hyalin, measured by a cell attachment assay.[41] Adjacent cells, which will not become mesenchyme, actually exhibit increased numbers of hyalin associated microvilli.[7] Once the cytoplasmic mass has shifted into the blastocoel, with the narrowing of the apical end, detachment from the apical side occurs rapidly, followed by retraction of the apical portion of the cell. This process results in the transposition of the primary mesenchyme cells into the interior as a round cell resting on the vegetal plate. The adjacent epithelial cells quickly fill the spaces left by the ingressing cells.[67] The factors that regulate the breakdown of the basal lamina, the onset of the blebbing behavior, and cell elongation, changes in the cell-cell and cell-ECM affinity and the stability of cell junctions are all likely to be involved in this morphogenetic process.

An average of between 32 to 64 PMCs form in the embryo of a given species depending on whether the progeny undergo three or four rounds of cell division following their formation at the sixteen cell stage. It should be noted that the micromeres that form at the 16-cell-stage divide again into large and small micromeres. It is the large micromeres that undergo EMT at the blastula stage, while the four small micromeres contribute to the coelom much later in development, and undergo EMT at that time. The number of large and small micromere descendants within a population is remarkably consistent. No further division or death of PMCs occurs following ingression, thus the numbers remain constant through the remainder of morphogenesis. At the time of ingression PMCs first being to express a new family of cell surface proteins (see for example Fig. 2).[11,20,81,92,100,108,117]

Ingression is followed by a period of relative quiescence during which the PMCs remain at the vegetal pole in a nonmotile state. This transient period may reflect changes that are occurring in the adhesive properties of the PMCs[41] and or reorganization of the cytoskeletal

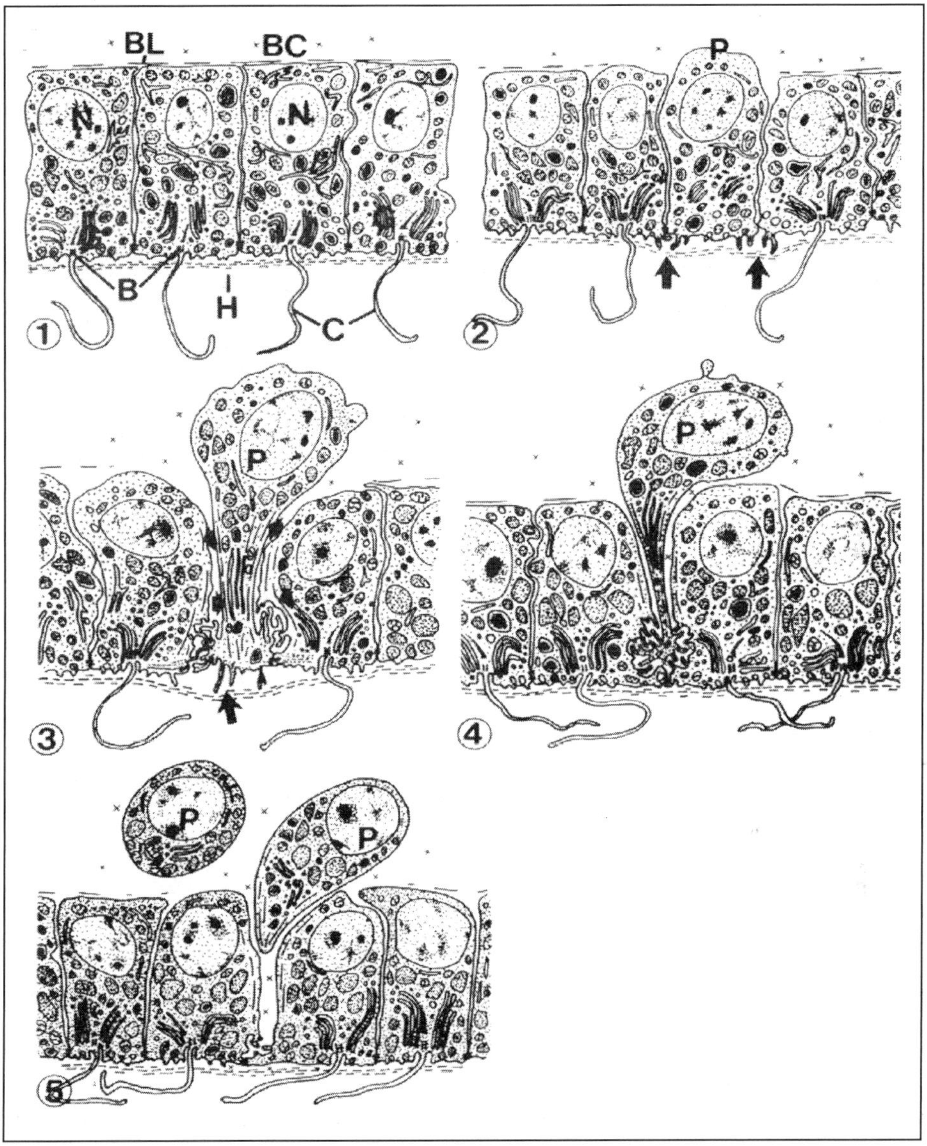

Figure 1. Diagram showing the PMC ingression process. 1) Presumptive PMCs are hardly distinctive with apical cilia (C) before thickening of vegetal plate occurs. B= basal body; BC= blastocoel; BL= basal lamina; G= Golgi complex; H= hyaline layer; N= nucleus. 2) During pulsatile movement, presumptive PMCs (P) lose apical cilia, and raise the hyaline layer, associated with apical elongation of microvilli (arrows). The basal surface disrupts the basal lamina. 3) Onset of basal elongation of presumptive PMC (P) with a bottle-neck feature in the ectoderm that is associated with the organization of apicobasal microtubules (arrowheads), and conspicuous apical elongation of cell processes (arrow). 4) Basal dislocation of the PMC (P) cell body by dissolving apical intercellular junctions. 5) Complete shifting of the cell body into the blastocoel, and quick sealing of the gap in the epithelium. The cell surface of the PMC (P) is characteristically smooth with few cell processes. Reprinted from J Exp Zool 1980; 213, Fig. 8.

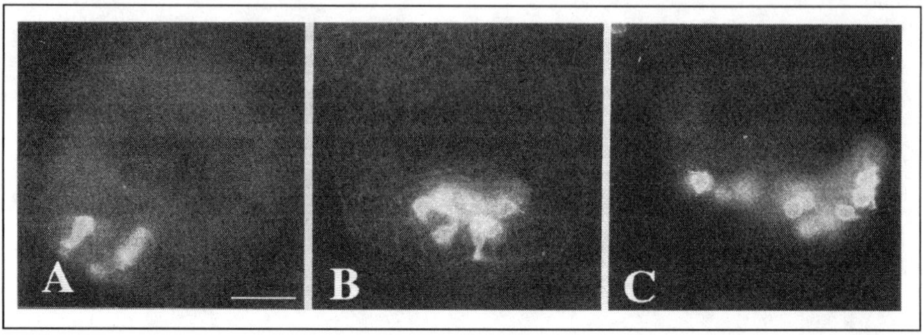

Figure 2. A) P4 antigen is expressed on the surface of presumptive PMCs before actual dislocation of the cell body into the blastocoel. B) PMCs shed on the vegetal ectoderm express P4. C) During migration PMCs express P4. Reprinted from Develop Growth Differ 1990; 32, Fig. 4A, B, and C. Bar shows 20 μm.

machinery required for protrusive activity and motility[10,45,76a] after as long as a few hours, the cells begin an almost spastic series of movements and migrate away from the vegetal pole along the inner surface of the blastocoel wall. The cells translocate by the continuous expression and retraction of slender filopodial cell processes (Figs. 3-5). Gradually the PMCs accumulate at specific target sites near the equator, forming a characteristic ring-like pattern known as the subequatorial PMC ring (Fig. 5). As the ring forms, filopodial processes of the PMCs fuse, forming an extensive network of syncytial cables (Figs. 6, 7). Within these cables, the cells secrete the larval skeleton, an array of crystalline rods composed of $CaCO_3$ and $MgCO_3$ embedded in a glycoprotein-rich matrix.[36,96]

Isolation and transplantation of the skeletogenic micromeres from early cleavage stage embryos show that these cells are autonomously programmed to give rise to skeletogenic cells and that all but the most terminal aspects of their differentiation program are insensitive to extrinsic cues.[9,17,38]

Prior to ingression, the PMCs are layered on both their apical and basal surfaces by two distinct ECMs. Each of these ECMs is critical in the regulation of EMT, and their analysis follows.

The Extracellular Matrix Involved in the EMT

Embryos of sea urchins and other echinoderms have two distinct extracellular matrices; an extra-embryonic matrix surrounding the entire embryo composed of the hyaline layer and apical lamina, and the basal lamina/blastocoel matrix upon which the epithelium resides, and through which the cells undergoing EMT must penetrate. These two ECMs have different origins, structures, and functions, which is especially apparent during gastrulation and EMT.

The Extra-Embryonic Matrices: The Hyaline Layer and the Apical Lamina

The hyaline layer and apical lamina are extracellular matrices that surround the embryo and larva from fertilization until metamorphosis. The hyaline layer is first formed at fertilization, when cortical granules undergo exocytosis and release molecules that assemble into a translucent layer on the external surface of the zygote. In addition to the cortical granules, the egg contains other classes of vesicles that undergo exocytosis shortly after fertilization and which comprise the apical lamina, an extracellular matrix distinct in origin and composition from the hyalin layer.[6,15,21,47,55,63,87-89]

The most abundant constituent of the hyaline layer is hyalin, a ~330 kDa, fibrillar glycoprotein that multimerizes in the presence of calcium.[110] The innermost region, referred to as the apical lamina, consists of a distinct layer of fibers that contains fibropellin proteins.[15,19] Cell adhesion assays have been used to measure binding of sea urchin embryo cells to several

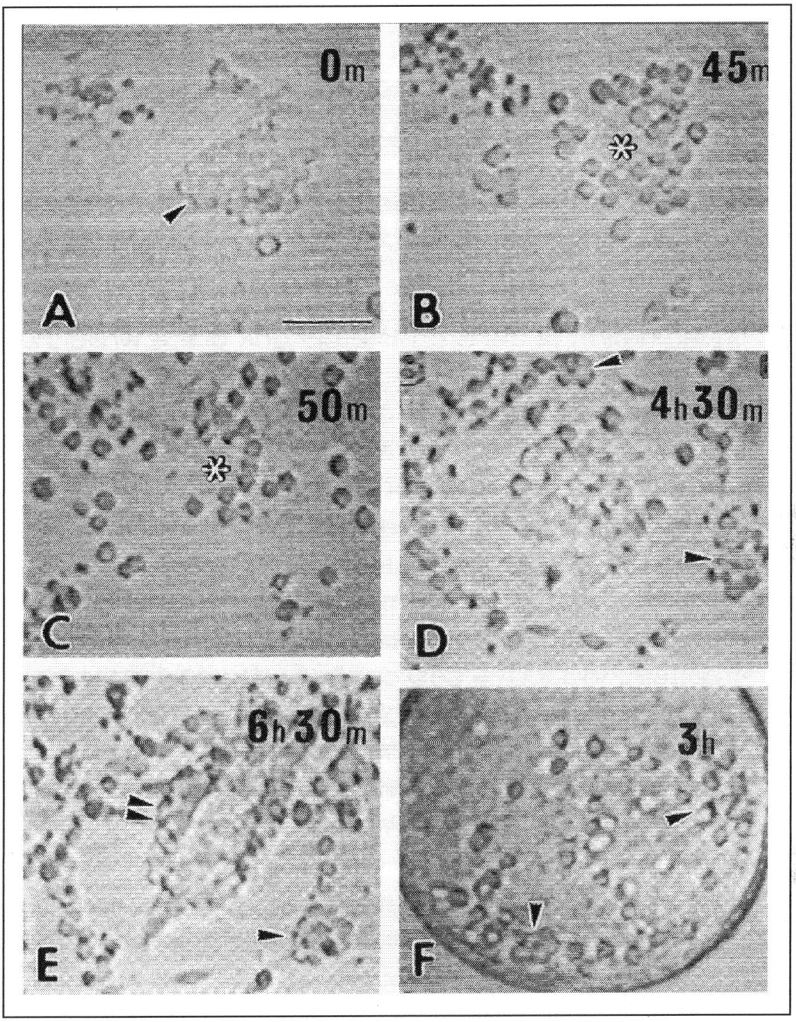

Figure 3. PMC ingression and pattern formation in 2D embryos. In spread blastulae, PMCs ingress (A), by leaving noningressing small micromere descendants at the center of a circle of PMCs (asterisk, B). Soon after ingression, PMCs spread on the epithelium (C). Four hours following ingression, PMCs formed a ring pattern around invaginating archenteron with two PMC aggregates (arrowheads) (D). Invagination occurs in 2 hours (double arrow heads) (E). In intact embryos, PMC pattern formation occurs similarly to the 2D embryos with two baso-lateral PMC aggregates (F). Bar in (A) shows 20 μM. Reprinted from Develop Growth Differ 2000; 42, Fig. 1.

hsyaline layer components, including hyalin, echinonectin, and fibropellins.[1,4,5,18,19,119] In several cases, cell-type specificity and/or temporal regulation of binding have been observed. In the case of hyalin, at least some of the cell binding activity of this protein has been mapped to a domain that includes multiple tandem repeats referred to as a hyalin repeat.[119]

The Basal Lamina and Blastocoel Matrix

The inner ECM of the embryo is composed of the basal lamina and blastocoel matrix (Fig. 1). The basal lamina is a thin but well-organized layer of fibers that lines the blastocoel (Fig. 8,

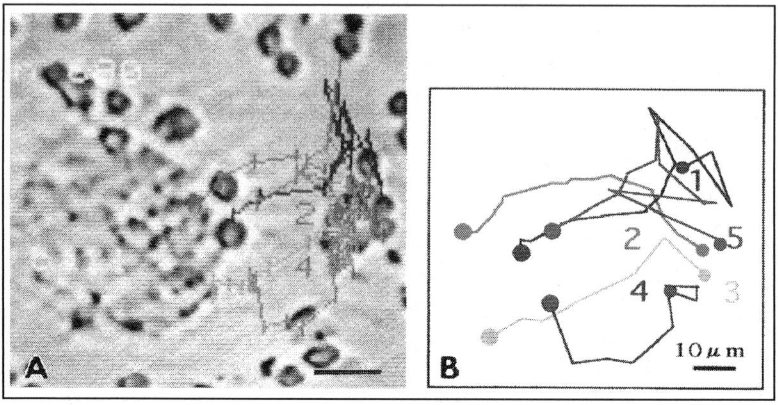

Figure 4. Computer-aided tracing of PMC migration pathways on 2D embryo. A) Each PMC migrated from the center of the photo frame to the right. B) Each line represents a single PMC. Extracted PMC migration pathway from (A). PMCs migrated from a large circle to a small circle where they aggregated. Reprinted from Develop Growth Differ 2000; 42, Fig. 2.

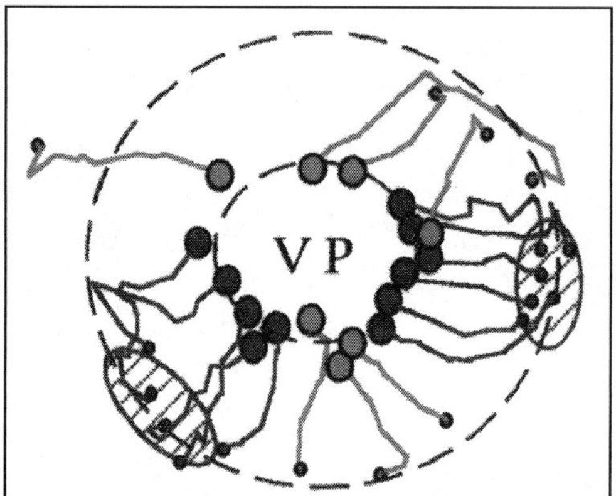

Figure 5. Summary of the PMC migration pathway on 2D embryo. Cell aggregates forming PMCs (dark gray) and cable forming PMCs (light gray) migrated straightforwardly to the respective targets. VP= vegetal plate. Reprinted from Develop Growth Differ 2000; 42, Fig. 4A.

9) and, after gastrulation begins, the wall of the archenteron. The blastocoel matrix is relatively sparse in organization and fills the blastocoel cavity at later stages of development. Like the hyaline layer, the basal lamina and blastocoel matrix first arise via the regulated exocytosis of vesicles stored in the unfertilized egg.[116] The basal lamina and blastocoel is formed early in development, during cleavage divisions. By the 4-8-cell stage, a space between the blastomeres is apparent and it already contains proteins indicative of the blastocoel of larvae. Ultrastructural studies show that a well-organized basal lamina is formed by the late blastula stage (Fig. 10).[43,66,68,116,122] The basal lamina of the vegetal plate is transiently disrupted during PMC ingression but reforms later in gastrulation. Like the hyaline layer, the molecular composition of the inner ECM changes progressively during development and many of the proteins are synthesized both maternally and zygotically. During gastrulation, the composition of the inner

Figure 6. Scanning electron microscopy of migrating PMCs. PMCs extend short cell processes in the direction of migration (small arrows), and leave thicker retraction protrusion behind (double arrows). Large arrows point the direction of migration. Bar shows 5 μm. Reprinted from Exp Cell Res 1981; 136, Fig. 2.

ECM changes as different populations of mesenchyme cells ingress and modify the matrix through the secretion of new molecules.[8,13,22,23,57,61,112,114,115,118]

Studies using antibodies against vertebrate ECM components suggested the sea urchin inner ECM might contain molecules, such as collagen, and a fibronectin-like protein.[29,59,61,69,109,116,117] A number of histochemical and biochemical studies suggested the presence of sulfated glycoproteins and proteoglycans of various kinds (reviewed by ref. 35). More recently, molecular approaches have led to the cloning and characterization of many components of the inner ECM, including laminin,[14,52,90] collagens,[12,16,22,23,118] fibronectin-like proteins, pamlin[63,72] and others (e.g., ECM3).[54] The ongoing sea urchin genome-sequencing project (see http://sugp.caltech.edu) has revealed many new components of the inner ECM. For example, sea urchin counterparts of fibrillin, perlecan, and nidogen/entactin have been identified in an EST sequencing project.[124] Because of its ease of isolation and conserved constituents, the blastocoel ECM from the sea urchin embryo has even been used in assays of invasion potential for mammalian metastatic tumor cells.[83]

EMT occurs at two occasions during embryogenesis in sea urchin. Initial EMT takes place at the vegetal plate in late mesenchyme blastulae to shed PMC into the blastocoel (Fig. 1). Subsequent EMT occurs at the tip of archenteron during gastrulation to shed SMC (in most sea urchins). PMC and SMC have different origins with common mechanisms.

Morphogenesis of the EMT

PMC Ingression

PMC formation occurs in indirectly developing sea urchins, but not in directly developing species. Morphologically, ingression process of PMC is divided into four stages:

1. Initial visual indication of presumptive PMCs by emergence of a cell body contour, resulting from loosening the adhesion between presumptive PMCs as well as between the surrounding epithelial cells,
2. Occurrence of pulsatile movement associating with thickening the vegetal plate as seen in time-lapse cinematography or video microscopy.
3. Entry of the presumptive PMCs into the blastocoelic space, and
4. Separation of the entire cell and initiation of PMC migration.

The entire ingression process does not require intact embryo, and occurs in vitro in embryos artificially spread as a sheet of ectoderm (Fig. 3),[64] implicating the process is autonomous. However, ultrastructural observations show the presence of shallow indentations on the lateral cell surfaces that face noningressing epithelial cells (Fig. 11), suggesting a squeezing

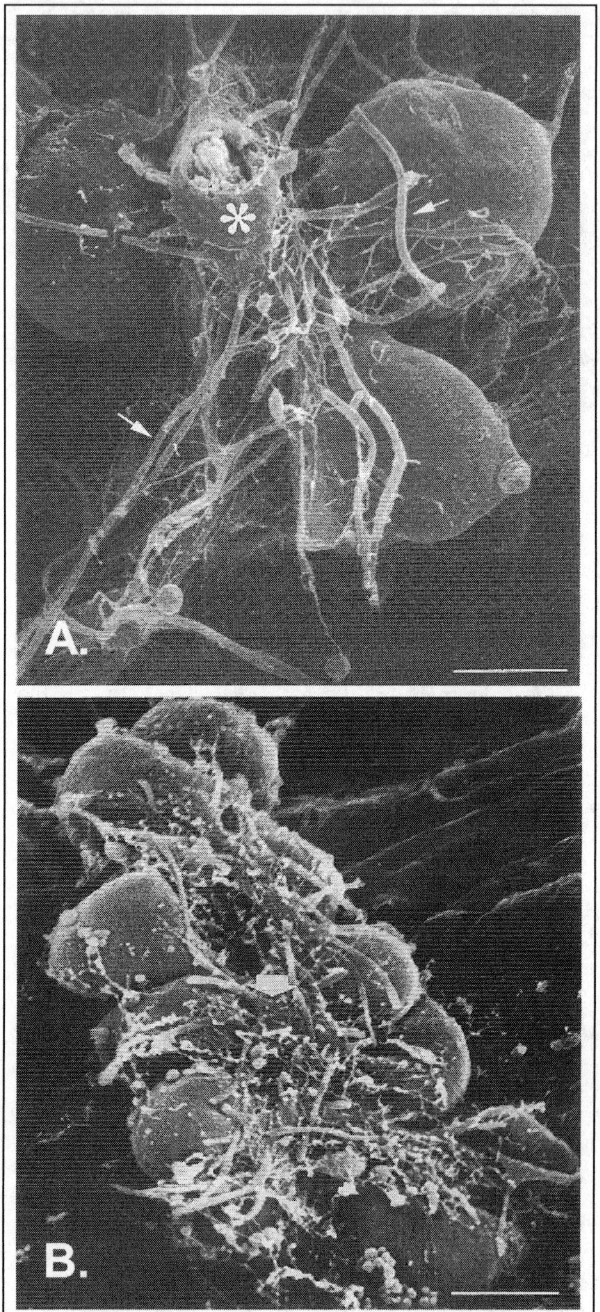

Figure 7. A) Scanning electron microscopy shows numerous long and thin cell processes extended from the PMC sessile site (arrows). Intercellular fusion takes place among these cell processes (asterisk). Bar shows 5 μm. (From Develop Growth Differ 1992; 34, Fig. 4.) B) Scanning electron microscopy of the surface beneath the PMC aggregate at its sessile site. Intercellular fusion occurs among a larger number of cell processes during development (arrow). Bar show 10 μm. Reprinted from Develop Growth Differ 1992; 34, Fig. 4.

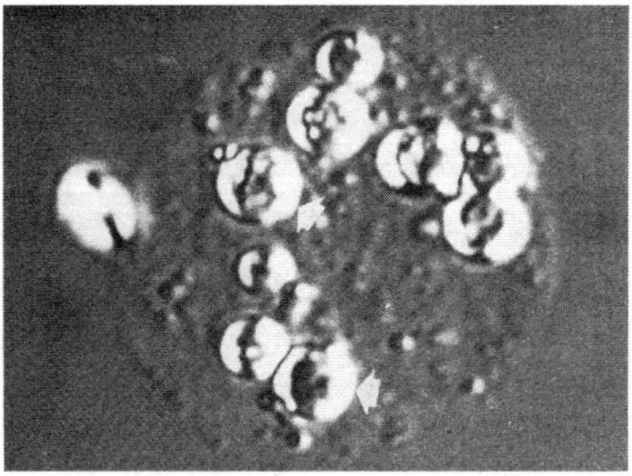

Figure 8. A bag isolated from mesenchyme blastula. PMCs in the blastocoel are now surrounded by the basal lamina (arrow). The diameter of the bag is considerably smaller than that of actual embryo.[69] Reprinted from Dev Biol 1982; 94, Fig. 2A.

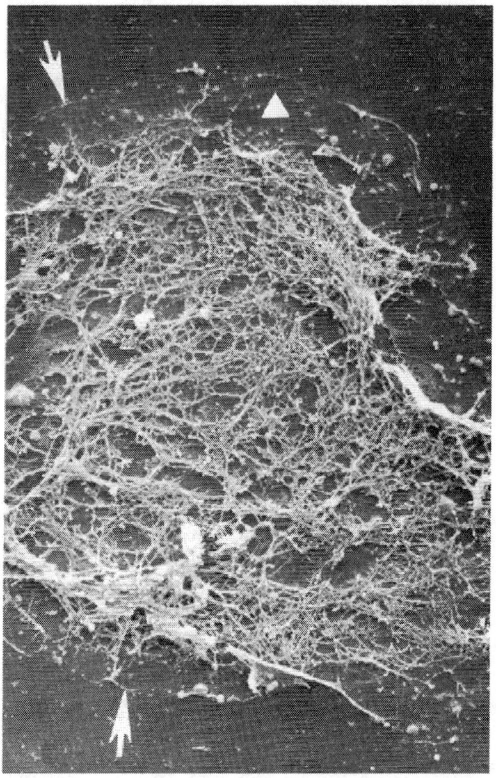

Figure 9. Scanning electron microscopy of an extracellular matrix bag, mesenchyme blastulae whose epithelium was removed, exposing the basal lamina and PMCs in the blastocoel. Fibrillar extracellular matrix is seen bound to the thin basal lamina (arrows) whose inner surface appears smooth (triangle). Reprinted from Dev Biol 1982; 94, Fig. 2C.

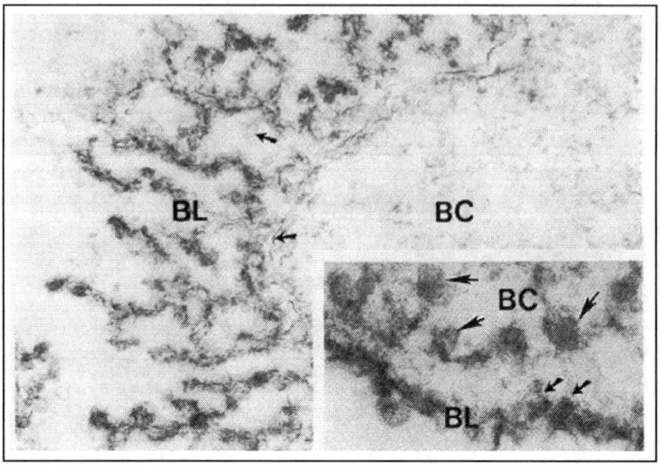

Figure 10. Transmission electron microscopy of the basal lamina in the bag. The basal lamina of the bag shrunk by showing considerable undulations (BL), and was filled with amorphous blastocoelic material (BC). Inset shows highly magnified basal lamina (BL), 25-30 nm diameter granules (large arrows) and fibers (small arrows). Reprinted from Dev Biol 1982; 94, Fig. 3.

force against presumptive PMCs by noningressing epithelial cells, and thus of such environmental cues near by as well.

The pulsatile movement is associated with the formation of apicobasally aligned bundle of microfilaments[10,11] and microtubules (Fig. 12)[67] in PMCs. Such characteristic microfilament development in PMC is correlated with acquisition of high sensitivity to cytochalasin B: treatment induces premature delamination resulting from depolymerization of the microfilaments in the cell cortex (Fig. 13).[60] This PMC-specific, microfilament-associated effect suggests that PMC ingression is associated with cytoskeleton organization and that repulsing forces may function between presumptive PMCs and the surrounding epithelial cells during ingression. This implicates that PMC ingression is not solely dependant on microfilament-generated cyoplasmic forces, but also includes cell surface-associated cues and signal transduction mechanisms (see below). Treatment with the calcium ionophore, A23187, separates presumptive PMCs from the adjacent epithelium cells on the basal side. The integrity of the apical epithelium though is retained on the apical side (Katow, unpublished data), implicating the involvement of calcium signal transduction pathway in early stages of EMT.

Cellular interactions with the ECM are dynamic during EMT.[79,80] The apical ECM bundles blastomeres together during early cleavage period, and provides adhesive substrate for epithelial cells thereafter. During ingression, however, presumptive PMCs alone loose their affinity to the apical ECM,[41,60,91] implicating the alteration of cell adhesion mechanisms (see below).

Different from the apical ECM, however, presumptive PMC needs to break through the basal ECM, the basal lamina, to complete the ingression to the blastocoel. Ultrastructurally, the basal lamina disappears on the surface of ingressing PMC (Figs. 14, 15),[67,70] and certain molecules are lost on these cell surfaces (e.g., pamlin; Fig. 16),[75] indicating the occurrence of local basal lamina disintegration. The pulsatile movement of ingresssing PMC is unlikely the major mechanism in disrupting the very elastic basal lamina (Fig. 10).[67] More likely, it is the involvement of matrix proteinase activity as is the case in metralloproteinase-2 (MMP-2) expression during mesenchyme formation from endocardium in avian embryo.[3,106] However, despite frequent reports on the presence of collagenase or hatching enzyme on the apical surface of sea urchin embryos,[82,89,123] no protease activity has been ascribed to this process in presumptive PMC so far.

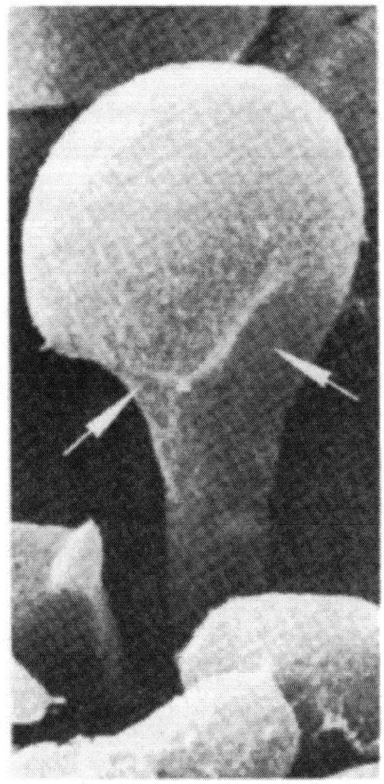

Figure 11. Scanning electron microscopy of PMC during ingression at the vegetal plate. The cell shows indentations on the lateral surfaces (arrows), implicating lateral squeezing forces by the surrounding cells. Reprinted from J Exp Zool 1980; 213, Fig. 5A.

Epithelial cells of the vegetal plate adhere also by intercellular junctions, including septate desmosomes on their apical side (Fig. 17).[67] During PMC ingression, these desmosomes are dissolved between noningressive epithelial cells and PMCs, as well as between the presumptive PMCs. This occurs slightly before visual recognition of presumptive PMCs in the vegetal ectoderm, and the PMC-specific loss of cilia.[67]

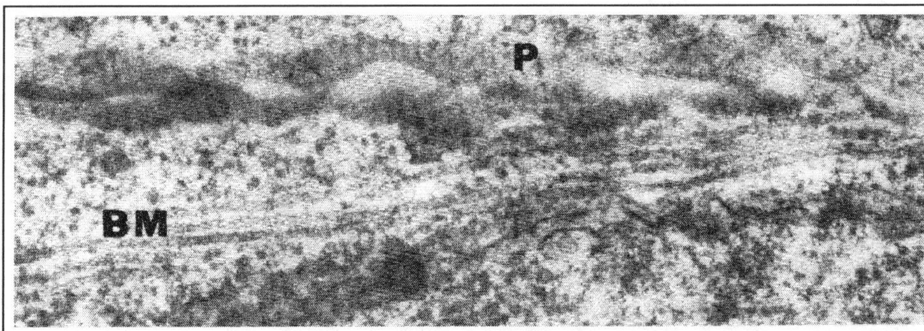

Figure 12. Blastomeres (BM) neighboring presumptive PMC (P) develops apicobasally aligned microtubules on the side that faces PMC. Reprinted from J Exp Zool 1980; 213, Fig. 12.

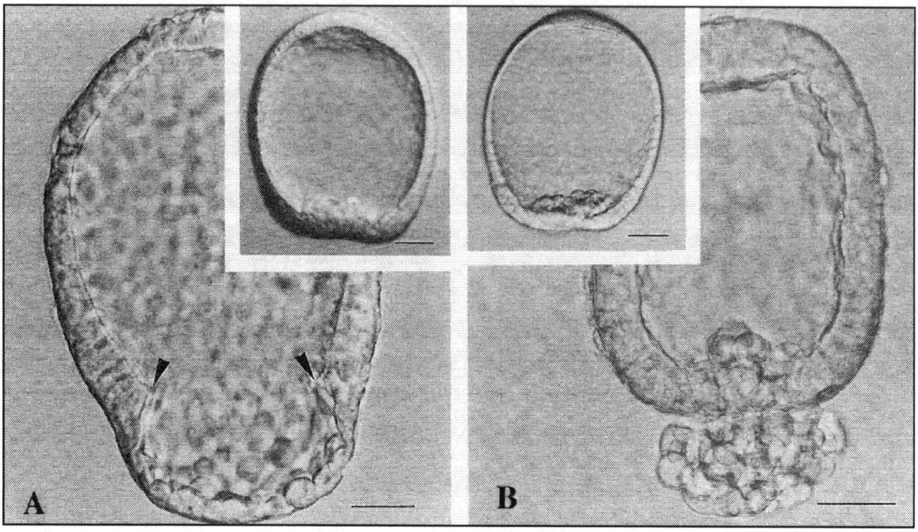

Figure 13. Vegetal plate cells are specifically sensitive to cytochalasin B treatment. Treated vegetal plate cells shorten at the apical surface with a sharp boundary between ectoderm cells (A, arrowheads). Inset shows control blastula. After the apical-ward shortening, these cells puffed out of the embryonic body (B).Inset shows control blastula that shows bulging vegetal plate cells. Bars show 20 μm. Reprinted from Develop Growth Differ 1989; 31, Fig. 2C, D.

SMC Ingression

SMCs are derived from a lineage distinct from the PMC. Whereas the PMCs originate from the micromeres at the 16-cell stage, SMCs are derived from veg2 blastomeres at 60-cell stage.[25,104] The veg2 lineage also contributes to the endoderm, and in most of sea urchins, the SMC are located at the tip of the invaginating gut, the archenteron and participate in direction of the invagination process.[24,49,51] Surprisingly, deletion of the archenteron at half invagination still yields SMC descendants.[48] In these embryos, even fully developed digestive tracts are still formed. During invagination, archen s teron forming cells are actively proliferate,[75] and even

Figure 14. Transmission electron microscopy of the basal lamina on the ectoderm (BL) in mesenchyme blastula which is comprised of fibrillar as well as amorphous materials. BC= blastocoel. Reprinted from J Exp Zool 1980; 213, Fig. 9A.

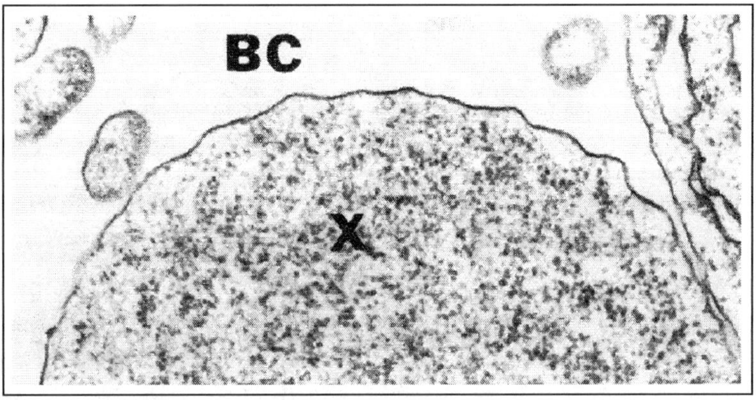

Figure 15. Transmission electron microscopy of the basal surface of a PMC during ingression (X). The surface lacks basal lamina and is characteristically smooth. BC= blastocoel. Reprinted from J Exp Zool 1980; 213, Fig. 9C.

by ubstantial deletion of the region by microsurgery, the archenteron still keeps elongating,[48] and produces SMC descendants, likely by lineage conversion. This highly conserved mechanism to keep supplying cells to form and archenteron is possible by the induced cell proliferation activity in the vegetal epithelium.[75] This implicates that any veg2-derived cells that participates in archenteron formation possess the potency to trans-differentiate into SMCs or a part of SMC descendants with high plasticity.

In *H. pulcherrimus* and *S. purpuratus*, SMCs extend long cellular processes toward the future mouth region of ectoderm, indicating distinctive movement among SMCs that is not seen in PMC ingression. The long cell process elongation, however, is not reported in *S. mirabilis* embryo.

Unlike the ingression of PMC, the site and stage in which SMCs ingress are different among species. In *Hemicentrotus pulcherrimus*[59] and *Strongylocentrotus purpuratus*,[94] SMCs ingress at the tip of archenteron in gastrulae. However, in *Scaphechinus mirabils* SMC ingression occurs at the vegetal plate soon after PMC ingression, but before gastrulation.[77] In this species, SMC even show pigment-cell properties prior to ingression, and its autofluorescence is retained even after formalin fixation. Surprisingly, these pigment cells spread within the ectodermal epithelium without first entering the blastocoel,[77] as they do in most species, implicating that SMC "ingression" is independent of crossing the basal lamina.

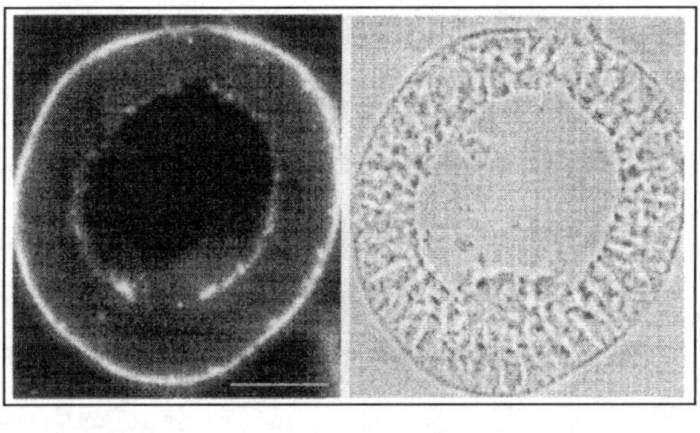

Figure 16. Vegetal plate ectoderm at the place where PMCs ingress into the blastocoel lack pamlin. Immunohistochemistry shows no positive staining at the vegetal plate (left). Left frame shows phase contrast micrograph of the right one. Bar shows 30 μm. Reprinted from Develop Growth Differ 2002; 44, Fig. 5C, D.

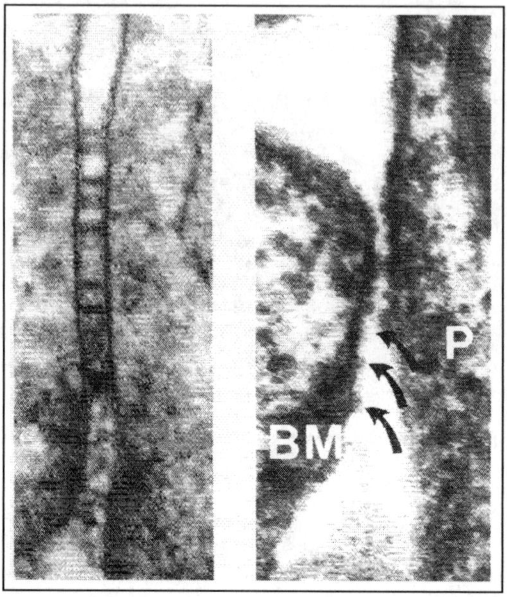

Figure 17. Epithelial cells are linked by apical septate desmosomes (A). The septate desmosomes are lost between neighboring blastomeres (BM) during PMC (P) ingression by leaving small granular debris on the cell surface (B, arrows). Reprinted from J Exp Zool 1980; 213, Fig. 10A, B.

Molecular Mechanisms Involved in EMT

PMC Formation

Before the occurrence of visual PMC ingression, molecular preludes are apparent in the vegetal epithelium. These include localized expression of the Ets transcription factor, which is down-regulated by Pmar1 in the earlier PMC lineage, and up-regulated of several other genes, including Sm50.[26] Besides these transcriptional activities that will be described in detail elsewhere in this chapter, the presumptive PMC also acquires sensitivity to calcium, develops bundles of microfilaments, and apicobasal alignments of microtubules (see above).

Along with the cytoskeletal reorganization, as the consequence of above transcription pathway, the cell surface properties are also modified. All of the epithelial cells of the early embryo express the lateral cell surface-specific glycoprotein, Epith-1, before gastrula stage. It is also expressed by both ectodermal and endodermal cells following gastrulation (Fig. 17).[59] This cell surface protein though is internalized by endocytosis selectively in PMCs (Fig. 18), thus altering their cell surface repertoire during ingression. Although biological role of Epith-1 is yet to be elucidated, its disappearance from the cell surface is temporally linked to the appearance of new surface proteins, including Meso 1[116] and the P4 antigen (Fig. 2). The link between dissociation of PMCs after ingression and Epith-1 internalization, appears to be downstream of a nonreceptor type protein tyrosine kinase (PTK) signal transduction pathway. Herbimycin A, a nonreceptor type PTK inhibitor, reversibly perturbs Epith-1 internalization and ingression (Figs. 19, 20).

In contrast, PMC acquires new cell surface protein, such as acidic msp130 that was found in *S. purpuratus* embryos[81] or its homologs in *H. pulcherrimus* embryos[62,107,108] during ingression associated with increasing negative cell surface charge.[105] Msp130 or P4 expression in the presumptive PMC occurs as accumulation of *HpEts* transcripts takes place in the same cell group. Both genes are down-regulated by Pmar1which is up-regulated by *Otx*.[26] In *H. pulcherrimus*, P4 expression on the cell surface occurs slightly before Epith-1 internalization, and is also reversibly down-regulated by herbimycin A.

The role of these cell surface modification in PMC-ingression is yet to be elucidated. Although a substantial number of integrins have been isolated and characterized so far in sea

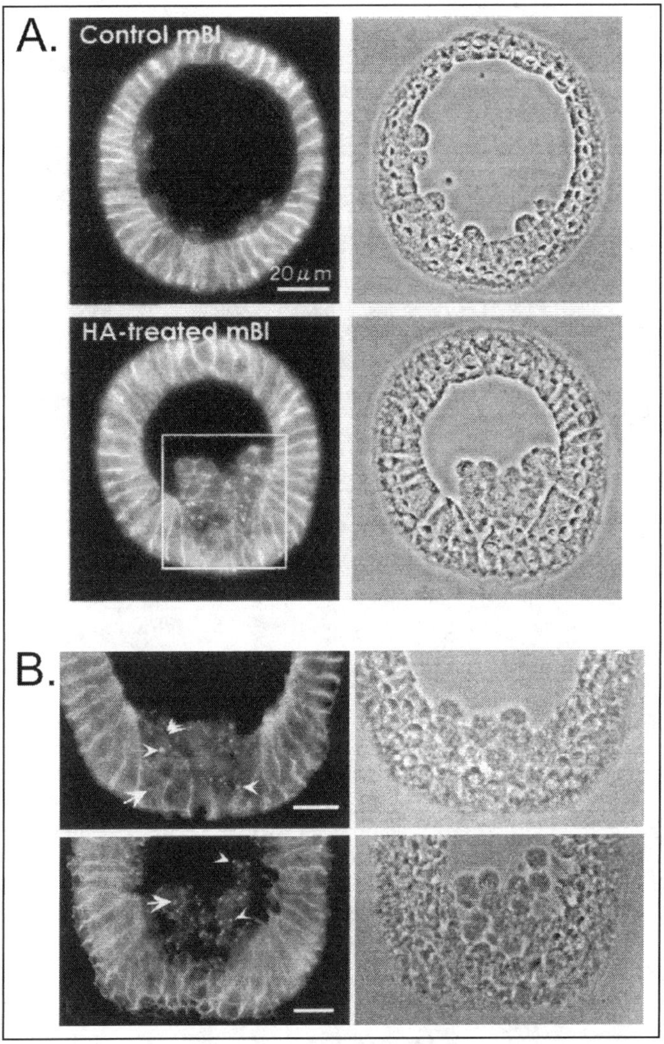

Figure 18. A) Herbimycin A, a nonreceptor type protein tyrosine kinase inhibitor, inhibits cytoplasmic transportation of cell surface protein. In control embryos, epithelium cell-specific cell surface protein (Epith-1) locates on the lateral surface in mesenchyme blastula. Epith-1 is internalized during PMC ingression and found with granular feature in cytoplasm. In herbimycin A-treated embryo, PMCs dislocate to the blastocoel, but never dissociate to single cells by retaining Epith-1 protein on their surface with incomplete internalization of the protein. Phase contrast micrographs are of control embryo and herbimycin A-treated one, respectively. B) Internalization of Epith-1 protein occurs while presumptive PMCs are in the vegetal ectoderm (upper), and shed PMCs completely lack Epith-1 afterdislocation to the blastocoel (bottom). Right column of photos is phase contrast micrograph of left column immunohistochemistry, respectively.

urchin embryos,[53,85,86,97,111] and they provided molecular basis of explaining the anchoring mechanism of basal lamina to the basal surface of the epithelium, none of them are localized on the apical surface. This hardly explains the molecular mechanism how PMCs acquire and loose their affinity to the hyaline layer.

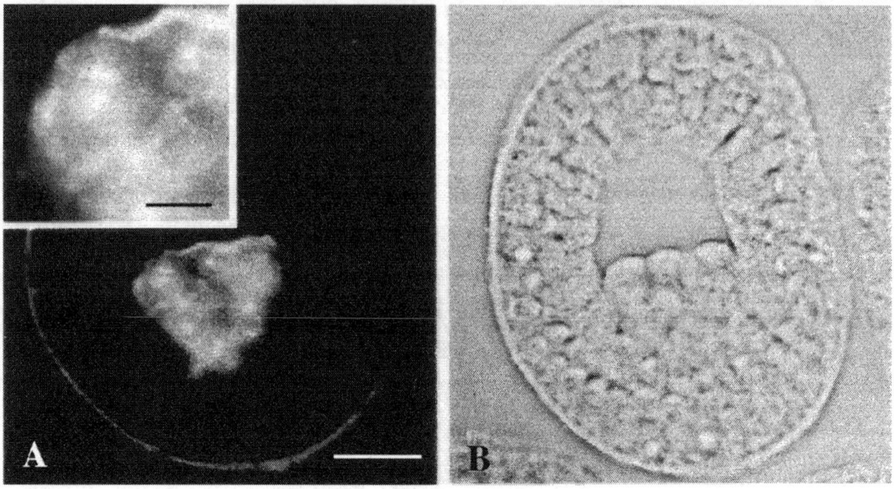

Figure 19. Herbimycin A perturbs cell surface transportation of PMC specific P4 protein. Although P4 protein was synthesized and detected in the Golgi of PMC, the protein was not present on the cell surface (A). Inset shows highly magnified basal surface of PMCs with slight positive cell surface P4, but not at all among PMCs. Bars in (A) an inset show 20 μm and 10 μm, respectively. B) phase contrast micrography of (A).

SMC Formation

SMC ingression from the tip of archenteron shares similar phenotypes to PMC ingression. Epith-1 protein is internalized[59] and pamlin on the basal surface of the invaginating archenteron is lost at the tip of archenterons.[63] No ultrastructural or immunohistochemical observations are available so far on the presence of the basal lamina on the invaginating archenteron, so the lack of pamlin on its basal side does not immediately mean the disappearance of entire basal lamina components. Unlike PMC though, so far no SMC surface-specific protein is known, despite numerous attempts by sea urchin biologists. Perhaps more detailed study of SMC formation in *S. mirabilis* will help in understanding SMC ingression mechanisms.

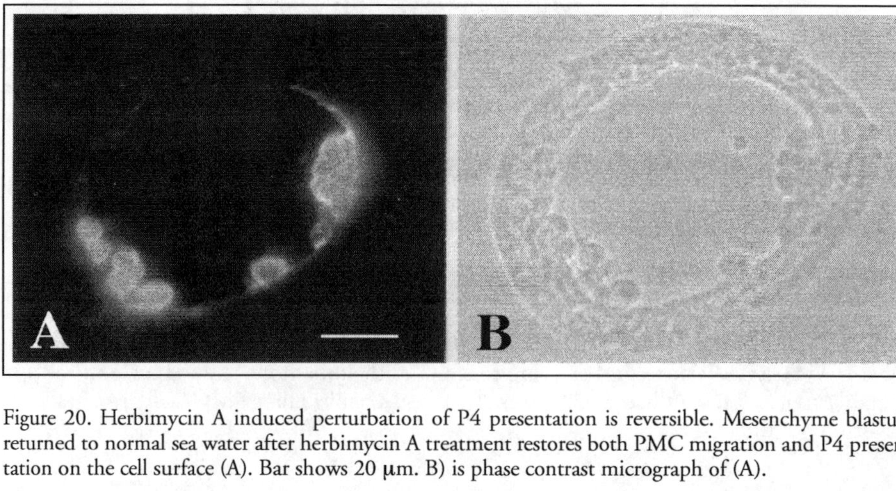

Figure 20. Herbimycin A induced perturbation of P4 presentation is reversible. Mesenchyme blastula returned to normal sea water after herbimycin A treatment restores both PMC migration and P4 presentation on the cell surface (A). Bar shows 20 μm. B) is phase contrast micrograph of (A).

Signal Transduction in PMC Cell Surface Modifications

The alteration of cell surface property among mesenchyme is regulated, at least in part, by PTK signal transduction pathways, including nonreceptor type PTKs. In herbimycin A-treated embryos, the internalization of Epith-1 and ingression of PMC is inhibited (Figs. 17, 18), and transportation of the cell surface-specific P4 protein to PMC surface was also inhibited (Figs. 19, 20). Furthermore, based on PTK activity it has been suggested that PMC ingression is comprised of, at least, two signal transduction stages. They are (1) the initial nonreceptor type PTK-independent dissociation of presumptive PMC from epithelial cells and (2) the nonreceptor type PTK-dependent dissociation among PMCs.

Genistein, a receptor type PTK inhibitor, perturbs neither PMC surface presentation of P4 nor Epith-I internalization. PMC dissociation soon after the ingression is also not affected by the PTK. This suggests that the second stage of ingression does not require extracellular signaling, and instead is regulated by autonomous cytoplasmic signal transduction pathways. This is consistent with the well used observation that micromeres separated from 16-cell embryos proliferate and differentiate spicules in vitro, independent of other embryonic cells. During this in vitro differentiation, they show characteristic morphological changes, dynamic cell surface changes, and migratory behavior in concert temporally with their intact sibling embryos.[98]

Genistein decreases phosphorylation of extracellular signal-regulated kinase 1 (ERK1) and myosin light chain (MLC) that is downstream of MAPK signal transduction pathway.[65,75] This indicates that PMC dissociation is up-regulated by MAP kinase signal transduction pathway through a MLC kinase that was isolated and characterized as a MAP kinase-activated protein kinase (MAPKAPK) in *H. pulcherrimus* embryos.[76] Microfilament activation also is involved in this ingression stage as has been implicated by immunohistochemical studies.[10,11] Furthermore, the observation that PD98059, an inhibitor of signal transduction from MEK1/2 to ERK1/2, inhibits PMC migration to some extent[74] agrees with the involvement of a MAPK pathway.

Suramin, a growth factor receptor inhibitor, blocks PMC dissociation, and exogenous supply of platelet-derived growth factor (PDGF) restores the deficiency.[75] This implicates the involvement of a signal transduction pathway that is initiated at growth factor receptor to IP_3 through G protein that is known to stimulate calcium ion release. PDGF receptor function in sea urchin embryos was implicated by a dominant/negative study in *Lytechinus pictus* embryos.[44,102,103]

PMCs begin migration in the blastocoel after the second stage of the nonreceptor type PTK-dependent PMC ingression process. This migration requires growth factor-like ligands as was shown by suramin-treated embryos. Suramin decreases phosphorylation of ERK1 and MLC, whereas exogenous PDGF restores both phosphorylation events,[75] implicating that PMC migration is downstream of growth factor-like ligand-initiated signal transduction pathway. Genistein does not perturb cell surface presentation of P4, but inhibits the early phase of PMC migration. This suggests the involvements of ECM receptor-generated signal transduction pathway in PMC migration. Consistent with these observations, the PMCs of suramin-treated embryos remain tightly clumped on the vegetal plate in early mesenchyme blastulae and on invaginating archenteron tip in early gastrulae, despite apparently normal PMC "dislocation" into the blastocoel (Fig. 21). In these embryos PMC dissociation never happens.[74] Thus, these observations are consistently implicating that PMC migration requires extracellular cues.

In herbimycin A-treated embryos, besides the failed internalization of Epith-1, delaminated PMCs do not dissociate from each other and remain clumped on the vegetal plate (Fig. 19). P4 in these embryos expressed in the cytoplasm of the clumped PMCs dissociate by expressing the protein on the surface after returning to normal seawater (Fig. 20). Herbimycin A also perturbs the phosphorylation of SUp62 protein, a homo-dimeric cytoplasmic protein, associated with hyper-phosphorylation at tyrosine residues of the protein.[64,74] In PMCs, phosphorylation of SUp62 increases during migration in vitro in the presence of pamlin, an endogenous PMC adhesion protein, whereas ectodermal cells do not during the same period.[74]

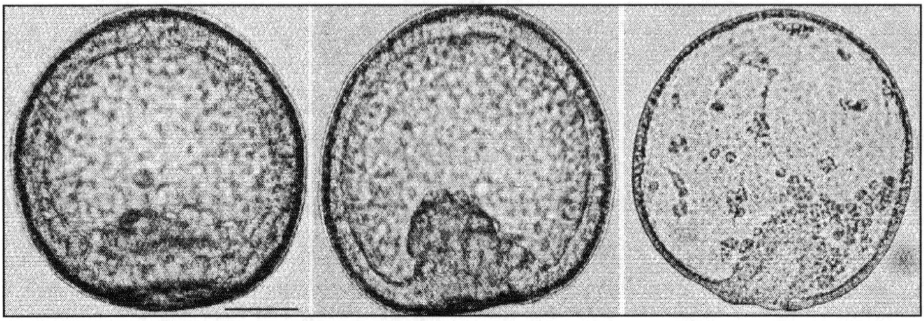

Figure 21. Suramin, a growth factor receptor inhibitor, perturbs PMC dissociation and migration. PMCs are shed to the blastocoel in mesenchyme blastula, but dissociation never occurs (far left). In early gastrula, PMCs sit on the tip of archenteron as a tightly packed cell mass with no visible single cell contour (middle). PMCs do not differentiate spicules, and granulated in prism larva stage of normal embryo (far right).Bar shows 30 μm. Reprinted from Develop Growth Differ 2002; 44, Fig. 2M, N, O.

Gene Regulation in Cells that Undergo EMT

Recent studies have begun to elucidate the gene regulatory network that underlies PMC specification (Fig. 22).[26] This information is critical to EMT morphogenesis as it is clear that new gene products must be transcribed in order for the PMCs to undergo EMT. The initial specification of the large micromere lineage is entrained by a patterning system linked to the animal-vegetal polarity of the unfertilized egg.[9,17,38] One important component of this system is beta-catenin.[84] Beta-catenin is localized in the nuclei of vegetal blastomeres during early cleavage and becomes concentrated at the highest levels in the micromere lineage.[84] Nuclearization of beta-catenin is required for all known aspects of mesoderm and endoderm formation, including large micromere specification.[26,30,84,120]

Alx1 is a member of the paired-type of homeodomain proteins. It is expressed selectively in the early micromere lineage and is required for at least two distinct morphogenetic processes: (1) ingression (epithelial-mesenchymal transition) and (2) skeletogenesis. The molecular changes required for ingression have not yet been identified, although this process is associated with changes in cell shape, protrusive activity, adhesive properties and cell surface architecture.[41,95] In contrast, a large number of terminal differentiation gene products have been identified that function in the formation of the biomineralized skeleton.[56,121] These gene products are expressed specifically in the large micromere lineage beginning at the mid-late blastula stage, prior to PMC ingression. The expression of four such markers was examined in Alx1-null cells, including SpMSP130, SpMSP130-related 2, SpP19 and SpSM50, and all four are regulated positively by Alx1. This suggests that Alx1 is a key regulator of the molecular subprogram that controls morphogenesis, including ingression and the terminal differentiation of skeletogenesis.

To determine the function of Alx1 in PMC specification,[39] injected morpholino antisense oligonucleotides (MOs) that were complementary to *alx1* mRNA, into fertilized sea urchin eggs. Development in the resultant embryos appeared normal during cleavage and blastula stages, and the injected embryos hatched within 1 hour of sibling controls. At the late blastula stage, however, a striking phenotype became apparent. PMCs did not ingress in the MO-injected embryos and invagination of the vegetal plate was delayed by several hours relative to control sibling embryos. Furthermore, MO-injected embryos failed to form visible skeletal elements even after extended periods of culture. Alx1 MOs also interfered with normal PMC specification as assayed by the expression of a battery of molecular markers: mAb 6a9 recognizes PMC-specific cell surface proteins of the MSP130 family,[34,56] *SpMSP130-related 2* (a MSP protein family member), *SpP19* and *SpFRP* (fibrinogen-related protein)[56,124] were all analyzed. Each showed highly down-regulated levels of expression coincident with the observed

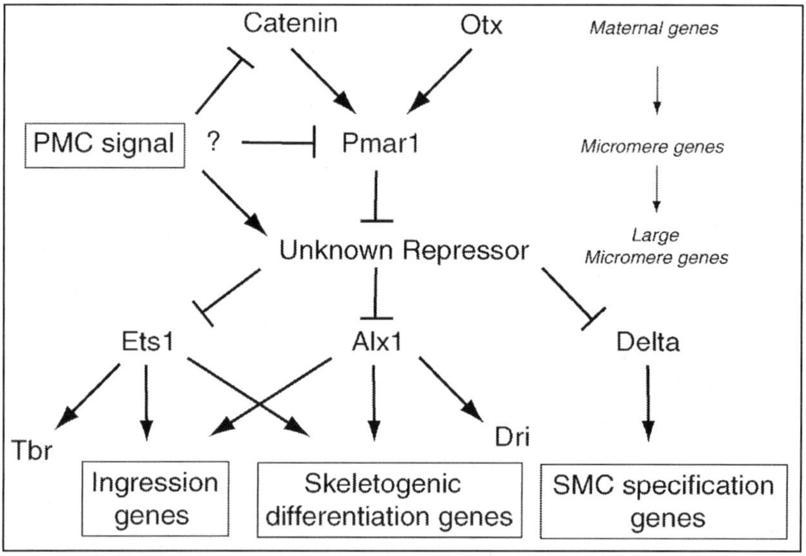

Figure 22. The micromerePMC gene regulatory network. The total developmental time represented in the diagram is from fertilization (top) to the blastula stage (bottom). Arrows and bars indicate positive and negative interactions, respectively. All genes shown encode transcription factors with the exception of *Delta*, which encodes a transmembrane protein. There is evidence for a direct interaction between Ets1 and *sm50*[78] but all other interactions may be indirect. beta-catenin and Otx are maternal proteins that become differentially enriched in micromere nuclei at the 16-cell stage.[21a,84,88] These two proteins are required for the activation of *pmar1*, which is expressed only by the micromeres and their progeny.[99] Pmar1 may block the expression of a putative repressor (Repressor X) specifically in the micromeres. This repressor (which may be several proteins) blocks PMC fate specification in all nonmicromere lineages (Oliveri et al, 2002). *Ets1*, *alx1* and *delta* are all regulated independently by pmar1 and the repressor. Ets1 regulates the *tbr* gene[42] and Alx1 regulates *dri* (this study). Alx1, Ets1 and Tbr are all expressed only by the large micromeres and their progeny. Alx1 and Ets1 both regulate genes involved in ingression and skeletogenesis.[39,78] Delta signaling activates genes involved in SMC specification, including *gcm*.[100a,112,113] PMC signals feed into the network upstream of *alx1* (this study); dashed bars and dashed arrow show possible inputs. Modified from Ettensohn et al, 2003.[39]

phenotype of the cells. In addition, the levels of nine different PMC markers were measured by QPCR. Four of the mRNAs tested, *dri* (G. Amore and E. Davidson, personal communication), *MSP130*,[100] *MSP130-related 2*[56] and *sm50*,[60a] were down-regulated in MO-injected embryos. On the other hand, SpAlx1 MO had no detectable effect on mRNA levels of four transcriptional elements of the PMCs: *tbr*,[42] *ets1*,[78] *delta*[113] and *pmar1*[99] when assessed either at 18-20 hours or 23-24 hours post-fertilization. The level of *Spalx1* mRNA was slightly elevated in MO-injected embryos, suggesting that Alx1 protein may act as a negative regulator of the *alx1* gene.

The Alx1 protein is closely related to the Cart1/Alx3/Alx4 family of vertebrate homeodomain proteins. In vertebrates, these proteins regulate the formation of skeletal elements of the limbs, face and neck. One conclusion from these findings then is that the ancestral deuterostome had a population of biomineral-forming mesenchyme cells that expressed an Alx1-like protein. Its subsequent function diversified to direct related biomineralization processes in different animals and in different tissues.

Pmar1 is a critical early transcriptional regulator in the gene network that controls PMC specification.[99] It predates expression of *Spalx1*, detectable by in situ hybridization analysis, by one cell cycle. That *Spalx1* might be regulated by *pmar1* is suggested by the overexpression of

Pmar1 or an engrailed-pmar1 fusion protein (EnHD)[99] resulting in a striking increase in levels of *Spalx1* mRNA expression (as assayed by QPCR). The fact that overexpression of wild-type Pmar1 and EnHD produced similar effects supports the view that Pmar1 normally acts as a repressor.[99] Moreover, overexpression of Pmar1 (or EnHD) could activate *Spalx1* expression to high levels even in cadherin mRNA-injected embryos.

Future Implications

The genome project of the sea urchin and its correlated EST sequences have already begun to yield important gene regulatory information for many aspects of the embryo, with particular interest in micromere/PMC specification. This is in part because it is critical in specifying adjacent cells and organizing the axis of the embryo. In this respect, it is being considered an "organizer" function, much like the dorsal lip of the blastopore classically defined by Spemann and Mangold. Several gene products involved in cell signaling and transcriptional regulation have recently been identified. These have a critical role in the development of the micromeres, including their morphogenesis during EMT.

References

1. Adelson DL, Humphreys T. Sea urchin morphogenesis and cell-hyalin adhesion are perturbed by a monoclonal antibody specific for hyalin. Development 1988; 104:391-402.
2. Akasaka K, Amemiya S, Terayama H. Scanning electron microscopical study of the inside of sea urchin embryos (Pseudocentotus depressus). Effects of aryl beta-xyloside, tunicamycin and deprivation of sulfate tions. Exp Cell Res 1980; 129:1-13.
3. Alexander SM, Jackson KJ, Bushnell KM et al. Spatial and temporal expression of the 72-kDa type IV collagenase (MMP-2) correlates with development and differentiation of valves in the embryonic avian heart. Dev Dyn 1997; 209:261-268.
4. Alliegro MC, Ettensohn CA, Burdsal CA et al. Echinonectin: A new embryonic substrate adhesion protein. J Cell Biol 1988; 107:2319-2327.
5. Alliegro MC, Burdsal CA, McClay DR. In vitro biological activities of echinonectin. Biochemistry 1990; 29:2135-2141.
6. Alliegro MC, Black SD, McClay DR. Deployment of extracellular matrix proteins in sea urchin embryogenesis. Microsc Res Tech 1992; 15:2-10.
7. Amemiya S. Electron microscopic studies on primary mesenchyme cell ingression and gastrulation in relation to vegetal pole cell behavior in sea urchin embryos. Exp Cell Res 1989; 183:453-62.
8. Angerer LM, Chambers SA, Yang Q et al. Expression of a collagen gene in mesenchyme lineages of the Strongylocentrotus purpuratus embryo. Genes Dev 1988; 2:239-246.
9. Angerer LM, Angerer RC. Patterning the sea urchin embryo: Gene regulatory networks, signaling pathways, and cellular interactions. In: Schatten G, ed. Current Topics in Developmental Biology. Academic Press, 2003:53:159-198.
10. Anstrom JA. Microfilaments, cell shape changes, and the formation of primary mesenchyme in sea urchin embryo. J Exp Zool 1992; 264:312-322.
11. Anstrom JA, Raff RA. Sea urchin primary mesenchyme cells: Relation of cell polarity to the epithelial-mesenchymal transformation. Dev Biol 1988; 130:57-66.
12. Benson S, Smith L, Wilt F et al. The synthesis and secretion of collagen by cultured sea urchin micromeres. Exp Cell Res 1990; 188:141-146.
13. Benson S, Rawson R, Killian C et al. Role of the extracellular matrix in tissue-specific gene expression in the sea urchin embryo. Mol Reprod Dev 1991; 29:220-226.
14. Benson S, Page L, Ingersoll E et al. Developmental characterization of the gene for laminin alpha-chain in sea urchin embryos. Mech Dev 1999; 81:37-49.
15. Bisgrove BW, Andrews ME, Raff RA. Fibropellins, products of an EGF repeat-containing gene, form a unique extracellular matrix structure that surrounds the sea urchin embryo. Dev Biol 1991; 146:89-99.
16. Blankenship J, Benson S. Collagen metabolism and spicule formation in sea urchin micromeres. Exp Cell Res 1984; 152:98-104.
17. Brandhorst B, Klein WH. Molecular patterning along the sea urchin animal-vegetal axis. Int Rev Cytol 2002; 213:183-232.
18. Burdsal CA, Alliegro MC, McClay DR. Tissue-specific, temporal changes in cell adhesion to echinonectin in the sea urchin embryo. Dev Biol 1991; 144:327-334.
19. Burke RD, Lail M, Nakajima Y. The apical lamina and its role in cell adhesion in sea urchin embryos. Cell Adhes Commun 1998; 5:97-108.

20. Carson DD, Farach MC, Earles DS et al. A monoclonal antibody inhibits calcium accumulation and skeleton formation in cultured embryonic cells of the sea urchin. Cell 1985; 41:639-48.
21a. Chuang CK, Wikramanayake AH, Mao CA et al. Transient appearance of Strongylocentrotus purpuratus Otx in micromere nuclei: Cytoplasmic retention of SpOtx possibly mediated through an alpha-actinin interaction. Dev Gene 1996; 19:231-237.
21. Coffman JA, McClay DR. A hyaline layer protein that becomes localized to the oral ectoderm and foregut of sea urchin embryos. Dev Biol 1990; 140:93-104.
22. D'Alessio M, Ramirez F, Suzuki HR et al. Structure and developmental expression of a sea urchin fibrillar collagen gene. Proc Natl Acad Sci USA 1989; 86:9303-9307.
23. D'Alessio M, Ramirez F, Suzuki HR et al. Cloning of a fibrillar collagen gene expressed in the mesenchymal cells of the developing sea urchin embryo. J Biol Chem 1990; 265:7050-7054.
24. Dan K, Okazaki K. Cyto-embryological studies of sea urchins. III. Role of the secondary mesenchyme cells in the formation of the primitive gut in sea urchin larvae. Biol Bull 1956; 110:29-42.
25. Davidson EH, Cameron RA, Ransick A. Specification of cell fate in the sea urchin embryo: Summary and some proposed mechanisms. Development 1998; 125:3269-3290.
26. Davidson EH et al. A provisional regulatory gene network for specification of endomesoderm in the sea urchin embryo. Dev Biol 2002; 246:162-190.
27. Davidson LA, Koehl MA, Keller R et al. How do sea urchins invaginate? Using biomechanics to distinguish between mechanisms of primary invagination. Development 1995; 121:2005-2018.
28. Davidson LA, Oster GF, Keller RE et al. Measurements of mechanical properties of the blastula wall reveal which hypothesized mechanisms of primary invagination are physically plausible in the sea urchin Strongylocentrotus purpuratus. Dev Biol 1999; 209:221-238.
29. DeSimone DW, Spiegel E, Spiegel M. The biochemical identification of fibronectin in the sea urchin embryo. Biochem. Biophys Res Comm 1985; 27:183-188.
30. Emily-Fenouil F, Ghiglione C, Lhomond G et al. GSK3beta/shaggy mediates patterning along the animal-vegetal axis of the sea urchin embryo. Development 1998; 125:2489-98.
31. Ettensohn CA. Gastrulation in the sea urchin embryo is accompanied by the rearrangement of invaginating epithelial cells. Dev Biol 1985; 112:383-390.
32. Ettensohn CA. Primary mesenchyme cell migration in the sea urchin embryo. In: Clark Jr RWH, Griffin F, eds. Gastrulation: Movements patterns and molecules. New York: Plenum Publishing Corp, 1991:289-304.
33. Ettensohn CA. Cell movements in the sea urchin embryo. Curr Opin Genet Dev 1999; 9:461-465.
34. Ettensohn CA, McClay DR. Cell lineage conversion in the sea urchin embryo. Dev Biol 1988; 125:396-409.
35. Ettensohn CA, Ingersoll EP. Morphogenesis of the sea urchin embryo. In: Rossomando EF Alexander S, eds. Morphogenesis: An Analysis of the Development of Biological Form. New York: Marcel Dekker Press, 1992:189-262.
36. Ettensohn CA, Guss KA, Hodor PG et al. The morphogenesis of the skeletal system of the sea urchin embryo. In: John Wiley and Sons, eds. Progress in Developmental Biology (Volume VIII, Reproductive Biology of Invertebrates). New York: 1997:225-265.
37. Ettensohn CA, Malinda KM, Sweet HC et al. The ontogeny of cell guidance information in an embryonic epithelium. In: Regulatory Processes in Development, Proceedings of a Wenner-Gren Foundation International Symposium. Portland Press, 1999:31-48.
38. Ettensohn CA, Sweet HC. Patterning the early sea urchin embryo. In: Schatten G, ed. Current Topics in Developmental Biology. Academic Press, 2000:50:1-44.
39. Ettensohn CA, Illes MR, Oliveri P et al. Alx1, a member of the Cart1/Alx3/Alx4 subfamily of paired-class homeodomain proteins, is an essential component of the gene network controlling skeletogenic fate specification in the sea urchin embryo. Development 2003; 130:2917-2928.
40. Ettensohn EA, Wray G, Wessel GM. Methods in Cell Biology: Experimental Analysis of the Development of Sea Urchins and Other NonVertebrate Deuterostomes. Academic Press, 2004:56 in press.
41. Fink RD, McClay DR. Three cell recognition changes accompany the ingression of sea urchin primary mesenchyme cells. Dev Biol 1985; 107:66-74.
42. Fuchikami T, Mitsunaga-Nakatsubo K, Amemiya S et al. T-brain homologue (HpTb) is involved in the archenteron induction signals of micromere descendant cells in the sea urchin embryo. Development 2002; 129:5205-5216.
43. Galileo DS, Morrill JB. Patterns of cells and extracellular material of the sea urchin Lytechinus variegatus (Echinodermata; Echinoidea) embryo, from hatched blastula to late gastrula. J Morph 1985; 185:387-402.
44. Govindarajan V, Ramachandran RK, George JM et al. An ECM-bound, PDGF-like growth factor and a TGF-alpha-like growth factor are required for gastrulation and spiculogenesis in the Lytechinus embryo. Dev Biol 1995; 172:541-551.

45. Gustafson T, Kinnander H. Microaquaria for time-lapse cinematographic studies of morphogenesis in swimming larvae and observations on sea urchin gastrulation. Exp Cell Res 1956; 11:36-51.
46. Gustafson T, Wolpert L. Cellular movement and contact in sea urchin morphogenesis. Biol Rev 1967; 42:442-498.
47. Hall HG, Vacquier VD. The apical lamina of the sea urchin embryo: Major glycoproteins associated with the hyaline layer. Dev Biol 1982; 89:168-78.
48. Hamada M, Kiyomoto M. Signals from primary mesenchyme cells proliferate endoderm differentiation in the sea urchin embryo. Develop Growth Differ 2003; 45:339-350.
49. Hardin J. The role of secondary mesenchyme cells during sea urchin gastrulation studied by laser ablation. Development 1988; 103:317-324.
50. Hardin J. Local shifts in position and polarized motility drive cell rearrangement during sea urchin gastrulation. Dev Biol 1989; 136:430-445.
51. Hardin J, McClay DR. Target recognition by the archenteron during sea urchin gastrulation. Dev Biol 1990; 142:86-102.
52. Hawkins RL, Fan J, Hille MB. Gastrulation in the sea urchin, Strongylocentrotus purpuratus, is disrupted by the small laminin peptides YIGSR and IKVAV. Cell Adhes Commun 1995; 3:163-177.
53. Hertzler PL, McClay DR et al. Alpha-SU2, an epithelial integrin that binds laminin in the sea urchin embryo. Dev Biol 1999; 207:1-13.
54. Hodor PG, Illies MR, Broadley S et al. Cell-substrate interactions during sea urchin gastrulation: Migrating primary mesenchyme cells interact with and align extracellular matrix fibers that contain ECM3, a molecule with NG2-like and multiple calcium-binding domains. Dev Biol 2000; 222:181-194.
55. Hylander BL, Summers RG. An ultrastructural immunocytochemical localization of hyalin in the sea urchin egg. Dev Biol 1982; 93:368-380.
56. Illies MR, Peeler MT, Dechtiaruk AM et al. Identification and developmental expression of new biomineralization proteins in the sea urchin Strongylocentrotus purpuratus. Dev Genes Evol 2002; 212:419-31.
57. Ingersoll EP, Ettensohn CA. An N-linked carbohydrate-containing extracellular matrix determinant plays a key role in sea urchin gastrulation. Dev Biol 1994; 163:351-366.
58. Iwata M, Nakano E. Characterization of sea-urchin fibronectin. Biochem J 1983; 215:205-208.
59. Kanoh K, Aizu G, Katow H. Disapperance of an epithelial cell surface-specific glycoprotein (Epith-1) associated with epithelial-mesenchymal conversion in sea urchin embryogenesis. Develop Growth Differ 2001; 43:83-95.
60a. Katoh-Fukui Y, Noce T, Ueda T et al. The corrected structure of the SM50 spicule matrix protein of Strongylocentrotus purpuratus. Dev Biol 1991; 145:201-202.
60. Katow H. Characterization of vegetal plate cells separated from cytochalasin B-treated blastulae of the sea urchin Clypeaster japonicus. Develop Growth Differ 1989; 31:439-446.
61. Katow H. A new technique for introducing anti-fibronectin antibodies and fibronectin-related peptides into blastulae of the sea urchin, Clypeaster japonicus. Develop Growth Differ 1990; 32:33-41.
62. Katow H. Role of primary mesenchyme cell surface specific antigen during early morphogenesis in sea urchin embryo. In: Yanagisawa, Yasumasu, Oguro, Suzuki and Motokawa, Balkema, eds. Biology of Echinodermata. Rotterdam: 1991:453-459.
63. Katow H. Pamlin, a primary mesenchyme cell adhesion protein, in the basal lamina of the sea urchin embryo. Exp Cell Res 1995; 218:469-478.
64. Katow H. Delamination and tyrosine phosphorylation of SUp62 during early embryogenesis of sea urchin. Zygote 2000; 8(Supplement):S59-40.
65. Katow H. Extracellular signal transdution in sea urchin embryogenesis: From extracellular matrix to MA kinase pathway. Current Topics in Peptide and Protein Research 2003; 5:149-160.
66. Katow H, Solursh M. Ultrastructure of blastocoel material in blastulae and gastrulae of the sea urchin Lytechinus pictus. J Exp Zool 1979; 210:561-567.
67. Katow H, Solursh M. Ultrastructure of primary mesenchyme cell ingression in the sea urchin Lytechinus pictus. J Exp Zool 1980; 213:231-246.
68. Katow H, Solursh M. Ultrastructural and time-lapse studies of primary mesenchyme cell behavior in normal and sulfate-deprived sea urchin embryos. Exp Cell Res 1981; 136:233-245.
69. Katow H, Yamada KM, Solursh M. Occurrence of fibronectin on the primary mesenchyme cell surface during migration in the sea urchin embryo. Differentiation 1982; 22:120-124.
70. Katow H, Amemiya S. Behavior of primary mesenchyme cells in situ associated with ultrastructural alteration of the blastocoelic material in the sea urchin, Anthocidaris crassispina. Develop Growth Differ 1986; 28:31-42.
71. Katow H, Yazawa S, Sofuku SA fibronectin-related synthetic peptide, Pro-Ala-Ser-Ser, inhibits fibronectin binding to the cell surface, fibronectin-promoted cell migration in vitro, and cell migration in vivo. Exp Cell Res 1990; 190:17-24.

72. Katow H, Komazaki, S. Spatio-temporal expression of pamlin during early embryogenesis in sea urchin and importance of N-glycosylation for the glycoprotein function. Roux's Arch Dev Biol 1996; 205:371-381.
73. Katow H, Uemura I, Nakajima Y. Primary mesenchyme cell-ring pattern formation in 2D-embryo of the sea urchin. Develop Growth Differ 2000; 42:9-17.
74. Katow H, Washio M. Pamlin induced tyrosine phosphorylation of SUp62 protein in primary mesenchyme cells during early embryogenesis in the sea urchin Hemicentrotus pulcherrimus. Develop Growth Differ 2000; 42:519-529.
75. Katow H, Aizu G. Essential role of growth factor receptor-mediated signal transduction through the mitogen-activated protein kinase pathyway in early embryogenesis of the echinoderm. Develop Growth Differ 2002; 44:3437-455.
76a. Kinnander H, Gustafson T. Further studies on the cellular basis of gastrulation in the sea urchin larva. Exp Cell Res 1960; 19:278-290.
76. Komatsu S, Murai N, Totsukawa G et al. Identification of MAPKAPK homologue (MAPKAPK-4) as a myosin II regulatory light chain kinase in sea urchin egg extracts. Arch Biochem Biophys 1997; 343:55-62.
77. Kominami T, Takata H, Takaichi M. Behavior of pigment cells in gastrula-stage embryos of Hemicentrotus pulcherrimus and Scaphechinus mirabilis. Develop Growth Differ 2001; 43:699-707.
78. Kurokawa D, Kitajima T, Mitsunaga-Nakatsubo K et al. HpEts, an ets-related transcription factor implicated in primary mesenchyme cell differentiation in the sea urchin embryo. Mech Dev 1999; 80:41-52.
79. Lane MC, Solursh M. Dependence of sea urchin primary mesenchyme cell migration on xyloside- and sulfate-sensitive cell surface-associated components. Dev Biol 1988; 127:78-87.
80. Lane MC, Solursh M. Primary mesenchyme cell migration requires a chondroitin sulfate/dermatan sulfate proteoglycan. Dev Biol 1991; 143:389-397.
81. Leaf DS, Anstrom JA, Chin JE et al. Antibodies to a fusion protein identify a cDNA clone encoding msp130, a primary mesenchyme-specific cell surface protein of the sea urchin embryo. Dev Biol 1987; 121:29-40.
82. Lepage T, Gache C. Early expression of a collagenase-like hatching enzyme gene in the sea urchin embryo. EMBO Journal 1990; 9:3003-3012.
83. Livant DL, Linn S, Markwart S et al. Invasion of selectively permeable sea urchin embryo basement membranes by metastatic tumor cells, but not by their normal counterparts. Cancer Res 1995; 55:5085-93.
84. Logan CY, Miller JR, Ferkowicz MJ et al. Nuclear beta-catenin is required to specify vegetal cell fates in the sea urchin embryo. Development 1999; 126:345-357.
85. Marsden M, Burke RD. Cloning and characterization of novel beta integrin subunits from a sea urchin. Dev Biol 1997; 181:234-245.
86. Marsden M, Burke RD. The alpha-L integrin subunit is necessary for gastrulation in sea urchin embryos. Dev Biol 1998; 203:134-148.
87. Matese JC, Black S, McClay DR. Regulated exocytosis and sequential construction of the extracellular matrix surrounding the sea urchin zygote. Dev Biol 1997; 186:16-26.
88. Mayne J, Robinson JJ. The sea urchin egg yolk granule is a storage compartment for HCL-32, an extracellular matrix protein. Biochem Cell Biol 1998; 76:83-88.
89. Mayne J, Robinson JJ. Localization and functional role of a 41 kDa collagenase/gelatinase activity expressed in the sea urchin embryo. Develop Growth Differ 2002; 44:345-56.
90. McCarthy RA, Beck K, Burger MM. Laminin is structurally conserved in the sea urchin basal lamina. EMBO J 1987; 6:1587-1593.
91. McClay DR, Fink RD. Sea urchin hyalin: Appearance and function in development. Dev Biol 1982; 92:285-293.
92. McClay DR, Cannon GW, Wessel GM et al. Patterns of Antigenic Expression in Early Sea Urchin Development. In: Jeffery WR, Raff RA, eds.Time, Space, and Pattern in Embryonic Development. AR Liss, 1983:157-169.
93. Miller J, Fraser SE, McClay D. Dynamics of thin filopodia during sea urchin gastrulation. Development 1995; 121:2501-2511.
94. McClay DR, Logan CY. Regulative capacity of the archenteron during gastrulation in the sea urchin. Development 1996; 122:607-616.
95. Miller JR, McClay DR. Characterization of the role of cadherin in regulating cell adhesion during sea urchin development. Dev Biol 1997; 192:323-339.
96. Mintz GR, DeFrancesco S, Lennarz WJ. Spicule formation by cultured embryonic cells from the sea urchin. J Biol Chem 1981; 256:13105-13111.
97. Murray G, Reed C, Marsden M et al. Institution Department of Biology, University of Victoria, British Columbia, Canada. Title The alphaBbetaC integrin is expressed on the surface of the sea urchin egg and removed at fertilization. Dev Biol 2000; 227:633-647.

98. Okazaki K. Normal development to metamorphosis. In: Czihak G ed. The Sea Urchin Embryo: Biochemistry and Morphogenesis. 1975:177-232.
99. Oliveri P, Davidson EH, McClay DR. Activation of pmar1 controls specification of micromeres in the sea urchin embryo. Dev Biol 2003; 258:32-43.
100a. Ransick A, Rast JP, Minokawa T et al. New early zygotic regulators in endomesoderm of sea urchin embryos discovered by differential array hybridization. Dev Biol 2002; 246:132-147.
100. Parr BA, Parks AL, Raff RA. Promoter structure and protein sequence of msp130, a lipid-anchored sea urchin glycoprotein. J Biol Chem 1990; 265:1408-13.
101. Peterson RE, McClay DR. Primary mesenchyme cell patterning during the early stages following ingression. Dev Biol 2003; 254:68-78.
102. Ramachandran RK, Seid CA, Lee H et al. PDGF-BB and TGF-alpha rescue gastrulation, spiculogenesis, and LpS1 expression in collagen-disrupted embryos of the sea urchin genus Lytechinus. Mech Dev 1993; 44:33-40.
103. Ramachandran Rk, Wikramanayake AH, Uzman JA et al. Disruption of gastrulation and oral ectoderm differentiatio in the Lytechinus pictus embryo by a dominant/negative PDGF receptor. Development 1997; 124:2355-2384.
104. Ruffins SW, Ettensohn CA. A fate map of the vegetal plate of the sea urchin (Lytechinus variegatus) mesenchyme blastula. Development 1996; 122:253-63.
105. Sano K. Changes in cell surface charges during differentiation of isolated micromeres and mesomeres from sea urchin embryos. Dev Biol 1977; 6:404-415.
106. Savagner P. Leaving the neighborhood: Molecular mechanisms involved during epithelial-mesenchymal transition. Bio Essays 2001; 23:912-923.
107. Shimizu K, Noro N, Matsuda R. Micromere differentiation in the sea urchin embryo. Expression of primary mesenchyme cell specific antigen during development. Develop Growth Differ 1988; 30:35-47.
108. Shimizu-Nishikawa K, Katow H, Matsuda R. Micromere differentiation in the sea urchin embryo: Immunochemical characterization of primary mesenchyme cell-specific antigen and its biological roles. Develop Growth Differ 1990; 32:629-636.
109. Spiegel E, Burger MM, Spiegel M. Fibronectin and laminin in the extracellular matrix and basement membrane of sea urchin embryos. Exp Cell Res 1983; 144:47-55.
110. Stephens RE, Kane RE. Some properties of hyalin: The calcium-insoluble protein of the hyaline layer of the sea urchin egg. J Cell Biol 1970; 44:611-617.
111. Susan JM, Just ML, Lennarz WJ. Cloning and characterization of alphaP integrin in embryos of the sea urchin Strongylocentrotus purpuratus. Biochem Biophys Res Com 2000; 272:929-935.
112. Sweet HC, Hodor PG, Ettensohn CA. The role of micromere signaling in Notch activation and mesoderm specification during sea urchin embryogenesis. Development 1999; 126:5255-5265.
113. Sweet HC, Gehring M, Ettensohn CA. LvDelta is a mesoderm-inducing signal in the sea urchin embryo and can endow blastomeres with organizer-like properties. Development 2002; 129:1945-55.
114. Tamboline CR, Burke RD. Ontogeny and characterization of mesenchyme antigens of the sea urchin embryo. Dev Biol 1989; 136:75-86.
115. Tesoro V, Zito F, Yokota Y et al. A protein of the basal lamina of the sea urchin embryo. Develop Growth Differ 1998; 40:527-535.
116. Wessel GM, Marchase RB, McClay DR. Ontogeny of the basal lamina in the sea urchin embryo. Dev Biol 1984; 103:235-245.
117. Wessel GM, McClay DR. Gastrulation in the sea urchin embryo requires the deposition of crosslinked collagen within the extracellular matrix. Dev Biol 1987; 121:149-165.
118. Wessel GM, Etkin M, Benson S. Primary mesenchyme cells of the sea urchin embryo require an autonomously produced, nonfibrillar collagen for spiculogenesis. Dev Biol 1991; 148:261-272.
119. Wessel GM, Berg L, Adelson DL et al. A molecular analysis of hyalin—a substrate for cell adhesion in the hyaline layer of the sea urchin embryo. Dev Biol 1998; 193:115-126.
120. Wikramanayake AH, Huang L, Klein WH. beta-Catenin is essential for patterning the maternally specified animal-vegetal axis in the sea urchin embryo. Proc Natl Acad Sci USA 1998; 95:9343-8.
121. Wilt FH. Biomineralization of the spicules of sea urchin embryos. Zoolog Sci 2002; 19:253-61.
122. Wolpert L, Mercer EH. An electron microscope study of the development of the blastula of the sea urchin and its radial polarity. Exp Cell Res 1963; 30:280-300.
123. Yasumasu S, Yamada K, Akasaka K et al. Isolation of cDNAs for LCE and HCE, two constituent proteases of the hatching enzyme of Oryzias latipes, and concurrent expression of their mRNAs during development. Dev. Biol. 1992; 153:250-8.
124. Zhu X, Mahairas G, Illies M et al. A large-scale analysis of mRNAs expressed by primary mesenchyme cells of the sea urchin embryo. Development 2001; 128:2615-2627.

CHAPTER 7

Change of Epithelial Fate: Lessons from Gastrulation in *Drosophila*

Atish Ganguly and Y. Tony Ip

Abstract

Change of epithelial cell fate occurs in developmental, physiological, and pathological processes. While the initiation and outcome of the various processes may differ, many of the cellular and molecular controls are common. Using *Drosophila* gastrulation as a model, we describe in this chapter the molecular genetics of cell shape changes of the epithelial cells in the blastoderm. The columnar epithelial cells in the ventral region, which are destined to become mesodermal precursors, behave differently from cells in other regions of the early embryo and invaginate to the appropriate position during gastrulation. Genetic studies point to the involvement of the region specific transcription factor Snail and the ubiquitous adherens junctions and cytoskeleton. How cell fate determinants turn on ubiquitous machineries to achieve the coordinated cell movements during gastrulation is being elucidated and is a fascinating area of research that attracts the attention of scientists from various disciplines.

Introduction

The cell movements of gastrulation are the first visible morphogenetic events during early embryogenesis.[4,19,28,30,32,42] During this developmental process, the single-cell-layered blastoderm undergoes a series of highly coordinated cell shape changes and movements to produce the three germ layers, the ectoderm, mesoderm and endoderm. Thus, gastrulation is not only a critical biological process, but also an excellent model for the study of cell shape changes and movements, as well as the conversion of epithelial cells to mesenchymal fates[24,44] (see also Chapter 2, Morali et al).

Drosophila melanogaster has proved to be a valuable model for studying the cellular and molecular control of gastrulation.[6,19,22,26,39] The techniques for forward and reverse genetics are well developed in this organism, and a large body of genetic information has been compiled.[13] Although the appearance of the morphogenetic process in various animals may differ, many of the molecules involved in gastrulation appear to be conserved throughout evolution. Therefore, the results obtained through studies in the fruit fly are relevant to similar processes in vertebrates. In this chapter, we will focus primarily on the conversion of the ventral epithelium into the mesodermal primordium during gastrulation, and on the zinc-finger transcription factor Snail as a key regulator of cell fate and movement in the process.

Gastrulation in the Fly Embryo

The fertilized *Drosophila* egg contains the zygotic nucleus surrounded by a large amount of yolk and protected by a plasma membrane, a vitelline membrane and a resilient outer chorion.[11] The yolk contains maternally deposited RNA and proteins, which enable the embryonic nuclei to divide in the absence of zygotic transcription. During the first two hours of development,

Rise and Fall of Epithelial Phenotype: Concepts of Epithelial-Mesenchymal Transition, edited by Pierre Savagner. ©2005 Eurekah.com and Kluwer Academic / Plenum Publishers.

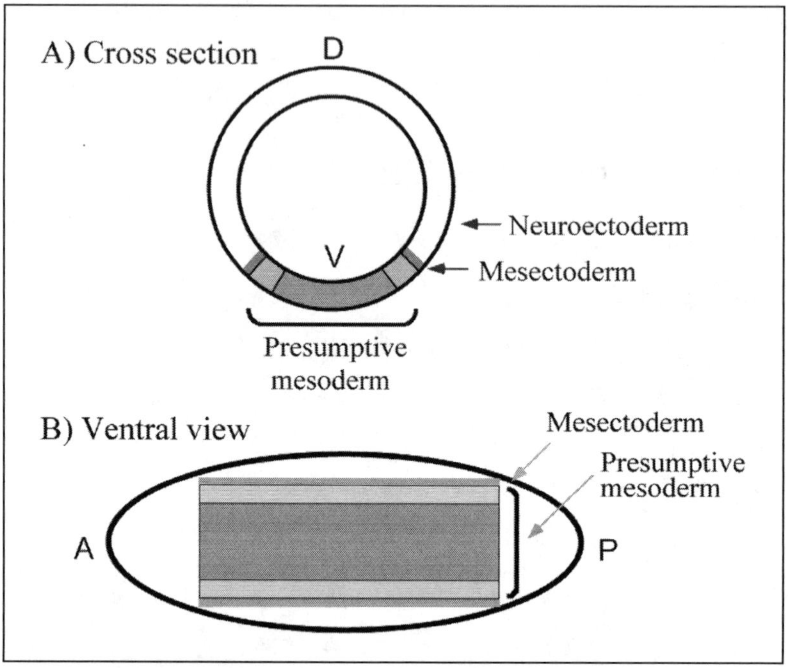

Figure 1. Schematic drawing of the blastoderm stage embryo. A) A cross-section at the middle of the embryo. The columnar epithelial cells are located at the periphery of the embryo, delimited by the two circles. Along the circumference there are approximately 100 cells, with 18 of these cells located in the ventral (V) region format the presumptive mesoderm. This area is subdivided into two regions. The blue region is 12 cells wide and is the midventral region where autonomous cell shape changes occur. The light blue regions on either side are 3 cells wide, and these cells passively follow the midventral cells during invagination. The mesectoderm (red line) is one cell wide along the circumference, and the gene *single-minded* is expressed specifically in this region. The neuroectoderm is approximately 10 cells wide along the two lateral regions of the embryo. D is dorsal side. B) A ventral view of the blastoderm. Anterior is A and posterior is P. The ventral presumptive mesoderm, blue and light blue together, is approximately 18 cells wide and 60 cells long. The mesectoderm (red line) constitutes the future midline cells and is one cell wide next to the mesoderm.

the zygotic nucleus undergoes 13 divisions without cytokinesis to form the syncytial blastoderm. Then cellularization begins and at its completion a single cell-layered epithelium surrounding the yolk forms the cellular blastoderm (Fig. 1).

The next developmental process is gastrulation. Gastrulation in *Drosophila* includes the ventral invagination of the presumptive mesoderm, invagination of the anterior and posterior endoderms, cephalic furrow formation, and germ band extension.[6] Formation of the ventral furrow occurs immediately after the ventral cells have completed cellularization. The furrow forms along the ventral midline of the embryo. An area around 60 cells long by 18 cells wide gets internalized as a result of the invagination. These cells go on to form the mesoderm of the embryo. Hence this ventral region is called the presumptive mesoderm or the ventral plate (Fig. 1). Around the same time, the cephalic furrow begins to form on the lateral sides of the embryo. Shortly after, the posterior midgut and the pole cells begin to invaginate. Germ band extension then commences from the ventral side of the embryo and pushes towards the dorsal side. This process carries the invaginating posterior midgut and the pole cells onto the dorsal side of the embryo. These various processes together take about one hour and 15 minutes. Gastrulation occurs rather independently of cell division, therefore cell shape changes and cell-cell interactions are the driving forces for the morphogenetic movements.

Cellular Changes in the Presumptive Mesoderm

At the blastoderm stage, every cell has a rounded apical surface (the surface facing outside the embryo). Just before the formation of the ventral furrow, the apical surface of every cell in the presumptive mesoderm becomes flat. Such apical flattening causes the cells to become more columnar in shape.[27,41] The next change is apical constriction, which occurs only in a 12 cell-wide region surrounding the ventral midline (Fig. 1). These cells are a subset of the presumptive mesoderm and are called the midventral cells. The constriction pushes the cytoplasm and nuclei, and the cells begin to elongate basally. At this stage, the constricting apical surfaces of the midventral cells appear to drag cells, 3 rows on each side, towards the ventral midline, and the first hint of the ventral furrow appears. The elongated midventral cells then begin to shorten. This causes their basal surfaces to expand laterally and exert force on their neighboring cells. As a result, the cells are displaced on their basal side while remaining bunched together on the apical side. An analogy would be a person holding a bunch of cylindrical balloons together at one end, and pushing the other end of the balloons towards the first. The balloons would then expand laterally and fan out. Similar arrangement and forces in the ventral epithelium promotes the formation of the ventral furrow. As the midventral cells contract further, the surrounding lateral cells continue to get dragged towards and into the furrow. Once the presumptive mesoderm is inside, these cells lose all attachment and collapse to form the mesoderm layer. They divide a total of three times, move to appropriate positions and differentiate to form mesodermal derivatives such as muscles, fat body and lymph gland.[16]

Dorsal-Ventral Patterning of the Blastoderm

Obviously, the ventral cells behave differently from other cells of the blastoderm stage embryo. The establishment of cell fate by the maternal Toll signaling pathway is the determining factor for the specific ventral cell behavior (Fig. 2). The Toll receptor, being maternally deposited, is expressed throughout the embryo.[1] As the result of early asymmetric cues tracing back to the oocyte, the ubiquitous Toll is activated only in the ventral side. The signal is transmitted through the adaptor protein Tube and the kinase Pelle, which regulate the IκB homologue Cactus.[5,45] As a result, Cactus is phosphorylated and targeted for degradation. The function of Cactus is to bind and inhibit Rel/NF-κB proteins. As cactus is degraded, Dorsal is released from the cytoplasm to enter the nucleus. Dorsal is also maternally deposited and hence present throughout the cytoplasm of the syncytial embryo. The strength of the Toll signal controls the amount of Dorsal nuclear transport. Thus, the Toll pathway sets up a gradient of nuclear Dorsal, with the ventral nuclei having the highest levels.

The Dorsal nuclear protein gradient activates and represses zygotic genes to establish patterning along the dorsal-ventral axis (Fig. 2). Two zygotic genes that Dorsal activates are *twist* and *snail*, which encodes a basic helix-loop-helix transcription activator and a zinc-finger transcriptional repressor, respectively. Both Twist and Snail are expressed on the ventral side of the embryo, and their patterns coincide with the presumptive mesoderm territory. While Dorsal directly activates Twist and Snail, Dorsal/Twist cooperation is also required for optimal expression of Snail.

Even though both *twist* and *snail* mutant embryos have defects of similar severity in gastrulation, there are a few important differences.[27,41] *twist* mutants are able to flatten the apical surfaces of the cells of the presumptive mesoderm but no apical constriction takes place. As a result, a very shallow furrow begins to form but no substantial invagination occurs. In *snail* mutants the ventral cells do express *twist* rather normally, but the embryos exhibit a more severe phenotype. The apices of the cells in the presumptive mesoderm do not flatten; consequently no apical constriction takes place in the midventral cells and no ventral furrow forms. Taken together, both Twist and Snail are important for ventral furrow formation, with Snail exerting a more direct influence on the process.[27,41,20] In addition, double mutants for *twist* and *snail* show no cell shape changes and no ventral furrow formation even at very late stages. This indicates that *twist* and *snail* have both distinct and overlapping functions during ventral furrow formation.

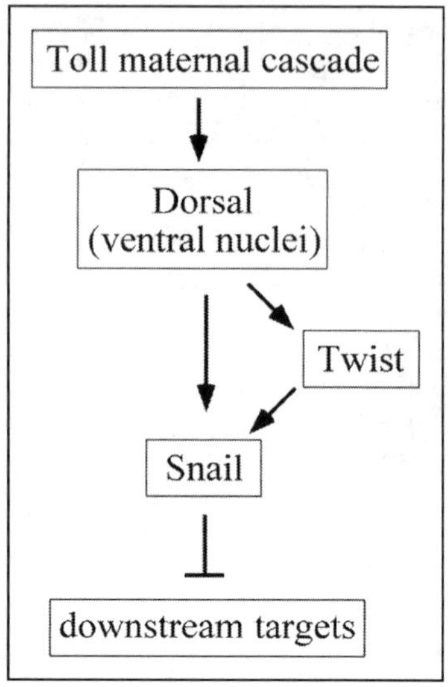

Figure 2. Genetic control of dorsal-ventral polarity and Snail expression. A maternal cascade, constituted of Toll receptor and other components including the adaptor Tube and the kinase Pelle, regulates Dorsal nuclear translocation. High levels of Dorsal protein in the ventral nuclei directly activate Twist and Snail. Dorsal cooperates with Twist for optimal expression of Snail in the cells of the presumptive mesoderm. Snail then represses downstream target genes to define the mesoderm territory.

Multifaceted Control of Gastrulation by Snail

A number of events are required for the coordinated invagination of the ventral cells. These include cell fate and polarity determination, cytoskeletal rearrangements, adherens junctions redistribution, and mitotic arrest before and during invagination. In *snail* mutants, many, if not all of these events do not occur; hence Snail has a direct or an instructive role in these cellular processes (Fig. 3).

Regulation of Cell Fate and Polarity

Snail functions as a sequence-specific DNA-binding protein and represses transcription of a number of target genes. Among them are *single-minded, rhomboid, lethal-of-scute, Delta, crumbs, shotgun* and genes in the *Enhancer-of-split* complex[21,23,25] (see also Chapter 11, Berx and Van Roy). These genes are normally expressed in the mesectodermal cells or in the neuroectodermal regions but are excluded from the presumptive mesoderm. Their expression is derepressed in the ventral region in *snail* mutant embryos. Many of these genes such as *single-minded* and *enhancer-of-split* are cell fate determinants for the mesectoderm and neuroectoderm. On the other hand, *crumbs* and *shotgun/E-cadherin* code for cell adhesion molecules. The repression of such a diverse array of target genes by Snail provides hints for the mechanisms involved in gastrulation. Since the ventral cells are committed to a mesodermal fate, they should not express other genes that will result in their taking up alternate fates. Expressing mesectodermal or neuroectodermal genes may push these cells towards a more neural fate. In fact, a hypomorphic allele of snail causes a derepression of *single-minded* and *rhomboid* in the presumptive mesoderm but still allows ventral furrow formation.[18] These ventral cells show the expression of both mesodermal and neural markers, indicating an abnormal fate specification. The repression of *Delta* may have a similar function as *Delta* may promote the formation of epidermoblasts at the expense of the mesoderm.

Crumbs is a cell adhesion molecule that is important for planar cell polarity in some epithelial cells of the *Drosophila* embryo.[2,22] Perhaps the presumptive mesodermal cells have to maintain apical-basal polarity in order to form the ventral furrow. Expression of Crumbs in the presumptive mesoderm might disrupt normal polarity cues in these cells, and thus needs to be repressed by Snail.

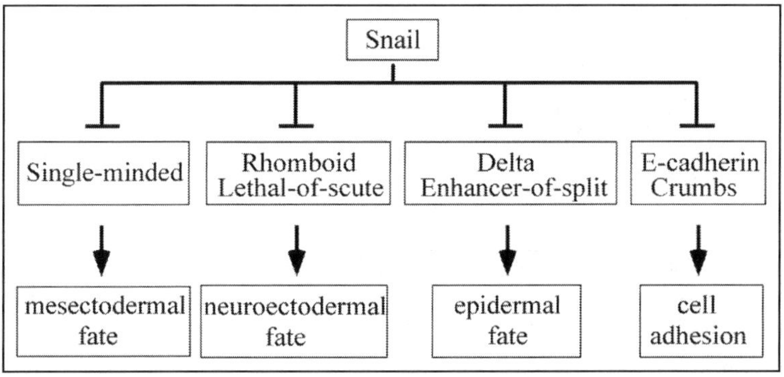

Figure 3. Snail is a transcriptional repressor that has an array of targets. Snail represses a variety of genes in the presumptive mesoderm. The repression of *single-minded, rhomboid,* and *lethal-of-scute* shuts off mesectodermal and neuroectodermal fates in the presumptive mesoderm. The repression of *Delta* presumably has similar effects on epidermal cell fate. E-cadherin and Crumbs are cell adhesion molecules, and the modulation of their activities and expression is needed to enable cell shape changes and cell movements during ventral furrow formation.

Regulation of the Cytoskeleton

The cells of the presumptive mesoderm go through many cell shape changes during ventral furrow formation. What causes this range of shape changes and movements? The most probable answer is through cytoskeletal rearrangements. Evidence comes from embryos mutant for the *flightless* gene.[40] These embryos show defects in gastrulation. *flightless* codes for a homologue of Gelsolin, a protein involved in actin depolymerization, thus implicating the actin-myosin complex in this process. In addition, Zipper, a nonmuscle myosin homologue, is enriched at the apical surface of the invaginating cells forming the ventral furrow.[33,47]

At this point, we can make two predictions. One, if the actin microfilament network is important for cell shape changes during ventral furrow formation, then disrupting the molecules required for the regulation of this network should have serious consequences. Two, there should be a mechanism by which these specific modulations of the ubiquitous actin-myosin network occur only in the ventral cells of the blastoderm embryo. The Rho family of GTPases plays a very important role in the regulation of the actin microfilament network, as do the associated molecules required for their proper function[10] (see also Chapter 17, Nakagawa et al). Disruption of the functions of RhoA and the GTP exchange factor RhoGEF2 causes severe defects in the formation of the ventral furrow.[3,15] The severity of the phenotype implies that these are essential molecules in the pathway and the cells have no alternate mechanisms to bypass them. As both molecules are expressed ubiquitously in the early embryo a question remains as to what provides the cue for their selective activation in the presumptive mesoderm.

The analysis of two more genes, *concertina* and *folded gastrulation (fog)*, has provided a partial solution to this problem.[7,31,37] *concertina* codes for a Gα- like protein and *fog* a putative ligand (Fig. 4). Concertina is expressed ubiquitously in the early embryo, being maternally deposited. On the other hand, Fog is a zygotic gene product regulated by Twist and Snail and is expressed in the ventral cells. Loss of function of either *concertina* or *fog* affects the flattening and the constriction of the apical surface of ventral plate cells and results in a delay, but not a total abrogation of ventral furrow formation. Misexpression studies with Fog have shown that it is capable of causing apical surface flattening and constriction in cells outside of the ventral domain.[31] If the same experiment is performed in a *rhoGEF2* mutant background, then these ectopic cell shape changes do not take place. This further confirms that RhoGEF2 is critical for

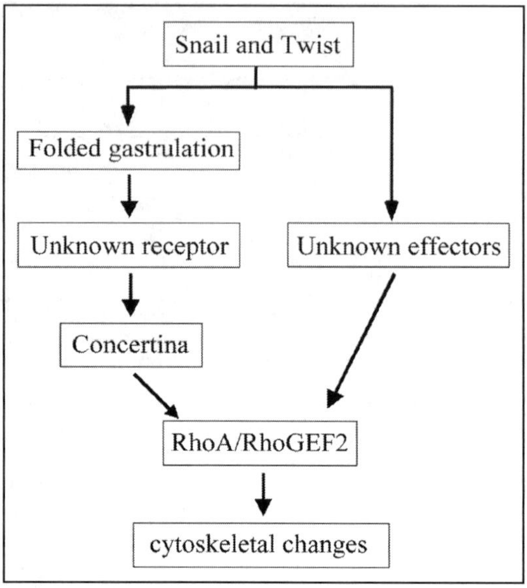

Figure 4. Regulation of cytoskeletal changes by Twist and Snail. Snail and Twist are involved in the ventral expression of the zygotic gene *folded gastrulation,* which codes for a putative ligand that binds to an unknown receptor to activate the maternal Gα-like protein Concertina. Activation of this pathway probably stimulates the ventral cells to turn on the ubiquitously expressed RhoA and RhoGEF2, which modulate the cytoskeletal rearrangement needed for cell shape changes and cell movements during the formation of ventral furrow. Based on the strength of phenotypes of various mutants, we postulate that there is an additional pathway functioning downstream of Twist and Snail to regulate RhoA/RhoGEF2 in the ventral cells.

these shape changes, and that cells outside the ventral domain contain all the necessary components for this process except Fog. These data point to a hypothesis where Fog is produced in the ventral cells of the early embryo under the control of Twist and Snail, and acts as a ligand for an as yet unknown Gα-coupled receptor. Activation of this receptor causes RhoA and RhoGEF2 mediated actin microfilament rearrangements in the presumptive mesoderm and also the associated cell shape changes, eventually leading to ventral furrow formation.

An interesting aspect is that mutants of *snail* and *rhoGEF2* do not form any ventral furrow, but mutants of *concertina* and *folded gastrulation* do, albeit delayed. This suggests that still missing is another critical pathway, which acts downstream of Snail but independently of Concertina and Fog to activate RhoGEF2 and the subsequent cytoskeletal rearrangements (Fig. 4).

Regulation of Intercellular Adhesion

The migration of the presumptive mesoderm occurs as a sheet of cells. Thus, the population is loose enough to migrate and adherent enough to remain an intact sheet. This points to a mechanism where intercellular adhesion is modified appropriately during the process to meet this specific requirement. Multi-protein complexes called adherens junctions are localized to plasma membranes of epithelial cells, and homotypic interactions of these complexes mediate intercellular adhesion.[2,22] The cell adhesion molecule that forms the core of adherens junctions is E-cadherin, which is encoded by the *shotgun* gene in *Drosophila.*[43,46] In wild type embryos, DE-cadherin is concentrated at the apicolateral surfaces of ventral cells during furrow formation. Such asymmetric localization may help the cells to remain attached together at the apical end during constriction, while leaving basal surfaces more loosely bound to facilitate the elongation, contraction, and lateral expansions.

There is a large maternal deposition of DE-cadherin in the embryo and hence the protein is ubiquitously expressed during the early stages of embryogenesis. This probably is sufficient for required functions and may explain why zygotic mutants of *shotgun* show no defects in gastrulation. Mosaic analysis shows that DE-cadherin is essential for oogenesis.[12,35] Therefore, ablating the maternal deposition interferes with the normal development of the oocyte, leaving no mutant embryos for examining the involvement of DE-cadherin in gastrulation. Another component of the adherens junction is β-catenin, encoded by the *armadillo* gene. This protein is bound to the cytoplasmic domain of DE-cadherin and serves as a link between DE-cadherin and the actin cytoskeleton through another protein, α-catenin. Null mutations of *armadillo* cause defects in oocytes, therefore embryos lacking maternal β-catenin cannot be obtained. Nonetheless, embryos derived from oocytes that are partial loss of function mutants of β-catenin exhibit defects in gastrulation.[8] Mutant cells lose their columnar shape and round up. Moreover, multiple cell layers are formed. This is probably due to the lack of coordinated cell movements during ventral furrow formation, thus indicating that adherens junctions are important for this process.

In *snail* mutants, the adherens junctions in the presumptive mesoderm are enriched in the lateral surfaces of the ventral cells as opposed to its apicolateral localization in wild type cells.[36] Therefore, Snail is somehow involved in this localization. One way for Snail to achieve this would be to regulate the actin cytoskeleton to which the adherens junctions are attached. However, the precise mechanisms by which adherens junctions are localized to the apicolateral surfaces of ventral cells remain unclear. Also, Snail is capable of repressing the RNA expression of *shotgun* in the cells of the presumptive mesoderm. How this could help the ventral furrow to form is unclear.

Regulation of Cell Division

Cell movements during gastrulation are driven by mechanisms autonomous to the ventral cells and do not depend on the increase in cell population. Early synchronized cell divisions stop at the end of the blastoderm stage and asynchronous mitosis begins in selected groups of cells in the embryo called mitotic domains.[11] The cells of the presumptive mesoderm start to divide only after they have completed invagination. It has also been shown that cell divisions actually need to be stopped during ventral invagination[14,29,38]

The gene *string* encodes the *Drosophila* ortholog of yeast Cdc25, a phosphatase that renders the Cdc2 kinase active and pushes the cell through the G2/M checkpoint (Fig. 5). Therefore, *Drosophila* Cdc25/String promotes mitosis and its expression in the cells of a mitotic domain usually precedes cell division in that domain.[9] In wild type embryos, *string* RNA is expressed in the presumptive mesoderm, but these cells remain arrested at G2. However, if String is expressed to higher levels in the presumptive mesoderm before ventral furrow formation, these cells enter premature mitosis and invagination is blocked.[14,29] Hence, while the RNA is present, String protein activity is normally inhibited in the presumptive mesoderm by post-transcriptional mechanisms. When the level of String activity exceeds a certain threshold, it overrides the inhibition and cells enter mitosis prematurely, thereby inhibiting the formation of the ventral furrow. Interestingly, in *snail* mutants, the ventral cells that do not invaginate also undergo mitosis. However, it is not clear whether the cell division is the cause or the effect of the lack of invagination.

Mutations in *tribbles* and *frühstart* show a phenotype of premature entry into mitosis similar to the loss of function of *snail* and the overexpression of String.[14,29,38] String RNA levels remain normal in the presumptive mesoderm of these mutants, thus establishing them as potential inhibitors of String protein activity. Tribbles has indeed been shown to promote degradation of String protein, while the function of *frühstart* remains undiscovered. Tribbles requires the Snail pathway to function optimally (Fig. 5). Overexpression of Tribbles in tissues where Snail is not normally expressed does not lead to any significant cell cycle block or developmental abnormality. In *snail* mutants, Tribbles activity in the presumptive

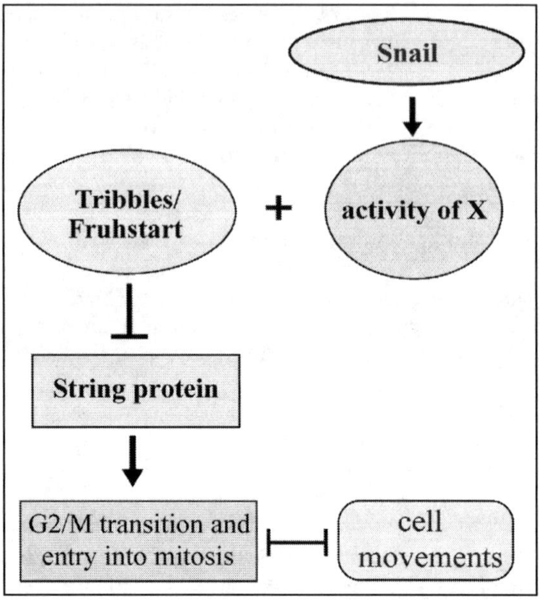

Figure 5. Coordination of cell division and cell movement. String protein is expressed in the ventral cells of the blastoderm but does not accumulate to high levels, therefore the cells do not enter mitosis. Tribbles and Frühstart are two proteins that mediate degradation of String in the ventral cells during invagination of the presumptive mesoderm. While the gene expression of *tribbles* (*frühstart* has not been molecularly cloned) does not depend on Snail, the activity of Tribbles is regulated by Snail. A possible function of Snail is to activate an unknown factor X that is required for the degradation of String protein by Tribbles.

mesoderm goes down and premature mitosis results, even though the expression of *tribbles* remains normal. However, no premature mitosis occurs in the presumptive mesoderm of *rhoGEF2* or *concertina* mutants, indicating that Snail is able to provide some unknown input to *Tribbles* for its optimal activity that is independent of the above genes. Since *tribbles* expression is normal in *snail* mutants, this input has to be at the post-transcriptional level (Fig. 5). Little else is known about this aspect of Snail's function during ventral furrow formation.

Summary

We have attempted to showcase the formation of the ventral furrow during gastrulation in the *Drosophila* embryo as a model for change in epithelial fate. The formation of ventral furrow requires coordinated regulation of cell fate, polarity, adhesion, cytoskeleton and division. Snail appears to be involved in each of these aspects and is thus, a key molecule controlling epithelial cell behavior. While the precise mechanisms vary in different species, the overall logic of rise and fall of epithelial cell fate is well conserved. It has been demonstrated that Snail homologues participate in vertebrate cell movements and cancer progression.[17,34] The relevance of epithelial fate change in human disease such as metastasis of malignant cells cannot be overstated. The study of model systems such as *Drosophila* gastrulation will not only increase the knowledge and understanding of a fascinating developmental process but also enhance the treatment of various diseases.

References

1. Anderson KV. Pinning down positional information: Dorsal-ventral polarity in the Drosophila embryo. Cell 1998; 95:439-442.
2. Bachmann A, Schneider M, Theilenberg E et al. Drosophila Stardust is a partner of Crumbs in the control of epithelial cell polarity. Nature 2001; 414:638-643.
3. Barrett K., Leptin M, Settleman J. The Rho GTPase and a putative RhoGEF mediate a signaling pathway for the cell shape changes in Drosophila gastrulation. Cell 1997; 91:905-915.
4. Beetschen JC. Amphibian gastrulation: History and evolution of a 125 year-old concept. Int J Dev Biol 2001; 45:771-795.
5. Belvin P, Jin Y, Anderson KV. Cactus protein degradation mediates Drosophila dorsal-ventral signaling. Genes Dev 1995; 9:783-793.
6. Costa M., Sweeton D, Wieschaus E. Gastrulation in Drosophila: Cellular mechanisms of morphogenetic movements. In: Bate Martinez Arias, eds. The development of Drosophila melanogaster. Cold Spring Harbor: Laboratory Press, 1993:425-465.
7. Costa M, Wilson ET, Wieschaus E. A putative cell signal encoded by the folded gastrulation gene coordinates cell shape changes during Drosophila gastrulation. Cell 1994; 76:1075-1089.
8. Cox RT, Kirkpatrick C, Peifer M. Armadillo is required for adherens junction assembly, cell polarity, and morphogenesis during Drosophila embryogenesis. J Cell Biol 1996; 134:133-148.
9. Edgar BA, Lehman DA, O'Farrell PH. Transcriptional regulation of string (cdc25): A link between developmental programming and the cell cycle. Development 1994; 120:3131-3143.
10. Etienne-Manneville S, Hall A. Rho GTPases in cell biology. Nature 2002; 420:629-935.
11. Foe VE, Odell GM, Edgar BA. Mitosis and morphogenesis in the Drosophila embryo: Point and counterpoint. In: Bate Martinez Arias, eds. The Development of Drosophila melanogaster. Cold Spring Harbor: Laboratory Press, 1993:149-300
12. Godt D, Tepass U. Drosophila oocyte localization is mediated by differential cadherin-based adhesion. Nature 1998; 395:387-391.
13. Gong WJ, Golic KG. Ends-out, or replacement, gene targeting in Drosophila. Proc Natl Acad Sci USA 2003; 100:2556-2561.
14. Grosshans J, Wieschaus E. A genetic link between morphogenesis and cell division during formation of the ventral furrow in Drosophila. Cell 2000; 101:523-531.
15. Hacker U, Perrimon N. DRhoGEF2 encodes a member of the Dbl family of oncogenes and controls cell shape changes during gastrulation in Drosophila. Genes Dev 1998; 12:274-284.
16. Hartenstein V, Lee A, Toga AW. A graphic digital database of Drosophila embryogenesis. Trends Genet 1995; 11:51-58.
17. Hemavathy K, Ashraf SI, Ip YT. Snail/slug family of repressors: Slowly going into the fast lane of development and cancer. Gene 2000; 257:1-12.
18. Hemavathy K, Meng X, Ip YT. Differential regulation of gastrulation and neuroectodermal gene expression by Snail in the Drosophila embryo. Development 1997; 124:3683-3691.
19. Ip YT, Gridley T. Cell movements during gastrulation: Snail dependent and independent pathways. Curr Opin Genet Dev 2002; 12:423-429.
20. Ip YT, Maggert K, Levine M. Uncoupling gastrulation and mesoderm differentiation in the Drosophila embryo. EMBO J 1994; 13:5826-5834.
21. Kasai Y, Nambu JR, Lieberman PM et al. Dorsal-ventral patterning in Drosophila: DNA binding of Snail protein to the single-minded gene. Proc Natl Acad Sci USA 1992; 89:3414-3418.
22. Knust E, Muller HJ. Drosophila morphogenesis: Orchestrating cell rearrangements. Curr Biol 1998; 8:R853-R855.
23. Kosman D, Ip YT, Levine M et al. Establishment of the mesoderm-neuroectoderm boundary in the Drosophila embryo. Science 1991; 254:118-122.
24. Kumar R. Another tie that binds the MTA family to breast cancer. Cell 2003; 113:142-3.
25. Leptin M. Twist and snail as positive and negative regulators during Drosophila mesoderm development. Genes Dev 1991; 5:1568-1576.
26. Leptin M. Gastrulation in Drosophila: The logic and the cellular mechanisms. EMBO J 1999; 18:3187-3192.
27. Leptin M, Grunewald B. Cell shape changes during gastrulation in Drosophila. Development 1990; 110:73-84.
28. Lu CC, Brennan J, Robertson EJ. From fertilization to gastrulation: Axis formation in the mouse embryo. Curr Opin Genet Dev 2001; 11:384-392.
29. Mata J, Curado S, Ephrussi A et al. Tribbles coordinates mitosis and morphogenesis in Drosophila by regulating string/CDC25 proteolysis. Cell 2000; 101:511-522.
30. Mlodzik M. Planar cell polarization: Do the same mechanisms regulate Drosophila tissue polarity and vertebrate gastrulation? Trends Genet 2002; 18:564-571.

31. Morize P, Christiansen AE, Costa M et al. Hyperactivation of the folded gastrulation pathway induces specific cell shape changes. Development 1998; 125:589-597.
32. Myers DC, Sepich DS, Solnica-Krezel L. Convergence and extension in vertebrate gastrulae: Cell movements according to or in search of identity? Trends Genet 2002; 18:447-455.
33. Nance J, Priess JR. Cell polarity and gastrulation in C. elegans. Development 2002; 129:387-397.
34. Nieto MA. The snail superfamily of zinc-finger transcription factors. Nat Rev Mol Cell Biol 2002; 3:155-166.
35. Niewiadomska P, Godt D, Tepass U. DE-Cadherin is required for intercellular motility during Drosophila oogenesis. J Cell Biol 1999; 144:533-547.
36. Oda H, Tsukita S. Real-time imaging of cell-cell adherens junctions reveals that Drosophila mesoderm invagination begins with two phases of apical constriction of cells. J Cell Sci 2001; 114:493-501.
37. Parks S, Wieschaus E. The Drosophila gastrulation gene concertina encodes a Gα-like protein. Cell 1991; 64:447-458.
38. Seher TC, Leptin M. Tribbles, a cell-cycle brake that coordinates proliferation and morphogenesis during Drosophila gastrulation. Curr Biol 2000; 10:623-629.
39. Stathopoulos A, Levine M. Dorsal gradient networks in the Drosophila embryo. Dev Biol 2002; 246:57-67.
40. Straub KL, Stella MC, Leptin M. The gelsolin-related flightless I protein is required for actin distribution during cellularisation in Drosophila. J Cell Sci 1996; 109:263-270.
41. Sweeton D, Parks S, Costa M et al. Gastrulation in Drosophila: The formation of the ventral furrow and posterior midgut invaginations. Development 1991; 112:775-789.
42. Tam PP, Behringer RR. Mouse gastrulation: The formation of a mammalian body plan. Mech Dev 1997; 68:3-25.
43. Tepass U, Gruszynski-DeFeo E, Haag TA et al. Shotgun encodes Drosophila E-cadherin and is preferentially required during cell rearrangement in the neurectoderm and other morphogenetically active epithelia. Genes Dev 1996; 10:672-685.
44. Thiery J-P. Epithelial-mesenchymal transitions in tumour progression. Nat Rev Cancer 2002; 2:442-454.
45. Towb P, Bergmann A, Wasserman SA. The protein kinase Pelle mediates feedback regulation in the Drosophila Toll signaling pathway. Development 2001; 128:4729-4736.
46. Uemura T, Oda H, Kraut R et al. Zygotic Drosophila E-cadherin expression is required for processes of dynamic epithelial cell rearrangement in the Drosophila embryo. Genes Dev 1996; 10:659-71.
47. Young PE, Pesacreta TC, Kiehart DP. Dynamic changes in the distribution of cytoplasmic myosin during Drosophila embryogenesis. Development 1991; 111:1-14.

CHAPTER 8

Cutaneous Wound Reepithelialization:
A Partial and Reversible EMT

Valérie Arnoux, Christophe Côme, Donna F. Kusewitt, Laurie G. Hudson and Pierre Savagner

Abstract

Successful cutaneous wound repair occurs in a series of tightly coordinated and overlapping steps: (1) inflammation and clot formation, (2) keratinocyte activation and migration, (3) remodeling of the basement membrane and extracellular matrix, and (4) dermal and epidermal maturation. During the final three stages of cutaneous wound healing, restoration of an intact epidermis occurs via a complex process termed reepithelialization. In this chapter, we focus on the process of wound reepithelialization, emphasizing the resemblance of reepithelialization to epithelial-mesenchymal transition (EMT) occurring during development and tumor progression. Based on the many morphologic and molecular similarities between the two processes, we propose that wound reepithelialization represents a partial and reversible EMT.

Introduction

EMT is essential for normal embryonic development and also occurs during some pathological conditions in adult tissues (see other chapters in this book). We propose that reepithelialization during wound healing represents an instance of physiologic EMT occurring during adult life, as opposed to developmental and pathologic EMT (see Table 1). In this chapter, we review those features of reepithelialization consistent with EMT.

Skin consists of the epidermis, a keratinized stratified squamous epithelium, overlying the collagen-rich dermis, which contains the skin's blood supply. The epidermis includes specialized structures such as hair follicles and annexae. The epidermis is composed primarily of keratinocytes (85%), with melanocytes, Langerhans cells, and Merkel cells. The basal layer of the epidermis (stratum basale) consists of mitotically active keratinocytes anchored to the basement membrane (BM) by hemidesmosomes. Suprabasal keratinocytes are organized into the stratum spinosum, the stratum granulosum, and the stratum corneum. Cells within the stratum corneum, the most external layer, have undergone nuclear degradation, cytokeratin aggregation, and replacement of the plasma membrane with a tough, insoluble cornified envelope that is cross-linked to extracellular lipid. The stratum corneum thus forms a highly impermeable barrier layer. Keratinocytes are tightly connected to each others by desmosomes and adherens junctions.

Keratinocytes differentiate physiologically as they migrate upward from the most basal layer toward the skin surface. Keratinocyte stem cells are located in the basal layer of the interfollicular dermis and in the bulge region of the hair follicle.[1,2] Proliferation and differentiation are mutually exclusive in the epidermis. Cells in the stratum basale proliferate but do not undergo terminal differentiation, while suprabasal cells differentiate but rarely divide. Differentiation of

Rise and Fall of Epithelial Phenotype: Concepts of Epithelial-Mesenchymal Transition, edited by Pierre Savagner.

Table 1. Distinct EMT types show similar characteristics

EMT Type	Examples	Intercellular Junctions	Motility	Invasiveness	Cytoskeleton	Reversibility
Developmental	Gastrulation Neural crest cell migration Heart morphogenesis Palatogenesis Digit formation	Complete loss	Increased	Enhanced by secretion of proteases	Change from cytokeratin to vimentin	Sometimes reversible by mesenchymal-epithelial transformation
Pathological	Cancer progression Renal fibrogenesis Cataractogenesis	Partial to complete loss	Increased	Enhanced by secretion of proteases	Change from cytokeratin to vimentin, sometimes to smooth muscle actin	Largely irreversible
Physiological (adult)	Wound healing	Partial loss	Increased	Enhanced by secretion of proteases	Modulation of cytokeratins	Completely reversible

stratified squamous epithelium is controlled by a variety of factors, including interactions with the BM,[3-6] calcium ion concentration gradients,[7] and vitamin D.[8,9]

Cutaneous Wound Healing, a Multistep Process

Embryonic wound healing proceeds without scarring[10] and will not be reviewed here. Cutaneous wounds in the adult include burn, blister, excisional, and incisional wounds. In adult wounds, the healing process is simpler and faster if the BM remains intact. Whether or not the BM is present, keratinocytes go through an activation process that initiates reepithelialization. Keratinocyte activation and subsequent stages in reepithelialization are characterized by events typical of EMT as described in other chapters of this book. Here, we will review these characteristics, emphasizing those wound healing processes occurring when the BM has been ruptured.

Studies using several animal models, including rabbit,[11] mouse[12] and pig,[13] have provided descriptions of in vivo cutaneous wound healing at the cellular level. In animals with loose skin, such as mouse and rabbit, contraction of the panniculus carnosus, a thin sheet of striated muscle lying beneath the epidermis, plays a prominent role during wound repair. In contrast, the human and pig have tight skin, lacking a panniculus carnosus. Despite these differences, Odland[14] has demonstrated that the timing of reepithelialization during wound healing is similar in man and experimental animals.

As illustrated in Figure 1, wound healing process may conveniently be divided into the following steps:

1. Clot formation and inflammation
2. Keratinocyte migration and proliferation
3. ECM remodeling
4. Dermal and epidermal maturation

Reepithelialization occurs during the final three stages of wound healing. Reepithelialization involves the migration of keratinocytes from the edges of the wound, followed by their proliferation, stratification, and redifferentiation to form an intact epithelium.[15]

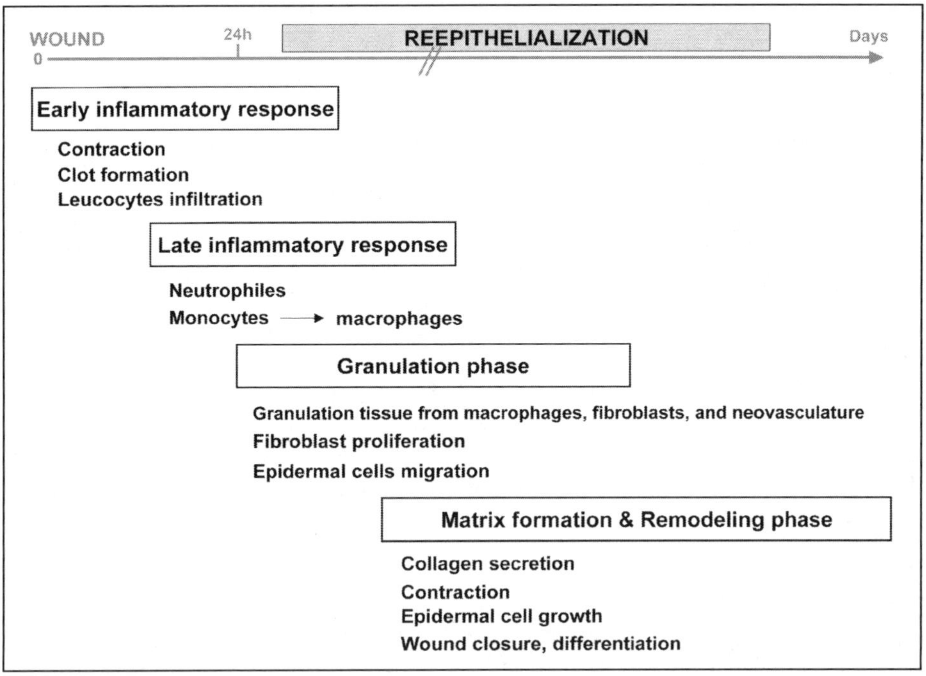

Figure 1. Wound repair is a multistep process with overlapping phases. Immediately following wound, local inflammation initiates the first phase during which the fibrin clot is formed, and contraction decreases wound extent. Leucocyte infiltration provides a source of neutrophils, monocytes, cytokines and growth factors which amplify the wound signal, supporting activation of downstream pathways. The granulation phase follows this phase and involves different cell types: macrophages, keratinocytes and fibroblasts whose activation results in the organization of the granulation tissue, a richly vascular connective tissue. Activated keratinocytes contribute to the ECM reconstitution through synthesis of new molecules and proteolytic activity. This remodeling phase is the ultimate step for complete reepithelialization and wound closure.

Keratinocyte Activation, a Partial EMT

Reepithelialization is initiated by activation of keratinocytes at wound margins.[16] This preparatory phase of keratinocyte activation precedes the onset of keratinocyte migration into the wound site[17] (Fig. 2). The original stimulus for activation is the injury itself which initiates the early signaling events. Using cultured corneal cells, it has been demonstrated that the stimulus for epithelial cell activation and migration is the availability of a denuded surface rather than mechanical damage to cells at the wound margin.[18]

Activated keratinocytes undergo morphologic modifications, develop new migratory capabilities, and express novel genes.[19] These alterations are reminiscent of "classic" EMT seen during development and described elsewhere in this book (Chapters 1: Hay and 2: Morali et al). Table 2 summarizes the changes in gene expression and their phenotypic effects that are shared by the processes of reepithelialization and EMT.

One of the hallmarks of EMT in vitro and in vivo is cell-cell dissociation. Likewise, keratinocytes at the leading edge of a wound undergo dramatic reorganization of the junctional complexes and associated cytoskeletal elements. Adhesion between keratinocytes is primarily mediated by desmosomes and adherens junctions. Desmosomes represent the dominant adhesive junction in the epidermis and are responsible for maintaining the architecture and structural integrity of the skin.[20] In addition, hemidesmosomes link basal keratinocytes to the BM at the junction of the epidermis and dermis. Desmosomes and hemidesmosomes are directly

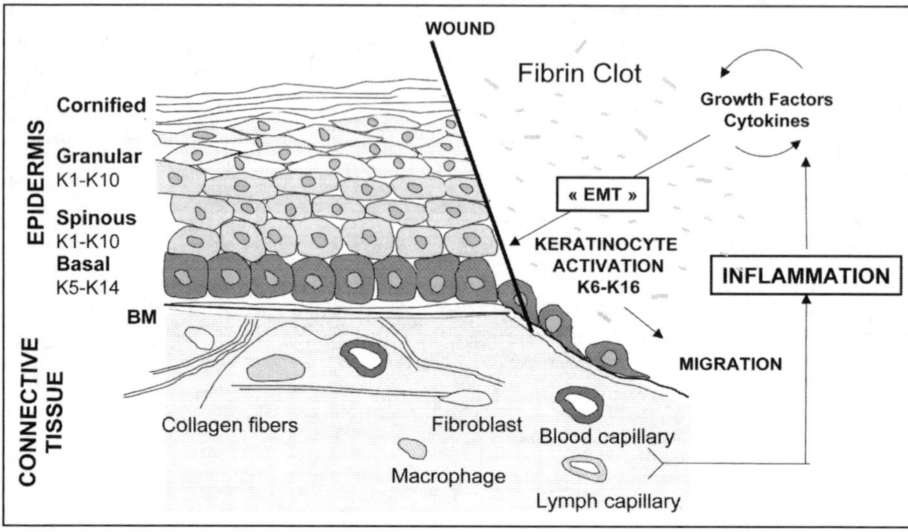

Figure 2. Keratinocyte activation is induced by early inflammatory response. The epidermis is organized into a keratinocyte multilayer. Keratinocytes express specific cytokeratins along their differentiation path, reflecting a vertical migration from the basal layer, attached to the basement membrane (BM), towards the cornified layer. Following a wound and local inflammation, cytokines and growth factors initiate combined pathways defining an activation state for the basal keratinocytes, associated with the specific expression of cytokeratins 6 and 16 (K6-K16). As a visible result of this activation, keratinocytes undergo a partial conversion of phenotype reminiscent of the EMT process and become flattened and motile.

linked to the cytokeratin cytoskeleton, while adherens junctions connect to actin cytoskeletal elements. At wound margins, keratinocytes change from polarized and stationary cuboidal cells to flattened cells with extended lamellipodia.[15] Cells undergo a reduction in the number of cell-cell junctions, disruption of the intermediate filament network and reorganization of the cytoskeleton; these alterations are associated with the appearance of gaps between keratinocytes.[16,21] At completion of reepithelialization, keratinocytes regain the epithelial phenotype, establish stable intercellular and cell-substrate contacts. It has been reported that desmosome formation is a critical event marking the end of migration.[22-24]

Our studies in HaCaT cells[25] showed that the internuclear distance between keratinocytes at wound margins increased by 60%. The flattened cells at the wound margins showed discontinuous staining for desmosomal components in regions of cell-cell contact, indicating decreased desmosome density but not complete loss of desmosomes.[26] Thus, although partially disconnected, cells at the wound margin remain part of a cohesive sheet. A generalized and dramatic remodeling of the cytokeratin cytoskeleton accompanies desmosomal dissolution. Morphological studies reveal that intermediate filaments retract from the keratinocyte cell surface as a subset of the desmosomes and hemidesmosomes in the cell undergo dissolution.[15] In unwounded epidermis, cytokeratins CK5 and CK14 are expressed in basal keratinocytes, which are the main participants in reepithelialization. Suprabasal keratinocytes express mostly CK1 and CK10.[27,28] In contrast, activated keratinocytes at wound margins[29] express cytokeratins CK6 and CK16, as well as CK5 and CK14.[21,30-32] Altered cytokeratin expression is also described as part of EMT occurring during development and malignant progression. During the formation of the secondary palate for example, K5/6 expression is upregulated and vimentin expressed prior to midline contact.[33] There are numerous instances of cytokeratin downregulation

Table 2. Similar molecules are involved in developmental EMT and wound healing

Molecules	Developmental EMT	References	Wound Healing (Activated Keratinocytes)
Collagen I	Neural crest cell migration	253	67
Fibronectin	Heart morphogenesis	254	62, 63, 255
Laminin	Heart morphogenesis Neural crest cell migration	253, 254	19
Tenascin	Embryo Neural crest cell migration	253, 256	64
β1 integrin	Heart morphogenesis	254, 257	203
MMP-1, -2, -8, -9, -10 plasminogen activator	Neural crest cell migration	258, Chapter 22	139, 143, 146, 148, 259
Keratins	Palate formation	33	21
EGF, TGF-α	Palate formation	260	161
EGF receptor	Palate formation	260	158
bFGF	Neural crest cell migration	176, Chapter 3	175, 261
TGF-β	Neural crest cell migration	262, 263	264
HGF		Chapter 13	170, 173
Snail, Slug	Gastrulation	219, 265	26
Ets	Neural crest cells	243, 266	248

during EMT occurring in transformed cell lines in vitro; in such cells, cytokeratins are often replaced by vimentin filaments (see other chapters in this book). Furthermore, during gastrulation, cytokeratins in cells undergoing EMT are replaced by vimentin filaments (see Chapter 2, Morali et al).

A major cell-cell adhesion molecule found in adherens junctions, E-cadherin, is also regulated during reepithelialization. Although no changes in E-cadherin mRNA or protein levels are detected in keratinocytes at wound margins in vitro[26] or in vivo,[34] E-cadherin vanishes from the free edges of keratinocytes forming the wound margin. This reflects structural changes in adherens junctions and actin cytoskeletal organization. A similar EMT-linked relocalization of E-cadherin is also observed in vivo during palate fusion[33] and in vitro in growth factor-induced EMT in NBT-II carcinoma cells.[35] In other in vitro situations and during EMT in gastrulation, E-cadherin is downregulated at the RNA and protein levels (see Chapter 3, Newgreen).

The relationship between E-cadherin expression and the EMT occurring as part of carcinoma progression is complex. Some breast cancers, such as infiltrating lobular carcinomas, have apparently undergone extensive EMT-like changes and have lost all cell-cell adhesion structures. These cells completely lack cell membrane-associated E-cadherin. In the case of breast invasive ductal carcinoma, E-cadherin expression is maintained but there is significant heterogeneity in the pattern of E-cadherin expression. Thus, there may be a total lack of expression in some limited regions of the tumor, possibly indicating regional variations in tumor phenotype. In general, a decrease in membrane-bound E-cadherin expression is associated with increased invasiveness in human tumors (see Chapter 9, Van Marck et al), and E-cadherin is overall considered to be a tumor suppressor gene.[36,37] On the other hand, maintaining some level of E-cadherin–mediated cell-cell adhesion in tumor metastases may be important for tumor progression,[38,39] probably because it could facilitate cell survival within metastatic deposits.

Migration of a Semi-Cohesive Sheet of Cells

Migration during EMT

Cell migration is a consistent feature of EMT and is also crucial for reepithelialization during cutaneous wound healing. During development, epithelial cells can move as sheets or, when they undergo the full panoply of EMT-associated changes, as individual cells.[40,41] Cancer cells can also invade either as multicellular strands or sheets or as individual cells.[42,43] Cell movement of multicellular structures resembles the wound healing responses described above for larger wounds, while invasion as single cells involves complete loss of cell-cell junctions and, often, replacement of cytokeratin intermediate filaments by vimentin. The latter mode of migration closely resembles the complete EMT occurring at some stages of development. Tumors can display both modes of invasion simultaneously, and it appears that the transition from collective to individual cell migration could mark an ominous change in the invasive and metastatic potential of a tumor. During some EMT-like processes, notably renal fibrogenesis, motility is associated with de novo synthesis of smooth muscle actin.[44]

Reepithelialization is characterized by the progressive extension of a sheet of keratinocytes from the wound margin across exposed dermis. This extension is the result of an active migratory process. From 12 to 48 hours after wounding, the epidermis migrates symmetrically over the wound bed from the free edge toward the center of the wound.[14,15] Advancing keratinocytes extend prominent cytoplasmic projections (filipodia and lamellipodia). Electron microscopy demonstrates a substantial increase in the diameter of cells at the advancing wound margin and sparse intermediate filaments. It also shows the persistence of at least some desmosomes in migrating keratinocytes, allowing coherent migration of motile but cohesive cells. In small circular wounds, epithelial cells at the wound margin develop bundles of actin filaments that insert into adherens junctions connecting the cells. These bundles form a circumferential "purse string" running around the wound margin; contraction of this structure reduces the wound.[45,46] In larger wounds, formation of a purse string is more difficult to demonstrate. In vitro, some epithelial cells at the wound margin develop into leader cells that extend large lamellae into the wound area. The cytoskeleton and cell-cell contacts of leader cells are modified to resemble those of fibroblasts and leader cells become partially dissociated from neighboring cells.[46,47] Movement of a coherent epithelial cell sheet during this process is not due simply to migration of a front row of motile cells that drags along a collection of trailing cells. Instead, there is dynamic reordering of the cells at the wound margin, which is facilitated by the lability of adherens junctions and desmosomes connecting the cells.

The leading edge of the wound is composed of a single layer of cells in vivo. The number of epithelial cell layers increases with distance away from the leading edge of the wound. It has been suggested that cells at the wound margin migrate in a leapfrog fashion, with migrating cells slightly distal to the wound edge moving over the leading cells to reach available substrate and cells located further from the wound edge migrating as a cohesive sheet. This hypothesis is based on observations in vivo and in reconstructed skin models in vitro.[48] However, this behavior is not observed in vitro two-dimensional wound healing models using keratinocytes grown on plastic or on a collagen substrate (P. Savagner, unpublished observation). In vivo observations may reflect the involvement of suprabasal keratinocytes in reepithelialization but the role of a "leapfrog mechanism" in wound healing has not been conclusively demonstrated. Other models for the generation of the mechanical forces implied in the migratory process involve (a) traction from the "leader cells" in vitro or their equivalent in vivo and (b) pressure from the cells proliferating behind the wound edge. The lack of tight adhesion between the cells at the wound margin and the following cells does not favor to this second hypothesis, although keratinocyte proliferation is certainly necessary for the overall success of wound healing process.

Epithelial cell migration during EMT and EMT-like processes is driven both by actin-myosin-based contraction and actin-polymerization-based protrusion. Branched actin networks located below the plasma membrane form a cortical actin ring. Bundles of actin filaments running through the cell cytoplasm constitute stress fibers, and highly branched networks of actin filaments are found in cellular protrusions like lamellipodia.[49] Actin polymerization, bundling, and branching are under the control of a variety of actin binding proteins such as cofilin, profilin, gelsolin, and tropomyosin.[49] Contraction of actin filaments in nonmuscle cells is mediated by the actin cross-linking protein myosin II.[50] Generation of traction forces by actin-myosin filaments requires interactions with the substrate or adjacent cells, thus the actin cytoskeleton is intimately connected to cell-substrate and cell-cell contacts.[51] Stress fibers insert into focal adhesion contacts that transiently anchor the cell to its substrate during migration and cortical actin cables may attach to adherens junctions connecting adjacent cells to form contractile structures involving multiple cells.[45,46] RhoA, rac1, and cdc42, members of the Rho family of small GTPases, link signals originating from cell surface receptors, including G-protein-coupled receptors, tyrosine kinase receptors, cytokine receptors, and adhesion receptors, with the actin cytoskeleton.[52] Effectors of Rho family GTPases are serine-threonine kinases, notably PAK1 and ROCK, that regulate the actin cytoskeleton.[53] Thus, the Rho family plays a critical role in cell migration as reviewed in Chapter 17 (Nakagawa et al). In general, rho activation stimulates formation of contractile stress fibers, while rac activation drives formation of lamellipodia and cdc42 controls filipodia development.[47] In addition to regulating changes in the actin cytoskeleton, these factors also modulate cell-cell and cell-substrate adhesion complexes, thus coordinating the two processes to promote cell migration.[54,51]

Migration versus Proliferation

Epidermal migration and proliferation both contribute to wound closure, but they appear to be controlled by independent mechanisms and to be somewhat mutually exclusive. Mitosis is rare in migrating cells.[55,56] During gastrulation in Drosophila, proliferation is suppressed while migration is occurring.[57,58] In corneal wounds, healing epithelium does not show more G2/M phase cells than unwounded control epithelium; however, cell division increases significantly by 12h 4mm distal to the wound.[59] In cutaneous wound healing, significant cell division occurs one day after wounding and appears to be restricted to cells distal from the leading edge of the wound.[12,15] Proliferation clearly contributes to the pool of migrating cells; however, migration of keratinocytes at the wound edge is an active process that cannot be explained by passive translocation of marginal cells due to proliferation.

Extracellular Matrix

Wound repair is initiated by aggregation of platelets, formation of a fibrin mesh,[60,61] and release of growth factors during activation of coagulation pathways. The fibrin-rich wound clot thus formed provides a temporary wound scaffolding. As granulation tissue forms, fibroblasts replace the plasmin- and MMP-digested fibrin clot with fibronectin,[62,63] tenascin[64,65] and collagen III.[66] Collagen type III is later replaced by collagen type I and these two types of collagen confer strength to the healing wound.[67]

The ECM controls keratinocyte migration. Typically, the BM is locally eliminated during the wounding, exposing migrating keratinocytes to the ECM of the underlying dermis. The fibronectin contribution to wound healing is reminiscent of its role in EMT during developmental processes including gastrulation and neural crest cell migration where it provides a migration substrate specifically enhanced within migratory pathways (see Chapter 2, Morali et al and Chapter 3, Newgreen).

The BM component laminin 5 (LM-5) (also called epiligrin, nicein, kalinin), secreted by epithelial cells, mediates stable attachment of epidermal cells to the BM in intact epithelium.[68-70] However, LM-5 also appears to play a significant role during keratinocyte migration by interacting with type VII collagen to link the epidermis to the dermis.[71-75] Immunohistochemical and in situ hybridization studies have shown increased LM-5 expression in migrating keratinocytes[76-81] and LM-5 deposition by keratinocytes is associated with persistent, directional migration.[78,79] Among other cell types involved in EMT-like events, LM-5 is also expressed at high levels by squamous cell carcinomas.[82,83] In these tumors, LM-5 is implicated as a mediator of invasion. The role played by LM-5 may depend on the integrity of the molecule since it is deposited as a precursor into the provisional BM.[84] In the mature BM, LM-5 is proteolytically processed and degraded into active peptide fragments reported to promote motility.[85]

Integrins

Keratinocytes interact with the ECM through cell surface receptors called integrins.[86-88] Integrins compose a large family of heterodimeric proteins that bind with varying affinity to different ECM components. Integrins participate in EMT occurring during developmental processes such as neural crest cells emergence (Chapter 3, Newgreen). The regulation of integrin expression is complex and species-dependent. In the epidermis, integrins are expressed mainly in basal keratinocytes; their expression is down-regulated during terminal differentiation. However, integrin expression can be detected in suprabasal keratinocytes in instances of benign hyperproliferation such as that occurring during wound healing.[5] Basal keratinocytes express a variety of integrins. The most abundant integrins in normal epidermis are α2β1 (a collagen receptor),[89] α3β1 (predominantly a LM-5 receptor),[90] and α6β4 (another laminin receptor).[91] The vitronectin receptor αvβ5 is also a normal epidermal integrin, but it is expressed at lower levels.[92] In their role as ECM receptors, integrins are directly involved in keratinocyte migration. However, as illustrated in Figure 3, they exhibit a surprisingly complex pattern of expression and regulation during wound healing.

Beginning early after wounding, several hours before reepithelialization begins, integrins are redistributed on the keratinocyte surface and keratinocytes expand their integrin repertoire. Hemidesmosomes detach from their laminin substrate and expression of integrins such as α5β1 (fibronectin receptor), αvβ6 (fibronectin and tenascin receptor), α9β1 (tenascin receptor), and αvβ5 is activated.[76,93,94] In addition, integrins α3β1 and α2β1 move from a lateral to a basal location where they come into direct contact with the ECM. Integrin α6β4 is redistributed on the surface of migrating keratinocytes lacking hemidesmosomes.[95] In vitro, α6β4 localizes to the leading lamellae of migrating keratinocytes.[96-98] Antibodies against α6β4 inhibit the migration of carcinoma cells by blocking the formation of filopodia and lamellipodia. Studies in knockout and transgenic mice demonstrate that α6β4 helps to maintain epithelial integrity by mediating strong keratinocyte adhesion to LM-5 via hemidesmosomes.[99-104]

Other studies demonstrate the importance of the ubiquitous β1 integrin subunit in wound healing. Loss of β1 integrin in keratinocytes causes a severe defect in wound healing in vivo. In vitro, β1-null keratinocytes are characterized by poor adhesion to various substrates, reduced proliferation, and impaired migratory capacity.[105] The integrin complex α2β1 has been shown to be necessary for keratinocyte migration[106] by mediating induction of collagenase-1 through its contact with native type I collagen[107] and by selecting LM-5 as its substrate once migration has been initiated.[106] The α2β1 integrin has also been found to be necessary for growth factor or collagen-induced EMT in carcinoma cells.[108]

The role of α3β1 integrin complex in cell migration during wound healing has been scrutinized by several laboratories. Integrin α3β1-deficient mice display BM disorganization at the junction between the epithelium and underlying connective tissue in skin,[109,110] kidney and lung,[111] suggesting a fundamental role for this integrin in ECM assembly and integrity during development. Integrin α3β1 is able to regulate LM-5 spatial organization within the ECM mesh.[112] Additionally, α3β1 integrin participates in the construction of hemidesmosomes by establishing molecular complexes with CD151.[113] However, neither α3β1 nor α6β4 complexes appear to be essential for epidermal morphogenesis during skin development.[114]

Xia et al have shown that α3β1 mediates initial attachment from epidermal cells to LM-5. In a later phase, α6β4 integrin takes over in binding preferentially LM-5.[115,116] Some groups have shown that α3β1 complex was involved into cell migration on laminin-5.[85,90,116,117] This is in agreement with the high level of expression of integrin α3β1in keratinocytes during

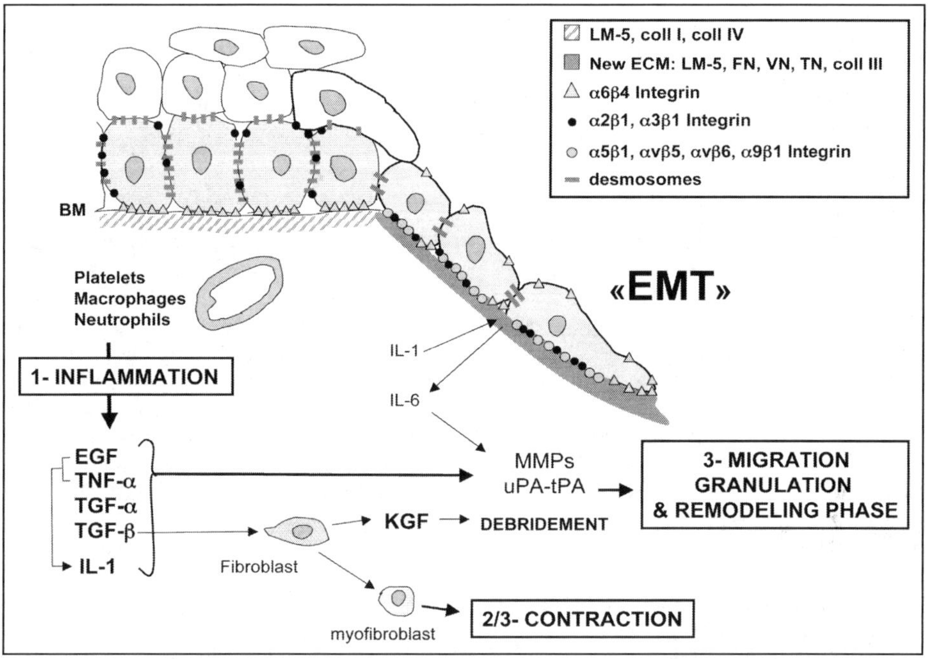

Figure 3. Keratinocyte EMT-like activation includes emergence of migratory and invasive potential. Cytokines and growth factors induced by inflammatory response control keratinocyte activation: Keratinocytes reorganize their cell-cell and cell-substrate adhesion structures, including integrin repertoire and localization, and initiate migratory process. Reepithelialization also includes debridement and renewal of the substrate, both by controlled proteolysis, facilitated by metalloproteinases (MMP) and plasminogen activator (PA), and synthesis of new extracellular matrix (ECM) components, including laminin chains (LN), collagens (coll), fibronectin (FN), vitronectin (VN) and tenascin (TN). In addition, the contraction, generated by myofibroblasts differentiating from local fibroblasts under TGF-β activation, participates in the early wound healing response.

cutaneous wound healing[76,87,118] and in various metastatic carcinomas.[119-121] Accordingly, inhibition of the α3β1-mediated adhesion to LM-5 can prevent formation of stable lamellipodia.[122]

However, others studies suggest that α3β1 complex actually controls negatively keratinocyte motility by providing enduring binding to newly synthesized LM-5 substrate.[106] In vitro studies based on α3β1 knockout keratinocytes suggest that α3β1 integrin may suppress fibronectin or collagen IV receptors function.[110] In current models,[106,110] it is suggested that after initial redistribution of α2β1 and α3β1 integrins to basal aspects of activated keratinocytes, it is predominantly α2β1 that initiates migration over dermal type I collagen[107] in link with MMP-1 activation. Accordingly, once migration has started, TGF-B stimulates LM-5 synthesis[123] and down-regulates functionally α3β1 integrin, allowing α2β1 to play a preferential role in binding LM-5 and support migration until the stationary phase. At this stage, when the basement membrane is partially reconstituted, α3β1 is functionally upregulated and supplants α2β1 in mediating LM-5 binding.

Additionally, a potential role for the β5 integrin subunit in keratinocyte motility was suggested by the reduced migratory activity of β5-null keratinocytes; however β5 knockout mice have no reported skin phenotype.[124]

During wound healing, keratinocyte integrins are dynamically regulated by numerous growth factors. For example, α2β1 is regulated through the EGF receptor pathway,[125] and β4 integrin subunit regulated through phosphorylation by EGF receptor,[126] MSP[127] or Fyn kinase[128] inducing hemidesmosome disruption. Members of the TGF-β family also regulate integrin expression, as detailed below. On the other hand, growth factor receptor phosphorylation and signaling pathway activation can, in some cases, be stimulated through integrins, even in the absence of specific ligand. This form of indirect activation appears to involve extracellular matrix molecules such as fibronectin.[129]

Proteases

Metalloproteinases (MMPs) cleave ECM components that impede keratinocyte migration.[130,131] MMPs contribute to keratinocyte migration and wound debridement by opening a path in the surrounding matrix. In addition, they also generate new ligands able to activate keratinocyte migration. For example, LM-5 digestion by MMP generate fragments that can directly stimulate cell motility.[132-135] MMP activity is controlled by physiological regulators, the tissue inhibitors of metalloproteinases (TIMPs).[136,137] The net proteolytic activity at any site depends on the local balance between MMPs and TIMPs. Growth factors such as KGF, EGF, TNF-α and TGF-β also regulate MMP production and activity, as discussed below.[138,139]

Expression of fibrinolytic MMPs and plasmin is increased in migrating keratinocytes,[140] prominently in migrating basal keratinocytes,[141,142] and during epidermal remodeling.[143] In human cutaneous wounds, MMP1 activity is increased about 100 fold. The highest level of expression is seen in dermal fibroblasts.[118] MMP1 expression is upregulated following integrin α2β1 activation by collagen I.[107]

Another member of the family, MMP-9, is expressed by basal keratinocytes[144] and activated during two stages: early after wound and later during maturation of newly formed epidermis.[143,145] Proteolysis appears to be essential for normal wound healing since pharmacological MMP inhibition delays epidermal regeneration in vivo.[146,147] Inhibition of MMP-2 activity drastically impairs keratinocyte migration, whereas inhibition of MMP-9 has no effect.[143] Impairment of wound healing is also observed in plasminogen-deficient mice,[148] indicating a crucial role for plasmin as well. Moreover, synergistic interactions between these two families of proteases are indicated by the complete lack of keratinocyte migration observed in plasminogen-deficient mice treated with MMP inhibitors.[147]

Protease expression by activated keratinocytes at wound margins is reminiscent of protease expression during developmental EMT, for instance during heart morphogenesis and neural crest cell migration (see Chapter 3, Newgreen; Chapter 4, Runyan et al; Chapter 20, Gilles et al). In vitro, MMP activation can induce EMT in mammary epithelial cells.[149] Moreover, proteomic studies have demonstrated increased MMP-2 and MMP-9 activity during renal fibrosis, a pathologic form of EMT.[140,150] These two MMPs are also highly expressed in invasive cancer cells and have been shown by cDNA microarray to be elevated in aggressive melanomas compared to nonaggressive forms 115 (see also Chapter 20, Gilles et al). The similarity in MMP expression between keratinocytes at wound margins and epithelial cells undergoing tumor-associated EMT is particularly interesting. Neither cell population usually expresses MMP, but both go through an activation process resulting in MMP expression.

Contribution of the Inflammatory Response to Reepithelialization

Role of Immune Cells

During migration, keratinocytes move over an inflammatory environment containing neutrophils and monocytes recruited from the circulating blood. Neutrophils are recruited within few minutes of wounding and remain activated for several days. They represent a source of pro-inflammatory cytokines (IL-1, IL-6, TNF-α) that, in turn, activate fibroblasts and keratinocytes.[151,152] Macrophages are essential for wound healing, since preventing macrophage infiltration severely impairs the process.[153] They phagocyte cell and matrix debris and recruit other inflammatory cells and local fibroblasts.[154] The role of macrophages in reepithelialization is indirect, mediated by numerous factors secreted locally. Overall, macrophages facilitate wound repair in the early stages of the process by the expression of embryonic fibronectin,[63] which is later supplied by fibroblasts.

Prostaglandins are lipidic signaling mediators that also play a major role in inflammation. Their synthesis from arachidonic acid is catalyzed by cyclooxygenases (COX-1, -2 or -3). Following an excisional injury in rat skin, COX-2 is induced in basal keratinocytes within 12h and remains elevated for 3 days. A COX-2 inhibitor delays reepithelialization in the early phase of wound healing and also inhibits angiogenesis.[155] In the same way, IL-6 knockout mice reveal a delayed wound healing and a decrease in inflammation and granulation tissue formation[156] that could be reverted by providing IL-6 locally. These studies based on COX-1, COX-2 as well as Il-6 knockout mice emphasize their importance in the wound healing process. However, they do not appear to involve keratinocyte phenotype and migration directly.

More generally, although no embryonic immune response can be involved in the EMT occurring during early development, immune cells are incriminated in pathological EMT processes such as cancer progression and kidney inflammation, mostly through the secretion of cytokines and growth factors.

Role of Growth Factors

Growth factors are released by keratinocytes, fibroblasts and inflammatory cells within and adjacent to the wound. The ECM limits diffusion of these growth factors, confining them to the local area. Growth factors are key regulators of keratinocyte migration and proliferation and also help restore the mature keratinocyte phenotype when wound healing has been achieved. EGF/TGF-α, KGF, FGF-2, TGF-β and IL-1 all promote cell movement in vitro and are therefore termed motogenic. Many growth factors have both mitogenic and motogenic effects on keratinocytes and other cell types. Growth factors play important roles in most EMT processes in vivo, as described in other chapters in this book. Growth factors contribute to EMT occurring during gastrulation (Chapter 2, Morali et al), neural crest cell migration (Chapter 3, Newgreen), and heart development (Chapter 4, Runyan et al). They are also incriminated in the pathologic EMT occurring during cancer progression (see Chapters 13-16) and kidney chronic degenerative pathologies.

Epidermal Growth Factor and Transforming Growth Factor-α

Epidermal growth factor (EGF) released by platelets and transforming growth factor a (TGF-α) produced by macrophages and keratinocytes are high-affinity polypeptide ligands for the EGF receptor.[157] The EGF receptor is transiently upregulated in wounded epidermis and activation of this receptor contributes significantly to the migratory and invasive potential of keratinocytes.[158] Application of EGF or TGF-α enhances reepithelialization in various animal models.[159,160] Moreover, overexpression of a human epidermal growth factor gene induces 20% acceleration of the healing process in vivo.[161]

TGF-α knockout mice exhibit normal cutaneous wound healing,[162] but this could reflect the presence of other autocrine or exogenous EGF receptor ligands in the environment.[163-165]

Expression of TGF-α under the control of cytokeratin 14 promoter doubles the proliferation rate of keratinocytes.[166,167] Overexpression of TGF-α in transgenic keratinocytes is also associated with enhanced migratory capacity.[166,168] Furthermore, genetic or pharmacological ablation of EGF receptor activity decreased reepithelialization and keratinocyte migration in vivo,[169] further illustrating the role of this protein in wound repair.

Hepatocyte Growth Factor

Hepatocyte growth factor (HGF) is a fibroblast-derived protein that modulates the proliferation, motility, and differentiation of epithelial cells, including epidermal keratinocytes (see also Chapter 13, Day et al). HGF enhances cutaneous wound healing processes including reepithelialization by multiple mechanisms including stimulation of keratinocyte proliferation and migration, neovascularization, and expansion of granulation tissue.[170-172] Expression of both HGF and its receptor (c-Met) are transiently increased in wounded skin,[173] and overexpression of HGF in human keratinocytes leads to hyperproliferation.[174]

Fibroblast Growth Factors

Fibroblast growth factors (FGF), an ever growing family of peptides includes several members directly involved in EMT situations in vitro[175] and in vivo[176](see Chapter 2, Morali) and also during wound healing and more specifically reepithelialization. FGFs induce mitogenesis and cell migration[177] underlying granulation tissue formation, and the production of MMPs.[178] FGF-1, FGF-2 and FGF-7 / KGF (keratinocyte growth factor) have been the most intensely studied. The role of FGF-2 has been confirmed in the FGF-2 null mouse which shows not only retarded epithelialization but also reduced collagen production.[179] In contrast, FGF1 null mouse does not show significant effect on wound healing. Compensation by FGF1 does not account for the mild phenotypic defects observed in FGF2 null mice.[180] KGF appears as the most specific FGF to be directly involved in wound healing process. Within 24 h after wounding, a 160-fold increase in KGF mRNA is observed in fibroblasts located at the wound edge in normal mice.[181] Wound healing studies in KGF knockout mice do not indicate a requirement for KGF during skin healing.[162] However, targeted expression of a dominant negative KGF receptor in basal keratinocytes in the mouse delays wound reepithelialization, leads to epidermal atrophy, and suppresses keratinocyte proliferation, indicating a role for KGF in wound repair.[182] Transgenic mice that express KGF in basal keratinocytes under control of the CK14 promoter have a thickened hyperproliferative epidermis.[183] In addition to having a mitogenic effect on keratinocytes, KGF also stimulates migration and plasminogen activator activity in human epidermal keratinocytes[184,185] and promotes wound repair in vivo through its impact on keratinocyte growth, motility, production of hyaluronan and inhibition of differentiation.[172,186-188] Interestingly, KGF plays a similar role in EMT in human stomach cancer cells.[189]

Transforming Growth Factor-β

TGF-β is released from platelets early in wound healing. TGF-β plays a dual role in cutaneous wound healing: it both induces keratinocyte migration and blocks cell proliferation. It also contributes to the wound healing process in a variety of other ways. At early stages of wound healing, TGF-β modulates ECM formation and remodeling by enhancing the expression of proteinases including MMPs,[190,191] urokinase type-plasminogen activator (uPA)[192] and collagenase.[193,194] At later stages, TGF-β stimulates formation of connective tissue by regulating other peptide growth factors[194] and modulating collagen synthesis by fibroblasts.[195] Finally, TGF-β can also accelerate wound healing in skin and increase the strength of repaired tissue by stimulating an increased influx of mononuclear cells and fibroblasts into the wound area.[196]

TGF-β signals are transduced by the Smad transcription factors, which play distinctive roles in different cell types. In HaCaT keratinocytes, the growth inhibitory effect of TGF-β1 is predominantly mediated by Smad3.[197] Overall, studies in transgenic mice suggest that inhibition of TGF-β signaling pathways accelerates cutaneous wound healing. Transgenic mice that express a dominant negative type II TGF-β receptor in keratinocytes[198] and Smad3 null-mice[199,200] have enhanced wound repair, characterized by an increased rate of reepithelialization. This increase probably results from increased epidermal proliferation and decreased apoptosis of keratinocytes bordering the wound rather than from a direct effect on keratinocyte motility. Furthermore, infiltration of monocytes into the wound area is significantly reduced in these mice, indicating another potential mechanism by which the effects of TGF-β on wound healing may be mediated.

TGF-β also decreases keratinocyte adhesion to the ECM and promotes keratinocyte migration by modulating the expression of integrin subunits and modifying their interaction with LM-5.[80] In vitro, TGF-β1 enhances the expression of α5, αv and β5 integrin subunits,[201] down-regulates α3 and α6 integrin subunit mRNA and protein production, and decreases expression of α3β1 and α6β4 integrins on the cell surface.[80] Downregulation of α3β1 integrins by TGF-β favors LM-5:α2β1 integrin interactions, thus supporting migration across newly synthesized LM-5.[106] Enhancement of motility also reflects the strong expression of integrins α5β1 and αvβ5[202] in link with TGF-β1 coexpression during reepithelialization.[76,203]

Activation of TGF-β pathways has been demonstrated in virtually all instances of developmental and pathologic EMT occurring in vivo (see Chapter 15, Vignais et al), notably during heart morphogenesis (see Chapter 4: Runyan et al) and kidney inflammation.[139]

Growth factors mediate many of the events occurring during reepithelialization in vivo. However these growth factors are multifunctional and may have synergistic or mutually antagonistic effects. Thus coordinated EMT-like behavior in keratinocytes during reepithelialization is probably centrally controlled at the level of the transcriptional response to these growth factors rather than at the level of growth factor production per se.

Transcriptional Control of Keratinocyte Migration

Through the wide variety of pathways described in Chapters 13 to 19, including the MAPK, Src, Ras, GTPases pathways, specific patterns of gene transcription are activated in keratinocytes during wound reepithelialization. These patterns of gene expression are coordinated by widely expressed transcription factors such as the AP-1 Jun complex and by more specifically expressed families of transcription factors such as the Snail and ETS families.

AP-1

The AP-1 transcription factor is a heterodimeric complex consisting of Fos and Jun proteins encoded by early stress response genes. The fos and jun gene families both include multiple members that can be variously combined to form distinct AP-1 heterodimers. Many signaling pathways, including some that control proliferation and migration, act by modulating AP-1 activity. AP-1 activity can be controlled both transcriptionally, by altering the amount or composition of the AP-1 subunits, or post-transcriptionally by phosphorylation or through combined mechanisms. Phosphorylation regulates DNA binding and transcriptional activity of these proteins. Wound-induced activation of the MAP kinase signaling pathways lead to AP-1 activation and upregulation of fos and jun in response to wounding.[204,205] Many genes identified as crucial to wound repair are AP-1 responsive genes.[204] AP-1 activity in keratinocytes is modulated both spatially and temporally.[206]

During reepithelialization, c-Jun is necessary for the organization of the keratinocytes at the wound margin.[207] AP-1 regulates many aspects of wound repair including the expression of genes involved in migration (TGF-β1, collagenase, stromelysin, gelatinase B, α2, α6, β4 integrins and laminin α3A) and keratinocyte differentiation,[208] by regulating promoters of the cytokeratins 1, 5, 14, filaggrin, loricrin, involucrin, and transglutaminase 1 genes.[209] Genetic ablation of c-jun impairs wound healing in vivo with a dominant impact on keratinocyte migration in vitro and reepithelialization in vivo.[207] This phenotype involved loss of HB-EGF production and EGF receptor activation in response to wounding and was rescued by exogenous HB-EGF.[207] Similarly, ERK and p38 are required for keratinocyte migration in vitro and epithelial outgrowth ex vivo.[205,210-213] In addition to wound healing, AP-1 activation occurs during other EMT events. Further evidence for the crucial role of AP-1 in epithelial sheet movements is based on studies of development and morphogenesis.[41] Another linkage between MAP kinase cascades and EMT-like events is based on the observation that Snail and Slug (see below) are both AP-1 regulated genes.[214] As discussed below, the Snail family of transcription factors is regulator of EMT during development and Slug expression is elevated at the margins of reepithelialization.[26]

Snail Family

The first member of the Snail family of transcription factors, Snail, was discovered twenty years ago in Drosophila. Since then, Snail homologs such as Slug have been identified in many animal groups, including diblastic organisms, nematodes, amphibians, fish and mammals.[215-217]

Analysis of forkhead and snail expression reveals epithelial-mesenchymal transitions during embryonic and larval development of *Nematostella vectensis*.[218] During embryonic development, Snail family transcription factors are involved in mesoderm differentiation, neural crest formation, specification of left-right asymmetry, neural development, apoptosis, cell division, and endoreduplication.[219] In mammals, Snail and Slug genes are expressed in the primitive streak, mesoderm, decondensing somites, neural plate, neural crest cells and mesenchymal tissues.[215,220] In the chicken, Slug seems to be required for gastrulation and neural crest migration.[217] However, in mice, it is Snail that is essential for early embryogenesis. Snail null mutants die at gastrulation,[221] whereas Slug null mice are viable and fertile despite a variety of abnormalities.[222] Because Snail and Slug play important roles in most developmental EMT processes, they are good candidates for master genes controlling EMT. Surprisingly, however, Snail transcription factors have been recently shown to be involved in a variety of functions unrelated to EMT including cell cycle regulation and anti-apoptotic activity in *C. elegans* and mammals.[223-226]

Several lines of evidence point to specific involvement of Slug in wound healing. Slug is expressed in keratinocytes at the wound edge in vivo and its expression coincides with keratinocyte activation and migration. In addition, stable or transient overexpression of Slug in keratinocytes causes desmosomal disruption and increased migratory activity but does not enhance mitogenesis. Finally, Slug null mutants exhibit deficient reepithelialization in ex vivo assays.[26]

In contrast to Slug, the other well studied member of the family, Snail was expressed at very low levels in immortalized and transformed keratinocyte cell lines. Moreover, Snail expression was not modulated at wound margins[26] making it a lesser candidate for involvement within wound healing process.

Snail factors act primarily as transcriptional repressors. They contain four to six C2H2 zinc-fingers that specifically bind the E-box sequence CAGGTG. This motif is also recognized by basic helix-loop-helix transcription factors and therefore is not specific to the Snail family. In mammalian cells, repression by Snail requires both the zinc finger region of the protein and a SNAG domain located in the N-terminal region. Although the SNAG domain is found in many members of the Snail family,[219] some members of the family, such as Drosophila Snail, lack this domain. In such cases, transcriptional repression results from interaction of Snail with the corepressor CtBP.[227] Recently, GSK-3β was found to mediate snail phosphorylation.[228] Snail phosphorylation appears to be critical for the stability of the protein[228] and for its subcellular location,[229] thus regulating activity of the transcriptional repressor Snail.

Only a few direct transcriptional targets of the Snail transcription factors have been identified. They include the transmembrane adhesion molecule E-cadherin,[230-236] aromatase,[237] aggrecan, collagen II,[238] Na/K-ATPase β1-subunit,[239] claudin and occludin.[240] Clearly, expression levels of target genes depend on other factors since several primary and transformed cell lines or population can simultaneously express Slug and E-cadherin at significant levels (Savagner et al, unpublished observations). This observation is in agreement with studies showing that, E-cadherin levels do not appear to be modulated by Slug during wound healing. Slug and Snail transcriptional repression may be mediated by histone deacetylases (HDACs).[241] Peinado and colleagues[242] found that E-caherin downregulation by Snail is due to chromatin modification. The transcription factor uses its SNAG domain to form a molecular complex with the effectors HDAC1, HDAC2 and mSin3A.

ETS Family

Several studies suggest that the Ets transcription factor family is involved in cell motility and invasion. Ets1 is involved in early organogenesis in the chick embryo in vivo and during the EMT-like scattering of MDCK epithelial cells in vitro.[243] A role for Ets1 in cell migration during mammary morphogenesis is suggested by studies using mutant forms in two or three dimensional culture systems.[244,245] In addition, there is a positive correlation between c-ets-1 expression and the invasive phenotype in carcinoma cells.[246]

At the present time, there is no direct evidence that Ets factors play a role in cutaneous wound healing. However, Ets factors are important in ECM remodeling,[247] and Ets targets genes such as tenascin-C[248] and collagen I are reexpressed during wound healing. Another possible role for Ets1 in wound healing is transcriptional control of uPA, a proteinase that, when activated, degrades extracellular matrix (ECM).[244,245] Moreover, binding sites for the ETS domain family of transcription factors are found in the promoters of several MMPs.[249,250] Finally, the localization of ESE-1 and ESE-2, two other members of the Ets family to the most differentiated layer of the epidermis indicates a role for the family in terminal differentiation of keratinocytes. ESE-1 and ESE-2 specifically control the transcription of the SPRR2A gene, a proline-rich protein linked with terminal differentiation of keratinocytes.[251,252]

Conclusion

Wound healing involves a complex and transient molecular reprogramming of keratinocytes involved in wound reepithelialization. This reprogramming is regulated by cytokines and growth factors that stimulate activation of keratinocytes. Keratinocyte activation results in phenotypic changes including enhanced migration, initiation of proteolytic cascades, profound alterations in the cytoskeleton and associated intercellular adhesion structures, reorganization of integrins on the keratinocyte surface, and substantial ECM remodeling. These changes recapitulate many aspects of developmental and pathologic EMT as described in other chapters of this book. However, keratinocytes at wound margins maintain some cell-cell cohesion, a situation not observed during typical developmental EMT. Like activated keratinocytes, carcinoma cells may also retain some intercellular junctions during EMT associated with tumor progression. Maintenance of a partially dissociated phenotype like that seen during wound healing may allow invading carcinoma cells to migrate as cords and clusters of cells, a behavior reminiscent of early organ morphogenesis. Because activated keratinocytes at wound margins and carcinoma cells both undergo partial phenotypic conversion to a more mesenchymal phenotype, wound healing is likely to be a relevant model of carcinoma progression. Understanding the molecular events that initiate and terminate keratinocyte activation during wound healing may provide important insights into mechanisms of tumor progression and ways to control the process.

Acknowledgments

Financial support is provided by the Fondation de France.

References

1. Moll I. Proliferative potential of different keratinocytes of plucked human hair follicles. J Invest Dermatol 1995; 105:14-21.
2. Christiano AM. Epithelial stem cells: Stepping out of their niche. Cell 2004; 118:530-2.
3. Green H. Terminal differentiation of cultured human epidermal cells. Cell 1977; 11:405-16.
4. Watt FM, Kubler MD, Hotchin NA et al. Regulation of keratinocyte terminal differentiation by integrin-extracellular matrix interactions. J Cell Sci 1993; 106(Pt 1):175-82.
5. Watt FM. Role of integrins in regulating epidermal adhesion, growth and differentiation. Embo J 2002; 21:3919-26.
6. Levy L, Broad S, Diekmann D et al. Beta1 integrins regulate keratinocyte adhesion and differentiation by distinct mechanisms. Mol Biol Cell 2000; 11:453-66.
7. Tu CL, Oda Y, Komuves L et al. The role of the calcium-sensing receptor in epidermal differentiation. Cell Calcium 2004; 35:265-73.
8. Bikle DD, Oda Y, Xie Z. Calcium and 1,25(OH)2D: Interacting drivers of epidermal differentiation. J Steroid Biochem Mol Biol 2004; 89-90:355-60.
9. Kira M, Kobayashi T, Yoshikawa K. Vitamin D and the skin. J Dermatol 2003; 30:429-37.
10. Martin P. Wound healing—aiming for perfect skin regeneration. Science 1997; 276:75-81.
11. Viziam CB, Matoltsy AG, Mescon H. Epithelialization of small wounds. J Invest Dermatol 1964; 43:499-507.
12. Krawczyk W. A pattern of epidermal cell migration during wound healing. J Cell Sci 1971; 49:247-263.
13. Miller SJ, Burke EM, Rader MD et al. Reepithelialization of porcine skin by the sweat apparatus. J Invest Dermatol 1998; 110:13-9.
14. Odland G, Ross R. Human wound repair. I. Epidermal regeneration. J Cell Biol 1968; 39:135-51.
15. Woodley DT et al. Reepithelialization. Human keratinocyte locomotion. Dermatol Clin 1993; 11:641-6.

16. Coulombe PA. Towards a molecular definition of keratinocyte activation after acute injury to stratified epithelia. Biochem Biophys Res Commun 1997; 236:231-8.
17. Grinnell F. Wound repair, keratinocyte activation and integrin modulation. J Cell Sci 1992; 101(Pt 1):1-5.
18. Block ER, Matela AR, SundarRaj N et al. Wounding induces motility in sheets of corneal epithelial cells through loss of spatial constraints: Role of heparin-binding epidermal growth factor-like growth factor signaling. J Biol Chem 2004; 279:24307-12.
19. Turchi L et al. Dynamic characterization of the molecular events during in vitro epidermal wound healing. J Invest Dermatol 2002; 119:56-63.
20. Getsios S, Huen AC, Green KJ. Working out the strength and flexibility of desmosomes. Nat Rev Mol Cell Biol 2004; 5:271-81.
21. Paladini RD, Takahashi K, Bravo NS et al. Onset of reepithelialization after skin injury correlates with a reorganization of keratin filaments in wound edge keratinocytes: Defining a potential role for keratin 16. J Cell Biol 1996; 132:381-97.
22. Krawczyk WS, Wilgram GF. Hemidesmosome and desmosome morphogenesis during epidermal wound healing. J Ultrastruct Res 1973; 45:93-101.
23. Sciubba JJ, Waterhouse JP, Meyer J. A fine structural comparison of the healing of incisional wounds of mucosa and skin. J Oral Pathol 1978; 7:214-27.
24. Stenn KS, Depalma L. The molecular and cellular biology of wound repair. In: Clark RAF HR, ed. New York: Plenum Press, 1988:321-335.
25. Boukamp P et al. Normal keratinization in a spontaneously immortalized aneuploid human keratinocyte cell line. J Cell Biol 1988; 106:761-71.
26. Savagner P et al. Involvement of the developmental transcription factor slug in reepithelialization by adult keratinocytes. J Cell Physiol 2005; In Press.
27. Fuchs E, Green H. Changes in keratin gene expression during terminal differentiation of the keratinocyte. Cell 1980; 19:1033-42.
28. Roop DR, Krieg TM, Mehrel T et al. Transcriptional control of high molecular weight keratin gene expression in multistage mouse skin carcinogenesis. Cancer Res 1988; 48:3245-52.
29. Nickoloff BJ, Turka LA. Keratinocytes: key immunocytes of the integument. Am J Pathol 1993; 143:325-31.
30. Weiss RA, Eichner R, Sun TT. Monoclonal antibody analysis of keratin expression in epidermal diseases: A 48- and 56-kdalton keratin as molecular markers for hyperproliferative keratinocytes. J Cell Biol 1984; 98:1397-406.
31. Jiang CK et al. Epidermal growth factor and transforming growth factor alpha specifically induce the activation- and hyperproliferation-associated keratins 6 and 16. Proc Natl Acad Sci USA 1993; 90:6786-90.
32. Freedberg IM, Tomic-Canic M, Komine M et al. Keratins and the keratinocyte activation cycle. J Invest Dermatol 2001; 116:633-40.
33. Gibbins JR et al. Midline fusion in the formation of the secondary palate anticipated by upregulation of keratin K5/6 and localized expression of vimentin mRNA in medial edge epithelium. Int J Dev Biol 1999; 43:237-44.
34. Kuwahara M, Hatoko M, Tada H et al. E-cadherin expression in wound healing of mouse skin. J Cutan Pathol 2001; 28:191-9.
35. Boyer B, Dufour S, Thiery JP. E-cadherin expression during the acidic FGF-induced dispersion of a rat bladder carcinoma cell line. Exp Cell Res 1992; 201:347-57.
36. Behrens J et al. The role of E-cadherin and scatter factor in tumor invasion and cell motility. Exs 1991; 59:109-26.
37. Birchmeier W. E-cadherin as a tumor (invasion) suppressor gene. Bioessays 1995; 17:97-9.
38. Ikeguchi M, Makino M, Kaibara N. Clinical significance of E-cadherin-catenin complex expression in metastatic foci of colorectal carcinoma. J Surg Oncol 2001; 77:201-7.
39. Graff JR, Gabrielson E, Fujii H et al. Methylation patterns of the E-cadherin 5' CpG island are unstable and reflect the dynamic, heterogeneous loss of E-cadherin expression during metastatic progression. J Biol Chem 2000; 275:2727-32.
40. Duband JL, Monier F, Delannet M et al. Epithelium-mesenchyme transition during neural crest development. Acta Anat (Basel) 1995; 154:63-78.
41. Martin P, Parkhurst SM. Parallels between tissue repair and embryo morphogenesis. Development 2004; 131:3021-34.
42. Friedl P, Wolf K. Tumour-cell invasion and migration: Diversity and escape mechanisms. Nat Rev Cancer 2003; 3:362-74.
43. Vasiliev JM. Cytoskeletal mechanisms responsible for invasive migration of neoplastic cells. Int J Dev Biol 2004; 48:425-39.

44. Liu Y. Epithelial to mesenchymal transition in renal fibrogenesis: Pathologic significance, molecular mechanism, and therapeutic intervention. J Am Soc Nephrol 2004; 15:1-12.
45. Bement WM, Forscher P, Mooseker MS. A novel cytoskeletal structure involved in purse string wound closure and cell polarity maintenance. J Cell Biol 1993; 121:565-78.
46. Omelchenko T, Vasiliev JM, Gelfand IM et al. Rho-dependent formation of epithelial "leader" cells during wound healing. Proc Natl Acad Sci USA 2003; 100:10788-93.
47. Ridley AJ, Allen WE, Peppelenbosch M et al. Rho family proteins and cell migration. Biochem Soc Symp 1999; 65:111-23.
48. Laplante AF, Germain L, Auger FA et al. Mechanisms of wound reepithelialization: Hints from a tissue-engineered reconstructed skin to long-standing questions. Faseb J 2001; 15:2377-89.
49. Revenu C, Athman R, Robine S et al. The coworkers of actin filaments: From cell structures to signals. Nat Rev Mol Cell Biol 2004; 5:635-46.
50. Fumoto K, Uchimura T, Iwasaki T et al. Phosphorylation of myosin II regulatory light chain is necessary for migration of HeLa cells but not for localization of myosin II at the leading edge. Biochem J 2003; 370:551-6.
51. Kamei T, Matozaki T, Takai Y. Mechanisms of cell adhesion and migration. Gan To Kagaku Ryoho 1999; 26:1359-66.
52. Kjoller L, Hall A. Signaling to Rho GTPases. Exp Cell Res 1999; 253:166-79.
53. Maruta H, Nheu TV, He H et al. Rho family-associated kinases PAK1 and rock. Prog Cell Cycle Res 2003; 5:203-10.
54. Frame MC, Brunton VG. Advances in Rho-dependent actin regulation and oncogenic transformation. Curr Opin Genet Dev 2002; 12:36-43.
55. Zagon IS, Sassani JW, Ruth TB et al. Cellular dynamics of corneal wound reepithelialization in the rat. III. Mitotic activity. Brain Res 2000; 882:169-79.
56. Chan KY, Patton DL, Cosgrove YT. Time-lapse videomicroscopic study of in vitro wound closure in rabbit corneal cells. Invest Ophthalmol Vis Sci 1989; 30:2488-98.
57. Mata J, Curado S, Ephrussi A et al. Tribbles coordinates mitosis and morphogenesis in Drosophila by regulating string/CDC25 proteolysis. Cell 2000; 101:511-22.
58. Grosshans J, Wieschaus E. A genetic link between morphogenesis and cell division during formation of the ventral furrow in Drosophila. Cell 2000; 101:523-31.
59. Thompson HW, Malter JS, Steinemann TL et al. Flow cytometry measurements of the DNA content of corneal epithelial cells during wound healing. Invest Ophthalmol Vis Sci 1991; 32:433-6.
60. Clark RA et al. Fibronectin and fibrin provide a provisional matrix for epidermal cell migration during wound reepithelialization. J Invest Dermatol 1982; 79:264-9.
61. Kubo M et al. Fibrinogen and fibrin are anti-adhesive for keratinocytes: A mechanism for fibrin eschar slough during wound repair. J Invest Dermatol 2001; 117:1369-81.
62. Grinnell F, Billingham RE, Burgess L. Distribution of fibronectin during wound healing in vivo. J Invest Dermatol 1981; 76:181-9.
63. Brown LF et al. Macrophages and fibroblasts express embryonic fibronectins during cutaneous wound healing. Am J Pathol 1993; 142:793-801.
64. Mackie EJ, Halfter W, Liverani D. Induction of tenascin in healing wounds. J Cell Biol 1988; 107:2757-67.
65. Latijnhouwers MA et al. Tenascin expression during wound healing in human skin. J Pathol 1996; 178:30-5.
66. Clark RA. Basics of cutaneous wound repair. J Dermatol Surg Oncol 1993; 19:693-706.
67. Vogel WF. Collagen-receptor signaling in health and disease. Eur J Dermatol 2001; 11:506-14.
68. Timpl R, Brown JC. The laminins. Matrix Biol 1994; 14:275-81.
69. Nguyen BP, Ryan MC, Gil SG et al. Deposition of laminin 5 in epidermal wounds regulates integrin signaling and adhesion. Curr Opin Cell Biol 2000; 12:554-62.
70. Kirtschig G, Marinkovich MP, Burgeson RE et al. Anti-basement membrane autoantibodies in patients with anti-epiligrin cicatricial pemphigoid bind the alpha subunit of laminin 5. J Invest Dermatol 1995; 105:543-8.
71. Rousselle P, Lunstrum GP, Keene DR et al. Kalinin: An epithelium-specific basement membrane adhesion molecule that is a component of anchoring filaments. J Cell Biol 1991; 114:567-76.
72. Rousselle P et al. Laminin 5 binds the NC-1 domain of type VII collagen. J Cell Biol 1997; 138:719-28.
73. Chen M et al. Interactions of the amino-terminal noncollagenous (NC1) domain of type VII collagen with extracellular matrix components. A potential role in epidermal-dermal adherence in human skin. J Biol Chem 1997; 272:14516-22.
74. Champliaud MF et al. Human amnion contains a novel laminin variant, laminin 7, which like laminin 6, covalently associates with laminin 5 to promote stable epithelial-stromal attachment. J Cell Biol 1996; 132:1189-98.

75. Uitto J, Pulkkinen L, Christiano AM. Molecular basis of the dystrophic and junctional forms of epidermolysis bullosa: Mutations in the type VII collagen and kalinin (laminin 5) genes. J Invest Dermatol 1994; 103:39S-46S.
76. Larjava H, Salo T, Haapasalmi K et al. Expression of integrins and basement membrane components by wound keratinocytes. J Clin Invest 1993; 92:1425-35.
77. Ryan MC, Tizard R, VanDevanter DR et al. Cloning of the LamA3 gene encoding the alpha 3 chain of the adhesive ligand epiligrin. Expression in wound repair. J Biol Chem 1994; 269:22779-87.
78. Frank DE, Carter WG. Laminin 5 deposition regulates keratinocyte polarization and persistent migration. J Cell Sci 2004; 117:1351-63.
79. Natarajan E et al. Coexpression of p16(INK4A) and laminin 5 gamma2 by microinvasive and superficial squamous cell carcinomas in vivo and by migrating wound and senescent keratinocytes in culture. Am J Pathol 2003; 163:477-91.
80. Decline F et al. Keratinocyte motility induced by TGF-beta1 is accompanied by dramatic changes in cellular interactions with laminin 5. Cell Motil Cytoskeleton 2003; 54:64-80.
81. Nguyen BP, Ren XD, Schwartz MA et al. Ligation of integrin alpha 3beta 1 by laminin 5 at the wound edge activates Rho-dependent adhesion of leading keratinocytes on collagen. J Biol Chem 2001; 276:43860-70.
82. Pyke C et al. Laminin-5 is a marker of invading cancer cells in some human carcinomas and is coexpressed with the receptor for urokinase plasminogen activator in budding cancer cells in colon adenocarcinomas. Cancer Res 1995; 55:4132-9.
83. Kainulainen T et al. Altered distribution and synthesis of laminin-5 (kalinin) in oral lichen planus, epithelial dysplasias and squamous cell carcinomas. Br J Dermatol 1997; 136:331-6.
84. Lampe PD et al. Cellular interaction of integrin alpha3beta1 with laminin 5 promotes gap junctional communication. J Cell Biol 1998; 143:1735-47.
85. Kikkawa Y, Umeda M, Miyazaki K. Marked stimulation of cell adhesion and motility by ladsin, a laminin-like scatter factor. J Biochem (Tokyo) 1994; 116:862-9.
86. Hynes RO. Integrins: Versatility, modulation, and signaling in cell adhesion. Cell 1992; 69:11-25.
87. Hertle MD, Adams JC, Watt FM. Integrin expression during human epidermal development in vivo and in vitro. Development 1991; 112:193-206.
88. Clark EA, Brugge JS. Integrins and signal transduction pathways: The road taken. Science 1995; 268:233-9.
89. Heino J. The collagen receptor integrins have distinct ligand recognition and signaling functions. Matrix Biol 2000; 19:319-23.
90. Carter WG, Ryan MC, Gahr PJ. Epiligrin, a new cell adhesion ligand for integrin alpha 3 beta 1 in epithelial basement membranes. Cell 1991; 65:599-610.
91. Niessen CM et al. The alpha 6 beta 4 integrin is a receptor for both laminin and kalinin. Exp Cell Res 1994; 211:360-7.
92. Adams JC, Watt FM. Expression of beta 1, beta 3, beta 4, and beta 5 integrins by human epidermal keratinocytes and nondifferentiating keratinocytes. J Cell Biol 1991; 115:829-41.
93. Haapasalmi K et al. Keratinocytes in human wounds express alpha v beta 6 integrin. J Invest Dermatol 1996; 106:42-8.
94. Hakkinen L, Hildebrand HC, Berndt A et al. Immunolocalization of tenascin-C, alpha9 integrin subunit, and alphavbeta6 integrin during wound healing in human oral mucosa. J Histochem Cytochem 2000; 48:985-98.
95. Kurpakus MA, Quaranta V, Jones JC. Surface relocation of alpha 6 beta 4 integrins and assembly of hemidesmosomes in an in vitro model of wound healing. J Cell Biol 1991; 115:1737-50.
96. Mercurio AM, Rabinovitz I, Shaw LM. The alpha 6 beta 4 integrin and epithelial cell migration. Curr Opin Cell Biol 2001; 13:541-5.
97. Stepp MA, Spurr-Michaud S, Tisdale A et al. Alpha 6 beta 4 integrin heterodimer is a component of hemidesmosomes. Proc Natl Acad Sci USA 1990; 87:8970-4.
98. Geuijen CA, Sonnenberg A. Dynamics of the alpha6beta4 integrin in keratinocytes. Mol Biol Cell 2002; 13:3845-58.
99. Baker SE et al. Laminin-5 and hemidesmosomes: Role of the alpha 3 chain subunit in hemidesmosome stability and assembly. J Cell Sci 1996; 109(Pt 10):2509-20.
100. Vidal F et al. Integrin beta 4 mutations associated with junctional epidermolysis bullosa with pyloric atresia. Nat Genet 1995; 10:229-34.
101. Dowling J, Yu QC, Fuchs E. Beta4 integrin is required for hemidesmosome formation, cell adhesion and cell survival. J Cell Biol 1996; 134:559-72.
102. Georges-Labouesse E et al. Absence of integrin alpha 6 leads to epidermolysis bullosa and neonatal death in mice. Nat Genet 1996; 13:370-3.

103. van der Neut R, Krimpenfort P, Calafat J et al. Epithelial detachment due to absence of hemidesmosomes in integrin beta 4 null mice. Nat Genet 1996; 13:366-9.
104. Pulkkinen L, Kurtz K, Xu Y et al. Genomic organization of the integrin beta 4 gene (ITGB4): A homozygous splice-site mutation in a patient with junctional epidermolysis bullosa associated with pyloric atresia. Lab Invest 1997; 76:823-33.
105. Grose R et al. A crucial role of beta 1 integrins for keratinocyte migration in vitro and during cutaneous wound repair. Development 2002; 129:2303-15.
106. Decline F, Rousselle P. Keratinocyte migration requires alpha2beta1 integrin-mediated interaction with the laminin 5 gamma2 chain. J Cell Sci 2001; 114:811-23.
107. Pilcher BK et al. The activity of collagenase-1 is required for keratinocyte migration on a type I collagen matrix. J Cell Biol 1997; 137:1445-57.
108. Valles AM, Boyer B, Tarone G et al. Alpha 2 beta 1 integrin is required for the collagen and FGF-1 induced cell dispersion in a rat bladder carcinoma cell line. Cell Adhes Commun 1996; 4:187-99.
109. DiPersio CM, Hodivala-Dilke KM, Jaenisch R et al. Alpha3beta1 Integrin is required for normal development of the epidermal basement membrane. J Cell Biol 1997; 137:729-42.
110. Hodivala-Dilke KM, DiPersio CM, Kreidberg JA et al. Novel roles for alpha3beta1 integrin as a regulator of cytoskeletal assembly and as a trans-dominant inhibitor of integrin receptor function in mouse keratinocytes. J Cell Biol 1998; 142:1357-69.
111. Kreidberg JA et al. Alpha 3 beta 1 integrin has a crucial role in kidney and lung organogenesis. Development 1996; 122:3537-47.
112. deHart GW, Healy KE, Jones JC. The role of alpha3beta1 integrin in determining the supramolecular organization of laminin-5 in the extracellular matrix of keratinocytes. Exp Cell Res 2003; 283:67-79.
113. Sterk LM et al. The tetraspan molecule CD151, a novel constituent of hemidesmosomes, associates with the integrin alpha6beta4 and may regulate the spatial organization of hemidesmosomes. J Cell Biol 2000; 149:969-82.
114. DiPersio CM et al. Alpha3beta1 and alpha6beta4 integrin receptors for laminin-5 are not essential for epidermal morphogenesis and homeostasis during skin development. J Cell Sci 2000; 113(Pt 17):3051-62.
115. Xia Y, Gil SG, Carter WG. Anchorage mediated by integrin alpha6beta4 to laminin 5 (epiligrin) regulates tyrosine phosphorylation of a membrane-associated 80-kD protein. J Cell Biol 1996; 132:727-40.
116. Goldfinger LE et al. The alpha3 laminin subunit, alpha6beta4 and alpha3beta1 integrin coordinately regulate wound healing in cultured epithelial cells and in the skin. J Cell Sci 1999; 112(Pt 16):2615-29.
117. Carter WG, Kaur P, Gil SG et al. Distinct functions for integrins alpha 3 beta 1 in focal adhesions and alpha 6 beta 4/bullous pemphigoid antigen in a new stable anchoring contact (SAC) of keratinocytes: Relation to hemidesmosomes. J Cell Biol 1990; 111:3141-54.
118. Salo T, Makela M, Kylmaniemi M et al. Expression of matrix metalloproteinase-2 and -9 during early human wound healing. Lab Invest 1994; 70:176-82.
119. Natali PG, Nicotra MR, Bartolazzi A et al. Integrin expression in cutaneous malignant melanoma: Association of the alpha 3/beta 1 heterodimer with tumor progression. Int J Cancer 1993; 54:68-72.
120. Bartolazzi A et al. Transformation and tumor progression are frequently associated with expression of the alpha 3/beta 1 heterodimer in solid tumors. Int J Cancer 1994; 58:488-91.
121. Patriarca C et al. Integrin laminin receptor profile of pulmonary squamous cell and adenocarcinomas. Hum Pathol 1998; 29:1208-15.
122. Choma DP, Pumiglia K, DiPersio CM. Integrin alpha3beta1 directs the stabilization of a polarized lamellipodium in epithelial cells through activation of Rac1. J Cell Sci 2004; 117:3947-59.
123. Korang K, Christiano AM, Uitto J et al. Differential cytokine modulation of the genes LAMA3, LAMB3, and LAMC2, encoding the constitutive polypeptides, alpha 3, beta 3, and gamma 2, of human laminin 5 in epidermal keratinocytes. FEBS Lett 1995; 368:556-8.
124. Huang X, Griffiths M, Wu J et al. Normal development, wound healing, and adenovirus susceptibility in beta5-deficient mice. Mol Cell Biol 2000; 20:755-9.
125. Chen JD et al. Epidermal growth factor (EGF) promotes human keratinocyte locomotion on collagen by increasing the alpha 2 integrin subunit. Exp Cell Res 1993; 209:216-23.
126. Mainiero F, Pepe A, Yeon M et al. The intracellular functions of alpha6beta4 integrin are regulated by EGF. J Cell Biol 1996; 134:241-53.
127. Santoro MM, Gaudino G, Marchisio PC. The MSP receptor regulates alpha6beta4 and alpha3beta1 integrins via 14-3-3 proteins in keratinocyte migration. Dev Cell 2003; 5:257-71.

128. Mariotti A et al. EGF-R signaling through Fyn kinase disrupts the function of integrin alpha6beta4 at hemidesmosomes: Role in epithelial cell migration and carcinoma invasion. J Cell Biol 2001; 155:447-58.
129. Yamada KM, Even-Ram S. Integrin regulation of growth factor receptors. Nat Cell Biol 2002; 4:E75-6.
130. Werb Z. ECM and cell surface proteolysis: Regulating cellular ecology. Cell 1997; 91:439-42.
131. Ravanti L, Kahari VM. Matrix metalloproteinases in wound repair (review). Int J Mol Med 2000; 6:391-407.
132. Udayakumar TS et al. Membrane type-1-matrix metalloproteinase expressed by prostate carcinoma cells cleaves human laminin-5 beta3 chain and induces cell migration. Cancer Res 2003; 63:2292-9.
133. Goldfinger LE, Stack MS, Jones JC. Processing of laminin-5 and its functional consequences: Role of plasmin and tissue-type plasminogen activator. J Cell Biol 1998; 141:255-65.
134. Giannelli G, Falk-Marzillier J, Schiraldi O et al. Induction of cell migration by matrix metalloprotease-2 cleavage of laminin-5. Science 1997; 277:225-8.
135. Franz CM, Jones GE, Ridley AJ. Cell migration in development and disease. Dev Cell 2002; 2:153-8.
136. Twining SS. Regulation of proteolytic activity in tissues. Crit Rev Biochem Mol Biol 1994; 29:315-83.
137. Dasu MR, Barrow RE, Spies M et al. Matrix metalloproteinase expression in cytokine stimulated human dermal fibroblasts. Burns 2003; 29:527-31.
138. Pilcher BK et al. Keratinocyte collagenase-1 expression requires an epidermal growth factor receptor autocrine mechanism. J Biol Chem 1999; 274:10372-81.
139. Madlener M et al. Regulation of the expression of stromelysin-2 by growth factors in keratinocytes: Implications for normal and impaired wound healing. Biochem J 1996; 320(Pt 2):659-64.
140. Kalluri R, Neilson EG. Epithelial-mesenchymal transition and its implications for fibrosis. J Clin Invest 2003; 112:1776-84.
141. Saarialho-Kere UK, Chang ES, Welgus HG et al. Distinct localization of collagenase and tissue inhibitor of metalloproteinases expression in wound healing associated with ulcerative pyogenic granuloma. J Clin Invest 1992; 90:1952-7.
142. Saarialho-Kere UK et al. Cell-matrix interactions modulate interstitial collagenase expression by human keratinocytes actively involved in wound healing. J Clin Invest 1993; 92:2858-66.
143. Makela M et al. Matrix metalloproteinase 2 (gelatinase A) is related to migration of keratinocytes. Exp Cell Res 1999; 251:67-78.
144. Sarret Y, Woodley DT, Goldberg GS et al. Constitutive synthesis of a 92-kDa keratinocyte-derived type IV collagenase is enhanced by type I collagen and decreased by type IV collagen matrices. J Invest Dermatol 1992; 99:836-41.
145. Inkinen K et al. Expression and activity of matrix metalloproteinase-2 and -9 in experimental granulation tissue. Apmis 2000; 108:318-28.
146. Agren MS, Mirastschijski U, Karlsmark T et al. Topical synthetic inhibitor of matrix metalloproteinases delays epidermal regeneration of human wounds. Exp Dermatol 2001; 10:337-48.
147. Lund LR et al. Functional overlap between two classes of matrix-degrading proteases in wound healing. Embo J 1999; 18:4645-56.
148. Romer J et al. Impaired wound healing in mice with a disrupted plasminogen gene. Nat Med 1996; 2:287-92.
149. Lochter A et al. Matrix metalloproteinase stromelysin-1 triggers a cascade of molecular alterations that leads to stable epithelial-to-mesenchymal conversion and a premalignant phenotype in mammary epithelial cells. J Cell Biol 1997; 139:1861-72.
150. Zeisberg M et al. Renal fibrosis: Collagen composition and assembly regulates epithelial-mesenchymal transdifferentiation. Am J Pathol 2001; 159:1313-21.
151. Hubner G et al. Differential regulation of pro-inflammatory cytokines during wound healing in normal and glucocorticoid-treated mice. Cytokine 1996; 8:548-56.
152. Ansel J et al. Cytokine modulation of keratinocyte cytokines. J Invest Dermatol 1990; 94:101S-107S.
153. Leibovich SJ, Ross R. The role of the macrophage in wound repair. A study with hydrocortisone and antimacrophage serum. Am J Pathol 1975; 78:71-100.
154. Loscalzo J. The macrophage and fibrinolysis. Semin Thromb Hemost 1996; 22:503-6.
155. Futagami A, Ishizaki M, Fukuda Y et al. Wound healing involves induction of cyclooxygenase-2 expression in rat skin. Lab Invest 2002; 82:1503-13.
156. Gallucci RM et al. Impaired cutaneous wound healing in interleukin-6-deficient and immunosuppressed mice. Faseb J 2000; 14:2525-31.
157. Cha D, O'Brien P, O'Toole EA et al. Enhanced modulation of keratinocyte motility by transforming growth factor-alpha (TGF-alpha) relative to epidermal growth factor (EGF). J Invest Dermatol 1996; 106:590-7.

158. Hudson LG, McCawley LJ. Contributions of the epidermal growth factor receptor to keratinocyte motility. Microsc Res Tech 1998; 43:444-55.
159. Brown GL et al. Enhancement of wound healing by topical treatment with epidermal growth factor. N Engl J Med 1989; 321:76-9.
160. Chernoff EA, Robertson S. Epidermal growth factor and the onset of epithelial epidermal wound healing. Tissue Cell 1990; 22:123-35.
161. Andree C et al. In vivo transfer and expression of a human epidermal growth factor gene accelerates wound repair. Proc Natl Acad Sci USA 1994; 91:12188-92.
162. Guo L, Degenstein L, Fuchs E. Keratinocyte growth factor is required for hair development but not for wound healing. Genes Dev 1996; 10:165-75.
163. Kim I, Mogford JE, Chao JD et al. Wound epithelialization deficits in the transforming growth factor-alpha knockout mouse. Wound Repair Regen 2001; 9:386-90.
164. Tokumaru S et al. Ectodomain shedding of epidermal growth factor receptor ligands is required for keratinocyte migration in cutaneous wound healing. J Cell Biol 2000; 151:209-20.
165. Kansra S, Stoll SW, Johnson JL et al. Autocrine extracellular signal-regulated kinase (ERK) activation in normal human keratinocytes: Metalloproteinase-mediated release of amphiregulin triggers signaling from ErbB1 to ERK. Mol Biol Cell 2004; 15:4299-309.
166. Vassar R, Fuchs E. Transgenic mice provide new insights into the role of TGF-alpha during epidermal development and differentiation. Genes Dev 1991; 5:714-27.
167. Dominey AM et al. Targeted overexpression of transforming growth factor alpha in the epidermis of transgenic mice elicits hyperplasia, hyperkeratosis, and spontaneous, squamous papillomas. Cell Growth Differ 1993; 4:1071-82.
168. Barrandon Y, Green H. Cell migration is essential for sustained growth of keratinocyte colonies: The roles of transforming growth factor-alpha and epidermal growth factor. Cell 1987; 50:1131-7.
169. Repertinger SK et al. EGFR enhances early healing after cutaneous incisional wounding. J Invest Dermatol 2004; 123:982-9.
170. Yoshida S et al. Neutralization of hepatocyte growth factor leads to retarded cutaneous wound healing associated with decreased neovascularization and granulation tissue formation. J Invest Dermatol 2003; 120:335-43.
171. Bevan D, Gherardi E, Fan TP et al. Diverse and potent activities of HGF/SF in skin wound repair. J Pathol 2004; 203:831-8.
172. Yoshida S et al. Recombinant hepatocyte growth factor accelerates cutaneous wound healing in a diabetic mouse model. Growth Factors 2004; 22:111-9.
173. Cowin AJ et al. Hepatocyte growth factor and macrophage-stimulating protein are upregulated during excisional wound repair in rats. Cell Tissue Res 2001; 306:239-50.
174. Hamoen KE, Morgan JR. Transient hyperproliferation of a transgenic human epidermis expressing hepatocyte growth factor. Cell Transplant 2002; 11:385-95.
175. Valles AM et al. Acidic fibroblast growth factor is a modulator of epithelial plasticity in a rat bladder carcinoma cell line. Proc Natl Acad Sci USA 1990; 87:1124-8.
176. Araki M et al. Epithelia-mesenchyme interaction plays an essential role in transdifferentiation of retinal pigment epithelium of silver mutant quail: Localization of FGF and related molecules and aberrant migration pattern of neural crest cells during eye rudiment formation. Dev Biol 2002; 244:358-71.
177. Ornitz DM et al. Receptor specificity of the fibroblast growth factor family. J Biol Chem 1996; 271:15292-7.
178. Powers CJ, McLeskey SW, Wellstein A. Fibroblast growth factors, their receptors and signaling. Endocr Relat Cancer 2000; 7:165-97.
179. Ortega S, Ittmann M, Tsang SH et al. Neuronal defects and delayed wound healing in mice lacking fibroblast growth factor 2. Proc Natl Acad Sci USA 1998; 95:5672-7.
180. Miller DL, Ortega S, Bashayan O et al. Compensation by fibroblast growth factor 1 (FGF1) does not account for the mild phenotypic defects observed in FGF2 null mice. Mol Cell Biol 2000; 20:2260-8.
181. Werner S et al. Large induction of keratinocyte growth factor expression in the dermis during wound healing. Proc Natl Acad Sci USA 1992; 89:6896-900.
182. Werner S et al. The function of KGF in morphogenesis of epithelium and reepithelialization of wounds. Science 1994; 266:819-22.
183. Guo L, Yu QC, Fuchs E. Targeting expression of keratinocyte growth factor to keratinocytes elicits striking changes in epithelial differentiation in transgenic mice. Embo J 1993; 12:973-86.
184. Tsuboi R et al. Keratinocyte growth factor (FGF-7) stimulates migration and plasminogen activator activity of normal human keratinocytes. J Invest Dermatol 1993; 101:49-53.

185. Sato C, Tsuboi R, Shi CM et al. Comparative study of hepatocyte growth factor/scatter factor and keratinocyte growth factor effects on human keratinocytes. J Invest Dermatol 1995; 104:958-63.
186. Marti G et al. Electroporative transfection with KGF-1 DNA improves wound healing in a diabetic mouse model. Gene Ther 2004.
187. Karvinen S et al. Keratinocyte growth factor stimulates migration and hyaluronan synthesis in the epidermis by activation of keratinocyte hyaluronan synthases 2 and 3. J Biol Chem 2003; 278:49495-504.
188. Jeschke MG et al. Nonviral liposomal keratinocyte growth factor (KGF) cDNA gene transfer improves dermal and epidermal regeneration through stimulation of epithelial and mesenchymal factors. Gene Ther 2002; 9:1065-74.
189. Shin EY, Ma EK, Kim CK et al. Src/ERK but not phospholipase D is involved in keratinocyte growth factor-stimulated secretion of matrix metalloprotease-9 and urokinase-type plasminogen activator in SNU-16 human stomach cancer cell. J Cancer Res Clin Oncol 2002; 128:596-602.
190. Kobayashi T, Hattori S, Shinkai H. Matrix metalloproteinases-2 and -9 are secreted from human fibroblasts. Acta Derm Venereol 2003; 83:105-7.
191. Wang WM et al. Transforming growth factor-beta induces secretion of activated ADAMTS-2. A procollagen III N-proteinase. J Biol Chem 2003; 278:19549-57.
192. Wikner NE, Elder JT, Persichitte KA et al. Transforming growth factor-beta modulates plasminogen activator activity and plasminogen activator inhibitor type-1 expression in human keratinocytes in vitro. J Invest Dermatol 1990; 95:607-13.
193. Salo T, Lyons JG, Rahemtulla F et al.Transforming growth factor-beta 1 up-regulates type IV collagenase expression in cultured human keratinocytes. J Biol Chem 1991; 266:11436-41.
194. Sporn MB, Roberts AB, Wakefield LM et al. Transforming growth factor-beta: Biological function and chemical structure. Science 1986; 233:532-4.
195. Pierce GF et al. Platelet-derived growth factor and transforming growth factor-beta enhance tissue repair activities by unique mechanisms. J Cell Biol 1989; 109:429-40.
196. Mustoe TA et al. Accelerated healing of incisional wounds in rats induced by transforming growth factor-beta. Science 1987; 237:1333-6.
197. Kretschmer A et al. Differential regulation of TGF-beta signaling through Smad2, Smad3 and Smad4. Oncogene 2003; 22:6748-63.
198. Amendt C, Mann A, Schirmacher P et al. Resistance of keratinocytes to TGFbeta-mediated growth restriction and apoptosis induction accelerates reepithelialization in skin wounds. J Cell Sci 2002; 115:2189-98.
199. Ashcroft GS et al. Mice lacking Smad3 show accelerated wound healing and an impaired local inflammatory response. Nat Cell Biol 1999; 1:260-6.
200. Ashcroft GS, Roberts AB. Loss of Smad3 modulates wound healing. Cytokine Growth Factor Rev 2000; 11:125-31.
201. Zambruno G et al. Transforming growth factor-beta 1 modulates beta 1 and beta 5 integrin receptors and induces the de novo expression of the alpha v beta 6 heterodimer in normal human keratinocytes: Implications for wound healing. J Cell Biol 1995; 129:853-65.
202. Gailit J, Welch MP, Clark RA. TGF-beta 1 stimulates expression of keratinocyte integrins during reepithelialization of cutaneous wounds. J Invest Dermatol 1994; 103:221-7.
203. Cavani A et al. Distinctive integrin expression in the newly forming epidermis during wound healing in humans. J Invest Dermatol 1993; 101:600-4.
204. Yates S, Rayner TE. Transcription factor activation in response to cutaneous injury: Role of AP-1 in reepithelialization. Wound Repair Regen 2002; 10:5-15.
205. Huang C, Jacobson K, Schaller MD. MAP kinases and cell migration. J Cell Sci 2004; 117:4619-28.
206. Rossi A, Jang SI, Ceci R et al. Effect of AP1 transcription factors on the regulation of transcription in normal human epidermal keratinocytes. J Invest Dermatol 1998; 110:34-40.
207. Li G et al. c-Jun is essential for organization of the epidermal leading edge. Dev Cell 2003; 4:865-77.
208. Eckert RL, Welter JF. Transcription factor regulation of epidermal keratinocyte gene expression. Mol Biol Rep 1996; 23:59-70.
209. Angel P, Szabowski A, Schorpp-Kistner M. Function and regulation of AP-1 subunits in skin physiology and pathology. Oncogene 2001; 20:2413-23.
210. Stoll SW, Kansra S, Elder JT. Keratinocyte outgrowth from human skin explant cultures is dependent upon p38 signaling. Wound Repair Regen 2003; 11:346-53.
211. Li W et al. The p38-MAPK/SAPK pathway is required for human keratinocyte migration on dermal collagen. J Invest Dermatol 2001; 117:1601-11.
212. Sharma GD, He J, Bazan HE. p38 and ERK1/2 coordinate cellular migration and proliferation in epithelial wound healing: Evidence of cross-talk activation between MAP kinase cascades. J Biol Chem 2003; 278:21989-97.

213. Matsubayashi Y, Ebisuya M, Honjoh S et al. ERK activation propagates in epithelial cell sheets and regulates their migration during wound healing. Curr Biol 2004; 14:731-5.
214. Barbera MJ et al. Regulation of Snail transcription during epithelial to mesenchymal transition of tumor cells. Oncogene 2004; 23:7345-54.
215. Smith DE, FdAF, Gridley T. Isolation of Sna, a mouse gene homologous to the Drosophila genes snail and escargot: Its expression pattern suggests multiple roles during postimplantation development. Development 1992; 116(4):1033-9. Erratum in: Development 1993; 117(3): P(precedi 1992).
216. Hemavathy K, Ashraf SI, Ip YT. Snail/slug family of repressors: Slowly going into the fast lane of development and cancer. Gene 2000; 257:1-12.
217. Nieto MA, Sargent MG, Wilkinson DG et al. Control of cell behavior during vertebrate development by Slug, a zinc finger gene. Science 1994; 264:835-839.
218. Fritzenwanker JH, Saina M, Technau U. Analysis of forkhead and snail expression reveals epithelial-mesenchymal transitions during embryonic and larval development of Nematostella vectensis. Dev Biol 2004; 275:389-402.
219. Nieto MA. The snail superfamily of zinc-finger transcription factors. Nat Rev Mol Cell Biol 2002; 3:155-166.
220. Savagner P, Karavanova I, Perantoni A et al. Slug mRNA is expressed by specific mesodermal derivatives during rodent organogenesis. Dev Dyn 1998; 213:182-7.
221. Carver EA, Jiang R, Lan Y et al. The mouse snail gene encodes a key regulator of the epithelial-mesenchymal transition. Mol Cell Biol 2001; 21:8184-8.
222. Jiang R, Lan Y, Norton CR et al. The Slug gene is not essential for mesoderm or neural crest development in mice. Dev Biol 1998; 198:277-285.
223. Inoue A et al. Slug, a highly conserved zinc finger transcriptional repressor, protects hematopoietic progenitor cells from radiation-induced apoptosis in vivo. Cancer Cell 2002; 2:279-88.
224. Vega S et al. Snail blocks the cell cycle and confers resistance to cell death. Genes Dev 2004; 18:1131-43.
225. Martinez-Alvarez C et al. Snail family members and cell survival in physiological and pathological cleft palates. Dev Biol 2004; 265:207-18.
226. Inukai T et al. SLUG, a ces-1-related zinc finger transcription factor gene with antiapoptotic activity, is a downstream target of the E2A-HLF oncoprotein. Mol Cell 1999; 4:343-52.
227. Nibu Y et al. dCtBP mediates transcriptional repression by Knirps, Kruppel and Snail in the Drosophila embryo. EMBO J 1998; 17:7009-20.
228. Zhou BP et al. Dual regulation of Snail by GSK-3beta-mediated phosphorylation in control of epithelial-mesenchymal transition. Nat Cell Biol 2004; 6:931-40.
229. Dominguez D et al. Phosphorylation regulates the subcellular location and activity of the snail transcriptional repressor. Mol Cell Biol 2003; 23:5078-89.
230. Cano A et al. The transcription factor snail controls epithelial-mesenchymal transitions by repressing E-cadherin expression. Nat Cell Biol 2000; 2:76-83.
231. Batlle E et al. The transcription factor snail is a repressor of E-cadherin gene expression in epithelial tumour cells. Nat Cell Biol 2000; 2:84-9.
232. Poser I et al. Loss of E-cadherin expression in melanoma cells involves up-regulation of the transcriptional repressor Snail. J Biol Chem 2001; 276:24661-6.
233. Yokoyama K et al. Reverse correlation of E-cadherin and snail expression in oral squamous cell carcinoma cells in vitro. Oral Oncol 2001; 37:65-71.
234. Jiao W, Miyazaki K, Kitajima Y. Inverse correlation between E-cadherin and Snail expression in hepatocellular carcinoma cell lines in vitro and in vivo. Br J Cancer 2002; 86:98-101.
235. Hajra KM, Chen DY, Fearon ER. The SLUG zinc-finger protein represses E-cadherin in breast cancer. Cancer Res 2002; 62:1613-8.
236. Bolos V et al. The transcription factor Slug represses E-cadherin expression and induces epithelial to mesenchymal transitions: A comparison with Snail and E47 repressors. J Cell Sci 2003; 116:499-511.
237. Chen S, Itoh T, Wu K et al. Transcriptional regulation of aromatase expression in human breast tissue. J Steroid Biochem Mol Biol 2002; 83:93-9.
238. Seki K et al. Mouse snail family transcription repressors regulate chondrocyte extracellular matrix, type II collagen and aggrecan. J Biol Chem 2003.
239. Espineda CE, Chang J, Twiss J et al. Repression of Na,K-ATPase {beta}1-subunit by the transcription factor Snail in carcinoma. Mol Biol Cell 2003.
240. Ikenouchi J, Matsuda M, Furuse M et al. Regulation of tight junctions during the epithelium-mesenchyme transition: Direct repression of the gene expression of claudins/occludin by Snail. J Cell Sci 2003; 116:1959-67.

241. Hemavathy K, Guru SC, Harris J. Human Slug is a repressor that localizes to sites of active transcription. Mol Cell Biol 2000; 20:5087-95.
242. Peinado H, Ballestar E, Esteller MA et al. Snail mediates E-cadherin repression by the recruitment of the Sin3A/histone deacetylase 1 (HDAC1)/HDAC2 complex. Mol Cell Biol 2004; 24:306-19.
243. Fafeur V et al. The ETS1 transcription factor is expressed during epithelial-mesenchymal transitions in the chick embryo and is activated in scatter factor-stimulated MDCK epithelial cells. Cell Growth Differ 1997; 8:655-65.
244. Delannoy-Courdent A et al. Expression of c-ets-1 and uPA genes is associated with mammary epithelial cell tubulogenesis or neoplastic scattering. Int J Dev Biol 1996; 40:1097-108.
245. Delannoy-Courdent A et al. The expression of an Ets1 transcription factor lacking its activation domain decreases uPA proteolytic activity and cell motility, and impairs normal tubulogenesis and cancerous scattering in mammary epithelial cells. J Cell Sci 1998; 111(Pt 11):1521-34.
246. Gilles C, Polette M, Birembaut P et al. Expression of c-ets-1 mRNA is associated with an invasive, EMT-derived phenotype in breast carcinoma cell lines. Clin Exp Metastasis 1997; 15:519-26.
247. Trojanowska M. Ets factors and regulation of the extracellular matrix. Oncogene 2000; 19:6464-71.
248. Shirasaki F, Makhluf HA, LeRoy C et al. Ets transcription factors cooperate with Sp1 to activate the human tenascin-C promoter. Oncogene 1999; 18:7755-64.
249. Westermarck J, Kahari VM. Regulation of matrix metalloproteinase expression in tumor invasion. Faseb J 1999; 13:781-92.
250. Singh S, Barrett J, Sakata K et al. ETS proteins and MMPs: Partners in invasion and metastasis. Curr Drug Targets 2002; 3:359-67.
251. Andreoli JM et al. The expression of a novel, epithelium-specific ets transcription factor is restricted to the most differentiated layers in the epidermis. Nucleic Acids Res 1997; 25:4287-95.
252. Oettgen P et al. Characterization of ESE-2, a novel ESE-1-related Ets transcription factor that is restricted to glandular epithelium and differentiated keratinocytes. J Biol Chem 1999; 274:29439-52.
253. McCarthy RA, Hay ED. Collagen I, laminin, and tenascin: Ultrastructure and correlation with avian neural crest formation. Int J Dev Biol 1991; 35:437-52.
254. Loeber CP, Runyan RB. A comparison of fibronectin, laminin, and galactosyltransferase adhesion mechanisms during embryonic cardiac mesenchymal cell migration in vitro. Dev Biol 1990; 140:401-12.
255. Ffrench-Constant C, Van de Water L, Dvorak HF et al. Reappearance of an embryonic pattern of fibronectin splicing during wound healing in the adult rat. J Cell Biol 1989; 109:903-14.
256. Erickson HP, Bourdon MA. Tenascin: An extracellular matrix protein prominent in specialized embryonic tissues and tumors. Annu Rev Cell Biol 1989; 5:71-92.
257. Zuk A, Hay ED. Expression of beta 1 integrins changes during transformation of avian lens epithelium to mesenchyme in collagen gels. Dev Dyn 1994; 201:378-93.
258. Giambernardi TA et al. Neutrophil collagenase (MMP-8) is expressed during early development in neural crest cells as well as in adult melanoma cells. Matrix Biol 2001; 20:577-87.
259. Oikarinen A, Kylmaniemi M, Autio-Harmainen H et al. Demonstration of 72-kDa and 92-kDa forms of type IV collagenase in human skin: Variable expression in various blistering diseases, induction during reepithelialization, and decrease by topical glucocorticoids. J Invest Dermatol 1993; 101:205-10.
260. Citterio HL, Gaillard DA. Expression of transforming growth factor alpha (TGF alpha), epidermal growth factor receptor (EGF-R) and cell proliferation during human palatogenesis: An immunohistochemical study. Int J Dev Biol 1994; 38:499-505.
261. Ishibashi T, Shinogami M, Ishimoto SI et al. Induction of KGF, basic FGF, and TGFalpha mRNA expression during healing of experimental TM perforations. Acta Otolaryngol 1998; 118:701-4.
262. Potts JD, Dagle JM, Walder JA et al. Epithelial-mesenchymal transformation of embryonic cardiac endothelial cells is inhibited by a modified antisense oligodeoxynucleotide to transforming growth factor beta 3. Proc Natl Acad Sci USA 1991; 88:1516-20.
263. Proetzel G et al. Transforming growth factor-beta 3 is required for secondary palate fusion. Nat Genet 1995; 11:409-14.
264. Frank S, Madlener M, Werner S. Transforming growth factors beta1, beta2, and beta3 and their receptors are differentially regulated during normal and impaired wound healing. J Biol Chem 1996; 271:10188-93.
265. Gammill LS, Bronner-Fraser M. Neural crest specification: Migrating into genomics. Nat Rev Neurosci 2003; 4:795-805.
266. Maroulakou IG, Papas TS, Green JE. Differential expression of ets-1 and ets-2 proto-oncogenes during murine embryogenesis. Oncogene 1994; 9:1551-65.

Chapter 9

Epithelial-Mesenchymal Transitions in Human Cancer

Veerle L. Van Marck and Marc E. Bracke

Abstract

Epithelial-mesenchymal transition (EMT) is a type of epithelial plasticity that is characterized by long-lasting morphological and molecular changes in epithelial cells as a result of transdifferentiation towards a mesenchymal cell type. To detect possible phenotypic transitions in human cancer, surgical pathology is a useful medical discipline, examining surgical or biopsy material at the microscopic and ultrastructural level. The expression in a particular tumor of epithelial and mesenchymal markers is evaluated by means of immunohistochemistry or in situ hybridization, and this could, besides directing to a correct diagnosis, substantiate a possible transdifferentiation. Whereas EMT occurs in several stages of embryonic development and can be readily induced in (cancer) cell lines in vitro, in human cancer the phenomenon is rarely encountered. Carcinosarcoma is the tumor best studied, in which monoclonality of both epithelial and mesenchymal cell components strongly favors an EMT. A challenging hypothesis considers EMT as a more general event, providing an additional survival advantage in all types of carcinoma. By means of EMT the epithelial tumor cells would transdifferentiate into myofibroblasts that lose their malignant phenotype but constitute the desmoplastic stroma which is essential for tumor growth, invasion and metastasis.

Introduction

Studying cancer cells in vitro or human tumors in the surgical pathology practice one can be confronted with morphological changes or an atypical appearance of cancer cells, which can be attributed to an epithelial-mesenchymal transition (EMT). In a broad sense, EMT encompasses all changes in cell morphology from epithelioid to mesenchymal/fibroblastoid/spindle-shaped. A more strict definition, that we will support, is formulated by Janda et al.[1] It considers EMT as a type of epithelial plasticity leading to morphological changes accompanied by 'drastic molecular changes' that 'persist'. Merely epithelial towards spindle cell transition is termed 'scattering'. 'Transdifferentiation' is a term used in the context of EMT to refer to the irreversible change in differentiation of one cell type to another.[2] In their clinical pratice pathologists use, rather than EMT, the terms 'sarcomatous/sarcomatoid dedifferentiation' ('resembling a sarcoma', a malignant mesenchymal tumor) or 'anaplasia' (literally meaning 'to form backward').[3] Instead of 'transdifferentiation' the term 'metaplasia' is generally used.[4] This implies that the distinction between both is not made in pathology practice, namely that 'transdifferentiation' is restricted to 'differentiated' cells, and 'metaplasia' includes both transdifferentiation and switches in stem cells.[2,5]

In this chapter we briefly review phenotypic cell transitions during embryonic development and the molecular basis of EMT in tumor biology. We then focus on the different features of

Rise and Fall of Epithelial Phenotype: Concepts of Epithelial-Mesenchymal Transition, edited by Pierre Savagner.

EMT in experimental and surgical pathology, with emphasis on the significance of differentiation markers. Examples of tumor types in which EMT presumably occurs are given and we conclude with recent insights into stromal-tumor interactions which might widen up the concept of EMT in the future.

The Molecular Basis of EMT

From a developmental point of view, EMT and other phenotypic transitions are important processes that ultimately lead to a complex organism with its different organs and tissue types. The precisely regulated switches between epithelial and mesenchymal differentiation programs serve the creation of specific migratory cells and the formation of the mesodermal cell layer and its differentiation into specific tissue types such as muscle, skin and kidney[6] (see also Chapters 1-4). In human embryo's after the second week of development, a mesodermal germ layer is formed by invagination and migration of altered epiblast cells at the primitive streak. Later on, several transitions occur in these mesodermal cells. They transiently retransform into epithelial structures, the somites, out of which migrating mesenchymal cell populations are formed again: the derma-, sclero- and myotome. Neural crest cells originate from the ectoderm and give rise to, amongst others, melanocytes, which spread over the whole body surface (see Chapter 3, Newgreen and McKeown). Other examples of EMT are the contribution of the epicardial cells to the formation of the coronary wall[7,8] (see also Chapter 4, Runyan et al) and the cytotrophoblast cells at the implantation site of the placenta which become spindle-shaped and get the capacity to infiltrate.[9] The best-studied example of the opposite, the mesenchymal-to-epithelial transition (MET), is the formation of the nephrons, the functional units in the kidney, from the metanephric mesenchyme.[10,11] Similarly, the cells arranged in an epithelium that secrete the dentin layer of a tooth, the odontoblasts, originate from neural crest-derived mesenchymal cells.[12] Together, these examples show that increased motility and changes in differentiation are two important consequences of EMT and that both EMT and MET may occur transiently.

Phenotypic plasticity has been observed in different cell lines in vitro. Some cell lines transdifferentiate simply through passaging. The chemically induced rat mammary carcinoma cells RANI for instance, undergo a polygonal-to-fusiform transition, accompanied by loss of some epithelial (keratin) and gain of some mesenchymal markers (Thy-1, vimentin, fibronectin).[13] Ovarian surface epithelial cells produce collagen type I and II when cultured in vitro and progressively lose cytokeratin expression with passaging.[14] The formation of networks and channels, together with the expression of endothelial markers is described in ovarian carcinoma cells grown in Matrigel or collagen type I, reminiscent of the 'vasculogenic mimicry' in melanomas.[15,16] When explanted in vitro, retinal pigment epithelium cells become fibroblastic.[17,18] In definitive embryonic avian lens epithelium a similar phenotypic conversion is observed, triggered by a collagen type I matrix through de novo expression of integrin-β1.[19] There are also reports on transdifferentiation from pancreas to liver cells,[20] from myoblasts to lipocytes[21] and from mesothelial cells to myoblasts.[22]

The cell-cell adhesion molecule E-cadherin plays a central role in the epithelial differentiation of cells. Counteracting E-cadherin expression is in most cells enough to direct them towards a mesenchymal differentiation program,[23] whereas E-cadherin transfection in some fibroblastic cells makes them epithelioid.[24,25] Much literature supports the idea that, besides promoting differentiation, E-cadherin also acts as an invasion suppressor molecule (see Chapter 11, Berx and Van Roy). Thus, from a clinical point of view, the significance of an E–cadherin downregulation and eventually EMT in tumors, is the associated increase in invasiveness and metastatic potential and the concomitantly decreased patient survival. Alterations in E-cadherin expression or functioning, either observed in tumors or induced in cell lines in vitro, occur at different levels (reviewed by Van Aken et al[26]). We will focus here on those modulations that, because of their persistent nature, could fit in the EMT concept. They are at the genomic and transcriptional level. First, the E-cadherin gene *CDH1* can be inactivated by mutations as has been described in lobular breast carcinoma,[27,28] diffuse gastric carcinoma,[29,30] endometrial,[31] and thyroid carcinoma.[32] Some of these are germline mutations. Second, loss of heterozygosity could be an alternative way to lose *CDH1* transcription.[27,28] A third level of regulation that is important in EMT, is the promoter level. Hypermethylation of the E-cadherin promoter is a well-described mechanism in tumors.[33-36] Time- and tissue-specific expression of different sets of transcriptional activators and repressors mediates the above mentioned phenotypic changes during embryogenesis. The Snail family of transcriptional repressors,[37] including Snail and Slug, together with SIP-1, another zinc-finger protein, play a key role in EMT during embryonic development,[38] and are often re-expressed in tumor conditions. By acting on the E-boxes of the E-cadherin promoter,[39-42] they downregulate E-cadherin expression, imposing an invasive and mesenchymal phenotype to the cells[43] (see also Chapter 11, Berx and Van Roy). Studies on tumor material demonstrated that Snail expression in invasive ductal breast carcinomas is inversely related to grade and correlates with lymph node metastasis,[44] whereas a role for SIP-1 has been shown in E-cadherin downregulation in the intestinal type of gastric carcinoma.[45] On the other hand, the transcription factor WT-1 (Wilms' tumor-1) which play a role in MET in the embryonic kidney,[10] induces epithelial differentiation and E-cadherin expression.[46,47]

An important issue regarding E-cadherin modulation is the counteracting role of another classical cadherin, the neural-type or N-cadherin. Aberrant de novo expression of N-cadherin has been noted in carcinomas from the breast,[48] prostate,[49] bladder[50] and the head and neck region,[51] where it parallels a downregulation of E-cadherin and a decrease in tumor differentiation. In breast carcinoma cell lines N-cadherin expression is associated with an increased invasion potential,[52-55] exerting a dominant influence on the cell's motility when co-expressed with E-cadherin.[53,54] An increased fibroblast growth factor 2 (FGF2) signaling through N-cadherin-mediated FGF receptor 1 (FGFR1) stabilization, results in an elevated matrix metalloproteinase-9 production, which could account for the invasive phenotype induced by N-cadherin.[55] Another proposed mechanism is the interaction of the tumor cells with N-cadherin-positive fibroblasts of the surrounding stroma.[52] Although the observed decrease in differentiation does not equal per se an EMT, further study of the putative active role of N-cadherin in EMT seems relevant since N-cadherin is mentioned in several studies as a marker coming up in the mesenchymal state.[1,56-58]

In their list of operational criteria of EMT, Janda et al[1] included the requirement for cooperation of transforming growth factor-β (TGF-β) and a hyperactive MAPkinase pathway. In Ha-Ras transformed mammary epithelial cells, they demonstrated that this pathway and not the PI3kinase pathway is responsible for the irreversible conversion.[1] An active integrin-β1 signaling through interaction with the extracellular matrix is required for the EMT induced by TGF-β/MAPK.[59] Others, however, found that the PI3K pathway is implicated in TGF-β-induced EMT.[60,61] According to the cell system and criteria applied, i.e., if reversibility is tolerated or not, other growth factors could be involved as well. The hepatocyte growth factor/c-met pathway,[62] but also the signaling pathways from other tyrosine kinase receptors, such as FGFR,[63] epidermal growth factor receptor (EGFR)[56,58] and insulin-like growth factor 1 receptor[64] have been studied in this context. Since discussing these would be beyond the scope of this chapter, we refer to recent reviews.[65-67]

Features of EMT in Pathology

To diagnose malignant tumors, surgical pathologists apply morphological criteria that are based on properties that these tumors share with their putative tissue of origin. Besides being classified, roughly spoken into an epithelial or mesenchymal type, the tumors are graded in a two- or three-tiered scheme, according to the extent of resemblance to their tissue of origin.[3] There is a high resemblance in so-called well-differentiated (grade I) tumors. As for well-differentiated carcinomas, which are malignant epithelial tumors, the tumor cells tend to be organized in epithelial structures (show tubular differentiation and an apical-basal polarity in the case of adenocarcinomas) and the nature of their cytoplasm refers to a presumed cell of origin (containing e.g., mucus, keratin, glycogen). In poorly differentiated /undifferentiated carcinomas, the tumor cells lose their epithelial properties. In some cases this is accompanied by the

Table 1. Features of EMT in vitro and in tumor pathology

In Vitro	In Vivo	
	Tumor Pathology	**Tools for Observation in Pathology**
Changes in morphology	Cell shape; organization cells in epithelial/ mesenchymal context; Differentiation	Microscopy: routine H&E staining IHC, ISH: differentiation markers, TEM: intercellular junctions, intermediate filaments
Synthesis of ECM	Increased extracellular matrix: collagenous or specialized (e.g., chondroid)	IHC: ECM typing, TEM
Degradation of ECM	Fragmentation of collagen fibers	TEM
Motility	Displaced cells: - Discontinued/absent basement membrane (invasion) - Ectopic presence of cells in tissue (invasion, metastasis)	IHC: - Laminin staining - Tumor-specific markers to visualize displaced cells

ECM= extracellular matrix; H&E= hematoxylin/eosin; IHC= immunohistochemistry; ISH= in situ hybridization; TEM= transmission electron microscopy

acquisition of mesenchymal characteristics in a morphological and molecular sense, thereby fulfilling the criteria of EMT. Malignant transformation also involves variation in size and shape, so-called pleomorphism of cells and their nuclei, hyperchromasia of nuclei due to hyperploidy, and an increased mitotic activity. These changes reflect the differentiation status of a tumor as well and thus are also taken into account for grading.

Table 1 summarizes the most important features of EMT in vitro, their reflection in tumor tissue and the tools that are applied to study them. In the surgical pathology practice, tumors are examined macroscopically ex vivo and after fixation at the microscopic, ultrastructural (by transmission electron microscopy) or molecular level (by immunohistochemistry/ in situ hybridization). As a result of the inherently 'static' nature of these practices, the observations made, i.e., in the case of EMT, can only be a reflection of the mechanisms described in vitro. Nevertheless, pathology can provide an important contribution to the study of EMT. By examining a particular tumor series or experimentally induced tumors in laboratory animals, observations in vitro can be extrapolated to the situation in vivo. Changes in morphology are readily appreciated microscopically at a routine hematoxylin/eosin (H&E) staining. Within a tumor exhibiting EMT, at least part of the tumor cells have lost their typical epithelial morphology. They lack epithelial organization in tubules, as for adenocarcinomas, or the typical organization of cells in squamous sheets, in the case of squamous cell carcinoma. Instead, the cells display either an undifferentiated mesenchymal phenotype, resembling spindle-shaped fibroblasts (hence 'fibroblastoid') and are then haphazardly oriented in a nondescript extracellular matrix, or they acquire characteristics of specialized mesenchymal tissue, like muscle, bone or cartilage. In the latter cases, cells tend to be organized in bundles or are surrounded by a specialized matrix. At the molecular level, tumor cells and stroma can be examined by immunohistochemistry or in situ hybridization for the presence/absence of proteins or their mRNA, that serve as markers of the epithelial or mesenchymal phenotype (see Table 2). In some cases

Table 2. Markers of the epithelial and mesenchymal phenotype

Type of Marker	Phenotype	
	Epithelial	**Mesenchymal**
Intermediate filaments	Cytokeratin[79]	Vimentin[78,84]
Cell-cell adhesion molecules	E-cadherin[26] desmoglein, desmocollin[121,240]	N-cadherin[55,201] cadherin-11[113-115]
Integrins	$\alpha 6\beta 4$, ($\alpha 2\beta 1$), ($\alpha 3\beta 1$)[241–243]	$\alpha 5\beta 1$, $\alpha v\beta 3$, ($\alpha 2\beta 1$) and ($\alpha 3\beta 1$)[19,123-125,242]
ECM molecules	Laminins, collagen type IV[244]	Collagen type I, III, V[212] fibronectin[245] tenascin-C[128,139]
Miscellaneous	Epithelial membrane antigen[142] E1a[143] matrilysin[246]	FSP1[145] stromelysin-1, -3, gelatinase A and MT1-MMP[95,148,151-153,247]

additional electron microscopic analysis is needed.[68-71] As is listed in Table 3, the examination focuses on the presence of cell junctions, the type of cytoplasmic intermediate filaments and the production of basement membrane or collagenous material. The latter feature brings us to the second aspect of EMT: cells that underwent EMT, synthesize a different set of extracellular matrix (ECM) molecules, such as different types of collagen, proteoglycans and hyaluronic acid. Although in general a routine staining suffices to appreciate the presence and extent of ECM in a tumor, immunohistochemistry is needed to identify the different types of matrix molecules. For instance, a chondroid matrix is rich in hyaluronic acid and shows immunospecificity for the matrix-associated proteins S100 and bone morphogenic protein.[72] Finally, an important feature of EMT is the increased cell motility. In a pathology perspective,

Table 3. Ultrastructural markers of the epithelial and mesenchymal phenotype

Structures	Phenotype	
	Epithelial	**Mesenchymal**
Intermediate filaments	Tonofilaments (keratin)	Non-keratinous filaments
Granules	Glycogen, mucin	
Organelles		Well-developed, dilated rough endoplasmic reticulum
Cell junctions	Tight junctions, desmosomes, adherens junctions, gap junctions	
Extracellular matrix (ECM)	Basement membrane ECM	Interstitial ECM: collagen fibres, mucin, calcification

this implies the abnormal presence of epithelium-derived cells in the connective tissue after crossing the basement membrane boundary (invasion), or in an organ at distance of the primary tumor mass (metastasis). Again, immunohistochemical markers can be of help. For instance, a laminin staining can show discontinuities of the basement membrane and metastatic tumor cells are better distinguished from their surrounding tissue when a tumor-specific panel of antibodies is applied. Carcinosarcomas, that very likely exemplify an EMT as we will discuss below, are in general highly infiltrative and metastasizing. The fact that these tumors usually behave similarly to poorly differentiated carcinomas, indicates once more that tumor grade rather than the EMT in se determines the tumor's potential to invade and metastasize and eventually the patient's survival. As extensively studied in breast carcinoma, some markers of mesenchymal differentiation such as vimentin, tenascin-C and stromelysin-3 may be predictive for an adverse outcome.[73-75]

Morphological Markers of Cell Differentiation

Table 2 gives an overview of the most common markers applied in the study of EMT and epithelial plasticity in general. Neither marker on itself has, when present in a cell, an absolute predictive value as to its reflection of an epithelial or mesenchymal phenotype.[76,77] Intermediate filaments, intracellular cytoskeleton proteins, have classically been used to distinguish between these phenotypes.[78] The cytokeratin subgroup is almost exclusively expressed in epithelia and consists of at least 20 different proteins varying in molecular weight (from 40 to 70kD) and showing some epithelium-, organ-, or tumor-specificity.[79-82] For routine immunohistochemistry purposes 'cocktails' of antibodies, such as AE1/AE3 covering most keratin subtypes, have been commercialized. On the other hand, CAM5.2 reactivity is restricted to 'simple' or ductal epithelia (as in adenocarcinomas).[83] Cytokeratins are relatively stable in vivo, as they remain expressed in most high-grade carcinomas and traces of it can even be detected in the sarcomatous component of a mixed tumor. Decreased differentiation however, is in some epithelial cell types like in breast epithelium, accompanied by the emergence of vimentin, a mesenchymal type of intermediate filaments.[84-89] In vitro studies indicated a link between vimentin and increased motility or invasion in e.g., breast,[90-93] prostate[94] and cervical[95,96] cancer cell lines. Sommers et al[97] suggested that vimentin is a marker but not an inducer of the fibroblastic phenotype in vitro, as transfection of MCF-7 mammary cancer cells did not alter their epithelial morphology, neither the expression of epithelial markers. In line with this is the study of Hendrix et al[92] showing that vimentin expression alone does not suffice for metastasis formation. Many reports discuss the prognostic significance of vimentin expression in carcinomas in vivo. In breast cancer, some groups found a correlation with a lower post-operative survival.[73,98-100] Other studies however, could not confirm this.[89,101,102] Remarkably, Thomas et al[103] suggested that a keratin/vimentin expression ratio is more predictive for a worse prognosis than vimentin expression alone; the tumors with a high keratin level in addition to vimentin were associated with the poorest survival. In cervical cancer, vimentin expression was noted in invasive carcinomas and in their lymph node metastases, but not in the intra-epithelial neoplasia precursor lesions.[104] That intermediate filament markers are in a sense 'promiscuous' is best illustrated by vimentin immunostaining. Vimentin/keratin co-expression is found in some carcinoma types without evidence of dedifferentiation or EMT, and even in normal epithelium. This is particularly true for epithelia that have evolved from mesenchymal precursors like the urogenital organs.[105,106] In a way, these epithelial cells seem to recall their mesenchymal roots. Ovarian surface epithelial cells for instance, consistently express vimentin.[14,107] Normal endometrial glands, like the ovary derived from the mesodermal celomic epithelium, also express vimentin, though mainly in their proliferative phase.[108] In series of renal cell carcinomas up to half of the tumors expressed vimentin.[109,110] Although a spindled morphology correlated with the vimentin expression, 32% of the well-differentiated carcinomas showed immunoreactivity as well.[109] Aberrant expression of keratins also occurs, e.g., in soft tissue tumors, though to a lesser extent than aberrant vimentin expression in carcinomas.[111,112]

Another class of differentially expressed markers is the cadherins. The switch from E- to N-cadherin has been discussed above. Another cadherin of interest is OB-cadherin or cadherin-11, belonging to the atypical class II cadherin family.[113] During embryonic development cadherin-11 is exclusively expressed in mesenchymal cells, in which it assists together with other cadherins in cell sorting.[114,115] In adult tissue, cadherin-11 is a quite specific marker for mesenchymal cells, with the exception of the endometrial epithelium and the trophoblast. Aberrant expression of cadherin-11 and its splice variant in breast carcinomas seems restricted to the more aggressive, invasive cell lines such as MDA-MB-231.[116] Transfection experiments with MCF-7[117] or BT-20[54] mammary carcinoma cells indicated an invasion promoter role, while in SKBR3 mammary carcinoma cells the induced epithelial morphology upon transfection would suggest a role in cell differentiation.[117] Of interest in the EMT context is the finding that decidualization of endometrial stromal cells coincides with upregulation of cadherin-11 expression.[118] Decidualization is the process of morphological changes in endometrial stromal cells towards an epithelioid cell type that occurs in normal women at the late secretory phase of the menstrual cycle, when progesteron levels are high. In the placenta, cadherin-11 is switched on in certain types of trophoblasts, that is, in the syncytiotrophoblasts (after cell fusion) and in the terminally differentiated cytotrophoblasts (after transdifferentiation towards stromal infiltrative cells).[119,120] The latter interact with the pregnancy-induced decidual stromal cells of the endometrium, possibly through cadherin-11.[119]

Changes in the expression level or pattern of desmosomal cadherins[121] and of other components of intercellular junctions, such as occludin[56,122] or ZO-1[60] are looked at to substantiate EMT-related loss of polarity. Desmosome dissociation in the NBT-II rat bladder carcinoma cell line is an initial step of the EMT induced by the zinc-finger protein Slug.[121]

An important role in EMT is played by integrins, as mentioned above. Switches have been described towards fibronectin- or collagen type I-interacting integrins.[123-125]

The type of extracellular proteins produced changes also during EMT. A decrease in the production of laminin and collagen type IV, components of the basement membrane, parallels an increase in collagen type I, III, and V, fibronectin and/or tenascin-C (TN-C). The latter molecule is a hexameric glycoprotein,[126-129] that is produced by (myo)fibroblasts at the stroma-epithelium interface.[130,131] During wound healing and in neoplasia an elevated TN-C expression is noted.[131-135] In these pathological conditions TN-C expression is also switched on in a small amount of the epithelial cells, particularly those bordering the stroma.[131-133,136-139] This expression pattern may relate to its known interference with (tumor) cell-matrix interactions. In a panel of mouse mammary epithelial cells TN-C expression was inversely correlated with the degree of epithelial differentiation, i.e., polarization, and could be induced by TGF-β.[140] An inverse correlation with survival was found in human breast tumors,[141] and TN-C expression in the tumor stroma at the invasive border was predictive for tumor recurrence, metastasis and decreased survival.[74]

The epithelial membrane antigen (EMA) is the mixture of antigenic glycoproteins from human mammary epithelial cells present in milk through reverse pinocytosis. Antibodies raised against these antigens are reactive in most epithelia and carcinomas and thus are routinely used in pathology together with keratin antibodies.[142]

The tumor suppressor gene adenovirus E1a induces an epithelial phenotype in mesenchymal tumor cell lines and fibroblasts, probably by globally reprogramming transcription.[143,144]

Fibroblast-specific protein-1 (FSP1) is a calcium-binding protein belonging to the S100 family, isolated by comparative transcript analysis from mouse renal interstitial fibroblasts and is characterized as highly specific for fibroblasts.[145] FSP1 is used as a marker for EMT in the mouse model of kidney fibrogenesis, which we will describe below.

Finally, 'stromal type' MMPs can be produced by tumor cells undergoing an EMT, i.e., stromelysin-1 (Str-1, MMP-3),[146,147] stromelysin-3 (Str-3, MMP-11),[148-150] gelatinase A (MMP-2),[95] and membrane type 1 (MT1)-MMP.[151] That these MMPs play an active role in EMT was demonstrated in the normal mouse mammary cell line SCp2, in which Str-1 expression was sufficient to trigger an EMT.[152] In transgenic mice, Str-1 promoted the development

Table 4. Examples of EMT in tumor pathology

Tumor Type	Localization	Tissue of Origin (Embryonic Source)	Ref.
Carcinosarcoma/ sarcomatoid carcinoma	Uterus, breast, lung, gastro-intestinal tract...	Epithelium anywhere in the body (endo-, meso-, ectoderm)	167
Malignant mesothelioma	Serosal surface of peritoneal/pleural cavity	Mesothelium (mesoderm)	176
Sex cord-stromal tumors	Ovary	Ovarian surface epithelium (mesoderm)	105
Ameloblastoma	Oral cavity, maxilla	Stellate reticulum cells are derived from ameloblastic epithelium (endoderm)	179, 180

of mammary tumors with a mesenchymal phenotype.[153] Str-1 expression in mouse skin carcinomas was observed in association with a squamous to spindle cell conversion.[154] As for Str-3, protein and mRNA were demonstrated in both the epithelial and mesenchymal compartment of metaplastic mammary carcinomas, which are examples of EMT,[148] but occasionally also in the tumor cells of common mammary ductal carcinomas.[155] The Str-3 level, either stromal[75,148] or epithelial,[149] is a prognostic parameter in breast cancer since a strong correlation has been shown with decreased postoperative (disease-free) survival.

We can conclude that assessing the expression of a panel of differentiation markers on tumor cells enables one in most cases to discriminate between the epithelial and mesenchymal phenotype. In the diagnostic pathology practice, mesenchymal differentiation of a tumor is rather documented by a loss of epithelial markers (keratins and EMA) or a gain of specific mesenchymal differentiation markers, e.g., α-smooth muscle actin (α-SMA), desmin, S100 and CD31 for smooth muscle, skeletal muscle, neuroid/chondroid and vascular differentiation respectively.

Examples of EMT

Carcinosarcoma/Sarcomatoid Carcinoma

First of all, it should be stressed that in the great majority of human tumors no EMT seems to occur. This is unlike the situation in vitro, in which phenotypic plasticity is common. One can argue that EMT is an artefact of in vitro culturing.[156] Alternatively, it may be that pathologists only detected the tip of the iceberg so far. The tumor stroma being an integral part of a malignant tumor could imply that virtually every carcinoma underwent an EMT while creating its own stroma. We will discuss this hypothesis further on. Yet, for the moment, the most convincing examples of EMT in tumor pathology are those neoplasms exhibiting both an epithelial and mesenchymal component (Table 4) (for a review see ref. 157). Together, they represent less than one percent of the total number of cancers.

Carcinosarcomas make up a rare group of malignant mixed tumors that can occur in the vicinity of all kinds of epithelium anywhere in the body.[158-164] The epithelial component is usually an adenocarcinoma. The mesenchymal component consists either of nondescript, undifferentiated spindle cells, or of pleomorphic cells featuring a particular line of differentiation such as rhabdoid (skeletal muscle),[165] osteoid (bone)[162,166] or chondroid[72] (cartilage) differentiation (Figs. 1, 2). The terminology of this group of tumors is diverse. When such a specific mesenchymal cell differentiation is present (in a large extent) the tumors are usually termed 'carcinosarcoma'. On the other hand, if transitions between the two cell populations are

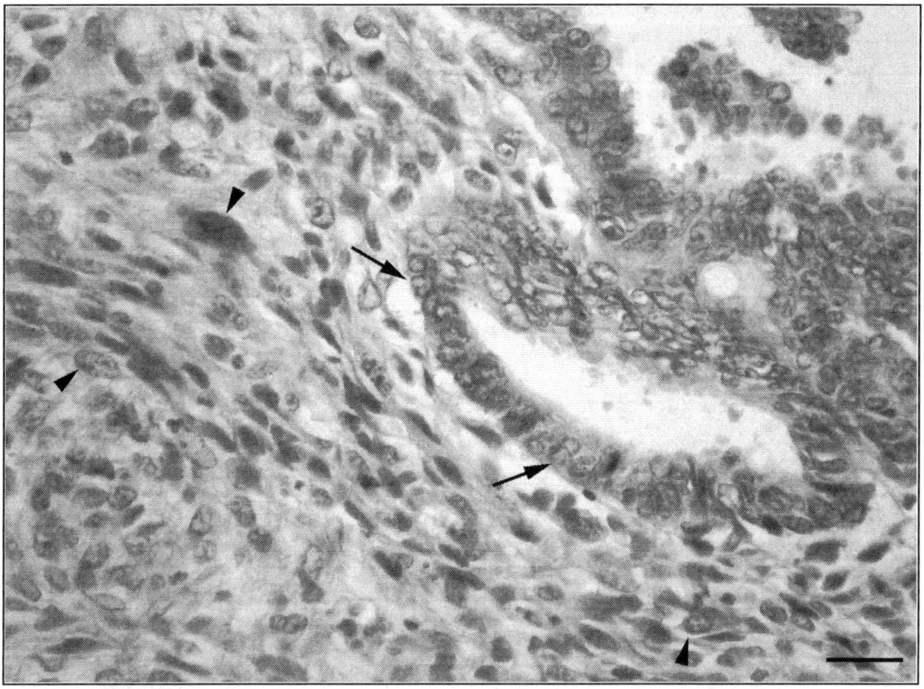

Figure 1. Malignant mixed Müllerian tumor (carcinosarcoma) of the ovary. The cells of the carcinomatous (arrows) and sarcomatous (arrowheads) component show malignant features: their nuclei are enlarged, pleomorphic and hyperchromatic. Formaldehyde fixed, paraffin embedded section. H&E staining. (Scale bar: 20 μm)

obvious, i.e., if there is a hybrid epithelial-mesenchymal cell type, the term 'sarcomatoid carcinoma' is often preferred. However, merely referring to the anatomical site, mixed tumors in the breast are usually termed 'spindle' or 'metaplastic' cell carcinoma, in the esophagus 'pseudosarcoma', and in the lining of the female genital tract 'malignant mixed Müllerian tumor'. Different opinions on the histogenesis of mixed tumors, which are nicely reviewed by Wick and Swanson,[167] to some extent account for this quite inconsistent classification. The debate focuses on two hypotheses. There is the convergence, polyclonal hypothesis in which mixed tumors consist of two separate, concurrent populations, hence the old term 'collision tumors'. Alternatively, the divergence, monoclonal hypothesis assumes that both epithelial and mesenchymal components are derived from a single uncommitted, totipotent cell which, after transformation, has proliferated towards separate differentiation lineages. Although rare cases of polyclonal mixed neoplasms might exist, most evidence in literature is favoring the divergence hypothesis.

First, at the light microscopic level, the epithelial and mesenchymal components are often not well demarcated. The tumor mass rather consists of a mixture of cells with smooth transitions in between the different cell types. Second, spindle cells frequently exhibit (at least minor) epithelial features, as is shown by immunohistochemical positivity for keratin and EMA, or by electron microscopy. Also, in some reports on the (lymph node) metastases of carcinosarcomas, the epithelial component predominates or is the only component,[168-170] which would not be expected in a true polyclonal neoplasm. In line with this is the fact that the clinical prognosis of mixed tumors in general equals that of poorly differentiated epithelial neoplasms.[170] Strong

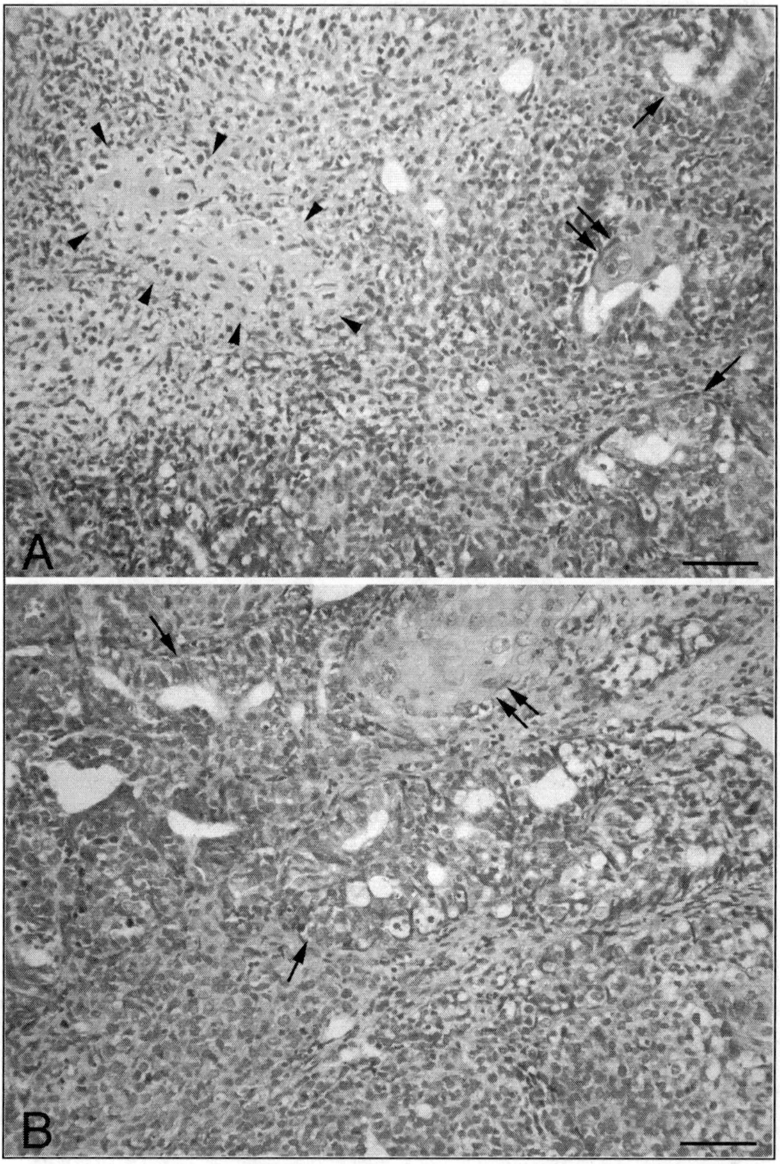

Figure 2. Carcinosarcoma of the bladder. A) area in which the mesenchymal (sarcomatous) component predominates. It is composed of disorganized spindle cells surrounded by a nondescript matrix, with an area of chondroid differentiation (arrowheads). Adjacent, the epithelial (carcinoma) component consists of tubules (single arrows) and squamous sheets (double arrows). B) area of carcinoma predominance. Formaldehyde fixed, paraffin embedded section. H&E staining. (Scale bar: 50 μm)

evidence of monoclonality is provided by the analysis of commonly mutated p53 exons,[171-173] allelic polymorphisms,[174] loss of heterozygosity[173] or the level of homology in chromosomal aberrations by comparative genomic hybridization.[175] These studies demonstrated identical genotypic changes in both components.

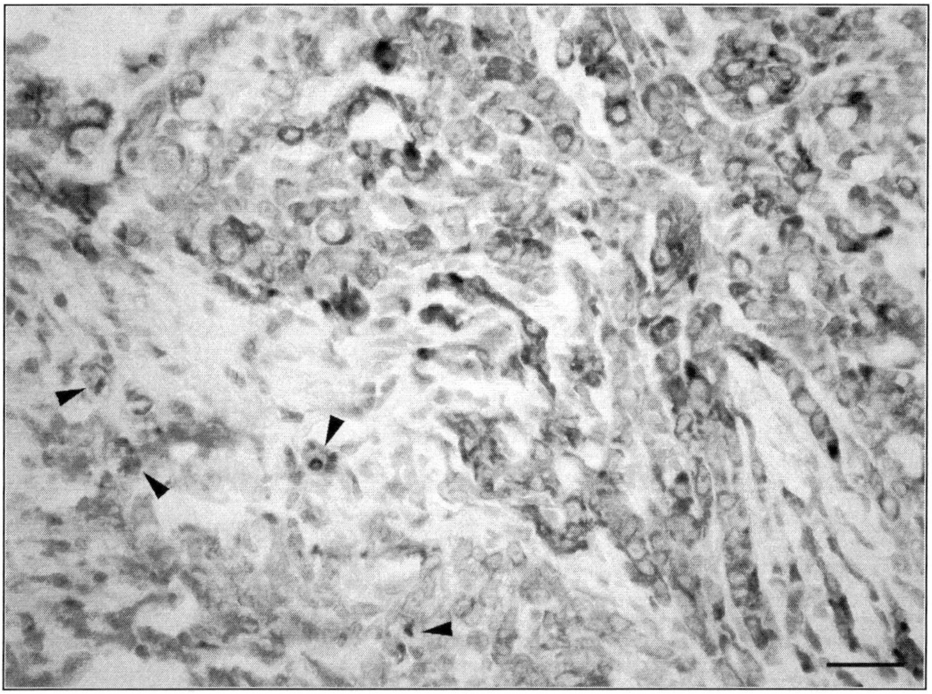

Figure 3. Biphasic malignant mesothelioma. Both the epithelial (upper right field) and mesenchymal (lower left field, arrowheads) component show intracytoplasmic immunoreactivity with the pan-cytokeratin AE1/AE3 antibody. Hematoxylin counterstaining. (Scale bar: 20 μm)

EMT in Other Human Tumors

Table 4 lists other examples of EMT in tumors. The most prevalent tumor is the malignant mesothelioma, a locally aggressive tumor that is related to asbestos exposure. It originates from the mesothelium which lines the pleural and peritoneal cavities. The tumor cells display either an epithelial or a mesenchymal phenotype and often, in so-called 'biphasic mesothelioma', both cell types are present (Fig. 3), probably representing cell plasticity.[176-178]

Sex cord-stromal tumors represent a group of rare ovarian stromal tumors probably derived from the ovarian surface epithelium.[105] This is a monolayered cuboidal epithelium, showing epithelial characteristics such as cytokeratin expression, desmosomes, and the production of the basement membrane molecules collagen type IV and laminin. However, concurrent vimentin and N-cadherin rather than E-cadherin expression suggests a dual epithelial-mesenchymal differentiation capacity,[105] which is also reflected by the plasticity of the ovarian surface epithelial cells in vitro, as mentioned above. This dual differentiation capacity might explain why as diverse tumor types as ovarian surface carcinomas and sex cord-stromal tumors could originate from them.

The last example of EMT illustrates that tumors might recapitulate embryonic development by showing differentiation along the same lineages. In the ameloblastoma, a tumor usually arising from remnants of the dental lamina or the enamel organ in the jaw, the so-called 'stellate reticulum' constitutes the tumor stroma adjacent to the ameloblastic tumor epithelium. These stromal cells share epithelial features with the ameloblastoma cells, as has been demonstrated ultrastructurally,[179] immunohistochemically and by in situ hybridization.[180] Both cell types are encountered in a similar context in the developing dental root.[181] In the more rare mixed odontogenic tumors, there is evidence for an EMT as well.[182]

Finally, Putz et al[183] investigated the marker profiles of cell lines that were established from bone marrow aspirates of stage M0 cancer patients and found that they were consistent with cancer cells that underwent an EMT, thereby underscoring the hypothesis that (micro)metastasized carcinoma cells have gone through an EMT.

EMT in Non-Tumoral Pathologies

Besides during embryogenesis and in tumors, EMT also occurs in certain non-tumoral pathologies, namely in some chronic inflammatory processes (see Table 5). Best studied is the transition in kidney fibrogenesis. Iwano et al induced a chronic inflammatory response in the kidney interstitium of transgenic mice by imposing a unilateral ureteral obstruction.[184] They demonstrated that a substantial number of the fibroblasts appearing in the induced fibrotic tissue were derived from the proximal tubular epithelium, which had been specifically LacZ-labeled through cell type specific gene targeting. Other reports focused on the underlying mechanisms; TGF-β, EGF and FGF2 secreted by inflammatory stromal cells, appeared inducers of this type of EMT.[63,185,186] In renal biopsies from patients suffering from tubulointerstitial fibrosis a switch was observed in the morphological markers, compatible with EMT.[187,188]

Another example of chronic inflammation is the peritoneal fibrosis that occurs during long-standing continuous ambulatory peritoneal dialysis of patients with renal failure. A recent study showed that in peritoneal biopsies of such patients the normal mesothelium and submesothelial connective tissue was replaced by fibrotic tissue and embedded in it, a population of elongated, fibroblast-like though still cytokeratin-positive cells.[189] In cultures of the mesothelial cells that were shedded into the dialysis fluid, a comparable phenotypic switch was found to be paralleled by an increase in vimentin expression and a downregulation of cytokeratin and E-cadherin. Similar cellular changes could be induced in vitro by treating normal mesothelium with TGF-β and IL1-β. Although in these cell cultures the transitions were reversible, so rather an example of scattering according to our definition, the persistent scar tissue observed in vivo could be attributed to an EMT.

A comparable transient phenotypic switch is described in repair mechanisms of human respiratory epithelial cells in vitro.[150] Motile, elongated cells emerging at the wound edge co-expressed cytokeratins and vimentin. The transition is possibly related to Str-3 (MMP-11) production in neighboring epithelial cells.

As mentioned above, retinal pigment epithelial cells show phenotypic plasticity in vitro. This cell type can be the key player in the 'proliferative vitreoretinal disorders', which are characterized by proliferation of scar-like tissue in response to inflammatory stimuli.[190]

Other Transitions

Examples of cell plasticity in adult tissue other than the epithelial-to-mesenchymal transition are listed in Table 5 and 6. Concerning the non-tumoral pathologies, Barrett esophagus, caused by chronic gastro-esophageal reflux, is clinically the most important example, since it imposes a high risk for developing adenocarcinoma.[191,192]

Blood vessels, when activated during inflammation, could become lined by an epithelioid instead of the normal flattened endothelium. An epithelioid morphology is also adapted by the macrophages in epithelioid granulomas, representing a type of chronic inflammation characterized by aggregates of activated macrophages. An electron microscopy study revealed that each epithelioid granuloma, irrespective of the underlying disease, consists of a heterogeneous population of macrophages. The cells have a large amount of eosinophilic cytoplasm due to the accumulation of different types of lysosomes and vesicles and the prominence of the rough endoplasmic reticulum and Golgi apparatus.[193] The cadherin expression detected in such epithelioid macrophages by immunohistochemistry, favors a process of true transdifferentiation.[194]

In contrast to the active changes in EMT, alterations in the epithelial cell shape can be the result of an accumulation of foreign material in the cytoplasm or the cytopathic effect of a viral infection. Besides nuclear alterations, cytoplasmic vacuolization and accumulation of keratohyaline granules are seen in Papilloma virus infections (cervix condylomata or warts of the skin), whereas epithelial cells transform to large, multinucleated cells in Herpes virus infections.

Table 5. Transitions in physiology and non-tumoral pathology

Physiologic Process / Disease	Transition	Refs.
	E to M	
– wound healing#,*	– epithelial cells adapt a motile, spindle-shaped phenotype	150
– chronic interstitial nephritis	– epithelium of renal tubules to myofibroblasts	184
– chronic ambulatory peritoneal dialysis	– mesothelial cells to fibroblastoid cells	189
– brain injury	– ependymal cells to astrocytes	248
– proliferative vitreoretinal disorders	– retinal pigment epithelial cells to myofibroblastoid cells	190
	M to E	
– decidual reaction endometrium#	– fibroblasts become epithelioid by gestagens during secretory phase of menstrual cycle	118, 249
– post-operative scar: e.g. post-operative spindle cell nodule of vaginal wall	– spindle cells become epithelioid and cytokeratin-reactive	250
	E to other	
– viral infections: – human Papilloma virus cervix, skin (wart); – cytomegalovirus, Herpes simplex virus	– cytopathic effect on epithelium: – cytoplasmic vacuolization, keratohyaline granules; – multinucleated, syncitial cells	
– metaplasia: e.g., Barrett esophagus	– squamous to intestinal type of columnar epithelium	192
– endosalpingiosis	– Müllerian differentiation of ovarian surface epithelium	251
	Other	
– placenta development#	– cytotrophoblast to infiltrative spindle cells, cyto- to syncytiotrophoblast (cell fusion)	119, 120
– wound healing#	– fibroblasts to myofibroblasts	230
– inflammation	– activated endothelium of small blood vessels becomes epithelioid	
– granuloma	– macrophages to epithelioid cells	193
– hepatic fibrogenesis	– activation of stellate cells	252

physiologic process; * note that in this case the transition is reversible, so rather 'scattering'

Concerning examples of cell plasticity in tumors (Table 6), epithelioid sarcoma is a clear example of a soft tissue tumor with a mixed marker expression pattern (EMA-, keratin- and vimentin-positive) but with an epithelioid cell morphology.[195,196] Synovial sarcoma is a 'biphasic' soft tissue tumor with a spindle-shaped and a variable epithelioid cell component.[197,198] Wilms' tumors are aggressive tumors of infancy and childhood exhibiting epithelial and stromal elements that are genetically related and that mimick immature renal structures.[199,200] Unlike low-grade lymphomas, anaplastic large cell lymphomas (ALCL) are characterized by large, apparently cohesive cells with abundant cytoplasm which are sometimes mistaken for carcinoma cells. Interestingly, strong pan-cadherin immunoreactivity was found in a series of ALCL, probably accounting for the epithelioid phenotype of the lymphoma cells.[194]

Table 6. Transitions in tumor pathology

Tumor	Transition	Refs.
	M to E	
– epithelioid sarcoma	– fibroblastoid to epithelioid	195,196
– synovial sarcoma	– mixed fibroblastoid, epithelioid	197,198
– gastro-intestinal stromal tumor (GIST)	– myoid to epithelioid	253
– anaplastic large cell lymphoma	– lymphoid (round cells) to epithelioid	194
– epithelioid hemangioma/hemangiosarcoma	– flattened endothelium to epithelioid	254
– Wilms' tumor	– mixed mesenchymal, epithelial	199
	E to other	
– carcinoma with neuro-endocrine features	– epithelial cells acquire neuro-endocrine secretory phenotype	
– carcinoma with clear cell, sebaceous cell differentiation; granular cell carcinoma	– epithelial cells varies in cell shape through changes in cytoplasmic content	
– mixed tumors with melanoma component	– epithelial cells transdifferentiate into melanoma cells	255
	Other	
– gliosarcoma	– neural to fibroblastoid	
– metatypical basal cell carcinoma	– basaloid epithelial to squamous epithelial cells	
– desmoplastic melanoma	– melanoma cells to fibroblastoid cells	256

Epithelium-Stroma Interactions

EMT in tumors should be considered as a possible phenotypic event occurring in a context of epithelial-mesenchymal interactions, which lead to homeostasis sustaining tumor progression. An important manifestation of these interactions in tumor pathology is the so-called 'desmoplastic stroma'. This is the reactive connective tissue found in most malignant tumors that surrounds to a variable extent the infiltrating tumor cells (Fig. 4). It mainly consists of activated myofibroblastic cells and an altered extracellular matrix (ECM). The stromal changes are the result of interactions during tumor progression between malignant cells, ECM, and the fibroblasts that normally populate the connective tissue, and involve also inflammatory cells and blood vessels.[201-204] In breast carcinoma, paracrine factors such as platelet derived growth factor[205] and TGF-β[206,207] secreted by the tumor cells, could account for the initiation of the desmoplastic response. Fibroblasts are converted to α-SMA-reactive myofibroblastic cells,[208] and produce a different set of ECM proteins, such as tenascin-C,[209] fibronectin,[210] hyaluronic acid,[211] as well as increased amounts of collagen type I, III and V.[212,213] The myofibroblastic cells also overexpress matrix metalloproteinases like MMP-9, MMP-3, MMP-2 and/or MMP-11,[214-216] and secrete growth factors such as FGF2, vascular endothelial growth factor-D and HGF,[217,218] which act on their turn on the tumor cells. Tumor-derived (myo)fibroblasts exerted a differential tumorigenic influence on breast[219] and prostate[220] cancer cells as compared to stromal cells from non-tumoral tissue. Benign stromal cells induced a differentiated, less tumorigenic phenotype in the rat Dunning prostatic adenocarcinoma cells.[221] A dominance of the stromal phenotype over the cancer cells' genotype was suggested in a mouse prostate model of ras+myc-induced carcinogenesis[222] and in the HMT-3522 breast cancer cell line, in which integrin-blocking antibodies were able to revert its malignant phenotype.[223] Serial analysis of gene expression (SAGE) and in situ hybridization studies described panels of genes that are differentially expressed in the desmoplastic stroma of breast and pancreas carcinoma, some of which appear to be organ-specific or restricted to specific regions within the tumors.[224,225]

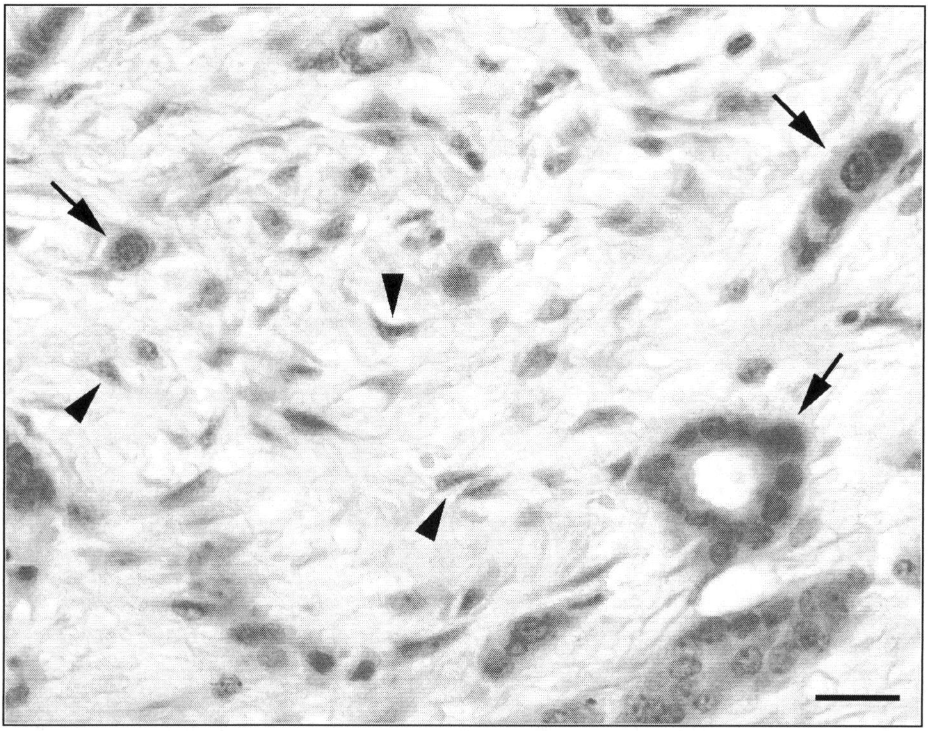

Figure 4. Desmoplastic stroma in an invasive ductal adenocarcinoma of the breast. It contains a high number of 'non-tumoral' myofibroblasts (arrowheads), of which the elongated nuclei are regular in size and shape compared to the pleomorphic nuclei of the sarcoma cells in Figure 1. Tumor cells are organized in tubules and solid cords or they seem to infiltrate individually (arrows). H&E staining. (Scale bar: 20 μm)

The interactions of the desmoplastic stroma with the tumor cells are reminiscent to the crosstalk observed in wound healing, in which the different components of the granulation tissue induce epithelial cell proliferation and migration necessary for wound closure.[204,226] Parallels can also be drawn with the epithelium-stroma interactions during the embryogenesis of e.g., breast or prostate.[227,228]

As for the myofibroblastic cells of the desmoplastic stroma, they most probably derive from quiescent stromal fibroblasts in response to TGF-β.[229-233] Concerning human prostate carcinoma however, Hayward et al suggests that the smooth muscle cells which make up the normal prostatic stroma are the cells of origin.[234] Since this transition would be accompanied by a loss of the smooth muscle markers myosin and desmin, it represents a type of dedifferentiation. In recent literature, the alternative hypothesis is brought up that the myofibroblasts in the desmoplastic stroma of a classical carcinoma, originate from the carcinoma cells themselves through a process of EMT.[235] A recent study by Petersen at al[236] addresses this subject indirectly. They examined two cell lines, established from the epithelial and mesenchymal component of a metaplastic breast carcinoma. Besides evidence for an EMT and for the clonality of both components, they showed that the fibroblastoid cell line behaved on itself in an indolent way. Although the cells were immortal, they were not tumorigenic in vivo and resembled normal fibroblasts in responding to TGF-β by de novo expression of α-SMA. The authors conclude that EMT-derived mesenchymal cells may differentiate into non-malignant myofibroblast-like cells. In a sense these results subscribe to a third theory concerning the origin of mixed tumors (besides the poly- and monoclonal theory discussed above): the stromal induction/metaplasia theory. According to this theory the stromal component of a mixed tumor is a non-malignant response to the growth of the epithelial malignant component. This particular case of a mixed breast carcinoma could also exemplify a tumor mechanism of carcinomas in general. One could hypothesize that in order to progress, the transformed epithelial cells of a carcinoma need the inductive signals of a surrounding stroma that they create themselves by undergoing a mesenchymal transition towards non-tumorigenic myofibroblastic cells (Fig. 4). In breast carcinomas, this could occur via an intermediate myoepithelial cell type. The fact that similar loci of loss of heterozygozity could be found in both mammary carcinoma and desmoplastic stromal cells from the same tumor, points to a similar genetic background, possibly indicating that both cell types derive from the same progenitor cell.[237-239]

Summary

In summary, epithelial-mesenchymal transitions are precisely regulated events. Reflected by changes in morphology (differentiation), cells exhibit altered functional properties in terms of motility, extracellular matrix remodeling and interactions with other cell types. In embryonic life, by means of EMT, the cells obtain the potential to create a wide range of different tissues, whereas in adult life, it helps maintaining a homeostasis, e.g., in adequate wound healing. In non-physiological conditions, EMT can also occur. Besides its suggested role in human chronic inflammatory disorders, EMT is described in malignant tumors. Much less numerous than the observations in vitro are the examples of EMT in human tumor pathology. In addition to the rare cases of mixed epithelial/mesenchymal tumors, we discussed the tempting hypothesis that EMT is a common event in carcinomas in general. By means of EMT the epithelial tumor cells would obtain the ability to transdifferentiate into myofibroblasts that lose their malignant phenotype but constitute the desmoplastic stroma which is essential for tumor growth, invasion and metastasis. Thus, by creating their own stroma, carcinoma cells would obtain an additional survival advantage.

Acknowledgements

We would like to thank M. Mareel, C. Stove and L. Derycke for critical reading of the manuscript, G. Debruyne for his help with the references and J. Roels and E. Van Marck for assistance with the illustrations. V. Van Marck is a research assistant of the Fund for Scientific Research- Flanders (F.W.O- Kom op Tegen Kanker, Belgium).

References

1. Janda E, Lehmann K, Killisch I et al. Ras and TGFβ cooperatively regulate epithelial cell plasticity and metastasis: dissection of Ras signaling pathways. J Cell Biol 2002; 156:299-313.
2. Slack JMW, Tosh D. Transdifferentiation and metaplasia-switching cell types. Curr Opin Genet Dev 2001; 11:581-586.
3. Cotran RS, Kumar V, Collins T (eds). Robbins Pathologic Basis of Disease. 6th Edition. WB Saunders Company, Philadelphia, 1999:264-267.
4. Cotran RS, Kumar V, Collins T (eds). Robbins Pathologic Basis of Disease. 6th Edition. WB Saunders Company, Philadelphia, 1999:36-38.
5. Tosh D, Slack JMW. How cells change their phenotype. Nat Rev Mol Cell Biol 2002; 3:187-194.
6. Hay ED, Zuk A. Transformations between epithelium and mesenchyme: normal, pathological, and experimentally induced. Am J Kidney Dis 1995; 26:678-690.
7. Vrancken Peeters M-PFM, Gittenberger-de Groot AC, Mentink MMT et al. Smooth muscle cells and fibroblasts of the coronary arteries derive from epithelial-mesenchymal transformation of the epicardium. Anat Embryol (Berl) 1999; 199:367-378.
8. Morabito CJ, Dettman RW, Kattan J et al. Positive and negative regulation of epicardial- mesenchymal transformation during avian heart development. Dev Biol 2001; 234:204-215.
9. Vicovac L, Aplin JD. Epithelial-mesenchymal transition during trophoblast differentiation. Acta Anat (Basel) 1996; 156:202-216.
10. Horster MF, Braun GS, Huber SM. Embryonic renal epithelia: induction, nephrogenesis, and cell differentiation. Physiol Rev 1999; 79:1157-1191.
11. Schedl A, Hastie ND. Cross-talk in kidney development. Curr Opin Genet Dev 2000; 10:543-549.

12. Vainio S, Karavanova I, Jowett A et al. Identification of BMP-4 as a signal mediating secondary induction between epithelial and mesenchymal tissues during early tooth development. Cell 1993; 75:45-58.
13. Dulbecco R, Henahan M, Bowman M et al. Generation of fibroblast-like cells from cloned epithelial mammary cells in vitro: a possible new cell type. Proc Natl Acad Sci USA 1981; 78:2345-2349.
14. Auersperg N, Maines-Bandiera SL, Dyck HG et al. Characterization of cultured human ovarian surface epithelial cells: phenotypic plasticity and premalignant changes. Lab Invest 1994; 71:510-518.
15. Sood AK, Seftor EA, Fletcher MS et al. Molecular determinants of ovarian cancer plasticity. Am J Pathol 2001; 158:1279-1288.
16. Maniotis AJ, Folberg R, Hess A et al. Vascular channel formation by human melanoma cells in vivo and in vitro: vasculogenic mimicry. Am J Pathol 1999; 155:739-752.
17. Burke JM, Skumatz CMB, Irving PE et al. Phenotypic heterogeneity of retinal pigment epithelial cells in vitro and in situ. Exp Eye Res 1996; 62:63-73.
18. Van Aken EH, De Wever O, Van Hoorde L et al. Invasion of retinal pigment epithelial cells: N-cadherin, hepatocyte growth factor, and focal adhesion kinase. Invest Ophthalmol Vis Sci 2003; 44:463-472.
19. Zuk A, Hay ED. Expression of β1 integrins changes during transformation of avian lens epithelium to mesenchyme in collagen gels. Dev Dyn 1994; 201:378-393.
20. Shen C-N, Slack JMW, Tosh D. Molecular basis of transdifferentiation of pancreas to liver. Nat Cell Biol 2000; 2:879-887.
21. Hu E, Tontonoz P, Spiegelman BM. Transdifferentiation of myoblasts by the adipogenic transcription factors PPARγ and C/EBPα. Proc Natl Acad Sci USA 1995; 92:9856-9860.
22. Donna A, Betta PG, Bianchi V et al. A new insight into the histogenesis of 'mesodermomas'-malignant mesotheliomas. Histopathology 1991; 19:239-244.
23. Vleminckx K, Vakaet L Jr, Mareel M et al. Genetic manipulation of E-cadherin expression by epithelial tumor cells reveals an invasion suppressor role. Cell 1991; 66:107-119.
24. Matsuzaki F, Mège RM, Jaffe SH et al. cDNAs of cell adhesion molecules of different specificity induce changes in cell shape and border formation in cultured S180 cells. J Cell Biol 1990; 110:1239-1252.
25. Nagafuchi A, Shirayoshi Y, Okazaki K et al. Transformation of cell adhesion properties by exogenously introduced E-cadherin cDNA. Nature 1987; 329:341-343.
26. Van Aken E, De Wever O, Correia da Rocha AS et al. Defective E-cadherin/catenin complexes in human cancer. Virchows Arch 2001; 439:725-751.
27. Berx G, Cleton-Jansen A-M, Nollet F et al. E-cadherin is a tumour/invasion suppressor gene mutated in human lobular breast cancers. EMBO J 1995; 14:6107-6115.
28. Berx G, Cleton-Jansen A-M, Strumane K et al. E-cadherin is inactivated in a majority of invasive human lobular breast cancers by truncation mutations throughout its extracellular domain. Oncogene 1996; 13:1919-1925.
29. Guilford P, Hopkins J, Harraway J et al. E-cadherin germline mutations in familial gastric cancer. Nature 1998; 392:402-405.
30. Humar B, Toro T, Graziano F et al. Novel germline CDH1 mutations in hereditary diffuse gastric cancer families. Hum Mutat 2002; 19:518-525.
31. Risinger JI, Berchuck A, Kohler MF et al. Mutations of the E-cadherin gene in human gynecologic cancers. Nat Genet 1994; 7:98-102.
32. Suriano G, Oliveira C, Ferreira P et al. Identification of CDH1 germline missense mutations associated with functional inactivation of the E-cadherin protein in young gastric cancer probands. Hum Mol Genet 2003; 12:575-582.
33. Grady WM, Willis J, Guilford PJ et al. Methylation of the CDH1 promoter as the second genetic hit in hereditary diffuse gastric cancer. Nat Genet 2000; 26:16-17.
34. Machado JC, Oliveira C, Carvalho R et al. E-cadherin gene (CDH1) promoter methylation as the second hit in sporadic diffuse gastric carcinoma. Oncogene 2001; 20:1525-1528.
35. Tamura G, Yin J, Wang S et al. E-cadherin gene promoter hypermethylation in primary human gastric carcinomas. J Natl Cancer Inst 2000; 92:569-573.
36. Yoshiura K, Kanai Y, Ochiai A et al. Silencing of the E-cadherin invasion-suppressor gene by CpG methylation in human carcinomas. Proc Natl Acad Sci USA 1995; 92:7416-7419.
37. Nieto MA. The Snail superfamily of zinc-finger transcription factors. Nat Rev Mol Cell Biol 2002; 3:155-166.
38. Nieto MA, Sargent MG, Wilkinson DG et al. Control of cell behavior during vertebrate development by Slug, a zinc finger gene. Science 1994; 264:835-839.
39. Batlle E, Sancho E, Franci C et al. The transcription factor Snail is a repressor of E-cadherin gene expression in epithelial tumour cells. Nat Cell Biol 2000; 2:84-89.

40. Cano A, Pérez-Moreno MA, Rodrigo I et al. The transcription factor Snail controls epithelial-mesenchymal transitions by repressing E-cadherin expression. Nature Cell Biol 2000; 2:76-83.
41. Bolós V, Peinado H, Pérez-Moreno MA et al. The transcription factor Slug represses E-cadherin expression and induces epithelial to mesenchymal transitions: a comparison with Snail and E47 repressors. J Cell Sci 2003; 116:499-511.
42. Comijn J, Berx G, Vermassen P et al. The two-handed E box binding zinc finger protein SIP1 downregulates E-cadherin and induces invasion. Mol Cell 2001; 7:1267-1278.
43. Yokoyama K, Kamata N, Hayashi E et al. Reverse correlation of E-cadherin and snail expression in oral squamous cell carcinoma cells in vitro. Oral Oncol 2001; 37:65-71.
44. Blanco MJ, Moreno-Bueno G, Sarrio D et al. Correlation of Snail expression with histological grade and lymph node status in breast carcinomas. Oncogene 2002; 21:3241-3246.
45. Rosivatz E, Becker I, Specht K et al. Differential expression of the epithelial-mesenchymal transition regulators Snail, SIP1, and Twist in gastric cancer. Am J Pathol 2002; 161:1881-1891.
46. Hosono S, Gross I, English MA et al. E-cadherin is a WT1 target gene. J Biol Chem 2000; 275:10943-10953.
47. Hosono S, Luo X, Hyink DP et al. WT1 expression induces features of renal epithelial differentiation in mesenchymal fibroblasts. Oncogene 1999; 18:417-427.
48. Han AC, Soler AP, Knudsen KA et al. Distinct cadherin profiles in special variant carcinomas and other tumors of the breast. Hum Pathol 1999; 30:1035-1039.
49. Tomita K, van Bokhoven A, van Leenders GJLH et al. Cadherin switching in human prostate cancer progression. Cancer Res 2000; 60:3650-3654.
50. Giroldi LA, Bringuier P-P, Shimazui T et al. Changes in cadherin-catenin complexes in the progression of human bladder carcinoma. Int J Cancer 1999; 82:70-76.
51. Islam S, Carey TE, Wolf GT et al. Expression of N-cadherin by human squamous carcinoma cells induces a scattered fibroblastic phenotype with disrupted cell-cell adhesion. J Cell Biol 1996; 135:1643-1654.
52. Hazan RB, Kang L, Whooley BP et al. N-cadherin promotes adhesion between invasive breast cancer cells and the stroma. Cell Adhesion Commun 1997; 4:399-411.
53. Hazan RB, Phillips GR, Fang Qiao R et al. Exogenous expression of N-cadherin in breast cancer cells induces cell migration, invasion, and metastasis. J Cell Biol 2000; 148:779-790.
54. Nieman MT, Prudoff RS, Johnson KR et al. N-cadherin promotes motility in human breast cancer cells regardless of their E-cadherin expression. J Cell Biol 1999; 147:631-643.
55. Suyama K, Shapiro I, Guttman M et al. A signaling pathway leading to metastasis is controlled by N-cadherin and the FGF receptor. Cancer Cell 2002; 2:301-314.
56. Grände M, Franzen A, Karlsson J-O et al. Transforming growth factor-β and epidermal growth factor synergistically stimulate epithelial to mesenchymal transition (EMT) through a MEK-dependent mechanism in primary cultured pig thyrocytes. J Cell Sci 2002; 115:4227-4236.
57. Bhowmick NA, Ghiassi M, Bakin A et al. Transforming growth factor-β1 mediates epithelial to mesenchymal transdifferentiation through a RhoA-dependent mechanism. Mol Biol Cell 2001; 12:27-36.
58. Ackland ML, Newgreen DF, Fridman M et al. Epidermal growth factor-induced epithelio-mesenchymal transition in human breast carcinoma cells. Lab Invest 2003; 83:435-448.
59. Bhowmick NA, Zent R, Ghiassi M et al. Integrin β_1 signaling is necessary for transforming growth factor-β activation of p38MAPK and epithelial plasticity. J Biol Chem 2001; 276:46707-46713.
60. Bakin AV, Tomlinson AK, Bhowmick NA et al. Phosphatidylinositol 3-kinase function is required for transforming growth factor β-mediated epithelial to mesenchymal transition and cell migration. J Biol Chem 2000; 275:36803-36810.
61. Gotzmann J, Huber H, Thallinger C et al. Hepatocytes convert to a fibroblastoid phenotype through the cooperation of TGF-β1 and Ha-Ras: steps towards invasiveness. J Cell Sci 2002; 115:1189-1202.
62. Potempa S, Ridley AJ. Activation of both MAP kinase and phosphatidylinositide 3-kinase by Ras is required for hepatocyte growth factor/scatter factor-induced adherens junction disassembly. Mol Biol Cell 1998; 9:2185-2200.
63. Strutz F, Zeisberg M, Ziyadeh FN et al. Role of basic fibroblast growth factor-2 in epithelial-mesenchymal transformation. Kidney Int 2002; 61:1714-1728.
64. Morali OG, Delmas V, Moore R et al. IGF-II induces rapid β-catenin relocation to the nucleus during epithelium to mesenchyme transition. Oncogene 2001; 20:4942-4950.
65. Thiery JP. Epithelial-mesenchymal transitions in tumour progression. Nat Rev 2002; 2:442-454.
66. Savagner P. Leaving the neighborhood: molecular mechanisms involved during epithelial-mesenchymal transition. BioEssays 2001; 23:912-923.
67. Boyer B, Valleé AM, Edme N. Induction and regulation of epithelial-mesenchymal transitions. Biochem Pharmacol 2000; 60:1091-1099.

68. Cheville NF (ed). Ultrastructural Pathology. An introduction to interpretation, Iowa State University Press, Ames, Iowa, 1994:234-259.
69. Lloreta-Trull J, Serrano S. The current role of electron microscopy in the diagnosis of epithelial and epithelioid tumors. Semin Diagn Pathol 2003; 20:46-59.
70. Suo Z, Nesland JM. Electron microscopy in diagnosis of spindle cell tumors. Semin Diagn Pathol 2003; 20:5-12.
71. Eyden B. Electron microscopy in the study of myofibroblastic lesions. Semin Diagn Pathol 2003; 20:13-24.
72. Sugai T, Oikawa M, Uesugi N et al. Esophageal squamous cell carcinoma characterized by extensive chondroid differentiation. Pathol Int 2000; 50:514-519.
73. Domagala W, Striker G, Szadowska A et al. p53 protein and vimentin in invasive ductal NOS breast carcinoma-relationship with survival and sites of metastases. Eur J Cancer 1994; 30A:1527-1534.
74. Jahkola T, Toivonen T, von Smitten K et al. Expression of tenascin in invasion border of early breast cancer correlates with higher risk of distant metastasis. Int J Cancer (Pred Oncol) 1996; 69:445-447.
75. Chenard M-P, O'Siorain L, Shering S et al. High levels of stromelysin-3 correlate with poor prognosis in patients with breast carcinoma. Int J Cancer (Pred Oncol) 1996; 69:448-451.
76. Davies JA, Garrod DR. Molecular aspects of the epithelial phenotype. BioEssays 1997; 19:699-704.
77. Sappino AP, Schürch W, Gabbiani G. Biology of disease. Differentiation repertoire of fibroblastic cells: expression of cytoskeletal proteins as marker of phenotypic modulations. Lab Invest 1990; 63:144-161.
78. Osborn M, Weber K. Intermediate filaments: cell-type-specific markers in differentiation and pathology. Cell 1982; 31:303-306.
79. Moll R, Franke WW, Schiller DL et al. The catalog of human cytokeratins: patterns of expression in normal epithelia, tumors and cultured cells. Cell 1982; 31:11-24.
80. Cooper D, Schermer A, Sun T-T. Classification of human epithelia and their neoplasms using monoclonal antibodies to keratins: strategies, applications, and limitations. Lab Invest 1985; 52:243-256.
81. Moll R, Löwe A, Laufer J et al. Cytokeratin 20 in human carcinomas. A new histodiagnostic marker detected by monoclonal antibodies. Am J Pathol 1992; 140:427-447.
82. Santini D, Ceccarelli C, Taffurelli M et al. Differentiation pathways in primary invasive breast carcinoma as suggested by intermediate filament and biopathological marker expression. J Pathol 1996; 179:386-391.
83. Goddard MJ, Wilson B, Grant JW. Comparison of commercially available cytokeratin antibodies in normal and neoplastic adult epithelial and non-epithelial tissues. J Clin Pathol 1991; 44:660-663.
84. Osborn M, Debus E, Weber K. Monoclonal antibodies specific for vimentin. Eur J Cell Biol 1984; 34:137-143.
85. Sommers CL, Walker-Jones D, Heckford SE et al. Vimentin rather than keratin expression in some hormone-independent breast cancer cell lines and in oncogene-transformed mammary epithelial cells. Cancer Res 1989; 49:4258-4263.
86. Raymond WA, Leong AS-Y. Vimentin-a new prognostic parameter in breast carcinoma ? J Pathol 1989; 158:107-114.
87. Domagala W, Wozniak L, Lasota J et al. Vimentin is preferentially expressed in high-grade ductal and medullary, but not in lobular breast carcinomas. Am J Pathol 1990; 137:1059-1064.
88. Heatley M, Whiteside C, Maxwell P et al. Vimentin expression in benign and malignant breast epithelium. J Clin Pathol 1993; 46:441-445.
89. Seshadri R, Raymond WA, Leong AS-Y et al. Vimentin expression is not associated with poor prognosis in breast cancer. Int J Cancer 1996; 67:353-356.
90. Sommers CL, Byers SW, Thompson EW et al. Differentiation state and invasiveness of human breast cancer cell lines. Breast Cancer Res Treat 1994; 31:325-335.
91. Thompson EW, Torri J, Sabol M et al. Oncogene-induced basement membrane invasiveness in human mammary epithelial cells. Clin Exp Metastasis 1994; 12:181-194.
92. Hendrix MJC, Seftor EA, Seftor REB et al. Experimental co-expression of vimentin and keratin intermediate filaments in human breast cancer cells results in phenotypic interconversion and increased invasive behavior. Am J Pathol 1997; 150:483-495.
93. Zajchowski DA, Bartholdi MF, Gong Y et al. Identification of gene expression profiles that predict the aggressive behavior of breast cancer cells. Cancer Res 2001; 61:5168-5178.
94. Ramaekers FC, Verhagen AP, Isaacs JT et al. Intermediate filament expression and the progression of prostatic cancer as studied in the Dunning R-3327 rat prostatic carcinoma system. Prostate 1989; 14:323-339.

95. Gilles C, Polette M, Piette J et al. Epithelial-to-mesenchymal transition in HPV-33-transfected cervical keratinocytes is associated with increased invasiveness and expression of gelatinase A. Int J Cancer 1994; 59:661-666.
96. Ebert AD, Wechselberger C, Nees M et al. Cripto-1-induced increase in vimentin expression is associated with enhanced migration of human Caski cervical carcinoma cells. Exp Cell Res 2000; 257:223-229.
97. Sommers CL, Heckford SE, Skerker JM et al. Loss of epithelial markers and acquisition of vimentin expression in adriamycin- and vinblastine-resistant human breast cancer cell lines. Cancer Res 1992; 52:5190-5197.
98. Domagala W, Lasota J, Dukowicz A et al. Vimentin expression appears to be associated with poor prognosis in node-negative ductal NOS breast carcinomas. Am J Pathol 1990; 137:1299-1304.
99. Fuchs IB, Lichtenegger W, Buehler H et al. The prognostic significance of epithelial-mesenchymal transition in breast cancer. Anticancer Res 2002; 22:3415-3419.
100. Domagala W, Markiewski M, Harezga B et al. Prognostic significance of tumor cell proliferation rate as determined by the MIB-1 antibody in breast carcinoma: its relationship with vimentin and p53 protein. Clin Cancer Res 1996; 2:147-154.
101. Bozcuk H, Uslu G, Pestereli E et al. Predictors of distant metastasis at presentation in breast cancer: a study also evaluating associations among common biological indicators. Breast Cancer Res Treat 2001; 68:239-248.
102. Heatley MK, Ewings P, Odling Smee W et al. Vimentin expression does not assist in predicting survival in ductal carcinoma of the breast. Pathology 2002; 34:230-232.
103. Thomas PA, Kirschmann DA, Cerhan JR et al. Association between keratin and vimentin expression, malignant phenotype, and survival in postmenopausal breast cancer patients. Clin Cancer Res 1999; 5:2698-2703.
104. Gilles C, Polette M, Piette J et al. Vimentin expression in cervical carcinomas: association with invasive and migratory potential. J Pathol 1996; 180:175-180.
105. Auersperg N, Wong AST, Choi K-C et al. Ovarian surface epithelium: biology, endocrinology, and pathology. Endocr Rev 2001; 22:255-288.
106. Holthöfer H. Immunohistology of renal carcinomas. Eur Urol 1990; 18 (Suppl 2):15-16.
107. Czernobilsky B, Moll R, Levy R et al. Co-expression of cytokeratin and vimentin filaments in mesothelial, granulosa and rete ovarii cells of the human ovary. Eur J Cell Biol 1985; 37:175-190.
108. Dabbs DJ, Geisinger KR, Norris HT. Intermediate filaments in endometrial and endocervical carcinomas. The diagnostic utility of vimentin patterns. Am J Surg Pathol 1986; 10:568-576.
109. Medeiros LJ, Michie SA, Johnson DE et al. An immunoperoxidase study of renal cell carcinomas. Correlation with nuclear grade, cell type, and histologic pattern. Hum Pathol 1988; 19:980-987.
110. Holthöfer H, Miettinen A, Paasivuo R et al. Cellular origin and differentiation of renal carcinomas. A fluorescence microscopic study with kidney-specific antibodies, antiintermediate filament antibodies, and lectins. Lab Invest 1983; 49:317-326.
111. Swanson PE. Heffalumps, jagulars, and cheshire cats. A commentary on cytokeratins and soft tissue sarcomas. Am J Clin Pathol 1991; 95 (Suppl 1):S2-S7.
112. Quentmeier H, Osborn M, Reinhardt J et al. Immunocytochemical analysis of cell lines derived from solid tumors. J Histochem Cytochem 2001; 49:1369-1378.
113. Okazaki M, Takeshita S, Kawai S et al. Molecular cloning and characterization of OB-cadherin, a new member of cadherin family expressed in osteoblasts. J Biol Chem 1994; 269:12092-12098.
114. Kimura Y, Matsunami H, Inoue T et al. Cadherin-11 expressed in association with mesenchymal morphogenesis in the head, somite, and limb bud of early mouse embryos. Dev Biol 1995; 169:347-358.
115. Simonneau L, Kitagawa M, Suzuki S et al. Cadherin 11 expression marks the mesenchymal phenotype: towards new functions for cadherins? Cell Adhesion Commun 1995; 3:115-130.
116. Pishvaian MJ, Feltes CM, Thompson P et al. Cadherin-11 is expressed in invasive breast cancer cell lines. Cancer Res 1999; 59:947-952.
117. Feltes CM, Kudo A, Blaschuk O et al. An alternatively spliced cadherin-11 enhances human breast cancer cell invasion. Cancer Res 2002; 62:6688-6697.
118. Getsios S, Chen GTC, Stephenson MD et al. Regulated expression of cadherin-6 and cadherin-11 in the glandular epithelial and stromal cells of the human endometrium. Dev Dyn 1998; 211:238-247.
119. MacCalman CD, Furth EE, Omigbodun A et al. Regulated expression of cadherin-11 in human epithelial cells: a role for cadherin-11 in trophoblast-endometrium interactions ? Dev Dyn 1996; 206:201-211.
120. Getsios S, Chen GTC, Huang DTK et al. Regulated expression of cadherin-11 in human extravillous cytotrophoblasts undergoing aggregation and fusion in response to transforming growth factor β1. J Reprod Fertil 1998; 114:357-363.

121. Savagner P, Yamada KM, Thiery JP. The zinc-finger protein slug causes desmosome dissociation, an initial and necessary step for growth factor-induced epithelial-mesenchymal transition. J Cell Biol 1997; 137:1403-1419.
122. Ikenouchi J, Matsuda M, Furuse M et al. Regulation of tight junctions during the epithelium-mesenchyme transition: direct repression of the gene expression of claudins/occludin by Snail. J Cell Sci 2003; 116:1959-1967.
123. Yi JY, Hur KC, Lee E et al. TGFβ1-mediated epithelial to mesenchymal transition is accompanied by invasion in the SiHa cell line. Eur J Cell Biol 2002; 81:457-468.
124. Kawano K, Kantak SS, Murai M et al. Integrin α3β1 engagement disrupts intercellular adhesion. Exp Cell Res 2001; 262:180-196.
125. Vallés AM, Boyer B, Tarone G et al. α2β1 integrin is required for the collagen and FGF-1 induced cell dispersion in a rat bladder carcinoma cell line. Cell Adhes Commun 1996; 4:187-199.
126. Erickson HP, Inglesias JL. A six-armed oligomer isolated from cell surface fibronectin preparations. Nature 1984; 311:267-269.
127. Bourdon MA, Wikstrand CJ, Furthmayr H et al. Human glioma-mesenchymal extracellular matrix antigen defined by monoclonal antibody. Cancer Res 1983; 43:2796-2805.
128. Jones FS, Jones PL. The tenascin family of ECM glycoproteins: structure, function, and regulation during embryonic development and tissue remodeling. Dev Dyn 2000; 218:235-259.
129. Vollmer G. Biologic and oncologic implications of tenascin-C/hexabrachion proteins. Crit Rev Oncol Hematol 1997; 25:187-210.
130. Howeedy AA, Virtanen I, Laitinen L et al. Differential distribution of tenascin in the normal, hyperplastic, and neoplastic breast. Lab Invest 1990; 63:798-806.
131. Hanamura N, Yoshida T, Matsumoto E-i et al. Expression of fibronectin and tenascin-c mRNA by myofibroblasts, vascular cells and epithelial cells in human colon adenomas and carcinomas. Int J Cancer 1997; 73:10-15.
132. Latijnhouwers M, Bergers M, Ponec M et al. Human epidermal keratinocytes are a source of tenascin-C during wound healing. J Invest Dermatol 1997; 108:776-783.
133. Yoshida T, Yoshimura E, Numata H et al. Involvement of tenascin-C in proliferation and migration of laryngeal carcinoma cells. Virchows Arch 1999; 435:496-500.
134. Schnyder B, Semadeni RO, Fischer RW et al. Distribution pattern of tenascin-C in normal and neoplastic mesenchymal tissues. Int J Cancer 1997; 72:217-224.
135. Riedl SE, Faissner A, Schlag P et al. Altered content and distribution of tenascin in colitis, colon adenoma, and colorectal carcinoma. Gastroenterology 1992; 103:400-406.
136. Ishihara A, Yoshida T, Tamaki H et al. Tenascin expression in cancer cells and stroma of human breast cancer and its prognostic significance. Clin Cancer Res 1995; 1:1035-1041.
137. Jahkola T, Toivonen T, Nordling S et al. Expression of tenascin-C in intraductal carcinoma of human breast: relationship to invasion. Eur J Cancer 1998; 34:1687-1692.
138. Tuominen H, Pöllänen R, Kallioinen M. Multicellular origin of tenascin in skin tumors-an in situ hybridization study. J Cutan Pathol 1997; 24:590-596.
139. Dandachi N, Hauser-Kronberger C, Moré E et al. Co-expression of tenascin-C and vimentin in human breast cancer cells indicates phenotypic transdifferentiation during tumour progression: correlation with histopathological parameters, hormone receptors, and oncoproteins. J Pathol 2001; 193:181-189.
140. Wirl G, Hermann M, Ekblom P et al. Mammary epithelial cell differentiation in vitro is regulated by an interplay of EGF action and tenascin-C downregulation. J Cell Sci 1995; 108:2445-2456.
141. Yoshida T, Ishihara A, Hirokawa Y et al. Tenascin in breast cancer development-is epithelial tenascin a marker for poor prognosis ? Cancer Lett 1995; 90:65-73.
142. Pinkus GS, Kurtin PJ. Epithelial membrane antigen-a diagnostic discriminant in surgical pathology: immunohistochemical profile in epithelial, mesenchymal, and hematopoietic neoplasms using paraffin sections and monoclonal antibodies. Hum Pathol 1985; 16:929-940.
143. Frisch SM. E1a induces the expression of epithelial characteristics. J Cell Biol 1994; 127:1085-1096.
144. Frisch SM. The epithelial cell default-phenotype hypothesis and its implications for cancer. BioEssays 1997; 19:705-709.
145. Strutz F, Okada H, Lo CW et al. Identification and characterization of a fibroblast marker: FSP1. J Cell Biol 1995; 130:393-405.
146. Hulboy DL, Matrisian LM, Crawford HC. Loss of JunB activity enhances stromelysin 1 expression in a model of the epithelial-to-mesenchymal transition of mouse skin tumors. Mol Cell Biol 2001; 21:5478-5487.
147. Martorana AM, Zheng G, Crowe TC et al. Epithelial cells up-regulate matrix metalloproteinases in cells within the same mammary carcinoma that have undergone an epithelial-mesenchymal transition. Cancer Res 1998; 58:4970-4979.

148. Ahmad A, Hanby A, Dublin E et al. Stromelysin 3: an independent prognostic factor for relapse-free survival in node-positive breast cancer and demonstration of novel breast carcinoma cell expression. Am J Pathol 1998; 152:721-728.
149. Nakopoulou L, Panayotopoulou EG, Giannopoulou I et al. Stromelysin-3 protein expression in invasive breast cancer: relation to proliferation, cell survival and patients' outcome. Mod Pathol 2002; 15:1154-1161.
150. Buisson A-C, Gilles C, Polette M et al. Wound repair-induced expression of a stromelysins is associated with the acquisition of a mesenchymal phenotype in human respiratory epithelial cells. Lab Invest 1996; 74:658-669.
151. Gilles C, Polette M, Seiki M et al. Implication of collagen type I-induced membrane-type 1-matrix metalloproteinase expression and matrix metalloproteinase-2 activation in the metastatic progression of breast carcinoma. Lab Invest 1997; 76:651-660.
152. Lochter A, Galosy S, Muschler J et al. Matrix metalloproteinase stromelysin-1 triggers a cascade of molecular alterations that leads to stable epithelial-to-mesenchymal conversion and a premalignant phenotype in mammary epithelial cells. J Cell Biol 1997; 139:1861-1872.
153. Sternlicht MD, Lochter A, Sympson CJ et al. The stromal proteinase MMP3/stromelysin-1 promotes mammary carcinogenesis. Cell 1999; 98:137-146.
154. Wright JH, McDonnell S, Portella G et al. A switch from stromal to tumor cell expression of stromelysin-1 mRNA associated with the conversion of squamous to spindle carcinomas during mouse skin tumor progression. Mol Carcinog 1994; 10:207-215.
155. Escot C, Zhao Y, Puech C et al. Cellular localisation by in situ hybridisation of cathepsin D, stromelysin 3, and urokinase plasminogen activator RNAs in breast cancer. Breast Cancer Res Treat 1996; 38:217-226.
156. Wells WA. Is transdifferentiation in trouble ? J Cell Biol 2002; 157:15-18.
157. Guarino M, Micheli P, Pallotti F et al. Pathological relevance of epithelial and mesenchymal phenotype plasticity. Pathol Res Pract 1999; 195:379-389.
158. George E, Manivel JC, Dehner LP et al. Malignant mixed müllerian tumors. An immunohistochemical study of 47 cases, with histogenetic considerations and clinical correlation. Hum Pathol 1991; 22:215-223.
159. Wick MR, Ritter JH, Humphrey PA. Sarcomatoid carcinomas of the lung. A clinicopathologic review. Am J Clin Pathol 1997; 108:40-53.
160. Torenbeek R, Blomjous CEM, de Bruin PC et al. Sarcomatoid carcinoma of the urinary bladder. Clinicopathologic analysis of 18 cases with immunohistochemical and electron microscopic findings. Am J Surg Pathol 1994; 18:241-249.
161. Wargotz ES, Norris HJ. Metaplastic carcinomas of the breast. III. Carcinosarcoma. Cancer 1989; 64:1490-1499.
162. Bleiweiss IJ, Huvos AG, Lara J et al. Carcinosarcoma of the submandibular salivary gland. Immunohistochemical findings. Cancer 1992; 69:2031-2035.
163. Ikegami H, Iwasaki H, Ohjimi Y et al. Sarcomatoid carcinoma of the urinary bladder: a clinicopathologic and immunohistochemical analysis of 14 patients. Hum Pathol 2000; 31:332-340.
164. Sneige N, Yaziji H, Mandavilli SR et al. Low-grade (fibromatosis-like) spindle cell carcinoma of the breast. Am J Surg Pathol 2001; 25:1009-1016.
165. Mount SL, Lee KR, Taatjes DJ. Carcinosarcoma (malignant mixed mullerian tumor) of the uterus with a rhabdoid tumor component. An immunohistochemical, ultrastructural, and immunoelectron microscopic case study. Am J Clin Pathol 1995; 103:235-239.
166. Ishida T, Tateishi M, Kaneko S et al. Carcinosarcoma and spindle cell carcinoma of the lung. Clinicopathologic and immunohistochemical studies. J Thorac Cardiovasc Surg 1990; 100:844-852.
167. Wick MR, Swanson PE. Carcinosarcomas: current perspectives and an historical review of nosological concepts. Semin Diagn Pathol 1993; 10:118-127.
168. Sreenan JJ, Hart WR. Carcinosarcomas of the female genital tract. A pathologic study of 29 metastatic tumors: further evidence for the dominant role of the epithelial component and the conversion theory of histogenesis. Am J Surg Pathol 1995; 19:666-674.
169. Bitterman P, Chun B, Kurman RJ. The significance of epithelial differentiation in mixed mesodermal tumors of the uterus. A clinicopathologic and immunohistochemical study. Am J Surg Pathol 1990; 14:317-328.
170. Silverberg SG, Major FJ, Blessing JA et al. Carcinosarcoma (malignant mixed mesodermal tumor) of the uterus. A Gynecologic Oncology Group pathologic study of 203 cases. Int J Gynecol Pathol 1990; 9:1-19.
171. Wang X, Mori I, Tang W et al. Metaplastic carcinoma of the breast: p53 analysis identified the same point mutation in the three histologic components. Mod Pathol 2001; 14:1183-1186.
172. Holst VA, Finkelstein S, Colby TV et al. p53 and K-ras mutational genotyping in pulmonary carcinosarcoma, spindle cell carcinoma, and pulmonary blastoma: implications for histogenesis. Am J Surg Pathol 1997; 21:801-811.

173. Abeln ECA, Smit VTHBM, Wessels JW et al. Molecular genetic evidence for the conversion hypothesis of the origin of malignant mixed müllerian tumours. J Pathol 1997; 183:424-431.
174. Thompson L, Chang B, Barsky SH. Monoclonal origins of malignant mixed tumors (carcinosarcomas). Evidence for a divergent histogenesis. Am J Surg Pathol 1996; 20:277-285.
175. Torenbeek R, Hermsen MAJA, Meijer GA et al. Analysis by comparative genomic hybridization of epithelial and spindle cell components in sarcomatoid carcinoma and carcinosarcoma: histogenetic aspects. J Pathol 1999; 189:338-343.
176. Leong AS-Y, Stevens MW, Mukherjee TM. Malignant mesothelioma: cytologic diagnosis with histologic, immunohistochemical, and ultrastructural correlation. Semin Diagn Pathol 1992; 9:141-150.
177. Oury TD, Hammar SP, Roggli VL. Ultrastructural features of diffuse malignant mesotheliomas. Hum Pathol 1998; 29:1382-1392.
178. Krismann M, Müller K-M, Jaworska M et al. Molecular cytogenetic differences between histological subtypes of malignant mesotheliomas: DNA cytometry and comparative genomic hybridization of 90 cases. J Pathol 2002; 197:363-371.
179. Nasu M, Ishikawa G. Ameloblastoma. Light and electron microscopic study. Virchows Arch A Pathol Anat Histopathol 1983; 399:163-175.
180. Heikinheimo K, Sandberg M, Happonen R-P et al. Cytoskeletal gene expression in normal and neoplastic human odontogenic epithelia. Lab Invest 1991; 65:688-701.
181. Thomas HF. Root formation. Int J Dev Biol 1995; 39:231-237.
182. Papagerakis P, Peuchmaur M, Hotton D et al. Aberrant gene expression in epithelial cells of mixed odontogenic tumors. J Dent Res 1999; 78:20-30.
183. Putz E, Witter K, Offner S et al. Phenotypic characteristics of cell lines derived from disseminated cancer cells in bone marrow of patients with solid epithelial tumors: establishment of working models for human micrometastases. Cancer Res 1999; 59:241-248.
184. Iwano M, Plieth D, Danoff TM et al. Evidence that fibroblasts derive from epithelium during tissue fibrosis. J Clin Invest 2002; 110:341-350.
185. Okada H, Danoff TM, Kalluri R et al. Early role of Fsp1 in epithelial-mesenchymal transformation. Am J Physiol 1997; 273:F563-F574.
186. Fan J-M, Huang X-R, Ng Y-Y et al. Interleukin-1 induces tubular epithelial-myofibroblast transdifferentiation through a transforming growth factor-β1-dependent mechanism in vitro. Am J Kidney Dis 2001; 37:820-831.
187. Jinde K, Nikolic-Paterson DJ, Huang XR et al. Tubular phenotypic change in progressive tubulointerstitial fibrosis in human glomerulonephritis. Am J Kidney Dis 2001; 38:761-769.
188. Rastaldi MP, Ferrario F, Giardino L et al. Epithelial-mesenchymal transition of tubular epithelial cells in human renal biopsies. Kidney Int 2002; 62:137-146.
189. Yáñez-Mo M, Lara-Pezzi E, Selgas R et al. Peritoneal dialysis and epithelial-to-mesenchymal transition of mesothelial cells. N Engl J Med 2003; 348:403-413.
190. Grisanti S, Guidry C. Transdifferentiation of retinal pigment epithelial cells from epithelial to mesenchymal phenotype. Invest Ophthalmol Vis Sci 1995; 36:391-405.
191. Morales CP, Souza RF, Spechler SJ. Hallmarks of cancer progression in Barrett's oesophagus. Lancet 2002; 360:1587-1589.
192. Seery JP. Stem cells of the oesophageal epithelium. J Cell Sci 2002; 115:1783-1789.
193. Epstein WL. Ultrastructural heterogeneity of epithelioid cells in cutaneous organized granulomas of diverse etiology. Arch Dermatol 1991; 127:821-826.
194. Ashton-Key M, Cowley GP, Smith MEF. Cadherins in reactive lymph nodes and lymphomas: high expression in anaplastic large cell lymphomas. Histopathology 1996; 28:55-59.
195. Kempson RL, Fletcher CDM, Evans HL et al. Tumors of the Soft Tissues. Atlas of Tumor Pathology, Third Series, Fascicle 30, Armed Forces Institute of Pathology, Washington, DC., 2001:484-492.
196. Chase DR, Enzinger FM, Weiss SW et al. Keratin in epithelioid sarcoma. An immunohistochemical study. Am J Surg Pathol 1984; 8:435-441.
197. Kempson RL, Fletcher CDM, Evans HL et al. Tumors of the Soft Tissues. Atlas of Tumor Pathology, Third Series, Fascicle 30, Armed Forces Institute of Pathology, Washington, DC., 2001:472-484.
198. Miettinen M, Limon J, Niezabitowski A et al. Patterns of keratin polypeptides in 110 biphasic, monophasic, and poorly differentiated synovial sarcomas. Virchows Arch 2000; 437:275-283.
199. Zhuang Z, Merino MJ, Vortmeyer AO et al. Identical genetic changes in different histologic components of Wilms' tumors. J Natl Cancer Inst 1997; 89:1148-1152.
200. Pritchard-Jones K. Malignant origin of the stromal component of Wilms' tumor. J Natl Cancer Inst 1997; 89:1089-1091.
201. De Wever O, Mareel M. Role of tissue stroma in cancer cell invasion. J Pathol 2003; 200:429-447.

202. Liotta LA, Kohn EC. The microenvironment of the tumour-host interface. Nature 2001; 411:375-379.
203. Rønnov-Jessen L, Petersen OW, Bissell MJ. Cellular changes involved in conversion of normal to malignant breast: importance of the stromal reaction. Physiol Rev 1996; 76:69-125.
204. Tuxhorn JA, Ayala GE, Rowley DR. Reactive stroma in prostate cancer progression. J Urol 2001; 166:2472-2483.
205. Shao Z-M, Nguyen M, Barsky SH. Human breast carcinoma desmoplasia is PDGF initiated. Oncogene 2000; 19:4337-4345.
206. Chiquet-Ehrismann R, Kalla P, Pearson CA. Participation of tenascin and transforming growth factor-β in reciprocal epithelial-mesenchymal interactions of MCF7 cells and fibroblasts. Cancer Res 1989; 49:4322-4325.
207. Walker RA, Dearing SJ, Gallacher B. Relationship of transforming growth factor β_1 to extracellular matrix and stromal infiltrates in invasive breast carcinoma. Br J Cancer 1994; 69:1160-1165.
208. Rønnov-Jessen L, Petersen OW. Induction of α-smooth muscle actin by transforming growth factor-β1 in quiescent human breast gland fibroblasts. Implications for myofibroblast generation in breast neoplasia. Lab Invest 1993; 68:696-707.
209. Yoshida T, Matsumoto E-I, Hanamura N et al. Co-expression of tenascin and fibronectin in epithelial and stromal cells of benign lesions and ductal carcinomas in the human breast. J Pathol 1997; 182:421-428.
210. Brown LF, Guidi AJ, Schnitt SJ et al. Vascular stroma formation in carcinoma in situ, invasive carcinoma, and metastatic carcinoma of the breast. Clin Cancer Res 1999; 5:1041-1056.
211. Bertrand P, Girard N, Delpech B et al. Hyaluronan (hyaluronic acid) and hyaluronectin in the extracellular matrix of human breast carcinomas: comparison between invasive and non-invasive areas. Int J Cancer 1992; 52:1-6.
212. Kauppila S, Stenbäck F, Risteli J et al. Aberrant type I and type III collagen gene expression in human breast cancer in vivo. J Pathol 1998; 186:262-268.
213. Lagacé R, Grimaud J-A, Schürch W et al. Myofibroblastic stromal reaction in carcinoma of the breast: variations of collagenous matrix and structural glycoproteins. Virchows Arch A Pathol Anat Histopathol 1985; 408:49-59.
214. Wood M, Fudge K, Mohler JL et al. In situ hybridization studies of metalloproteinases 2 and 9 and TIMP-1 and TIMP-2 expression in human prostate cancer. Clin Exp Metastasis 1997; 15:246-258.
215. Poulsom R, Pignatelli M, Stetler-Stevenson WG et al. Stromal expression of 72 kda type IV collagenase (MMP-2) and TIMP-2 mRNAs in colorectal neoplasia. Am J Pathol 1992; 141:389-396.
216. Masson R, Lefebvre O, Noël A et al. In vivo evidence that the stromelysin-3 metalloproteinase contributes in a paracrine manner to epithelial cell malignancy. J Cell Biol 1998; 140:1535-1541.
217. Orlandini M, Oliveiro S. In fibroblasts Vegf-D expression is induced by cell-cell contact mediated by cadherin-11. J Biol Chem 2001; 276:6576-6581.
218. Nakamura T, Matsumoto K, Kiritoshi A et al. Induction of hepatocyte growth factor in fibroblasts by tumor-derived factors affects invasive growth of tumor cells: in vitro analysis of tumor-stromal interactions. Cancer Res 1997; 57:3305-3313.
219. Brouty-Boyé D, Mainguené C, Magnien V et al. Fibroblast-mediated differentiation in human breast carcinoma cells (MCF-7) grown as nodules in vitro. Int J Cancer 1994; 56:731-735.
220. Olumi AF, Grossfeld GD, Hayward SW et al. Carcinoma-associated fibroblasts direct tumor progression of initiated human prostatic epithelium. Cancer Res 1999; 59:5002-5011.
221. Hayashi N, Cunha GR, Wong YC. Influence of male genital tract mesenchymes on differentiation of Dunning prostatic adenocarcinoma. Cancer Res 1990; 50:4747-4754.
222. Thompson TC, Timme TL, Kadmon D et al. Genetic predisposition and mesenchymal-epithelial interactions in ras+myc-induced carcinogenesis in reconstituted mouse prostate. Mol Carcinog 1993; 7:165-179.
223. Weaver VM, Petersen OW, Wang F et al. Reversion of the malignant phenotype of human breast cells in three-dimensional culture and in vivo by integrin blocking antibodies. J Cell Biol 1997; 137:231-245.
224. Iacobuzio-Donahue CA, Ryu B, Hruban RH et al. Exploring the host desmoplastic response to pancreatic carcinoma. Gene expression of stromal and neoplastic cells at the site of primary invasion. Am J Pathol 2002; 160:91-99.
225. Iacobuzio-Donahue CA, Argani P, Hempen PM et al. The desmoplastic response to infiltrating breast carcinoma: gene expression at the site of primary invasion and implications for comparisons between tumor types. Cancer Res 2002; 62:5351-5357.
226. Dvorak HF. Tumors: wounds that do not heal. N Engl J Med 1986; 315:1650-1659.
227. Wiseman BS, Werb Z. Stromal effects on mammary gland development and breast cancer. Science 2002; 296:1046-1049.

228. Cunha GR, Hayward SW, Wang YZ. Role of stroma in carcinogenesis of the prostate. Differentiation 2002; 70:473-485.
229. De Wever O, Mareel M. Role of myofibroblasts at the invasion front. Biol Chem 2002; 383:55-67.
230. Schmitt-Gräff A, Desmoulière A, Gabbiani G. Heterogeneity of myofibroblast phenotypic features: an example of fibroblastic cell plasticity. Virchows Arch 1994; 425:3-24.
231. Rønnov-Jessen L, Petersen OW, Koteliansky VE et al. The origin of the myofibroblasts in breast cancer. Recapitulation of tumor environment in culture unravels diversity and implicates converted fibroblasts and recruited smooth muscle cells. J Clin Invest 1995; 95:859-873.
232. Martin M, Pujuguet P, Martin F. Role of stromal myofibroblasts infiltrating colon cancer in tumor invasion. Pathol Res Pract 1996; 192:712-717.
233. Tuxhorn JA, Ayala GE, Smith MJ et al. Reactive stroma in human prostate cancer: induction of myofibroblast phenotype and extracellular matrix remodeling. Clin Cancer Res 2002; 8:2912-2923.
234. Hayward SW, Cunha GR, Dahiya R. Normal development and carcinogenesis of the prostate. A unifying hypothesis. Ann NY Acad Sci 1996; 784:50-62.
235. Petersen OW, Lind Nielsen H, Gudjonsson T et al. The plasticity of human breast carcinoma cells is more than epithelial to mesenchymal conversion. Breast Cancer Res 2001; 3:213-217.
236. Petersen OW, Nielsen HL, Gudjonsson T et al. Epithelial to mesenchymal transition in human breast cancer can provide a nonmalignant stroma. Am J Pathol 2003; 162:391-402.
237. Moinfar F, Man YG, Arnould L et al. Concurrent and independent genetic alterations in the stromal and epithelial cells of mammary carcinoma: implications for tumorigenesis. Cancer Res 2000; 60:2562-2566.
238. Kurose K, Hoshaw-Woodard S, Adeyinka A et al. Genetic model of multi-step breast carcinogenesis involving the epithelium and stroma: clues to tumour-microenvironment interactions. Hum Mol Genet 2001; 10:1907-1913.
239. Wernert N, Löcherbach C, Wellmann A et al. Presence of genetic alterations in microdissected stroma of human colon and breast cancers. Anticancer Res 2001; 21:2259-2264.
240. Schmidt A, Heid HW, Schäfer S et al. Desmosomes and cytoskeletal architecture in epithelial differentiation: cell type-specific plaque components and intermediate filament anchorage. Eur J Cell Biol 1994; 65:229-245.
241. Hynes RO. Integrins: versatility, modulation, and signaling in cell adhesion. Cell 1992; 69:11-25.
242. Sun H, Santoro SA, Zutter MM. Downstream events in mammary gland morphogenesis mediated by reexpression of the $\alpha_2\beta_1$ integrin: the role of the α_6 and β_4 integrin subunits. Cancer Res 1998; 58:2224-2233.
243. Wang Z, Symons JM, Goldstein SL et al. $\alpha3\beta1$ integrin regulates epithelial cytoskeletal organization. J Cell Sci 1999; 112:2925-2935.
244. Timpl R, Brown JC. Supramolecular assembly of basement membranes. BioEssays 1996; 18:123-132.
245. Kosmehl H, Berndt A, Katenkamp D. Molecular variants of fibronectin and laminin: structure, physiological occurrence and histopathological aspects. Virchows Arch 1996; 429:311-322.
246. Tanimoto H, Shigemasa K, Sasaki M et al. Differential expression of matrix metalloprotease-7 in each component of uterine carcinosarcoma. Oncol Rep 2000; 7:1209-1212.
247. Chambers AF, Matrisian LM. Changing views of the role of matrix metalloproteinases in metastasis. J Natl Cancer Inst 1997; 89:1260-1270.
248. Johansson CB, Momma S, Clarke DL et al. Identification of a neural stem cell in the adult mammalian central nervous system. Cell 1999; 96:25-34.
249. Heatley MK, Maxwell P, Toner PG. The immunophenotype of human decidua and extra-uterine decidual reactions. Histopathology 1996; 29:437-442.
250. Proppe KH, Scully RE, Rosai J. Postoperative spindle cell nodules of genitourinary tract resembling sarcomas. A report of eight cases. Am J Surg Pathol 1984; 8:101-108.
251. Dubeau L. The cell of origin of ovarian epithelial tumors and the ovarian surface epithelium dogma: does the emperor have no clothes ? Gynecol Oncol 1999; 72:437-742.
252. Friedman SL. Molecular regulation of hepatic fibrosis, an integrated cellular response to tissue injury. J Biol Chem 2000; 275:2247-2250.
253. Miettinen M, Lasota J. Gastrointestinal stromal tumors-definition, clinical, histological, immunohistochemical, and molecular genetic features and differential diagnosis. Virchows Arch 2001; 438:1-12.
254. Gray MH, Rosenberg AE, Dickersin GR et al. Cytokeratin expression in epithelioid vascular neoplasms. Hum Pathol 1990; 21:212-217.
255. Nobukawa B, Fujii H, Hirai S et al. Breast carcinoma diverging to aberrant melanocytic differentiation. A case report with histopathologic and loss of heterozygosity analyses. Am J Surg Pathol 1999; 23:1280-1287.
256. From L, Hanna W, Kahn HJ et al. Origin of the desmoplasia in desmoplastic malignant melanoma. Hum Pathol 1983; 14:1072-1080.

Chapter 10

Structural and Functional Regulation of Desmosomes

Spiro Getsios, Lisa M. Godsel and Kathleen J. Green

Introduction

Intercellular adhesion and communication in mammalian epithelial cells occurs via specialized junctional complexes, which include tight junctions, adherens junctions, desmosomes, and gap junctions (Fig. 1).[1] Desmosomes are unique among these junctions, as they are coupled to the intermediate filament (IF)-based cytoskeleton. These form stable complexes that facilitated the ultrastructural and biochemical characterization of desmosomes but also made progress in the molecular organization of these junctions more challenging.[2-5] Although the individual components that make up a desmosome are now fairly well established, we are just beginning to appreciate how desmosomes are assembled into highly regulated and dynamic adhesive units.

Desmosomes have been best studied in epithelial cells and are most prevalent in tissues that undergo extensive mechanical stress, such as the skin.[6] However, these junctions are also present in myocardial and Purkinje fiber cells of the heart, follicular dendritic cells of lymph nodes, and the arachnoid plexus of the brain meninges.[7] An expanding list of autoimmune and genetic diseases that target the different junctional proteins indicates that desmosomes are key mediators of epidermal integrity but also play important in these other organs (Table 1).

Molecular Components of the Desmosome

The basic structure of desmosomes is determined by proteins from three independent gene families (Fig. 2). The transmembrane component of desmosomes includes members of the gene superfamily of calcium-dependent cell adhesion molecules (CAMs), known as the cadherins.[8] Structurally-related proteins of the armadillo (*arm*) family, including plakoglobin (Pg; also known as γ-catenin) and the plakophilins (PKP), interact directly with the desmosomal cadherin cytoplasmic domains.[9] Finally, plakin family proteins and, in particular desmoplakin (DP), provide a structural link between the desmosomes and IFs.[10] Several other proteins have been localized to desmosomes but whose functions are not as well characterized in these junctional complexes.

The Desmosomal Cadherins

Sequence comparison of the desmosomal cadherins has further subdivided these into the desmogleins (Dsg) and desmocollins (Dsc). To date, four Dsg and three Dsc isoforms have been identified in humans that are the products of separate but closely linked genes.[8,11,12] Differential splicing of the *DSC* genes results in two distinct isoforms: an 'a' and a 'b' form that contains a shorter cytoplasmic domain with unique sequences.[13,14] Recently, two genes related to Dsg1, and thus termed *Dsg1-β* and *Dsg1–γ* have been identified in mice.[15,16] Whether these

Rise and Fall of Epithelial Phenotype: Concepts of Epithelial-Mesenchymal Transition, edited by Pierre Savagner.

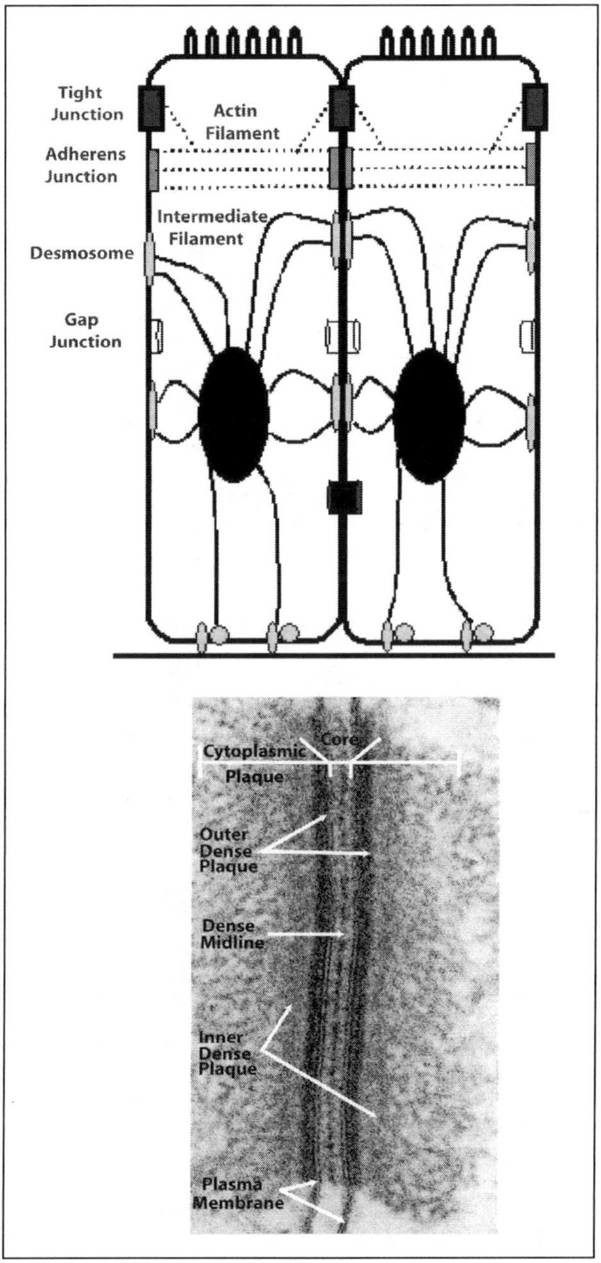

Figure 1. Two epithelial cells are represented (top panel) along with the major adhesive and communicating junctions. Tight junctions and adherens junctions are connected to microfilaments, whereas desmosomes interact with intermediate filaments. Gap junctions serve to directly connect the cytoplasm of adjacent cells by forming intrercellular channels. The bottom panel shows an electron micrograph of a single bovine tongue desmosome.[168] Distinct desmosomal regions have been identified, including a plasma membrane-associated core region that contains an extracellular electron dense midline as well as a cytoplasmic region that contains an outer and inner dense plaque.

Table 1. Genetic diseases involving desmosomal proteins

Mutated Gene/Protein	Disease/Phenotype
***DSG1* (Desmoglein 1)** N-terminal deletion, possible haploinsufficiency[52,53,169]	**Striate Palmoplantar Keratoderma** Epidermal thickening inpalms and soles of feet
***DSG4* (Desmoglein 4)** Homozygous deletion in extracellular domain[11]	**Localized Autosomal Recessive Hypotrichosis** Hair loss restricted to scalp, chest, arms, and legs
***CDSN* (Corneodesmosin)** Nonsense mutation, prematurely truncated protein[57]	**Hypotrichosis Simplex of the Scalp** Scalp-specific hair loss with childhood onset
***JUP* (Plakoglobin)** Autosomal recessive frameshift mutation, C-terminal domain truncation[75]	**Naxos Disease** Arrythmogenic right cardiomyopathy with palmoplantar keratoderma and woolly hair
***PKP1* (Plakophilin 1)** Autosomal recessive null mutation[80,170]	**Ectodermal Dysplasia/Skin Fragility Syndrome** with hair and nail defects
***DSP* (Desmoplakin)** N-terminal deletion, haploinsufficiency[106,107]	**Striate Palmoplantar Keratoderma**
Autosomal recessive frameshift mutation, C-terminal truncation[108]	**Striate Palmoplantar Keratoderma** with left ventricular cardiomyopathy and woolly hair
Compound heterozygosity of nonsense (C-terminal deletion) and missense (N-terminal) mutations[171]	**Striate Palmoplantar Keratoderma** with hair defects
Autosomal dominant N-terminal missense mutation[109]	**Arrhythmogenic Right Ventricular Cardiomyopathy**
Autosomal recessive C-terminal missense mutation[172]	**Arrhythmogenic Right Ventricular Dysplasia** with woolly hair and skin blisters

additional murine *DSG* genes have been lost in the human as a result of functional redundancy between desmosomal cadherins remains to be established.

E-cadherin (E-cad) is the best-studied member of the classical cadherins and localizes to adherens junctions in epithelial cells.[8] In general, the classical and desmosomal cadherins possess several common structural characteristics (Fig. 3). For example, the mature extracellular region of these cadherins is composed of five ectodomains with conserved calcium binding sites: four cadherin repeats (EC1-EC4) and a more divergent membrane proximal subdomain (EA). Short peptides generated against a cell adhesion recognition (CAR) sequence present in the EC1 subdomain of the classical cadherins disrupt cadherin-mediated homophilic adhesion.[17,18] In contrast, interfering with desmosomal adhesion requires the presence of CAR peptides specific for both Dsg and Dsc isoforms, suggesting that these CAMs interact in a heterophilic manner.[19,20] Dsg/Dsc heterophilic complexes have been identified in cultured cells and using recombinant EC1-2 subdomains of the desmosomal cadherins in vitro.[21-23] Collectively, these studies emphasize important differences between the organization of classical and desmosomal cadherin-based junctions.

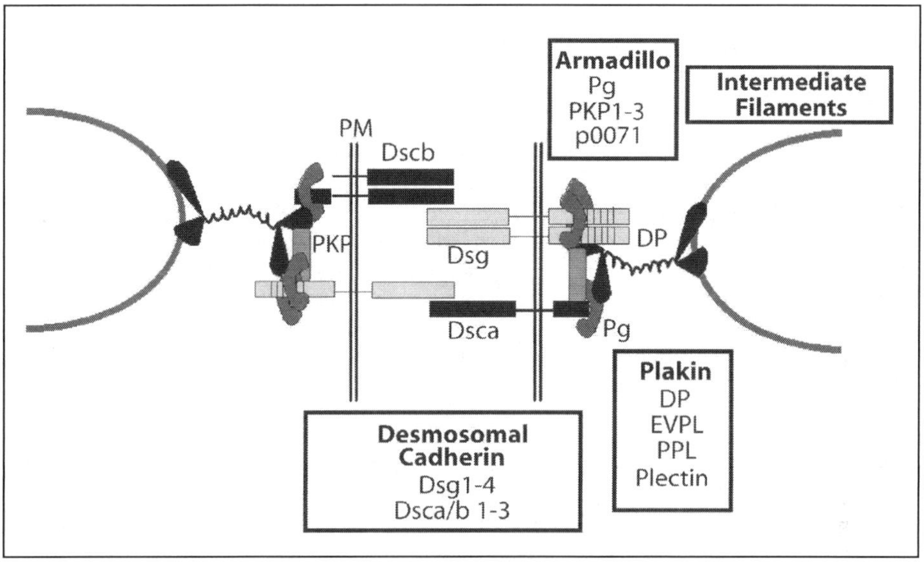

Figure 2. A molecular model of the desmosome. The desmosomal cadherins interact with one another and via their cytoplasmic domains with members of the armadillo family of proteins. Plakins are capable of linking these membrane complexes to the intermediate filament based cytoskeleton.

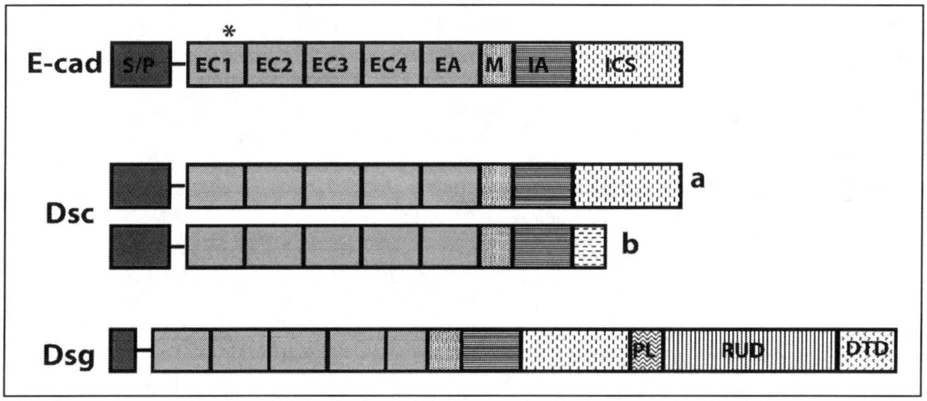

Figure 3. Comparison of the cadherin molecular domain structure. Classical and desmosomal cadherins are synthesized as precursor proteins containing a signal and pro-peptide region (S/P), five extracellular domains (EC1-4 and EA), a transmembrane domain (M), and an intracellular anchor domain (IA). The EC1 subdomain harbors a subtype-specific cell adhesion recognition sequence (indicated by an asterisk). The only cadherin subtypes that lack an intracellular catenin sequence (ICS) are the alternatively spliced Dsc 'b' isoforms. The Dsg isoforms further possess a proline-rich linker region (PL), a repeating unit domain (RUD), and a desmoglein terminal domain (DTD).

The desmosomal cadherin cytoplasmic domains are extensively divergent compared to the classical cadherins and between the Dsg and Dsc subfamilies. All cadherins possess an intracellular anchor (IA) subdomain that is proximal to the plasma membrane and is known to modulate classical cadherin-based adhesion but whose function is less clear for the desmosomal cadherins.[24,25] The intracellular catenin segment (ICS) of classical cadherins is capable of interacting with the highly related *arm* proteins, Pg and β-catenin, in a mutually exclusive manner.[26,27] In contrast, the ICS of desmosomal cadherins primarily mediates binding to Pg and contributes to its junction localization.[28-34] The shorter Dsc 'b' isoforms lack an ICS and subsequently fail to bind Pg but Dsc3b has recently been shown to interact with another desmosomal *arm* protein, PKP3.[35] The PKPs are equally likely to contribute to Dsg incorporation into junctions, although different Dsg cytoplasmic subdomains are responsible for these interactions.[35,36]

The Dsg isoforms also distinguish themselves from other cadherins by an extended region beyond the ICS which has been further subdivided into an intracellular proline-rich linker region (PL), a variable number of repeated unit domains (RUD), and a Dsg terminal domain (DTD). Although rotary shadowed preparations of recombinant RUD peptides suggest that these might serve as homodimerization domains, the functions or cellular proteins that interact with these repeats remain completely unknown.[37]

Changes in the expression of the classical cadherins provide the molecular basis for cell segregation and sorting events critical for embryonic development and the same principle appears to hold true for the desmosomal cadherins. For example, CAR peptides specific for the desmosomal cadherins can inhibit alveolar morphogenesis in a 3-dimensional model for mammary gland lumen formation.[20] In the stratified squamous epithelium of the epidermis, Dsg2/Dsc2 and Dsg3/Dsc3 are expressed in the basal keratinocytes.[38,39] In contrast, Dsg1/Dsc1 and Dsg4 are increased in the upper, more differentiated keratinocyte layers of the skin.[11,38,39] The extent of overlap between these desmosomal cadherin isoforms varies in different squamous epithelial tissues, such as the oral mucosa and hair follicle, and distinct desmosomal cadherin isoforms can even be detected within the same epidermal desmosome.[39-44] Nevertheless, transgenic mice containing an extracellular domain-deleted Dsg3 mutant in the basal keratinocytes or aberrantly overexpressing wild-type Dsg3 throughout the cellular layers of the epidermis exhibit defects in epidermal differentiation or barrier formation, suggesting that the ratio of different Dsg isoforms is an important determinant in skin morphogenesis.[45-47]

Evidence that individual desmosomal cadherins maintain the integrity of the epidermis comes from two related autoimmune diseases that result in skin blistering. In pemphigus vulgaris (PV) and pemphigus foliaceus (PF), autoantibodies that target Dsg3 and Dsg1, respectively, promote acantholysis in these patients.[48] The epidermal layers that form blisters correlate with the cellular localization of the Dsg subtype targeted in these disease; PV blisters form in the deeper layers of the skin whereas PF blisters are present in the more superficial epidermal layers. In addition, the extracellular domain of Dsg1, but not Dsg3, is cleaved by the exfoliative toxins produced by *Staphylococcus aureus* and promote blister formation in the upper layers of skin in bullous impetigo and staphylococcal scalded skin syndrome.[49-51] Finally, genetic mutations in the human *DSG1* gene cause an epidermal thickening disease, known as striate palmoplantar keratoderma (SPPK), whereas *DSG4* mutations are involved in defective hair follicle differentiation.[11,52,53] These studies emphasize important differences in the morphogenetic roles of the distinct Dsg isoforms.

Although the desmosomal cadherins appear to be the predominant transmembrane components of desmosomes, other glycoproteins have been localized to these junctional complexes in epithelial cells. For example, corneodesmosin is a secreted glycoprotein that can promote homophilic adhesion and associates with the cross-linked desmosomal complexes found in the upper layers of the skin, termed corneodesmosomes.[54-56] Nonsense mutations in the gene that encodes for corneodesmosin have been identified in patients with a scalp-specific hair loss disease.[57] In addition, polycystin-1 is implicated in polycystic kidney disease and has been reported in the desmosomes of Madine Darby canine kidney (MDCK) epithelial cells.[58] However, the importance of these proteins in desmosome function remains to be firmly established.

The Armadillo Proteins

The *arm* proteins serve as the primary binding partners for the desmosomal cadherins.[9] Members of this family share a series of structural motifs, known as *arm* repeats, that are

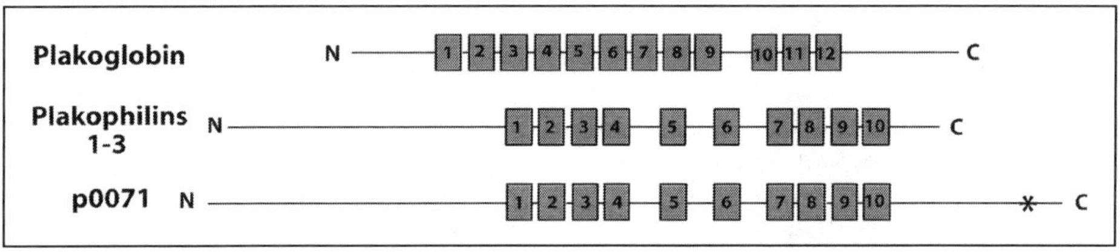

Figure 4. Comparison of the desmosomal armadillo molecular domain structure. These proteins comprise a variable number of central armadillo repeat motifs and more divergent N- and C-terminal domains. Plakoglobin contains 12 armadillo repeats whereas the plakophilins and p0071 contain only 10 of these motifs. p0071 also harbors a PDZ binding motif within its C-terminus, indicated by the asterisk.

flanked by more divergent N- and C-terminal domains (Fig. 4). Although the Armadillo protein was initially identified as a target of the *Wnt* signaling cascade during *Drosophila* development, the vertebrate homologue turned out to be the adherens junction protein, β-catenin (see also Chapter 12, Conacci-Sorrell and Ben-Ze'ev).[59,60] The desmosomal subset of *arm* proteins are thought to play an important role in clustering these junctional complexes at the plasma membrane and providing an indirect link to the IF-based cytoskeleton, but may also be involved in intracellular signaling pathways.

Plakoglobin

Based on high-resolution crystallography studies of β-catenin, Pg is thought to possess 12 central *arm* repeats that are likely organized into a single superhelical structure containing a positively charged groove.[61] The *arm* repeats of Pg are capable of binding desmosomal and classical cadherins and are thus responsible for its localization to both desmosomes and adherens junctions (see also Chapter 12, Conacci-Sorrell and Ben-Ze'ev).[62-64] Pg is further able to link the desmosomal cadherins to the IFs by interacting with DP as well as other *arm* proteins, like PKP2 and 3.[33,35,65] The flanking domains of Pg also contribute to junction assembly. For example, the N-terminal domain of Pg is involved in binding α-catenin, an actin-associated protein present in adherens junctions, but not desmosomal cadherins.[66,67] In contrast, the deletion of the Pg C-terminus alters the length of these junctions in the human epidermoid A431 cell line.[67]

Desmosome formation is influenced by the earlier assembly of classical cadherin complexes and Pg is a good candidate for regulating cross-talk between these two junctions. For example, carcinoma cells that lack endogenous Pg are capable of forming desmosomes when this *arm* protein is reintroduced, but only if classical cadherins are present.[68,69] Pg may only facilitate desmosome assembly, as these junctions can form in the epidermis of null mutant mice.[70,71] Although these animals exhibit skin defects, many embryos die of cardiac failure. β-catenin probably compensates for some Pg adhesive functions in the skin of these mutant mice, as this catenin subtype is mislocalized to desmosomes in Pg null keratinocytes.[72] Nevertheless, keratinocytes lacking Pg fail to respond to PV sera, not only implicating this catenin subtype in the pathogenesis of blistering diseases but also in functions beyond junction cross-talk.[73] Alterations in cellular proliferation were observed when an N-terminal deletion of Pg was overexpressed in the epidermis of transgenic mice, providing further evidence for its role in signaling.[74]

Recently, a homozygous mutation in the gene that encodes for Pg was identified in patients with Naxos disease.[75] This frameshift mutation results in a premature truncation of Pg in the C-terminus and is implicated in the autosomal recessive arrhythmic right ventricular cardiomyopathy (ARVC), SPPK, and woolly hair phenotypes observed in these patients. The fact that the truncation of Pg in these patients occurs within a region that is important for its transactivation activity raises the possibility that defects in Pg signaling may contribute to this disease.

Plakophilins and P0071

The three PKPs (PKP1-3) and p0071 (also known as PKP4) are members of the p120^{ctn} subfamily of proteins that are characterized by 10 central *arm* repeats.[9] In contrast to Pg, the N-terminus of the PKPs is critical for interacting with desmosomal proteins.[35,36,65,76-79] For example, PKP1 is capable of binding Dsg1, DP, IFs, but not Pg, although these two *arm* members are coordinated to promote efficient clustering of desmosomal complexes. Similarly, PKP2 and 3 associate with the desmosomal cadherins and DP but can be further detected in a complex with Pg.

The ability of PKP1 to interact with desmosomal proteins might explain its role in maintaining the integrity of skin. PKP1 is present in the desmosomes of the suprabasal epidermis and genetic defects are involved in a skin fragility disease that affects newborns.[80,81] Alterations in the desmosomes present in the epidermis and keratinocytes isolated from these patients disrupted adhesion but also increased cell motility.[82] Interestingly, DP is redistributed to the cytoplasm of PKP1 affected patients and the expression of several desmosomal proteins is reduced in PKP1 null keratinocytes. A role for the PKPs in recruiting DP to desmosomes and increasing desmosomal protein expression has since been demonstrated in cells transfected with this *arm* protein.[77,78,82,83]

PKP1 increases the expression of desmosomal proteins primarily from the post-translational stabilization of desmosomal complexes.[82] Nevertheless, all three PKP isoforms have also been localized to the nucleus and might be involved in intracellular signaling. Recently, the phosphorylation of PKP2 by C-TAK-1 has been shown to regulate its nuclear localization but the functional consequences of this remain unknown.[84] PKP2 has also been found in a complex with the RNA polymerase III holoenzyme and modulates β-catenin signaling activity in a colon cancer cell line.[65,85]

p0071 is more closely related to p120^{ctn} compared to the other PKPs and localizes to both adherens and desmosomal junctions.[86,87] Similar to p120,^{ctn} the central *arm* domains of p0071 mediate its interaction with the classical cadherins.[87,88] In contrast, the p0071 N-terminus is capable of interacting with Dsc3a, Pg, and DP. The C-terminal domain of p0071 harbors a PDZ binding motif that is absent in p120^{ctn} and the other PKPs and this protein can be further found in complexes with the PDZ-motif containing proteins, papin and Erbin.[89-91] As these proteins have been implicated in the establishment of epithelial polarity, it would seem likely that p0071 serves as a molecular scaffold for recruiting signaling molecules to junctional complexes.

The Plakins

The molecular mediators of IF-binding in desmosomes include members of an emerging family of cytolinker proteins, known as the plakins.[10,92] The plakin family is defined by the presence of a plakin domain (PD) and/or plakin repeat domain (PRD) that are implicated in junction targeting and IF-binding, respectively (Fig. 5). Several other subdomains, including an actin-binding domain, a coiled-coil rod domain, a spectrin-repeat-containing rod, and a microtubule-binding domain, are present in some members but not in others. As such, these proteins are well suited for linking junctional complexes to the cytoskeleton.

Desmoplakin

DP exists as two distinct isoforms derived from the alternative splicing of a single gene and is thought to be the most abundant component of desmosomes.[5,93] Biochemical and rotary shadow studies suggest that DP forms homodimers which comprise a central α-helical coiled-coil rod domain with two globular ends.[94] The parallel orientation of these DP dimers adopt higher ordered oligomeric structures with the N-terminus being more proximal to the plasma membrane compared to the C-terminal domain, and this topology corresponds nicely to the defined functions for these respective ends of DP.[10,95] For example, the N-terminus of DP harbors the PD that interacts with the desmosomal *arm* proteins but can also directly bind the desmosomal cadherins at the cell surface.[32,33,35,65,76-78,87,88] In contrast, the DP C-terminus

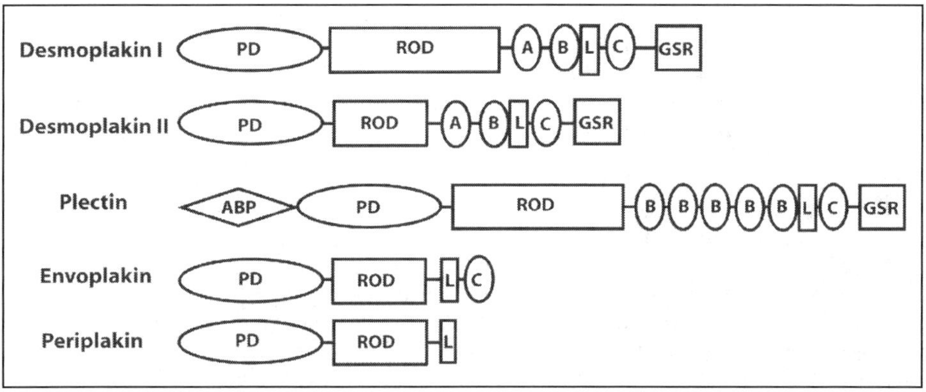

Figure 5. Comparison of the desmosomal plakin molecular domain structure. The two isoforms of desmoplakin, plectin, envoplakin, and periplakin all harbor an N-terminal plakin domain (PD) and a central coiled-coil rod domain (ROD). Plectin contains an additional actin-binding domain (ABD) at its N-terminal end. A variable number of plakin repeat domains, termed A, B, and C, are present in all the desmosomal plakin members except for periplakin and have been implicated in intermediate filament binding. The linker subdomain (L) sequence present in the C-terminus also mediates interactions with intermediate filaments. Desmoplakin and plectin further possess a glycine-serine-arginine (GSR) domain at the C-terminal end that is not present in either envoplakin or periplakin.

contributes to IF binding; the molecular basis of these interactions has been the focus of several studies.

The C-terminal domain of DP is comprised of three interspersed PRDs, termed A, B, and C, followed by downstream sequences and a specialized linker sequence (L) found between the last two repeats. High resolution crystallography of the DP PRDs has demonstrated that the B and C subdomains adopt a globular structure, consisting of a short β-hairpin followed by two anti-parallel α-helices, and contains a basic groove that might be involved in IF interactions.[96] Although this study showed that the individual PRDs interact directly with IFs, a combination of the B and C subdomains did so with higher efficiency. Other work suggests that the L subdomain and amino acid sequences downstream of the PRDs are sufficient for these cytoskeletal interactions.[97-99] Collectively, these results suggest that multiple, low-affinity interactions are involved in regulating the strength of DP-IF interactions.

The relationship of IF attachment to desmosome function has been examined by introducing a C-terminal truncated DP construct into A431 cells.[100,101] This dominant negative construct not only inhibited the ability of IFs to couple with the desmosomal cadherins at the cell surface but also reduced the adhesive strength of these cells. DP is critical for maintaining the mechanical integrity of tissues in vivo. Targeted deletion of the DP gene in mice resulted in early embryonic lethality.[102] Rescuing the extraembryonic tissue defects in these animals further revealed cardiac, epidermal, and neuroepithelial abnormalities at later stages of development, but these animals eventually succumbed to a failure in microvasculature development.[103] This latter finding emphasized a less appreciated role of DP in endothelial cell junctions, where it complexes with classical cadherins.[104] Finally, the targeted deletion of DP in the epidermis resulted in neonatal lethality as a consequence of severe adhesive defects in the skin.[105] Although desmosomes were capable of forming in the epidermis of these animals, the incorporation of IFs into these structures was severely compromised. A lack of DP was also associated with adherens junction abnormalities, suggesting that the processes regulated by DP in the desmosome have a direct influence on the maturation of other junctional complexes.

Several human diseases have been identified that target the DP gene. For example, autosomal dominant mutations in the DP gene cause epidermal defects in a form of SPPK.[106,107]

Similarly, autosomal recessive mutations that result in the premature deletion of the C PRD in DP cause SPPK, woolly hair, and left ventricular cardiomyopathy.[108] A point mutation in a DP N-terminal serine residue implicated in binding Pg has been identified in patients with cardiomyopathy, leading the authors to speculate that the disruption of desmosomal protein interactions might contribute to this disease.[109]

Plectin

Plectin is a versatile member of the plakin family that is able to interact with all three components of the cytoskeleton and comprises an N-terminal actin-binding domain, a PD, a coiled-coil rod domain, and six C-terminal PRDs (five B and one C subdomain), followed by a microtubule-binding domain.[110-112] The L sequence between the fifth and sixth PRD is capable of binding IFs.[97] As this plakin family member directly interacts with DP, it can be localized to desmosomes but appears to play a more primary role within the context of integrin-containing hemidesmosomes that mediate interactions between epithelial cells and the basement membrane.[113,114] In particular, targeted deletion of the plectin gene in mice results in decreased hemidesmosomes in the basal epidermis.[115] Furthermore, homozygous mutations in the human gene cause skin blisters in epidermolysis bullosa simplex with late-onset muscular dystrophy.[116]

Envoplakin and Periplakin

EVPL and PPL serve as substrates for the transglutaminase enzyme responsible for cross-linking protein complexes in the cornified envelope of the skin.[117-119] Both of these plakins contain an N-terminal PD and a coiled-coil rod domain but EVPL contains a single C PRD downstream of an L sequence and PPL contains only the L sequence. As with DP and plectin, the L sequence has been implicated in regulating the interaction with IFs.[98,120] Both EVPL and PPL localize to epidermal desmosomes, probably as a heterodimeric complex. However, a role for these plakins in directly linking desmosomes to IFs has not been firmly established. Instead, a supportive role for EVPL in maintaining the barrier function of skin has recently been demonstrated, suggesting that these plakins might regulate the assembly of the epidermal cornified envelope.[121]

Other Molecular Components of Desmosomes

Several other proteins have been localized to desmosomes but their roles in this junctional complex remain poorly resolved. For example, pinin is a phosphoprotein present in the nucleus but also in regions surrounding epithelial desmosomes.[122,123] Overexpression of pinin inhibits corneal epithelial cell migration and binds directly to IFs, suggesting a structural role in desmosomes.[124,125] On the other hand, pinin mediates alternative splicing in COS cells.[126] Much less is know about the desmosomal role of the microtubule-binding cytoplasmic linker protein (CLIP)-170, beyond its localization to the desmosomal plaque in MDCK cells.[127] Similarly, an IF-binding protein, known as desmocalmin, and a calmodulin-binding protein, known as keratocalmin, have been localized to epidermal desmosomes.[128,129] Finally, the AHNAK gene encodes for a protein that was originally localized to desmosomes and termed desmoyokin, but was independently identified as a nuclear phosphoprotein in several cancer cell lines.[130,131] Desmoyokin/AHNAK is also present in nondesmosomal regions at the cell surface, where it might be involved in membrane repair and expansion processes.[132,133]

Functional Regulation of Desmosome Assembly

The expanding repertoire of desmosomal proteins suggests that intricate cellular processes coordinate the integration of these components into highly ordered complexes. Alterations in cell adhesion that occur during embryonic development, tissue remodeling, and wound healing require these junctional complexes be dynamic and under tight regulation. Recent studies have provided important insight into the molecular and cellular mechanisms that control desmosome dynamics.

One approach to understanding desmosome assembly has been to functionally reconstitute these junctions in cells that are not normally adhesive, namely fibroblastic L cells. The introduction of a single classical cadherin subtype into these cells promotes calcium-dependent homophilic adhesion.[134] In contrast, several desmosomal components, including Dsg, Dsc, and Pg, are required for pronounced adhesion in this culture system.[19,135-138] Although the overexpression of Dsg1, Dsc2a, and Pg failed to promote aggregation in one study, subsequent work in which Dsc2a was introduced under the control of an inducible promoter, along with Dsg1 and Pg, resulted in strong calcium-dependent adhesion.[137,138] Furthermore, the coordinate expression of Dsc1a, Dsc1b, Dsg1, and Pg not only enhanced L cell aggregation but also inhibited the invasive capacity of these cells.[19] These studies support the idea that desmosomes require multiple components to be coordinated at the cell surface in order to regulate adhesion.

Desmosomal-adhesion might not require DP in fibroblasts but this plakin is a key structural component of all desmosomes in vivo. The primary importance of DP is highlighted in a series of studies attempting to reconstitute ultrastructural plaques in cultured cells. Fibroblastic L cells transfected with Pg and a chimeric construct containing the extracellular domains of E-cad and the cytoplasmic domains of Dsg1 only assemble electron dense plaques when the N-terminal domain of DP is also present.[33] In COS cells expressing the same E-cad/Dsg1 chimera, the further addition of DP, Pg, and PKP1 is needed for efficient desmosomal plaque formation.[77] Similarly, endogenous Dsg2 clustering into the desmosomes of a fibrosarcoma-derived cell line was dependent on DP, Pg, and PKP2.[83] PKP2 was capable of promoting desmosome-like plaques in these cells even in the presence of a DP mutant construct lacking the C-terminal IF binding domain. Thus, the PKPs appear to play a key role in clustering desmosomal complexes at the cell surface but our understanding of the signals that mediate this assembly process remains limited.

Switching cultured epithelial cells from low (< 0.1mM) to high (> 1.2mM) calcium containing media allows for desmosome formation, as assessed by an increase in the stability and insolubility of desmosomal proteins, recruitment to cell borders, and attachment of IFs.[139-152] Although desmosomal proteins are still synthesized in low calcium media, their organization within the cytoplasm is complex. For example, cytoplasmic clusters of DP in association with IFs have been proposed as precursors for desmosome assembly triggered by an increase in extracellular calcium.[141] This is consistent with the biochemical and spatial dissociation of desmosomal cadherins and DP during a calcium switch in MDCK cells.[153] In contrast, membrane-associated half-desmosomes that are constantly endocytosed have been described in keratinocytes cultured in low calcium for extended periods of time, but these are only stabilized to form intact desmosomes once cell-cell contacts are initiated.[146,152] The tracking of fluorescently-tagged desmosomal proteins in living cells should help sort out how these distinct precursors are assembled and transported to the cell surface during junction formation.

Calcium directly influences classical and desmosomal cadherin extracellular interactions but might also trigger intracellular signaling pathways in cultured epithelial cells.[22,23,154] For example, the activation of protein kinase C (PKC) by calcium has been described in human keratinocyte cultures.[155] PKC activation is capable of promoting desmosome assembly even in low calcium conditions and influences the desmosomes present in confluent epithelial sheets, where these junctions become calcium-independent.[140,156,157] After wounding, the desmosomes in confluent MDCK cultures revert to a calcium-dependent state but this transition could be blocked by the inhibition of PKC.[158] Identifying the desmosomal targets that are influenced by these intracellular signaling pathways remains an active area of investigation.

The posttranslational regulation of desmosomal proteins can reorganize desmosomes and alter the adhesive characteristics of epithelial cells. For example, the phosphorylation of a specific serine residue present in the C-terminal domain of DP disrupts the ability of this plakin to interact with cytoplasmic IFs and promotes the incorporation of DP into desmosomes (Godsel and Green, manuscript in preparation).[159] Okadaic acid is an inhibitor of serine phosphatases that is capable of interfering with desmosome assembly in MDCK cells, providing further support for the hypothesis that desmosomal protein phosphorylation regulates the organization of these junctions.[160] Desmosomal proteins also interact directly with signaling molecules and this might negatively regulate desmosome assembly. For example, Pg has been found in a complex with the epidermal growth factor (EGF)-receptor and the protein tyrosine phosphatases, LAR and PTPκ.[161-163] Stimulation of A431 cells with EGF not only resulted in the tyrosine phosphorylation of Pg but this *arm* protein was no longer detectable in a complex with DP.[164]

Understanding how desmosomes are disassembled is likely to yield important information regarding the maintenance of these junctions. Although most studies have focused on the posttranslational regulation of desmosome dynamics, transcription factors have also been identified as candidates for disrupting these complexes. For example, Slug is a zinc-finger protein capable of inducing cell separation in a rat bladder cancer cell line, at least in part, through a loss of desmosomal proteins.[165] Desmosomal proteins are also targeted for cleavage by caspases and matrix metalloproteinases as cells undergo apoptosis (Dusek and Green, unpublished observations).[166,167] A similar role for proteases in regulating desmosomal organization might accompany cellular processes that require alterations in cell adhesion, such as tumor invasion, wound healing, and differentiation.

Conclusions

Desmosomes have classically been defined as adhesive units that are critical for maintaining tissue integrity and genetic studies of desmosomal proteins in humans and mice has borne out this hypothesis. Recent work has emphasized that these junctions are not merely static structures that bind cells together but are highly dynamic complexes that vary in composition and activity in different cell types. The further characterization of novel proteins within desmosomes, some of which act as components or downstream targets of intracellular signaling pathways, will not only help us understand how these junctions are regulated at the cell surface but also provides insight into their ability to instruct cells to respond to the extracellular environment in developmental and diseased states.

Acknowledgements

The authors would like to thank Rachel L. Dusek for the critical reading of the manuscript. The authors are supported by grants from the NIH, RO1AR43380, RO1AR41836, project #4 of PO1 DE12328 (K.G.), and NIH T32 AR07593 (L.M.G.), and a CIHR postdoctoral fellowship (S.G.).

References

1. Farquhar MG, Palade GE. Junctional complexes in various epithelia. J Cell Biol 1963; 17:375-412.
2. Skerrow CJ, Matoltsy AG. Chemical characterization of isolated epidermal desmosomes. J Cell Biol 1974; 63(2 Pt 1):524-530.
3. Drochmans P, Freudenstein C, Wanson JC et al. Structure and biochemical composition of desmosomes and tonofilaments isolated from calf muzzle epidermis. J Cell Biol 1978; 79(2 Pt 1):427-443.
4. Gorbsky G, Steinberg MS. Isolation of the intercellular glycoproteins of desmosomes. J Cell Biol 1981; 90(1):243-248.
5. Mueller H, Franke WW. Biochemical and immunological characterization of desmoplakins I and II, the major polypeptides of the desmosomal plaque. J Mol Biol 1983; 163(4):647-671.
6. White FH, Gohari K. Desmosomes in hamster cheek pouch epithelium: Their quantitative characterization during epithelial differentiation. J Cell Sci 1984; 66:411-429.
7. Borrmann CM, Mertens C, Schmidt A et al. Molecular diversity of plaques of epithelial-adhering junctions. Ann NY Acad Sci 2000; 915:144-150.
8. Nollet F, Kools P, van Roy F. Phylogenetic analysis of the cadherin superfamily allows identification of six major subfamilies besides several solitary members. J Mol Biol 2000; 299(3):551-572.
9. Hatzfeld M. The armadillo family of structural proteins. Int Rev Cytol 1999; 186:179-224.
10. Leung CL, Green KJ, Liem RK. Plakins: A family of versatile cytolinker proteins. Trends Cell Biol 2002; 12(1):37-45.

11. Kljuic A, Bazzi H, Sundberg JP et al. Desmoglein 4 in hair follicle differentiation and epidermal adhesion: Evidence from inherited hypotrichosis and acquired pemphigus vulgaris. Cell 2003; 113(2):249-260.
12. Whittock NV, Bower C. Genetic evidence for a novel human desmosomal cadherin, desmoglein 4. J Invest Dermatol 2003; 120(4):523-530.
13. Collins JE, Legan PK, Kenny TP et al. Cloning and sequence analysis of desmosomal glycoproteins 2 and 3 (desmocollins): Cadherin-like desmosomal adhesion molecules with heterogeneous cytoplasmic domains. J Cell Biol 1991; 113(2):381-391.
14. Parker AE, Wheeler GN, Arnemann J et al. Desmosomal glycoproteins II and III. Cadherin-like junctional molecules generated by alternative splicing. J Biol Chem 1991; 266(16):10438-10445.
15. Kljuic A, Christiano AM. A novel mouse desmosomal cadherin family member, desmoglein 1gamma. Exp Dermatol 2003; 12(1):20-29.
16. Pulkkinen L, Choi YW, Kljuic A et al. Novel member of the mouse desmoglein gene family: Dsg1-beta. Exp Dermatol 2003; 12(1):11-19.
17. Blaschuk OW, Sullivan R, David S et al. Identification of a cadherin cell adhesion recognition sequence. Dev Biol 1990; 139(1):227-229.
18. Noe V, Fingleton B, Jacobs K et al. Release of an invasion promoter E-cadherin fragment by matrilysin and stromelysin-1. J Cell Sci 2001; 114(Pt 1):111-118.
19. Tselepis C, Chidgey M, North A et al. Desmosomal adhesion inhibits invasive behavior. Proc Natl Acad Sci USA 1998; 95(14):8064-8069.
20. Runswick SK, O'Hare MJ, Jones L et al. Desmosomal adhesion regulates epithelial morphogenesis and cell positioning. Nat Cell Biol 2001; 3(9):823-830.
21. Chitaev NA, Troyanovsky SM. Direct Ca2+-dependent heterophilic interaction between desmosomal cadherins, desmoglein and desmocollin, contributes to cell-cell adhesion. J Cell Biol 1997; 138(1):193-201.
22. Troyanovsky RB, Klingelhofer J, Troyanovsky S. Removal of calcium ions triggers a novel type of intercadherin interaction. J Cell Sci 1999; 112(Pt 23):4379-4387.
23. Syed SE, Trinnaman B, Martin S et al. Molecular interactions between desmosomal cadherins. Biochem J 2002; 362(Pt 2):317-327.
24. Yap AS, Niessen CM, Gumbiner BM. The juxtamembrane region of the cadherin cytoplasmic tail supports lateral clustering, adhesive strengthening, and interaction with p120ctn. J Cell Biol 1998; 141(3):779-789.
25. Thoreson MA, Anastasiadis PZ, Daniel JM et al. Selective uncoupling of p120(ctn) from E-cadherin disrupts strong adhesion. J Cell Biol 2000; 148(1):189-202.
26. Butz S, Kemler R. Distinct cadherin-catenin complexes in Ca(2+)-dependent cell-cell adhesion. FEBS Lett 1994; 355(2):195-200.
27. Nathke IS, Hinck L, Swedlow JR et al. Defining interactions and distributions of cadherin and catenin complexes in polarized epithelial cells. J Cell Biol 1994; 125(6):1341-1352.
28. Korman NJ, Eyre RW, Klaus-Kovtun V et al. Demonstration of an adhering-junction molecule (plakoglobin) in the autoantigens of pemphigus foliaceus and pemphigus vulgaris. N Engl J Med 1989; 321(10):631-635.
29. Troyanovsky SM, Eshkind LG, Troyanovsky RB et al. Contributions of cytoplasmic domains of desmosomal cadherins to desmosome assembly and intermediate filament anchorage. Cell 1993; 72(4):561-574.
30. Mathur M, Goodwin L, Cowin P. Interactions of the cytoplasmic domain of the desmosomal cadherin Dsg1 with plakoglobin. J Biol Chem 1994; 269(19):14075-14080.
31. Troyanovsky SM, Troyanovsky RB, Eshkind LG et al. Identification of the plakoglobin-binding domain in desmoglein and its role in plaque assembly and intermediate filament anchorage. J Cell Biol 1994; 127(1):151-160.
32. Troyanovsky SM, Troyanovsky RB, Eshkind LG et al. Identification of amino acid sequence motifs in desmocollin, a desmosomal glycoprotein, that are required for plakoglobin binding and plaque formation. Proc Natl Acad Sci USA 1994; 91(23):10790-10794.
33. Kowalczyk AP, Bornslaeger EA, Borgwardt JE et al. The amino-terminal domain of desmoplakin binds to plakoglobin and clusters desmosomal cadherin-plakoglobin complexes. J Cell Biol 1997; 139(3):773-784.
34. Andl CD, Stanley JR. Central role of the plakoglobin-binding domain for desmoglein 3 incorporation into desmosomes. J Invest Dermatol 2001; 117(5):1068-1074.
35. Bonne S, Gilbert B, Hatzfeld M et al. Defining desmosomal plakophilin-3 interactions. J Cell Biol 2003; 161(2):403-416.
36. Hatzfeld M, Haffner C, Schulze K et al. The function of plakophilin 1 in desmosome assembly and actin filament organization. J Cell Biol 2000; 149(1):209-222.

37. Rutman AJ, Buxton RS, Burdett ID. Visualisation by electron microscopy of the unique part of the cytoplasmic domain of a desmoglein, a cadherin-like protein of the desmosome type of cell junction. FEBS Lett 1994; 353(2):194-196.
38. Garrod D, Chidgey M, North A. Desmosomes: Differentiation, development, dynamics and disease. Curr Opin Cell Biol 1996; 8(5):670-678.
39. Ishii K, Green KJ. Cadherin function: Breaking the barrier. Curr Biol Jul 24 2001; 11(14):R569-572.
40. North AJ, Chidgey MA, Clarke JP et al. Distinct desmocollin isoforms occur in the same desmosomes and show reciprocally graded distributions in bovine nasal epidermis. Proc Natl Acad Sci USA 1996; 93(15):7701-7705.
41. Kurzen H, Moll I, Moll R et al. Compositionally different desmosomes in the various compartments of the human hair follicle. Differentiation 1998; 63(5):295-304.
42. Wan H, Dopping-Hepenstal PJ, Gratian MJ et al. Desmosomes exhibit site-specific features in human palm skin. Exp Dermatol 2003; 12(4):378-388.
43. Wan H, Stone MG, Simpson C et al. Desmosomal proteins, including desmoglein 3, serve as novel negative markers for epidermal stem cell-containing population of keratinocytes. J Cell Sci 2003;Pt.
44. Wu H, Stanley JR, Cotsarelis G. Desmoglein isotype expression in the hair follicle and its cysts correlates with type of keratinization and degree of differentiation. J Invest Dermatol 2003; 120(6):1052-1057.
45. Allen E, Yu QC, Fuchs E. Mice expressing a mutant desmosomal cadherin exhibit abnormalities in desmosomes, proliferation, and epidermal differentiation. J Cell Biol 1996; 133(6):1367-1382.
46. Elias PM, Matsuyoshi N, Wu H et al. Desmoglein isoform distribution affects stratum corneum structure and function. J Cell Biol 2001; 153(2):243-249.
47. Merritt AJ, Berika MY, Zhai W et al. Suprabasal desmoglein 3 expression in the epidermis of transgenic mice results in hyperproliferation and abnormal differentiation. Mol Cell Biol 2002; 22(16):5846-5858.
48. Amagai M. Autoimmunity against desmosomal cadherins in pemphigus. J Dermatol Sci 1999; 20(2):92-102.
49. Amagai M, Matsuyoshi N, Wang ZH et al. Toxin in bullous impetigo and staphylococcal scalded-skin syndrome targets desmoglein 1. Nat Med 2000; 6(11):1275-1277.
50. Amagai M, Yamaguchi T, Hanakawa Y et al. Staphylococcal exfoliative toxin B specifically cleaves desmoglein 1. J Invest Dermatol 2002; 118(5):845-850.
51. Hanakawa Y, Schechter NM, Lin C et al. Molecular mechanisms of blister formation in bullous impetigo and staphylococcal scalded skin syndrome. J Clin Invest 2002; 110(1):53-60.
52. Rickman L, Simrak D, Stevens HP et al. N-terminal deletion in a desmosomal cadherin causes the autosomal dominant skin disease striate palmoplantar keratoderma. Hum Mol Genet 1999; 8(6):971-976.
53. Hunt DM, Rickman L, Whittock NV et al. Spectrum of dominant mutations in the desmosomal cadherin desmoglein 1, causing the skin disease striate palmoplantar keratoderma. Eur J Hum Genet 2001; 9(3):197-203.
54. Serre G, Mils V, Haftek M et al. Identification of late differentiation antigens of human cornified epithelia, expressed in reorganized desmosomes and bound to cross-linked envelope. J Invest Dermatol 1991; 97(6):1061-1072.
55. Simon M, Montezin M, Guerrin M et al. Characterization and purification of human corneodesmosin, an epidermal basic glycoprotein associated with corneocyte-specific modified desmosomes. J Biol Chem 1997; 272(50):31770-31776.
56. Jonca N, Guerrin M, Hadjiolova K et al. Corneodesmosin, a component of epidermal corneocyte desmosomes, displays homophilic adhesive properties. J Biol Chem 2002; 277(7):5024-5029.
57. Levy-Nissenbaum E, Betz RC, Frydman M et al. Hypotrichosis simplex of the scalp is associated with nonsense mutations in CDSN encoding corneodesmosin. Nat Genet 2003; 34(2):151-153.
58. Scheffers MS, van der Bent P, Prins F et al. Polycystin-1, the product of the polycystic kidney disease 1 gene, colocalizes with desmosomes in MDCK cells. Hum Mol Genet 2000; 9(18):2743-2750.
59. Riggleman B, Wieschaus E, Schedl P. Molecular analysis of the armadillo locus: Uniformly distributed transcripts and a protein with novel internal repeats are associated with a Drosophila segment polarity gene. Genes Dev 1989; 3(1):96-113.
60. Peifer M, Polakis P. Wnt signaling in oncogenesis and embryogenesis—a look outside the nucleus. Science 2000; 287(5458):1606-1609.
61. Huber AH, Nelson WJ, Weis WI. Three-dimensional structure of the armadillo repeat region of beta-catenin. Cell 1997; 90(5):871-882.

62. Chitaev NA, Leube RE, Troyanovsky RB et al. Troyanovsky SM. The binding of plakoglobin to desmosomal cadherins: Patterns of binding sites and topogenic potential. J Cell Biol 1996; 133(2):359-369.
63. Wahl JK, Sacco PA, McGranahan-Sadler TM et al. Plakoglobin domains that define its association with the desmosomal cadherins and the classical cadherins: Identification of unique and shared domains. J Cell Sci 1996; 109 (Pt 5):1143-1154.
64. Witcher LL, Collins R, Puttagunta S et al. Desmosomal cadherin binding domains of plakoglobin. J Biol Chem 1996; 271(18):10904-10909.
65. Chen X, Bonne S, Hatzfeld M et al. Protein binding and functional characterization of plakophilin 2. Evidence for its diverse roles in desmosomes and beta -catenin signaling. J Biol Chem 2002; 277(12):10512-10522.
66. Sacco PA, McGranahan TM, Wheelock MJ et al. Identification of plakoglobin domains required for association with N-cadherin and alpha-catenin. J Biol Chem 1995; 270(34):20201-20206.
67. Palka HL, Green KJ. Roles of plakoglobin end domains in desmosome assembly. J Cell Sci 1997; 110 (Pt 19):2359-2371.
68. Lewis JE, Wahl 3rd, Sass KM et al. Cross-talk between adherens junctions and desmosomes depends on plakoglobin. J Cell Biol 1997; 136(4):919-934.
69. Parker HR, Li Z, Sheinin H, Lauzon G et al. Plakoglobin induces desmosome formation and epidermoid phenotype in N-cadherin-expressing squamous carcinoma cells deficient in plakoglobin and E-cadherin. Cell Motil Cytoskeleton 1998; 40(1):87-100.
70. Bierkamp C, McLaughlin KJ, Schwarz H et al. Embryonic heart and skin defects in mice lacking plakoglobin. Dev Biol 1996; 180(2):780-785.
71. Ruiz P, Brinkmann V, Ledermann B et al. Targeted mutation of plakoglobin in mice reveals essential functions of desmosomes in the embryonic heart. J Cell Biol 1996; 135(1):215-225.
72. Bierkamp C, Schwarz H, Huber O et al. Desmosomal localization of beta-catenin in the skin of plakoglobin null-mutant mice. Development 1999; 126(2):371-381.
73. Caldelari R, de Bruin A, Baumann D et al. A central role for the armadillo protein plakoglobin in the autoimmune disease pemphigus vulgaris. J Cell Biol 2001; 153(4):823-834.
74. Charpentier E, Lavker RM, Acquista E et al. Plakoglobin suppresses epithelial proliferation and hair growth in vivo. J Cell Biol 2000; 149(2):503-520.
75. McKoy G, Protonotarios N, Crosby A et al. Identification of a deletion in plakoglobin in arrhythmogenic right ventricular cardiomyopathy with palmoplantar keratoderma and woolly hair (Naxos disease). Lancet 2000; 355(9221):2119-2124.
76. Smith EA, Fuchs E. Defining the interactions between intermediate filaments and desmosomes. J Cell Biol 1998; 141(5):1229-1241.
77. Bornslaeger EA, Godsel LM, Corcoran CM et al. Plakophilin 1 interferes with plakoglobin binding to desmoplakin, yet together with plakoglobin promotes clustering of desmosomal plaque complexes at cell-cell borders. J Cell Sci 2001; 114(Pt 4):727-738.
78. Kowalczyk AP, Hatzfeld M, Bornslaeger EA et al. The head domain of plakophilin-1 binds to desmoplakin and enhances its recruitment to desmosomes. Implications for cutaneous disease. J Biol Chem 1999; 274(26):18145-18148.
79. Hofmann I, Mertens C, Brettel M et al. Interaction of plakophilins with desmoplakin and intermediate filament proteins: An in vitro analysis. J Cell Sci 2000; 113(Pt 13):2471-2483.
80. McGrath JA, McMillan JR, Shemanko CS et al. Mutations in the plakophilin 1 gene result in ectodermal dysplasia/skin fragility syndrome. Nat Genet 1997; 17(2):240-244.
81. Moll I, Kurzen H, Langbein L et al. The distribution of the desmosomal protein, plakophilin 1, in human skin and skin tumors. J Invest Dermatol 1997; 108(2):139-146.
82. South AP, Wan H, Stone MG et al. Lack of plakophilin 1 increases keratinocyte migration and reduces desmosome stability. J Cell Sci 2003;116(Pt 16):3303-14.
83. Koeser J, Troyanovsky SM, Grund C et al. De novo formation of desmosomes in cultured cells upon transfection of genes encoding specific desmosomal components. Exp Cell Res 2003; 285(1):114-130.
84. Muller J, Ritt DA, Copeland TD et al. Functional analysis of C-TAK1 substrate binding and identification of PKP2 as a new C-TAK1 substrate. Embo J 2003; 22(17):4431-4442.
85. Mertens C, Hofmann I, Wang Z et al. Nuclear particles containing RNA polymerase III complexes associated with the junctional plaque protein plakophilin 2. Proc Natl Acad Sci USA 2001; 98(14):7795-7800.
86. Hatzfeld M, Nachtsheim C. Cloning and characterization of a new armadillo family member, p0071, associated with the junctional plaque: Evidence for a subfamily of closely related proteins. J Cell Sci 1996; 109 (Pt 11):2767-2778.
87. Hatzfeld M, Green KJ, Sauter H. Targeting of p0071 to desmosomes and adherens junctions is mediated by different protein domains. J Cell Sci 2003; 116(Pt 7):1219-1233.

88. Calkins CC, Hoepner BL, Law CM et al. The Armadillo family protein p0071 is a VE-cadherin- and desmoplakin-binding protein. J Biol Chem 2003; 278(3):1774-1783.
89. Deguchi M, Iizuka T, Hata Y et al. PAPIN. A novel multiple PSD-95/Dlg-A/ZO-1 protein interacting with neural plakophilin-related armadillo repeat protein/delta-catenin and p0071. J Biol Chem 2000; 275(38):29875-29880.
90. Izawa I, Nishizawa M, Tomono Y et al. ERBIN associates with p0071, an armadillo protein, at cell-cell junctions of epithelial cells. Genes Cells 2002; 7(5):475-485.
91. Jaulin-Bastard F, Arsanto JP, Le Bivic A et al. Interaction between Erbin and a Catenin-related protein in epithelial cells. J Biol Chem 2002; 277(4):2869-2875.
92. Ruhrberg C, Watt FM. The plakin family: Versatile organizers of cytoskeletal architecture. Curr Opin Genet Dev 1997; 7(3):392-397.
93. Angst BD, Nilles LA, Green KJ. Desmoplakin II expression is not restricted to stratified epithelia. J Cell Sci 1990; 97(Pt 2):247-257.
94. O'Keefe EJ, Erickson HP, Bennett V. Desmoplakin I and desmoplakin II. Purification and characterization. J Biol Chem 1989; 264(14):8310-8318.
95. North AJ, Bardsley WG, Hyam J et al. Molecular map of the desmosomal plaque. J Cell Sci 1999; 112(Pt 23):4325-4336.
96. Choi HJ, Park-Snyder S, Pascoe LT et al. Structures of two intermediate filament-binding fragments of desmoplakin reveal a unique repeat motif structure. Nat Struct Biol 2002; 9(8):612-620.
97. Nikolic B, Mac Nulty E, Mir B et al. Basic amino acid residue cluster within nuclear targeting sequence motif is essential for cytoplasmic plectin-vimentin network junctions. J Cell Biol 1996; 134(6):1455-1467.
98. DiColandrea T, Karashima T, Maatta A et al. Subcellular distribution of envoplakin and periplakin: Insights into their role as precursors of the epidermal cornified envelope. J Cell Biol 2000; 151(3):573-586.
99. Fontao L, Favre B, Riou S et al. Interaction of the Bullous Pemphigoid Antigen 1 (BP230) and Desmoplakin with Intermediate Filaments Is Mediated by Distinct Sequences within Their COOH Terminus. Mol Biol Cell 2003; 14(5):1978-1992.
100. Bornslaeger EA, Corcoran CM, Stappenbeck TS et al. Breaking the connection: Displacement of the desmosomal plaque protein desmoplakin from cell-cell interfaces disrupts anchorage of intermediate filament bundles and alters intercellular junction assembly. J Cell Biol 1996; 134(4):985-1001.
101. Huen AC, Park JK, Godsel LM et al. Intermediate filament-membrane attachments function synergistically with actin-dependent contacts to regulate intercellular adhesive strength. J Cell Biol 2002; 159(6):1005-1017.
102. Gallicano GI, Kouklis P, Bauer C et al. Desmoplakin is required early in development for assembly of desmosomes and cytoskeletal linkage. J Cell Biol 1998; 143(7):2009-2022.
103. Gallicano GI, Bauer C, Fuchs E. Rescuing desmoplakin function in extra-embryonic ectoderm reveals the importance of this protein in embryonic heart, neuroepithelium, skin and vasculature. Development 2001; 128(6):929-941.
104. Kowalczyk AP, Navarro P, Dejana E et al. VE-cadherin and desmoplakin are assembled into dermal microvascular endothelial intercellular junctions: A pivotal role for plakoglobin in the recruitment of desmoplakin to intercellular junctions. J Cell Sci 1998; 111(Pt 20):3045-3057.
105. Vasioukhin V, Bowers E, Bauer C et al. Desmoplakin is essential in epidermal sheet formation. Nat Cell Biol 2001; 3(12):1076-1085.
106. Armstrong DK, McKenna KE, Purkis PE et al. Haploinsufficiency of desmoplakin causes a striate subtype of palmoplantar keratoderma. Hum Mol Genet 1999; 8(1):143-148.
107. Whittock NV, Ashton GH, Dopping-Hepenstal PJ et al. Striate palmoplantar keratoderma resulting from desmoplakin haploinsufficiency. J Invest Dermatol 1999; 113(6):940-946.
108. Norgett EE, Hatsell SJ, Carvajal-Huerta L et al. Recessive mutation in desmoplakin disrupts desmoplakin-intermediate filament interactions and causes dilated cardiomyopathy, woolly hair and keratoderma. Hum Mol Genet 2000; 9(18):2761-2766.
109. Rampazzo A, Nava A, Malacrida S et al. Mutation in human desmoplakin domain binding to plakoglobin causes a dominant form of arrhythmogenic right ventricular cardiomyopathy. Am J Hum Genet 2002; 71(5):1200-1206.
110. Svitkina TM, Verkhovsky AB, Borisy GG. Plectin sidearms mediate interaction of intermediate filaments with microtubules and other components of the cytoskeleton. J Cell Biol 1996; 135(4):991-1007.
111. Fuchs P, Zorer M, Rezniczek GA et al. Unusual 5' transcript complexity of plectin isoforms: Novel tissue-specific exons modulate actin binding activity. Hum Mol Genet 1999; 8(13):2461-2472.

112. Sun D, Leung CL, Liem RK. Characterization of the microtubule binding domain of microtubule actin crosslinking factor (MACF): Identification of a novel group of microtubule associated proteins. J Cell Sci 2001; 114(Pt 1):161-172.
113. Eger A, Stockinger A, Wiche G et al. Polarisation-dependent association of plectin with desmoplakin and the lateral submembrane skeleton in MDCK cells. J Cell Sci 1997; 110(Pt 11):1307-1316.
114. Nievers MG, Kuikman I, Geerts D et al. Formation of hemidesmosome-like structures in the absence of ligand binding by the (alpha)6(beta)4 integrin requires binding of HD1/plectin to the cytoplasmic domain of the (beta)4 integrin subunit. J Cell Sci 2000; 113(Pt 6):963-973.
115. Andra K, Lassmann H, Bittner R et al. Targeted inactivation of plectin reveals essential function in maintaining the integrity of skin, muscle, and heart cytoarchitecture. Genes Dev 1997; 11(23):3143-3156.
116. Pulkkinen L, Smith FJ, Shimizu H et al. Homozygous deletion mutations in the plectin gene (PLEC1) in patients with epidermolysis bullosa simplex associated with late-onset muscular dystrophy. Hum Mol Genet 1996; 5(10):1539-1546.
117. Ruhrberg C, Hajibagheri MA, Simon M et al. Envoplakin, a novel precursor of the cornified envelope that has homology to desmoplakin. J Cell Biol 1996; 134(3):715-729.
118. Ruhrberg C, Hajibagheri MA, Parry DA et al. Periplakin, a novel component of cornified envelopes and desmosomes that belongs to the plakin family and forms complexes with envoplakin. J Cell Biol 1997; 139(7):1835-1849.
119. Marekov LN, Steinert PM. Ceramides are bound to structural proteins of the human foreskin epidermal cornified cell envelope. J Biol Chem 1998; 273(28):17763-17770.
120. Karashima T, Watt FM. Interaction of periplakin and envoplakin with intermediate filaments. J Cell Sci 2002; 115(Pt 24):5027-5037.
121. Maatta A, DiColandrea T, Groot K et al. Gene targeting of envoplakin, a cytoskeletal linker protein and precursor of the epidermal cornified envelope. Mol Cell Biol 2001; 21(20):7047-7053.
122. Ouyang P, Sugrue SP. Identification of an epithelial protein related to the desmosome and intermediate filament network. J Cell Biol 1992; 118(6):1477-1488.
123. Ouyang P, Sugrue SP. Characterization of pinin, a novel protein associated with the desmosome-intermediate filament complex. J Cell Biol Nov 1996; 135(4):1027-1042.
124. Shi J, Sugrue SP. Dissection of protein linkage between keratins and pinin, a protein with dual location at desmosome-intermediate filament complex and in the nucleus. J Biol Chem 2000; 275(20):14910-14915.
125. Shi Y, Tabesh M, Sugrue SP. Role of cell adhesion-associated protein, pinin (DRS/memA), in corneal epithelial migration. Invest Ophthalmol Vis Sci 2000; 41(6):1337-1345.
126. Wang P, Lou PJ, Leu S et al. Modulation of alternative premRNA splicing in vivo by pinin. Biochem Biophys Res Commun 2002; 294(2):448-455.
127. Wacker IU, Rickard JE, De Mey JR et al. Accumulation of a microtubule-binding protein, pp170, at desmosomal plaques. J Cell Biol 1992; 117(4):813-824.
128. Tsukita S. Desmocalmin: A calmodulin-binding high molecular weight protein isolated from desmosomes. J Cell Biol 1985; 101(6):2070-2080.
129. Fairley JA, Scott GA, Jensen KD et al. Characterization of keratocalmin, a calmodulin-binding protein from human epidermis. J Clin Invest 1991; 88(1):315-322.
130. Hieda Y, Tsukita S. A new high molecular mass protein showing unique localization in desmosomal plaque. J Cell Biol 1989; 109(4 Pt 1):1511-1518.
131. Shtivelman E, Bishop JM. The human gene AHNAK encodes a large phosphoprotein located primarily in the nucleus. J Cell Biol 1993; 120(3):625-630.
132. Masunaga T, Shimizu H, Ishiko A et al. Desmoyokin/AHNAK protein localizes to the nondesmosomal keratinocyte cell surface of human epidermis. J Invest Dermatol 1995; 104(6):941-945.
133. Borgonovo B, Cocucci E, Racchetti G et al. Regulated exocytosis: A novel, widely expressed system. Nat Cell Biol 2002; 4(12):955-962.
134. Nagafuchi A, Shirayoshi Y, Okazaki K et al. Transformation of cell adhesion properties by exogenously introduced E-cadherin cDNA. Nature 1987; 329(6137):341-343.
135. Amagai M, Karpati S, Klaus-Kovtun V et al. Extracellular domain of pemphigus vulgaris antigen (desmoglein 3) mediates weak homophilic adhesion. J Invest Dermatol 1994; 102(4):402-408.
136. Chidgey MA, Clarke JP, Garrod DR. Expression of full-length desmosomol glycoproteins (desmocollins) is not sufficient to confer strong adhesion on transfected L929 cells. J Invest Dermatol 1996; 106(4):689-695.
137. Kowalczyk AP, Borgwardt JE, Green KJ. Analysis of desmosomal cadherin-adhesive function and stoichiometry of desmosomal cadherin-plakoglobin complexes. J Invest Dermatol 1996; 107(3):293-300.

138. Marcozzi C, Burdett ID, Buxton RS et al. Coexpression of both types of desmosomal cadherin and plakoglobin confers strong intercellular adhesion. J Cell Sci 1998; 111 (Pt 4):495-509.
139. Hennings H, Michael D, Cheng C et al. Calcium regulation of growth and differentiation of mouse epidermal cells in culture. Cell 1980; 19(1):245-254.
140. Watt FM, Mattey DL, Garrod DR. Calcium-induced reorganization of desmosomal components in cultured human keratinocytes. J Cell Biol 1984; 99(6):2211-2215.
141. Jones JC, Goldman RD. Intermediate filaments and the initiation of desmosome assembly. J Cell Biol 1985; 101(2):506-517.
142. Mattey DL, Garrod DR. Calcium-induced desmosome formation in cultured kidney epithelial cells. J Cell Sci 1986; 85:95-111.
143. Mattey DL, Garrod DR. Splitting and internalization of the desmosomes of cultured kidney epithelial cells by reduction in calcium concentration. J Cell Sci 1986; 85:113-124.
144. Penn EJ, Burdett ID, Hobson C et al. Structure and assembly of desmosome junctions: Biosynthesis and turnover of the major desmosome components of Madin-Darby canine kidney cells in low calcium medium. J Cell Biol 1987; 105(5):2327-2334.
145. Penn EJ, Hobson C, Rees DA et al. Structure and assembly of desmosome junctions: Biosynthesis, processing, and transport of the major protein and glycoprotein components in cultured epithelial cells. J Cell Biol 1987; 105(1):57-68.
146. Duden R, Franke WW. Organization of desmosomal plaque proteins in cells growing at low calcium concentrations. J Cell Biol 1988; 107(3):1049-1063.
147. Pasdar M, Nelson WJ. Kinetics of desmosome assembly in Madin-Darby canine kidney epithelial cells: Temporal and spatial regulation of desmoplakin organization and stabilization upon cell-cell contact. I. Biochemical analysis. J Cell Biol 1988; 106(3):677-685.
148. Pasdar M, Nelson WJ. Kinetics of desmosome assembly in Madin-Darby canine kidney epithelial cells: Temporal and spatial regulation of desmoplakin organization and stabilization upon cell-cell contact. II. Morphological analysis. J Cell Biol 1988; 106(3):687-695.
149. Pasdar M, Nelson WJ. Regulation of desmosome assembly in epithelial cells: Kinetics of synthesis, transport, and stabilization of desmoglein I, a major protein of the membrane core domain. J Cell Biol 1989; 109(1):163-177.
150. Penn EJ, Hobson C, Rees DA et al. The assembly of the major desmosome glycoproteins of Madin-Darby canine kidney cells. FEBS Lett 1989; 247(1):13-16.
151. Mattey DL, Burdge G, Garrod DR. Development of desmosomal adhesion between MDCK cells following calcium switching. J Cell Sci 1990; 97(Pt 4):689-704.
152. Demlehner MP, Schafer S, Grund C et al. Continual assembly of half-desmosomal structures in the absence of cell contacts and their frustrated endocytosis: A coordinated Sisyphus cycle. J Cell Biol 1995; 131(3):745-760.
153. Pasdar M, Krzeminski KA, Nelson WJ. Regulation of desmosome assembly in MDCK epithelial cells: Coordination of membrane core and cytoplasmic plaque domain assembly at the plasma membrane. J Cell Biol 1991; 113(3):645-655.
154. Leckband D. The structure of the C-cadherin ectodomain resolved. Structure (Camb) 2002; 10(6):739-740.
155. Kitajima Y, Aoyama Y, Seishima M. Transmembrane signaling for adhesive regulation of desmosomes and hemidesmosomes, and for cell-cell datachment induced by pemphigus IgG in cultured keratinocytes: Involvement of protein kinase C. J Investig Dermatol Symp Proc 1999; 4(2):137-144.
156. Sheu HM, Kitajima Y, Yaoita H. Involvement of protein kinase C in translocation of desmoplakins from cytosol to plasma membrane during desmosome formation in human squamous cell carcinoma cells grown in low to normal calcium concentration. Exp Cell Res 1989; 185(1):176-190.
157. van Hengel J, Gohon L, Bruyneel E et al. Protein kinase C activation upregulates intercellular adhesion of alpha-catenin-negative human colon cancer cell variants via induction of desmosomes. J Cell Biol 1997; 137(5):1103-1116.
158. Wallis S, Lloyd S, Wise I et al. The alpha isoform of protein kinase C is involved in signaling the response of desmosomes to wounding in cultured epithelial cells. Mol Biol Cell 2000; 11(3):1077-1092.
159. Stappenbeck TS, Lamb JA, Corcoran CM et al. Phosphorylation of the desmoplakin COOH terminus negatively regulates its interaction with keratin intermediate filament networks. J Biol Chem 1994; 269(47):29351-29354.
160. Pasdar M, Li Z, Chan H. Desmosome assembly and disassembly are regulated by reversible protein phosphorylation in cultured epithelial cells. Cell Motil Cytoskeleton 1995; 30(2):108-121.
161. Hoschuetzky H, Aberle H, Kemler R. Beta-catenin mediates the interaction of the cadherin-catenin complex with epidermal growth factor receptor. J Cell Biol 1994; 127(5):1375-1380.
162. Fuchs M, Muller T, Lerch MM et al. Association of human protein-tyrosine phosphatase kappa with members of the armadillo family. J Biol Chem 1996; 271(28):16712-16719.

163. Muller T, Choidas A, Reichmann E et al. Phosphorylation and free pool of beta-catenin are regulated by tyrosine kinases and tyrosine phosphatases during epithelial cell migration. J Biol Chem 1999; 274(15):10173-10183.
164. Gaudry CA, Palka HL, Dusek RL et al. Tyrosine-phosphorylated plakoglobin is associated with desmogleins but not desmoplakin after epidermal growth factor receptor activation. J Biol Chem 2001; 276(27):24871-24880.
165. Savagner P, Yamada KM, Thiery JP. The zinc-finger protein slug causes desmosome dissociation, an initial and necessary step for growth factor-induced epithelial-mesenchymal transition. J Cell Biol 1997; 137(6):1403-1419.
166. Brancolini C, Sgorbissa A, Schneider C. Proteolytic processing of the adherens junctions components beta-catenin and gamma-catenin/plakoglobin during apoptosis. Cell Death Differ 1998; 5(12):1042-1050.
167. Weiske J, Schoneberg T, Schroder W et al. The fate of desmosomal proteins in apoptotic cells. J Biol Chem 2001; 276(44):41175-41181.
168. Kowalczyk AP, Stappenbeck TS, Parry DA et al. Structure and function of desmosomal transmembrane core and plaque molecules. Biophys Chem 1994; 50(1-2):97-112.
169. Kljuic A, Gilead L, Martinez-Mir A et al. A nonsense mutation in the desmoglein 1 gene underlies striate keratoderma. Exp Dermatol 2003; 12(4):523-527.
170. McGrath JA, Hoeger PH, Christiano AM et al. Skin fragility and hypohidrotic ectodermal dysplasia resulting from ablation of plakophilin 1. Br J Dermatol 1999; 140(2):297-307.
171. Whittock NV, Wan H, Morley SM et al. Compound heterozygosity for nonsense and mis-sense mutations in desmoplakin underlies skin fragility/woolly hair syndrome. J Invest Dermatol 2002; 118(2):232-238.
172. Alcalai R, Metzger S, Rosenheck S et al. A recessive mutation in desmoplakin causes arrhythmogenic right ventricular dysplasia, skin disorder, and woolly hair. J Am Coll Cardiol 2003;42(2):319-327.

Chapter 11

Epithelial Cell Plasticity by Dynamic Transcriptional Regulation of E-Cadherin

Geert Berx and Frans Van Roy

Abstract

Epithelial cell plasticity is associated with coordinated changes in cell adhesion and migratory behavior. The cell-cell adhesion protein E-cadherin regulates the functional integrity of epithelia by mediating specific intercellular adhesion. E-cadherin is a well-established invasion/tumor suppressor. Dynamic transcriptional regulation is a major mechanism of controlling E-cadherin expression during embryogenesis and malignant progression of various epithelial tumors. A variety of transcription regulators implicated in both embryonic development and tumorigenesis have been described to regulate E-cadherin expression in a reversible way.

Introduction

Epithelial-mesenchymal transition (EMT) is a normal developmental process in which epithelial cells, attached to the underlying extracellular matrix, lose apical-basal polarity and migrate as separate cells into the matrix. The transition from an epithelial to a mesenchymal phenotype must be strictly controlled after embryonic development in order to avoid chaotic cell behavior in the adult organism. It is also well documented that EMT occurs to a variable degree during tumor progression in man, resulting in mesenchymal-like cells that invade and metastasize (see Chapter 9, Van Marck and Bracke). Thus, the invasive behavior of these cells can be largely explained by repression of the epithelial state. The calcium-dependent cell adhesion molecule E-cadherin is most important in the formation and maintenance of epithelial cell junctions. E-cadherin is a transmembrane protein consisting of an extracellular domain containing five repeat domains and a cytoplasmic domain that interacts with catenins to establish functional cell-cell adhesion. In normal epithelia, the E-cadherin/catenin complex is typically present in the adherens junctions between adjacent cells. The human E-cadherin gene (*CDH1*) is located on chromosome 16q22.1 and consists of 16 exons.[1] Recent research has demonstrated that disturbance of E-cadherin functionality has a causative role in modulation of the epithelial differentiation during both developmental processes and cancer progression. In embryonic development there is a clear spatiotemporal regulation of E-cadherin expression, which allows cell migration and morphogenesis. During gastrulation this dynamic regulation is exemplified with a decreasing E-cadherin expression in epiblast cells entering the primitive streak to form mesoderm (see also Chapter 2, Morali et al).[2] This EMT is associated with loss of epithelial differentiation and gain of migratory characteristics.[3,4] In a similar way, E-cadherin is reversibly downregulated during epithelial tissue healing, possibly allowing cells to migrate over areas of ulceration.[5] Later mesoderm derivatives remain negative for E-cadherin, except for those cells that undergo the reciprocal mesenchymal-epithelial transition (MET) where

Rise and Fall of Epithelial Phenotype: Concepts of Epithelial-Mesenchymal Transition, edited by Pierre Savagner.

E-cadherin is reexpressed. Such a conversion with upregulation of E-cadherin is taking place during for instance metanephric kidney development.[6] There are multifactorial mechanisms by which the expression and function of E-cadherin are regulated (for a review see ref. 7). This review specifically focuses on the transcriptional regulation of the E-cadherin gene and its aberrations during the process of tumor progression.

E-Cadherin Transcriptional Regulation

E-Cadherin Promoter Regulation

Our actual understanding of E-cadherin transcriptional regulation started with the identification of specific transcriptional regulatory elements in the E-cadherin promoter sequences of different species.

Transcription regulation of the gene encoding L-CAM (the chicken homologue of E-cadherin) has been studied first. The intergenic region between the K-CAM and L-CAM genes was proven to have low transcriptional activity without clear tissue specificity.[8] So far, no further detailed characterization of this promoter has been reached to explain the particular expression activity. Later, the upstream regulatory sequences of the mouse, dog and human E-cadherin genes have been identified.[9-11] These promoters show a high degree of sequence homology (Fig. 1). Detailed analysis of the mouse E-cadherin promoter revealed a modular structure with different critical transcription factor binding regions, which determine the regulation of the E-cadherin promoter.[9,12,13] The positive regulatory elements within the 5' proximal E-cadherin promoter include a CCAAT box and a GC box. Mutational inactivation of these factor-binding sequences drastically reduced the transcriptional activity of the E-cadherin promoter.[12]

An important regulatory element determining the epithelial specificity was confined to the E-pal element. This 12-bp long, tandemly duplicated palindromic E-box (-86,-81/-80,-75) was originally identified in the mouse E-cadherin gene.[9] Only the 3' E-box (-80,-75) from this mouse E-pal element is conserved in both the human and the canine promoter (Fig. 1).[10,11] In addition, a third (-30,-25) and the fourth E-box (+22, +27),[14,15] both located more downstream, are highly conserved across species and were shown to be functional in the human E-cadherin promoter. The most downstream fourth E-box is, however, absent from the mouse promoter.[9] Mutational inactivation of these different sequence elements results in upregulation of the E-cadherin promoter in mesenchymal cells and dedifferentiated tumor cell lines, whereas the wild-type promoter shows low activity in such cells.[9-11,14-16] In this context it should be noted that the divergent importance in epithelial specific expression, assigned in different studies to particular E-boxes, is most likely a reflection of difference in species and promoter sequence context.

In vivo footprinting analysis of the human E-cadherin promoter identified a second, more downstream GC box, which is involved in transcription factor binding although so far no function has been assigned to this element.[12]

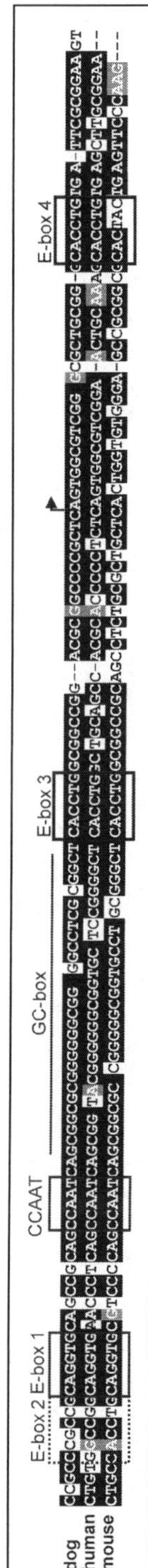

Figure 1. E-cadherin promoter homology across species. Alignment of the E-cadherin promoter sequences of dog, man and mouse. Conserved regulatory elements are indicated: E2-boxes 1, and 3, CCAAT box, and GC box. The E2-box 2 has been described as part of a palindromic E-pal sequence in the mouse E-cadherin promoter, but is conserved neither in canine nor in human sequences. E2-box 4 is conserved in the dog and human *CDH1* promoter but not in the mouse sequence.

Enhancer Regulation of the E-Cadherin Gene

Despite the detailed knowledge of the E-cadherin promoter and its epithelial specificity, it is doubtful that the complex expression regulation of the E-cadherin gene during embryonic development can be fully explained by the different elements identified so far in the *CDH1* promoter. Little is known about the role of transcriptional enhancer and silencer elements in the dynamic modulation of E-cadherin expression. The first evidence for the involvement of such transregulatory elements was the identification of an enhancer sequence in intron 2 of the L-CAM gene.[17] This enhancer element harbors putative binding sites for the transcription factors SP1, E2A and AP2, as well as a consensus binding site for hepatocyte nuclear factor-1 (HNF-1), which is a liver-enriched POU (Pit-Oct-Unc) homeodomain transcription factor.[8,18] In vitro analysis of the L-CAM promoter in combination with the intron-2 enhancer clearly showed that HNF-1 and HOXD9 can stimulate transcriptional activity in E-cadherin-negative fibroblasts through binding with the HNF-1 regulatory sequence in this enhancer sequence. This HNF-1 consensus binding site was also proven to be bound by transcription factors in vivo.[18] In conclusion, the low nonspecific promoter activity of the L-CAM gene was found to gain elevated specific expression during epithelial liver differentiation when combined with an intronic enhancer.

The existence of such intronically located regulatory elements in the large mouse E-cadherin gene appeared from the localization of hypersensitive sites in introns 1 and 2.[12] Indeed, in intron 1 a 149-bp long enhancer element was proven to be active. As this element exclusively activates transcription in epithelial cells, it was designated ESE (epithelial specific enhancer). It consists of three GC-rich subregions (EI, EII and EIII), which show differential factor binding in function of both the E-cadherin expression status and the cell context.[13] Importantly, the integrity of the whole ESE is necessary to confer enhancer activity. The enhancer regions EII and EIII contain consensus sequences for AP-2. Although the involvement of AP-2-related factors was demonstrated by footprint analysis, this cannot explain the epithelial specific expression of the ESE. So far, no further evidence is available on regulatory sequences in intron 2 of the mouse gene. On the other hand, we identified an enhancer-like element in the large intron 2 (65 kbp) of the human E-cadherin gene (G. Berx, P. Vermassen and F. Van Roy, manuscript in preparation).

Transcriptional Up-Regulation of E-Cadherin

The GC region in the mouse E-cadherin promoter contains two AP-2 consensus-binding sites and binds recombinant AP-2 (or an AP-2-like factor) in a tandem arrangement (Figs. 1 and 2).[13] Dominant-negative AP-2 lacking the transactivation domain was found to inhibit *CDH1* promoter activity in E-cadherin-positive cells.[13] Interestingly, the retinoblastoma protein (Rb) and c-Myc seem to act as coactivators for AP-2 binding to the E-cadherin promoter, and activate the E-cadherin expression exclusively in epithelial cells (Table 1 and Fig. 2).[19] Conversely, inactivation of Rb in MDCK cells causes downregulation of the proto-oncogene c-*myc* and induces EMT and invasiveness associated with loss of epithelial markers, including E-cadherin.[20] In conflict with an important role of AP2 in E-cadherin transcriptional regulation is the finding that only one AP-2 consensus sequence is present in the minimal human E-cadherin promoter and that this binding site is apparently not essential for high transcriptional activity.[21] Moreover, AP2 expression is not able to induce E-cadherin promoter activity in E-cadherin-negative breast cancer cells.[21] The latter phenomenon could be explained, since the breast cancer cell lines mentioned show a high coexpression of different transrepressors, which have the potential to overrule the activating effect of AP-2.[11,15]

Another candidate transcription factor interacting with the GC box of the E-cadherin promoter is the zinc finger protein WT1, encoded by the tumor suppressor gene *Wt1* affected in Wilms' tumor of the kidney. GC box binding by WT1 has been shown to activate the E-cadherin promoter (Table 1 and Fig. 2) and to induce epithelial differentiation of fibroblasts upon ectopic expression.[22] CP-2 and C/EBP-related nuclear factors seem to be responsible for the specific binding of the CCAAT box of the E-cadherin promoter.[23]

Table 1. Factors known to repress or to activate E-cadherin transcription

E-Cadherin Activators	Refs.	E-Cadherin Repressors	Refs.
AP2	13	COX-2	65
CP2	23	E12/E47	29
C/EBP	23	δEF1	40
E1A	13,40,41	ErbB2	61
HNF4	24	c-Fos	28
LIF1	27	ILK	50,53,109
c-Myc	19	SIP1	11
PAX2	26	Slug	15
Rb	19,20	Snail	14,16
WT1	22	TGF-β	48,49
		TNF-α	67

The transcription factor hepatocyte nuclear factor 4 (HNF4), member of the steroid hormone receptor superfamily, has been demonstrated to be an epithelial morphogen inducing E-cadherin gene expression.[24] Moreover, transforming growth factor (TGF)-β has been shown to upregulate Snail (see also below) and to downregulate HNF4 expression in murine hepatocytes.[25] During the formation of the urogenital system, epithelia arise from mesenchymal cells (MET). PAX2, an important factor for MET of kidney mesenchyme, induces an increase in WT-1 and E-cadherin expression in HEK293 human fetal kidney epithelial cells.[26] Finally, stimulation with leukemia inhibitory factor (LIF) induces epithelialization and tubulogenesis in metanephric mesenchymes, also coinciding with WT1 and E-cadherin mRNA expression.[27]

Transcriptional Down-Regulation of E-Cadherin

Transcriptional Repressors

The different E-boxes in the mouse and human E-cadherin promoter sequences were demonstrated to play a crucial role in determining the epithelial specific expression of the E-cadherin gene.[9,10] The most potent repressor effect of the E-boxes was found in fibroblasts and dedifferentiated cancer cells. Mutation of these sequence elements resulted in upregulation of the E-cadherin promoter in such cells. This epithelial discriminative role of the E-boxes suggested the existence of transcription factors able to downregulate E-cadherin and possibly playing an important role in epithelial tumor progression.

Fos oncogene activation in the mouse mammary epithelial cell line EP-FosER induces EMT and invasiveness with loss of E-cadherin mRNA expression (Table 1).[28] This downregulation of E-cadherin expression is accompanied by the loss of transcription factor binding in the GC region and the CCAAT box.[12] E-pal mutation results in increased promoter activity in the EP-FosER cell line, indicating that an E-box related silencing mechanism is involved.[13]

A yeast one-hybrid screening with the E-pal sequence as a bait identified Snail, Slug and E12/E47 as transcription factors directly binding to the proximal E-cadherin promoter (Table 1 and Fig. 2).[16,29] Snail and Slug are members of the Snail family of zinc finger DNA-binding transcription factors, which play critical roles in neural crest formation and mesoderm specification[30,31] Snail and Slug contain at their C-terminal end, respectively, four and five zinc finger domains of the C_2H_2 type. The N-terminal SNAG (Snail/Gfi) domain of Snail family members is important for the repressor activity of these proteins in mammalian cells.[32,33] The well-established role of Snail in EMT during embryogenesis reflects its possiíje participation during cancer progression.[31] Expression of Snail in MDCK cells leads to loss of differentiation, concomitantly with the loss of E-cadherin, and promotes the acquisition of invasive and

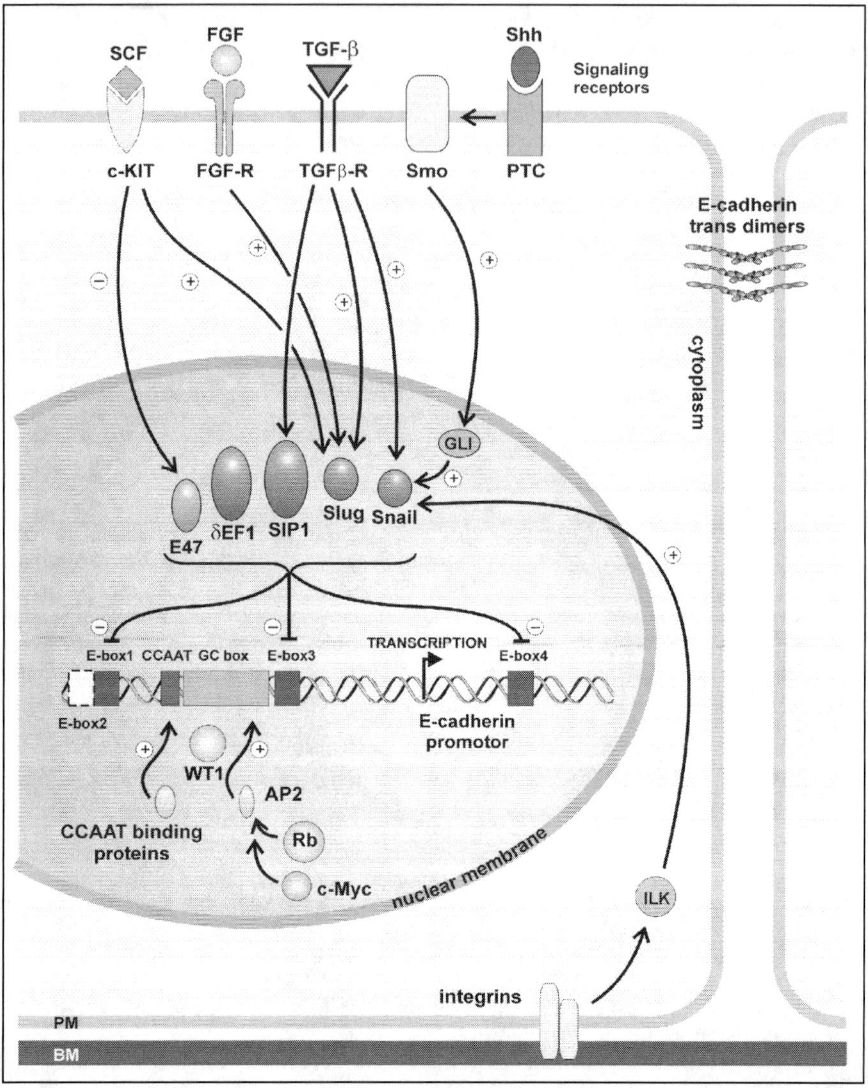

Figure 2. Overview of E-cadherin transcriptional regulation mechanisms. Different signaling pathways in mammalian cells have been implicated in influencing cell-cell adhesion by controlling E-cadherin transcriptional regulation. For the sake of simplicity the various cytoplasmic signal transduction steps leading to nuclear action of the different transcriptional repressors are not shown. The direct action mechanism and physiological relevance of the inductive (+) and repressing (-) signals of E-cadherin modulators remain to be further established. BM= basement membrane; ILK= integrin linked kinase; FGF= fibroblast growth factor; FGF-R= fibroblast growth factor receptor; PM= plasma membrane; PTC= Patched; Rb= Retinoblastoma tumor suppressor protein; SCF= stem cell factor; Shh= sonic hedgehog protein; SIP1= Smad interacting protein 1; Smo= Smoothened; TGF-β1= transforming growth factor-β1; TGF-βR= transforming growth factor receptor; WT1= Wilms' tumor 1 gene product.

migratory behaviors.[14,16] In addition, a Snail retrogene (Snail-like) was proven to have maintained its potential to induce EMT and to modulate invasive and migratory properties of epithelial cells.[34] Compared to Snail, the ability of Slug to suppress E-cadherin seems to be more dependent on the cellular context.[35,36] Ectopic expression of Slug in MDCK cells induces EMT with complete loss of E-cadherin expression and gain of migratory behavior.[37] This is in agreement with the finding that Slug downregulates E-cadherin efficiently in breast cancer cells.[15] In contrast, overexpression of Slug in NBT-II bladder carcinoma cells triggers desmosome dissociation with loss of desmoplakin and desmoglein expression, but is unable to downregulate E-cadherin in these cells and also unable to promote the expression of mesenchymal markers or the acquisition of cell motility.[36]

SIP1 (ZEB2) belongs to the δEF-1 or ZEB protein family. These proteins are characterized by a homeodomain flanked by two separated, highly conserved zinc fingers: an N-terminal and a C-terminal one, which contain four and three zinc fingers, respectively.[38] Each zinc finger can bind independently to CACCT sequences.[39] SIP1 shows specific binding to the E2-boxes of the E-cadherin promoter (Fig. 2). Conditional expression of SIP1 in E-cadherin-positive cells abrogates E-cadherin-mediated intercellular adhesion and simultaneously induces invasion.[11] δEF1 (ZEB1), a homologue of SIP1, has also been shown to be a potent repressor of E-cadherin transcription (Table 1 and Fig. 2).[40] In this context it is important to mention that mesenchymal to epithelial conversions can be induced by adenovirus E1A.[12,41] As δEF-1 (ZEB1) is known to suppress transcription by recruitment of the transcription corepressor protein CtBP,[42] the underlying mechanism for the E1A-mediated upregulation of E-cadherin becomes clear. Indeed, the E1A protein interacts with CtBP and in this way activates epithelial gene expression.[40]

Also the class-I basic helix-loop-helix transcription factor E12/E47, encoded by the *E2A* gene, is a potent repressor of E-cadherin transcription through binding of the E-pal element in the E-cadherin promoter (Table 1 and Fig. 2).[29] Expression of E47 in MDCK cells results in EMT and induction of a migratory phenotype. Another bHLH factor binding the E-pal sequence is the product Immunoglobulin Transcription Factor-2 of the *ITF-2* gene.[29]

The surprisingly high redundancy of different transcription factors so far characterized as repressors of E-cadherin transcription is, however, reasonable when one takes into account the variety of different tissues that arise from embryonic epithelial cells. The existence of the different transcriptional repressors could therefore be a reflection of the need for epithelial gene repression at different points in time and space during various developmental and physiological processes. For instance, the transcription factor Snail is expressed in areas undergoing active EMT, whereas Slug and E12/E47 are excluded from those areas, but are expressed in effectively migrating cells.[16,29,43] Homozygous SIP1 gene disruption results in loss of normal E-cadherin downregulation in the neuro-epithelium, which strongly argues that SIP1 is indeed a physiologically relevant repressor of the E-cadherin gene at particular stages of development.[44]

It is intriguing to observe that several embryonic tissues and many human cancer cell lines show simultaneous coexpression of different *CDH1* transcriptional repressors having overlapping binding sites. Regarding this particular coexpression it is noteworthy that so far no comparable protein expression data have been reported, due to lack of highly specific antibodies. However, cooperation between the different transcription factors cannot be excluded at this stage of research. It is therefore feasible that expression of one transcriptional repressor needs additional expression of another to reach a critical level of repression for complete loss of E-cadherin expression. The existence of such complementary simultaneous repression is suggested by the mouse Snail knockout, where E-cadherin expression in mesodermal cells is elevated, but still lower than in ectodermal cells of the same mutant embryos; this indicates that other repressors play a supporting role during gastrulation.[45] The expression pattern of the *E2a* gene in early mouse embryos suggests that this transcription factor might be involved in the maintenance of the mesenchymal state, whereas other transcriptional repressors, like Snail, initiate the EMT process.[29] In addition, it was shown that Snail expression can induce δEF-1 (ZEB1) in HT29 M6 cells and that this coexpression was necessary to obtain strong downregulation of E-cadherin.[46] On the other hand, the recently proven differences in binding affinities of Snail, E47 and Slug for the mouse E-pal sequence suggest that the different overexpression studies could be misleading as to *CDH1* being a primary target for all E-cadherin repressor proteins.[36] It could therefore well be that under coexpressing circumstances, some E-box binding factors have complementary but specific target gene sets that are important in epithelial differentiation. The basis for this could be differences in affinity for E-boxes in another promoter sequence context. Identification of the different transcriptional repressors of other target genes in addition to *CDH1* will clarify this. Under coexpressing circumstances, the different E-box binding factors may compete directly with each other for the same binding sequences, but they may also synergize because of a broader spectrum of genes efficiently targeted.

E-Cadherin Repressing Signaling Pathways

Although the upstream signals are largely unknown today, there is clear evidence that various signaling pathways converge in the different abovementioned transcriptional repressors of E-cadherin to modulate epithelial cell plasticity. During embryonic development, a number of growth factors, such as TGF-β1, TGF-β2, bone morphogenetic protein (BMP) and fibroblast growth factor (FGF), can induce Snail and Slug upregulation (Fig. 2).[31] Similarly, during reepithelialization of a skin wound, an interface of epithelial-mesenchymal cell status arises where epidermal cells transiently downregulate intercellular adhesion and dramatically rearrange their cytoskeleton. In this context, different factors comprising epidermal growth factor (EGF), TGF-α, FGFs and TGF-β play key roles during efficient epidermal repair (see Chapter 8: Arnoux et al). Some of these growth factors, e.g., FGF and TGF-β, have a strong inducing effect in keratinocytes on the transcriptional factors Snail and Slug.[36,47] TGF-β is able to repress E-cadherin transcription.[48,49] SIP1, initially isolated on the basis of its interaction with TGF-β regulated Smad proteins, can effectively be induced by TGF-β treatment.[11]

Overexpression of integrin-linked kinase (ILK), a component of focal adhesion plaques in epithelial cells, downregulates E-cadherin concomitantly with stimulation of lymphocyte-enhancing factor (LEF-1)/β-catenin signaling, induction of anchorage-independent growth and oncogenic transformation.[50-52] Inhibition of ILK expression in colon cancer cells results in repression of Snail transcription and induction of E-cadherin expression (Fig. 2).[53] Likewise, in HT-144 melanoma cells TGF-β1 specifically induces expression of an ILK isoform, which also results in loss of E-cadherin mRNA expression.[54]

Recently, it became clear that the transcription factor oncogene *GLI* is a potent inducer of Snail expression (Fig. 2).[55] This oncogene encodes a zinc finger protein activated by Sonic Hedgehog signaling. Mutations in the latter pathway induce *GLI* expression in basal cell carcinoma. E-cadherin shows a frequently decreased expression in human basal cell carcinoma.[56,57] GLI induction of Snail, which mediates EMT and downregulates E-cadherin, is consistent with the locally invasive nature of these tumors.

Elevated signaling via the tyrosine kinase receptor c-KIT induces Slug expression in leukemia cells.[58] Constitutive activation of c-KIT by mutations in solid tumors could therefore confer invasive properties on tumor cells.[59,60] Overexpression of the proto-oncogene encoding the transmembrane tyrosine kinase receptor ErbB2 (HER-2/neu) causes a decrease in E-cadherin gene transcription.[61] However, this correlation could not be confirmed when large series of breast cancer cell lines and tumors were investigated.[62]

Cyclooxygenase 2 or prostaglandin endoperoxide synthase 2 (COX-2) is markedly increased in human colorectal adenocarcinomas,[63] as well as in invasive breast cancers and adjacent ductal carcinomas in situ.[64] Increased COX-2 expression in rat intestinal epithelial cells results in decreased expression of E-cadherin (Table 1),[65] and has been shown to increase the metastatic potential of human colon cancer cells.[66] Stimulation of Caco2 colon cancer cells with TNF-α results in a reversible transcriptional downregulation of E-cadherin and also in a modest reduction in β-catenin expression.[67] Sex steroids have an impact on the E-cadherin expression status, but this seems to be critically dependent on the cellular system used. Administration of estrogens to endometrial cancer cells disturbs proper cell-cell adhesion and reduces steady-state mRNA levels of E-cadherin, whereas anti-estrogenic compounds, such as progesterone or danazol, reverse this effect.[68]

E-Cadherin Transrepression in Human Tumors

E-cadherin is a well-established invasion/tumor suppressor.[69-73] Extensive analyses made clear that aberrant E-cadherin expression, resulting from somatic inactivating mutations of both E-cadherin alleles, is rare and so far largely confined to diffuse gastric carcinomas and infiltrative lobular breast carcinomas.[74-76]

RT-PCR analysis, Northern hybridization and RNA in situ hybridization analysis revealed that reduced E-cadherin immunoreactivity, frequently reported for thyroid, colon, ductal breast and prostate carcinomas, correlates with decreased mRNA levels.[70,77-85] Thorough analysis of expression data on primary carcinomas has to reveal whether the transcription factors reported to downregulate *CDH1* indeed contribute to decreased E-cadherin expression in particular tumor types. An inverse correlation between E-cadherin and Snail expression could be shown for different hepatocellular carcinoma cell lines in vitro and in vivo.[86] Also in ductal breast carcinomas, increased Snail expression has been found to be common and to be strongly associated with reduced E-cadherin expression.[85] High-grade breast tumors and lymph-node positive tumors consistently show strong Snail expression.[87] High expression of SIP1, associated with loss of E-cadherin expression, was reported in gastric cancers of the intestinal type, whereas Snail does not seem to be involved.[88] In contrast, diffuse gastric cancer, a tumor subtype frequently affected by E-cadherin inactivating mutations, shows upregulated Snail.[88] Strikingly, all these correlative expression studies of different transcriptional repressors versus E-cadherin were performed at the mRNA level due to lack of available specific antibodies. In this context, it should be taken into account that contaminating fibroblasts in the tumor stroma could contribute to the identified expression status of the different transcriptional repressors.

In ductal breast carcinomas, a negative correlation has been found between expression of E-cadherin/catenin complexes and ErbB2 receptor overexpression.[89] In a set of 210 breast cancers, p53 accumulation associated with both ErbB2 expression and reduced E-cadherin expression.[62] However, no statistically significant association was seen when E-cadherin expression and ErbB2 overexpression were compared in this analysis. The highly aggressive inflammatory breast cancer form, which is exceptional due to its persistent E-cadherin expression, shows no association either between E-cadherin and ErbB2 expression.[90] Also in cholangiocarcinomas of the liver, E-cadherin downregulation does not correlate with ErbB2 expression, whereas β-catenin downregulation does.[91]

E-Cadherin Gene Silencing by CpG Hypermethylation

Besides transrepression of the E-cadherin gene, hypermethylation has been postulated to result in loss of E-cadherin expression. Methylation analysis using methylation-sensitive enzymes, methylation-specific PCR and in vivo footprinting revealed that loss of E-cadherin expression in tumoral tissue is correlated with methylation within the promoter or the 5' CpG island of the E-cadherin gene.[1,80,81,92,93] Interesting, though not understood, is the remarkable variable distribution of methylated CpGs in the CpG islands of silenced cell lines.[94]

So far, CpG hypermethylation in the E-cadherin promoter region concomitant with decreased E-cadherin expression has been reported for primary invasive ductal breast carcinomas,[85,95] lobular breast carcinomas,[96] diffuse and intestinal gastric carcinomas,[97-100] hepatocellular carcinomas,[93] oral squamous-cell carcinomas,[101,102] bladder neoplasms,[103] and also leukemia.[104,105] This methylation seems to be dynamic and unstable, reflecting the heterogeneous loss of E-cadherin protein expression during malignant cancer progression. Such dynamic increase in E-cadherin promoter CpG methylation has been documented during hepatocarcinogenesis, more particularly during transition from precancerous chronic hepatitis and cirrhosis to hepatocellular carcinomas,[93] and during malignant progression from ductal carcinoma in situ to metastatic lesions.[95,106] In addition, these epigenetic changes can be mimicked in vitro, depending on microenvironmental conditions favoring either homotypic cell adhesion or in vitro invasion.[106]

DNA methylation in combination with loss of heterozygocity (LOH) could be a major inactivation mechanism for sporadic cancers. Remarkably, in patients with hereditary diffuse gastric cancers (HDGC) LOH was consistently undetected in tumor cells; in contrast, *CDH1* germline mutations were accompanied with methylation of the E-cadherin promoter as second hit.[107] Interestingly, in E-cadherin-silenced cell lines, methylated CpGs present in the *CDH1* CpG island are bound by the methyl CpG-binding proteins MeCP2 and MBD2.[94] These methyl-binding proteins recruit histone deacetylase (HDAC) activity, which could explain the proven deacetylated histone H3 associated with the methylated *CDH1* CpG island in E-cadherin-silenced cell lines.[94] Direct involvement of hypermethylation in suppressing E-cadherin gene expression is further supported by the observation that E-cadherin expression can be reactivated by treatment with the demethylating agent 5-aza-2'-deoxycytidine (5AzaC) in some carcinoma cell lines.[80,81] Conversely, E-cadherin expression could not be restored in somatic cell hybrids resulting from fusions between E-cadherin-positive cell lines and cell lines with a methylated inactive E-cadherin promoter.[108] In the latter hybrids, transacting repression might explain the inability to reactivate E-cadherin expression by treatment 5AzaC. Thus, loss of E-cadherin mRNA is definitely not only attributable to hypermethylation.

Further research has to reveal whether E-cadherin gene silencing results from cooperation between transcriptional (dys)regulations and DNA methylation, or whether these are independent mechanisms.

Acknowledgments

G.B. is supported by the Fund for Scientific Research (FWO) – Flanders. Research of the authors has been supported by the Association for International Cancer Research (AICR - UK), the Geconcerteerde Onderzoeksacties of Ghent University, FWO - Flanders, Fortis Verzekeringen (Belgium) and Interuniversitaire attractiepolen (Belgian Goverment).

References

1. Berx G, Staes K, van Hengel J et al. Cloning and characterization of the human invasion suppressor gene E-cadherin (CDH1). Genomics 1995; 26:281-289.
2. Takeichi M, Matsunami H, Inoue T et al. Roles for cadherins in patterning of the developing brain. Dev Neurosci 1997; 19(1):86-87.
3. Butz S, Larue L. Expression of catenins during mouse embryonic development and in adult tissues. Cell Adhes Commun 1995; 3(4):337.
4. Hatta K, Takeichi M. Expression of N-Cadherin adhesion molecules associated with early morphogenetic events in chick development. Nature 1986; 320:447-449.
5. Dogan A, Wang ZD, Spencer J. E-cadherin expression in intestinal epithelium. J Clin Pathol 1995; 48(2):143-146.
6. Kuure S, Vuolteenaho R, Vainio S. Kidney morphogenesis: Cellular and molecular regulation. Mech Develop 2000; 92(1):31-45.
7. Strumane K, van Roy F, Berx G. The role of E-cadherin in epithelial differentiation and cancer progression. In: Pandalai SG, ed. Recent Research Development in Cellular Biochemistry. Kerala, India: Transworld Research Network; 2003; in press.
8. Sorkin BC, Jones FS, Cunningham BA et al. Identification of the promoter and a transcriptional enhancer of the gene encoding L-CAM, a calcium-dependent cell adhesion molecule. Proc Natl Acad Sci USA. 1993; 90(23):11356-11360.
9. Behrens J, Löwrick O, Klein-Hitpass L et al. The E-cadherin promoter: functional analysis of a G. C-rich region and an epithelial cell-specific palindromic regulatory element. Proc Natl Acad Sci USA 1991; 88(24):11495-11499.
10. Giroldi LA, Bringuier PP, de Weijert M et al. Role of E boxes in the repression of E-cadherin expression. Biochem Biophys Res Commun 1997; 241:453-458.
11. Comijn J, Berx G, Vermassen P et al. The two-handed E-box-binding zinc finger protein SIP1 downregulates E-cadherin and induces invasion. Mol Cell 2001; 7:1267-1278.
12. Hennig G, Behrens J, Truss M et al. Progression of carcinoma cells is associated with alterations in chromatin structure and factor binding at the E-cadherin promoter in vivo. Oncogene 1995; 11(3):475-484.
13. Hennig G, Lowrick O, Birchmeier W et al. Mechanisms identified in the transcriptional control of epithelial gene expression. J Biol Chem 1996; 271(1):595-602.

14. Batlle E, Sancho E, Franci C et al. The transcription factor Snail is a repressor of E-cadherin gene expression in epithelial tumour cells. Nat Cell Biol 2000; 2(2):84-89.
15. Hajra KM, Chen DY, Fearon ER. The SLUG zinc-finger protein represses E-cadherin in breast cancer. Cancer Res 2002; 62(6):1613-1618.
16. Cano A, Perez-Moreno MA, Rodrigo I et al. The transcription factor Snail controls epithelial-mesenchymal transitions by repressing E-cadherin expression. Nat Cell Biol 2000; 2(2):76-83.
17. Begemann M, Tan S S, Cunningham BA et al. Expression of chicken liver cell adhesion molecule fusion genes in transgenic mice. Proc Natl Acad Sci USA 1990; 87:9042-9046.
18. Goomer RS, Holst BD, Wood IC et al. Regulation in vitro of an L-CAM enhancer by homeobox genes HoxD9 and HNF-1. Proc Natl Acad Sci USA 1994; 91(17):7985-7989.
19. Batsche E, Muchardt C, Behrens J et al. RB and c-Myc activate expression of the E-cadherin gene in epithelial cells through interaction with transcription factor AP-2. Mol Cell Biol 1998; 18(7):1-12.
20. Martel C, Harper F, Cereghini S et al. Inactivation of retinoblastoma family proteins by SV40 T antigen results in creation of a hepatocyte growth factor/scatter factor autocrine loop associated with an epithelial-fibroblastoid conversion and invasiveness. Cell Growth Differ 1997; 8(2):165-178.
21. Ji XD, Woodard AS, Rimm DL et al. Transcriptional defects underlie loss of E-cadherin expression in breast cancer. Cell Growth & Differentiation 1997; 8(7):773-778.
22. Hosono S, Gross I, English MA et al. E-cadherin is a WT1 target gene. J Biol Chem 2000; 275(15):10943-10953.
23. Faraldo MLM, Rodrigo I, Behrens J et al. Analysis of the E-cadherin and P-cadherin promoters in murine keratinocyte cell lines from different stages of mouse skin carcinogenesis. Mol Carcinogen 1997; 20(1):33-47.
24. Spath GF, Weiss MC. Hepatocyte nuclear factor 4 provokes expression of epithelial marker genes, acting as a morphogen in dedifferentiated hepatoma cells. J Cell Biol 1998; 140(4):935-946.
25. Spagnoli FM, Cicchini C, Tripodi M et al. Inhibition of MMH (Met murine hepatocyte) cell differentiation by TGF(beta) is abrogated by pretreatment with the heritable differentiation effector FGF1. J Cell Sci 2000; 113(Pt 20):3639-3647.
26. Torban E, Goodyer PR. Effects of PAX2 expression in a human fetal kidney (HEK293) cell line. Biochim Biophys Acta - Mol Cell Res 1998; 1401(1):53-62.
27. Barasch J, Yang J, Ware CB et al. Mesenchymal to epithelial conversion in rat metanephros is induced by LIF. Cell 1999; 99(4):377-386.
28. Reichmann E, Schwarz H, Deiner EM et al. Activation of an inducible c-FosER fusion protein causes loss of epithelial polarity and triggers epithelial-fibroblastoid cell conversion. Cell 1992; 71:1103-1116.
29. Perez-Moreno MA, Locascio A, Rodrigo I et al. A new role for E12/E47 in the repression of E-cadherin expression and epithelial-mesenchymal transitions. J Biol Chem 2001; 276(29):27424-27431.
30. Leptin M. Morphogenesis - control of epithelial cell shape changes. Curr Biol 1994; 4(8):709-712.
31. Nieto MA. The snail superfamily of zinc finger transcription factors. Nat Rev Mol Cell Biol. 2002; 3:155-166.
32. Grimes HL, Chan TO, Zweidler-McKay PA et al. The Gfi-1 proto-oncoprotein contains a novel transcriptional repressor domain, SNAG, and inhibits G1 arrest induced by interleukin-2 withdrawal. Mol Cell Biol 1996; 16(11):6263-6272.
33. Nakayama H, Scott IC, Cross JC. The transition to endoreduplication in trophoblast giant cells is regulated by the mSNA zinc finger transcription factor. Dev Biol 1998; 199(1):150-163.
34. Locascio A, Vega S, De Frutos CA et al. Biological potential of a human SNAIL retrogene. J Biol Chem 2002; 31:31.
35. Ros MA, Sefton M, Nieto MA. Slug, a zinc finger gene previously implicated in the early patterning of the mesoderm and the neural crest, is also involved in chick limb development. Development 1997; 124(9):1821-1829.
36. Savagner P, Yamada KM, Thiery JP. The zinc-finger protein slug causes desmosome dissociation, an initial and necessary step for growth factor-induced epithelial-mesenchymal transition. J Cell Biol 1997; 137(6):1403-1419.
37. Bolos V, Peinado H, Perez-Moreno MA et al. The transcription factor Slug represses E-cadherin expression and induces epithelial to mesenchymal transitions: A comparison with Snail and E47 repressors. J Cell Sci 2003; 116(Pt 3):499-511.
38. Verschueren K, Remacle JE, Collart C et al. SIP1, a novel zinc finger/homeodomain repressor, interacts with Smad proteins and binds to 5'-CACCT sequences in candidate target genes. J Biol Chem 1999; 274(29):20489-20498.
39. Remacle JE, Kraft H, Lerchner W et al. New mode of DNA binding of multi-zinc finger transcription factors: δEF1 family members bind with two hands to two target sites. EMBO J 1999; 18(18):5073-5084.

40. Grooteclaes ML, Frisch SM. Evidence for a function of CtBP in epithelial gene regulation and anoikis. Oncogene 2000; 19(33):3823-3828.
41. Frisch SM. E1a induces the expression of epithelial characteristics. J Cell Biol 1994; 127(4):1085-1096.
42. Postigo AA, Dean DC. ZEB represses transcription through interaction with the corepressor CtBP. Proc Natl Acad Sci USA 1999; 96(12):6683-6688.
43. Sefton M, Sanchez S, Nieto MA. Conserved and divergent roles for members of the Snail family of transcription factors in the chick and mouse embryo. Development 1998; 125:3111-3121.
44. Van de Putte T, Maruhashi M, Francis A et al. Mice lacking Zfhx1b, the gene that codes for Smad- interacting protein-1, reveal a role for multiple neural crest cell defects in the etiology of Hirschsprung disease- mental retardation syndrome. Am J Hum Genet 2003; 72(2):465-470.
45. Carver EA, Jiang R, Lan Y et al. The mouse snail gene encodes a key regulator of the epithelial-mesenchymal transition. Mol Cell Biol 2001; 21(23):8184-8188.
46. Guaita S, Puig I, Franci C et al. Snail induction of epithelial to mesenchymal transition in tumor cells is accompanied by MUC1 repression and ZEB1 expression. J Biol Chem 2002; 277(42):39209-39216.
47. Zavadil J, Bitzer M, Liang D et al. Genetic programs of epithelial cell plasticity directed by transforming growth factor-beta. Proc Natl Acad Sci USA 2001; 98(12):6686-6691.
48. Miettinen PJ, Ebner R, Lopez AR et al. TGF-beta induced transdifferentiation of mammary epithelial cells to mesenchymal cells: Involvement of type I receptors. J Cell Biol 1994; 127(6):2021-2036.
49. Piek E, Moustakas A, Heldin CH et al. TGF-β type I receptor/ALK-5 and Smad proteins mediate epithelial to mesenchymal transdifferentiation in NMuMG breast epithelial cells. J Cell Sci 1999; 112:4557-4568.
50. Wu CY, Keightley SY, Leung-Hagesteijn C et al. Integrin-linked protein kinase regulates fibronectin matrix assembly, E-cadherin expression, and tumorigenicity. J Biol Chem 1998; 273(1):528-536.
51. Novak A, Hsu SC, Leung Hagensteijn C et al. Cell adhesion and the integrin-linked kinase regulate the LEF-1 and beta-catenin signaling pathways. Proc Natl Acad Sci USA 1998; 95(8):4374-4379.
52. Wu C, Dedhar S. Integrin-linked kinase (ILK) and its interactors: A new paradigm for the coupling of extracellular matrix to actin cytoskeleton and signaling complexes. J Cell Biol 2001; 155(4):505-510.
53. Tan C, Costello P, Sanghera J et al. Inhibition of integrin linked kinase (ILK) suppresses beta-catenin-Lef/Tcf-dependent transcription and expression of the E-cadherin repressor, snail, in APC-/- human colon carcinoma cells. Oncogene. 2001; 20(1):133-140.
54. Janji B, Melchior C, Vallar L et al. Cloning of an isoform of integrin-linked kinase (ILK) that is upregulated in HT-144 melanoma cells following TGF-beta1 stimulation. Oncogene 2000; 19(27):3069-3077.
55. Louro ID, Bailey EC, Li XN et al. Comparative gene expression profile analysis of GLI and c-MYC in an epithelial model of malignant transformation. Cancer Res 2002; 62(20):5867-5873.
56. Pizarro A, Benito N, Navarro P et al. E-cadherin expression in basal cell carcinoma. Br J Cancer 1994; 69(1):157-162.
57. Kooy AJW, Tank B, deJong AAW et al. Expression of E-cadherin, alpha- & &BETA;-catenin, and CD44V(6) and the subcellular localization of E-cadherin and CD44V(6) in normal epidermis and basal cell carcinoma. Hum Pathol 1999; 30(11):1328-1335.
58. Perez-Losada J, Sanchez-Martin M, Rodriguez-Garcia A et al. Zinc-finger transcription factor Slug contributes to the function of the stem cell factor c-kit signaling pathway. Blood. 2002;100(4):1274-1286.
59. Matsuda R, Takahashi T, Nakamura S et al. Expression of the c-kit protein in human solid tumors and in corresponding fetal and adult normal tissues. Am J Pathol 1993; 142(1):339-346.
60. Inoue M, Kyo S, Fujita M et al. Coexpression of the c-kit receptor and the stem cell factor in gynecological tumors. Cancer Res 1994; 54(11):3049-3053.
61. D'Souza B, Taylor-Papadimitriou J. Overexpression of erb-B2 in human mammary epithelial cells signals inhibition of transcription of the E-cadherin gene. Proc Natl Acad Sci USA 1994; 91(15):7202-7206.
62. Bukholm IK, Nesland JM, Karesen R et al. Expression of E-cadherin and its relation to the p53 protein status in human breast carcinomas. Virchows Arch - Int J Pathol 1997; 431(5):317-321.
63. Eberhart CE, Coffey RJ, Radhika A et al. Up-regulation of cyclooxygenase 2 gene expression in human colorectal adenomas and adenocarcinomas. Gastroenterol 1994; 107(4):1183-1188.
64. Half E, Tang XM, Gwyn K et al. Cyclooxygenase-2 expression in human breast cancers and adjacent ductal carcinoma in situ. Cancer Res 2002; 62(6):1676-1681.

65. Tsujii M, DuBois RN. Alterations in cellular adhesion and apoptosis in epithelial cells overexpressing prostaglandin endoperoxide synthase 2. Cell 1995; 83(3):493-501.
66. Tsujii M, Kawano S, DuBois RN. Cyclooxygenase-2 expression in human colon cancer cells increases metastatic potential. Proc Natl Acad Sci USA 1997; 94:3336-3340.
67. Perry I, Tselepis C, Hoyland J et al. Reduced cadherin/catenin complex expression in celiac disease can be reproduced in vitro by cytokine stimulation. Lab Invest 1999; 79(12):1489-1499.
68. Fujimoto J, Ichigo S, Hori M et al. Progestins and danazol effect on cell-to-cell adhesion, and E-cadherin and alpha- and beta-catenin mRNA expressions. J Steroid Biochem Mol Biol 1996; 57(5-6):275-282.
69. Behrens J, Mareel MM, van Roy FM et al. Dissecting tumor cell invasion: Epithelial cells acquire invasive properties after the loss of uvomorulin-mediated cell-cell adhesion. J Cell Biol 1989; 108:2435-2447.
70. Frixen UH, Behrens J, Sachs M et al. E-cadherin-mediated cell-cell adhesion prevents invasiveness of human carcinoma cells. J Cell Biol 1991; 113(1):173-185.
71. Vleminckx K, Vakaet Jr L, Mareel M et al. Genetic manipulation of E-cadherin expression by epithelial tumor cells reveals an invasion suppressor role. Cell.1991; 66:107-119.
72. Meiners S, Brinkmann V, Naundorf H et al. Role of morphogenetic factors in metastasis of mammary carcinoma cells. Oncogene 1998; 16:9-20.
73. Perl AK, Wilgenbus P, Dahl U et al. A causal role for E-cadherin in the transition from adenoma to carcinoma. Nature 1998; 392(6672):190-193.
74. Becker KF, Atkinson MJ, Reich U et al. E-cadherin gene mutations provide clues to diffuse type gastric carcinomas. Cancer Res 1994; 54(14):3845-3852.
75. Berx G, Cleton-Jansen A M, Nollet F et al. E-cadherin is a tumor/invasion suppressor gene mutated in human lobular breast cancers. EMBO J 1995; 14:6107-6115.
76. Berx G, Becker K-F, Höfler H et al. Mutation Update: Mutations of the human E-cadherin (CDH1) gene. Hum Mutat 1998; 12(4):226-237.
77. Sommers CL, Thompson EW, Torri JA et al. Cell adhesion molecule uvomorulin expression in human breast cancer cell lines: Relationship to morphology and invasive capacities. Cell Growth & Differentiation 1991; 2(8):365-372.
78. Brabant G, Hoangvu C, Cetin Y et al. E-cadherin - a differentiation marker in thyroid malignancies. Cancer Res 1993; 53(20):4987-4993.
79. Dorudi S, Hanby AM, Poulsom R et al. Level of expression of E-cadherin mRNA in colorectal cancer correlates with clinical outcome. Br J Cancer 1995; 71(3):614-616.
80. Graff JR, Herman JG, Lapidus RG et al. E-cadherin expression is silenced by DNA hypermethylation in human breast and prostate carcinomas. Cancer Res 1995; 55(22):5195-5199.
81. Yoshiura K, Kanai Y, Ochiai A et al. Silencing of the E-cadherin invasion-suppressor gene by CpG methylation in human carcinomas. Proc Natl Acad Sci USA 1995; 92(16):7416-7419.
82. Munro SB, Turner IM, Farookhi R et al. E-cadherin and OB-cadherin mRNA levels in normal human colon and colon carcinoma. Exp Mol Pathol 1995; 62(2):118-122.
83. Guriec N, Marcellin L, Gairard B et al. E-cadherin mRNA expression in breast carcinomas correlates with overall and disease-free survival. Invasion & Metastasis 1996; 16(1):19-26.
84. Wang JZ, Krill D, Torbenson M et al. Expression of cadherins and catenins in paired tumor and non neoplastic primary prostate cultures and corresponding prostatectomy specimens. Urol Res 2000; 28(5):308-315.
85. Cheng CW, Wu PE, Yu JC et al. Mechanisms of inactivation of E-cadherin in breast carcinoma: Modification of the two-hit hypothesis of tumor suppressor gene. Oncogene 2001; 20(29):3814-3823.
86. Jiao W, Miyazaki K, Kitajima Y. Inverse correlation between E-cadherin and Snail expression in hepatocellular carcinoma cell lines in vitro and in vivo. Br J Cancer 2002; 86(1):98-101.
87. Blanco MJ, Moreno-Bueno G, Sarrio D et al. Correlation of Snail expression with histological grade and lymph node status in breast carcinomas. Oncogene 2002; 21(20):3241-3246.
88. Rosivatz E, Becker I, Specht K et al. Differential expression of the epithelial-mesenchymal transition regulators Snail, SIP1, and Twist in gastric cancer. Am J Pathol 2002; 161(5):1881-1891.
89. Schönborn I, Zschiesche W, Behrens J et al. Expression of E-cadherin/catenin complexes in breast cancer: Orrelation with favourable prognostic factors and survival. Int J Oncol 1997; 11(6):1327-1334.
90. Kleer CG, vanGolen KL, Braun T et al. Persistent E-cadherin expression in inflammatory breast cancer. Modern Pathol 2001; 14(5):458-464.
91. Ashida K, Terada T, Kitamura Y et al. Expression of E-cadherin, alpha-catenin, beta-catenin, and CD44 (Standard and variant isoforms) in human cholangiocarcinoma: An immunohistochemical study. Hepatology 1998; 27(4):974-982.

92. Herman JG, Jen J, Merlo A et al. Hypermethylation-associated inactivation indicates a tumor suppressor role for p15(INK4B1). Cancer Res 1996; 56(4):722-727.
93. Kanai Y, Ushijima S, Hui AM et al. The E-cadherin gene is silenced by CpG methylation in human hepatocellular carcinomas. Int J Cancer 1997; 71(3):355-359.
94. Koizume S, Tachibana K, Sekiya T et al. Heterogeneity in the modification and involvement of chromatin components of the CpG island of the silenced human CDH1 gene in cancer cells. Nucleic Acids Res 2002; 30(21):4770-4780.
95. Nass SJ, Herman JG, Gabrielson E et al. Aberrant methylation of the estrogen receptor and E-cadherin 5 ‘ CpG islands increases with malignant progression in human breast cancer. Cancer Res 2000; 60(16):4346-4348.
96. Droufakou S, Deshmane V, Roylance R et al. Multiple ways of silencing E-cadherin gene expression in lobular carcinoma of the breast. Int J Cancer 2001; 92(3):404-408.
97. Leung WK, Yu J, Ng EKW et al. Concurrent hypermethylation of multiple tumor-related genes in gastric carcinoma and adjacent normal tissues. Cancer 2001; 91(12):2294-2301.
98. Mingchao, Devereux TR, Stockton P et al. Loss of E-cadherin expression in gastric intestinal metaplasia and later stage p53 altered expression in gastric carcinogenesis. Exp Toxicol Pathol 2001; 53(4):237-246.
99. Tamura G, Yin J, Wang S et al. E-cadherin gene promoter hypermethylation in primary human gastric carcinomas. Journal of the National Cancer Institute 2000; 92(7):569-573.
100. Tamura G, Sato K, Akiyama S et al. Molecular characterization of undifferentiated-type gastric carcinoma. Lab Invest 2001; 81(4):593-598.
101. Nakayama S, Sasaki A, Mese H et al. The E-cadherin gene is silenced by CpG methylation in human oral squamous cell carcinomas. Int J Cancer 2001; 93(5):667-673.
102. Saito Y, Takazawa H, Uzawa K et al. Reduced expression of E-cadherin in oral squamous cell carcinoma: Relationship with DNA methylation of 5' CpG island. Int J Oncol 1998; 12(2):293-298.
103. Bornman DM, Mathew S, Alsruhe J et al. Methylation of the E-cadherin gene in bladder neoplasia and in normal urothelial epithelium from elderly individuals. Am J Pathol 2001; 159(3):831-835.
104. Melki JR, Vincent PC, Clark SJ. Concurrent DNA hypermethylation of multiple genes in acute myeloid leukemia. Cancer Res 1999; 59:3730-3740.
105. Melki JR, Vincent PC, Brown RD et al. Hypermethylation of E-cadherin in leukemia. Blood 2000; 95(10):3208-3213.
106. Graff JR, Gabrielson E, Fujii H et al. Methylation patterns of the E-cadherin 5 ‘ CpG island are unstable and reflect the dynamic, heterogeneous loss of E-cadherin expression during metastatic progression. J Biol Chem 2000; 275(4):2727-2732.
107. Grady WM, Willis J, Guilford PJ et al. Methylation of the CDH1 promoter as the second genetic hit in hereditary diffuse gastric cancer. Nat Genet 2000; 26(1):16-17.
108. Hajra KM, Ji XD, Fearon ER. Extinction of E-cadherin expression in breast cancer via a dominant repression pathway acting on proximal promoter elements. Oncogene 1999; 18(51):7274-7279.
109. Somasiri A, Howarth A, Goswami D et al. Overexpression of the integrin-linked kinase mesenchymally transforms mammary epithelial cells. J Cell Sci 2001; 114(Pt 6):1125-1136.

CHAPTER 12

The Regulation of Catenins in Cancer

Maralice Conacci-Sorrell and Avri Ben-Ze'ev

Abstract

The cadherin-catenin system of cell-cell adhesion molecules plays a key role in determining cellular and tissue morphogenesis. The normal function of this molecular complex is indispensable during various stages throughout development, not only for determining the proper adhesive interactions between neighboring cells, but also for transducing the signals elicited by the Wnt pathway, mostly by β-catenin. The dual role of β-catenin in the assembly of cell-cell adherens junctions and its role as a cotranscriptional activator of target genes in the Wnt pathway is often disrupted in cancer cells. In this perspective, we discuss the interplay between the adhesive and signaling roles of the cadherin-catenin system, its regulation by various mechanisms, with special emphasis on its role in the development of cancer.

Introduction

The adhesion of cells to their neighbors determines cellular and tissue morphogenesis and regulates major cellular processes including motility, growth, differentiation and survival. Cell-cell adherens junctions (AJ), the most ubiquitous type of intercellular adhesions, are important for maintaining tissue architecture, cell polarity, limiting cell movement and proliferation.[1,2] AJ are assembled via homophylic interactions between the extracellular domains of transmembrane receptors of the cadherin family on the surface of neighboring cells. The cytoplasmic domains of cadherins bind to the submembranal plaque proteins β-catenin or plakoglobin (γ-catenin) that are linked to the actin-cytoskeleton via α-catenin (Fig. 1 and refs. 3,4). The binding of cadherin receptors to the cytoskeleton is necessary for the stabilization of cell-cell adhesions that maintain normal cell architecture and physiology. Malignant transformation is often characterized by major changes in the organization of the cytoskeleton, decreased adhesion and aberrant adhesion-mediated signaling.[5] Disruption of normal cell-cell adhesion in transformed cells promotes enhanced cell migration and proliferation, leading to invasion and metastasis.[6] This disruption is caused by various defects that alter the assembly of cadherin-catenin complexes in cancer cells, including loss of function mutations and other defects in the expression or stability of cadherins and/or catenins.[7]

In addition to their structural role in AJ-organization, catenins function as key transducers of adhesion-mediated signaling. Deregulation of their signaling function is characteristic of cancer progression. In this perspective, we discuss the molecular mechanisms underlying the role of the cadherin-catenin system (especially β-catenin) in the regulation of cell proliferation, invasion and its signaling roles in transformed cells.

Gene Deletion or Inactivation of Catenins and E-Cadherin in Cancer Cells

The loss of E-cadherin expression, causing disruption of AJ, is a frequent event in many types of sporadic and hereditary carcinomas.[8,9] Cancer cells employ reversible (transcriptional

Rise and Fall of Epithelial Phenotype: Concepts of Epithelial-Mesenchymal Transition, edited by Pierre Savagner.

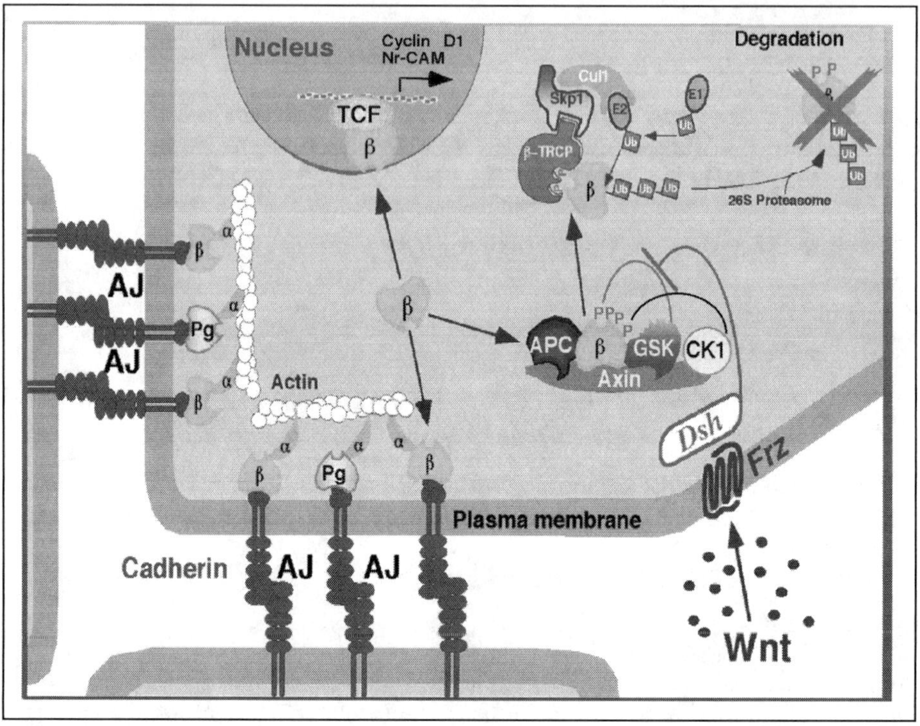

Figure 1. A dual role for the cadherin-catenin system in cell adhesion and transcriptional activation. β-catenin (β) and plakoglobin (Pg) bind to the cytoplasmic part of cadherin transmembrane adhesion receptors, and via α-catenin (α), associate with the actin cytoskeleton to form adherens junctions (AJ). When the Wnt signaling pathway is inactive, the cytoplamsic pool of β-catenin is degraded by a complex including casein kinase 1 (CK1), glycogen synthase kinase 3β (GSK), APC and Axin, that phosphorylate β-catenin (PPPP). Phosphorylated β-catenin is recognized by the ubiquitin ligase component β-TrCP, which together with Skp1, Cul1 and the E1 and E2 ubiquitination components, mediates the ubiquitylation (Ub) of β-catenin and directs it to degradation by the 26S proteasome. The binding of Wnt to Frizzled (Frz) receptors activates Wnt signaling, that induces a disheveled (Dsh) -mediated inhibition of β-catenin phosphorylation by GSK. This results in β-catenin accumulation and its complexing in the nucleus with T-cell factor (TCF), followed by activation of target genes such as Cyclin D1 and Nr-CAM. (Modified from ref. 7.)

repression and promoter methylation) or irreversible (gene deletion) mechanisms to silence the E-cadherin gene that eliminates AJ formation. These changes in E-cadherin expression are associated with the transition from adenoma to carcinoma and acquisition of metastatic capacity[6] (also see Chapter 9, Van Marck and Bracke, and Chapter 11, Berx and Van Roy). Reestablishment of AJ in cancer cells by restoration of cadherin expression[10] exerts tumor suppressive effects, including decreased proliferation and motility. The cytoplasmic tail of the cadherin receptors binds to β- or γ-catenin (plakoglobin),[11] which are linked to the actin cytoskeleton by α-catenin (Fig. 1 and ref. 4). This interaction is critical for the establishment of stable and functional AJ.[12] The juxtamembranal region of the E-cadherin cytoplasmic tail binds exclusively p120, another member of the catenin family. However the exact role of p120 in the control of AJ assembly is still controversial since both positive and negative roles for p120 in the regulation of AJ assembly were reported.[13]

Mutations in β- or α-catenin genes that prevent the binding of cadherin receptors to the actin cytoskeleton were reported in lung, prostate, ovarian, and colon cancer cell lines that lack normal AJ.[14] A homozygous deletion of the α-catenin gene (CTNNA1), leading to loss of cell-cell adhesion, was reported in a human lung cancer cell line[15,16] and the introduction of wild-type α-catenin into these cells restored normal adhesion. In an ovarian carcinoma cell line harboring a deletion in α-catenin the binding to β-catenin and the formation of AJ was blocked.[17] Expression of full-length α-catenin in these cells restored epithelial morphology, reduced the growth rate and decreased the tumorigenic capacity of these cells.[17] Moreover, in mice in which the α-catenin gene was eliminated in the epidermis, AJ were disrupted and the epithelial skin cells displayed hyperproliferation[18] that was attributed to aberrant adhesion-mediated signaling, indicating that disruption of AJ assembly can directly contribute to oncogenesis. Deletions of the α-catenin binding site in the β-catenin gene (CTNNB1) were detected in signet ring cell carcinoma of the stomach,[19] where β-catenin-cadherin complexes are formed, but the lack of α-catenin binding prevented their linkage to the actin cytoskeleton and the assembly of AJ.[19] Except for these examples, deletions or inactivating mutations in the genes encoding for β-catenin, plakoglobin, or p120 catenin, are rarely seen.[14] In contrast, mutations in β-catenin or in its cytosolic partners that interfere with its stability and/or enhance its signaling abilities (without directly disrupting AJ assembly) were extensively reported in many types of cancer[20] and are discussed below.

The Wnt Signaling Pathway

In addition to providing a physical link between cells, the mode of AJ assembly can influence various signaling pathways. A major route for signal transduction by AJ involves the regulation of signaling by β-catenin. Besides its structural role in AJ, β-catenin is a key component of the Wnt signaling pathway, acting as a transcription factor in the nucleus by being a coactivator of LEF/TCF family of DNA binding proteins (Fig. 1 and ref. 21, see also Chapter 18, Klymkowsky). Consistent with its dual function, there are two cellular pools of β-catenin; one insoluble associated with AJ, and the second soluble that, if not degraded, can be engaged in transcriptional activation. Under homeostatic conditions, the soluble pool of β-catenin is efficiently targeted to degradation after its phosphorylation on three serines and one threonine residue in the N-terminal region of β-catenin (Fig. 1 and ref. 22). These phosphorylations occur in a large multiprotein complex consisting of the tumor suppressor adenomatous polyposis coli (APC), the serine threonine kinase glycogen synthase kinase 3β GSK3β and the scaffolding protein axin (Fig. 1 and refs. 23-25). In this complex, the priming phosphorylation of β-catenin on serine 45 is carried out by casein kinase I (CKI)[26-28] and the subsequent phosphorylations on threonine 41, serines 37 and 33, are executed by GSK3β (Fig. 1 and ref. 29). Phosphoserines 33 and 37 of β-catenin are recognized by the E3 ubiquitin ligase component β-TrCP that targets β-catenin to ubiquitination and rapid proteasomal degradation (Fig. 1 and ref. 22). In the presence of soluble glycoproteins of the Wnt family, the proteasomal degradation of β-catenin is inhibited due to inactivation of GSK3β and relocalization of axin from the degradation complex to the cell membrane.[30] This results in the accumulation of β-catenin in the cytoplasm and in the nucleus where it forms a complex with transcription factors of the LEF/TCF family (Fig. 1 and ref. 31). In this complex, β-catenin provides the transactivation domain while the LEF/TCF factors contribute the DNA binding domain, together inducing the transcription of genes that have LEF/TCF binding sequences in their promoters. In addition, β-catenin can also interact with components of the basal transcriptional machinery such as the TATA-box binding protein (TBP),[32] the CREB-binding protein (CBP/p300),[33] and with nuclear factors that antagonize its transcriptional abilities, for example Chibby.[34]

β-catenin-mediated transcription induced by activation of the Wnt pathway is crucial during various stages of embryonic development,[35] and aberrant activation of β-catenin signaling is often involved in cancer progression.[7,20] Mutations in the N-terminus of β-catenin that compromise its degradation were reported in various cancers, including colorectal carcinoma, melanoma, medulloblastoma, endometrial, ovarian and prostate carcinomas, and many other types of cancer.[20] Mutations in components of the β-catenin degradation complex such as APC and axin that result in β-catenin accumulation and enhanced β-catenin-mediated signaling, were also reported.[22] In about 70-80% of colorectal carcinoma, and a subset of other cancers including pancreatic, hepatocellular, breast and ovarian carcinoma, inactivating mutations of APC were detected,[36,37] while axin mutations were detected in hepatocellular,[38] colorectal[39] and ovarian[40] carcinoma.

Transcriptional Targets of β-Catenin in Cancer

Stabilizing mutations in β-catenin, or inactivating mutations in APC or axin can lead, in cancer cells, to accumulation of β-catenin in the nucleus and elevated transcription of a variety of target genes. These genes contribute to both the first steps in cell transformation, as well as to the acquisition of invasive and metastatic capacity. They include positive regulators of the cell cycle, anti-apoptotic genes, regulators of cell motility, extravasation, angiogenesis and drug resistance. *Cyclin D1* and *C-MYC* that promote cell cycle progression, and can also act as oncogenes when aberrantly overexpressed, were the first transcriptional targets of β-catenin-TCF signaling detected in colon cancer cells.[41-43] WISP-1, an oncogene induced by Wnt-1 and β-catenin,[44,45] was recently shown to induce cell growth and transformation, at least in part, by suppressing apoptosis.[46] The neural cell adhesion molecule Nr-CAM is also a transcriptional target of β-catenin in melanoma and colon carcinoma conferring cell transformation and motility when expressed in fibroblasts.[47] Other β-catenin target genes that can induce colon cancer cell motility are the extracellular matrix (ECM) proteins fibronectin,[48] laminin γ2[49] and various metalloproteases,[50-52] that in addition to digesting the ECM and promoting extravasation of tumor cells, can also cleave and thereby inactivate transmembrane or secreted growth inhibitory proteins.[53] The expression of vascular endothelial growth factor (VEGF)[54] that promotes angiogenesis, and the multi-drug resistance gene 1 (Mdr1) conferring resistance to stress and drug treatment[55] are also targets of β-catenin signaling. In a positive feedback loop, the LEF/TCF family genes *LEF1* and *TCF1* were shown to be induced by β-catenin[56,57] and thus potentially facilitate the expression of LEF/TCF-regulated genes. β-catenin can therefore enhance the expression of genes that confer a variety of advantages to cells in the course of malignant transformation.

Besides playing a key role in colon carcinoma development, recent studies suggest that activation of β-catenin signaling most probably regulates also the proliferation of normal colon cells by inducing specific target genes.[58] This view is supported by the observation that mice with inactivated TCF4 gene die early after birth since they fail in maintaining the normal proliferation of intestinal cells.[59] In normal colon cells Wnt signaling induces cell proliferation by inducing myc, which represses a p21-induced growth arrest.[60] β-catenin signaling is activated in the stem cell compartment that is responsible for maintaining the proliferative pool of intestinal cells. Cells that exit this compartment are not exposed to Wnt signaling, stop dividing and differentiate into specialized intestinal cells. This differentiation was shown to also involve β-catenin-TCF4 signaling that activates the ephrin gene receptors and represses the expression of ephrin ligands.[61] The inverse regulation by β-catenin/TCF signaling of ephrins and their receptors in normal colon epithelium is grossly disrupted in colon carcinoma.[61] In addition to intestinal cells, Wnt signaling was recently shown to also play a key role in hematopoietic stem cell self-renewal by activating the transcription of HoxB4 and Notch1 genes.[62,63]

Antiproliferative Responses to Oncogenic β-Catenin that Involve p53 and PML

In nontransformed cells, the aberrant activation of oncogenes, including *c-myc, ras* and *E2F1*, can trigger the activation of the tumor suppressor p53 as a protective mechanism against the development of cancer.[64,65] The best understood mechanism of p53 activation by oncogenes involves ARF, the alternative reading frame product of the INK4A tumor suppressor locus ($p19^{ARF}$ in mouse and $p14^{ARF}$ in human cells). ARF binds Mdm2 and inhibits its ability to function in the degradation of p53.[66] The expression of deregulated β-catenin, as shown for other oncogenes, also induces p53.[67] This involves the induction of ARF transcription that leads to inhibition of Mdm2 and consequently, to the accumulation of active p53.[67] Retroviral

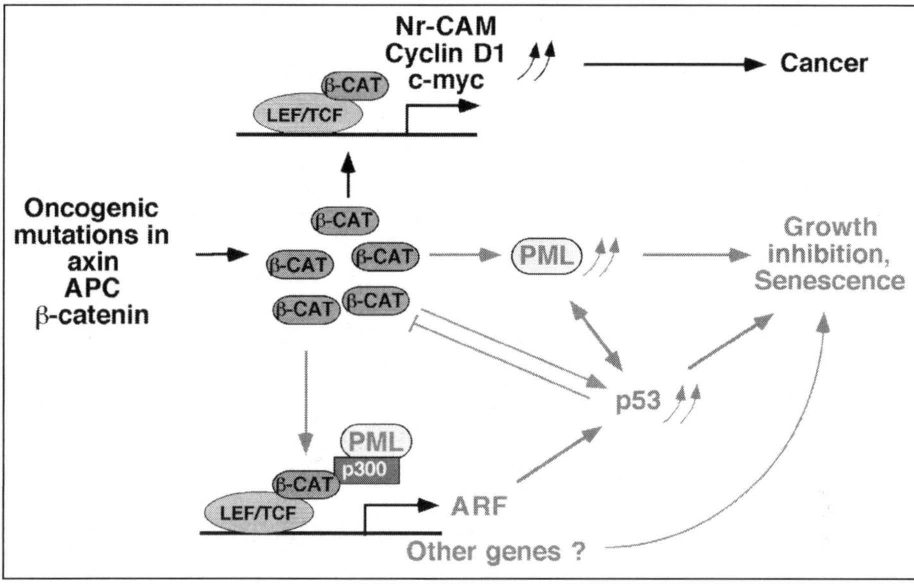

Figure 2. Molecular mechanisms involved in the oncogenic and tumor suppressive effects elicited by activation of β-catenin signaling. Oncogenic mutations in components of the Wnt pathway lead to accumulation of β-catenin (β-CAT) and aberrant activation (by β-catenin together with LEF/TCF) of target genes that promote cell proliferation and motility, including *Cyclin D1* and *Nr-CAM*, thus leading to development of cancer. β-catenin also induces the expression of the *PML* gene that promotes growth inhibition or senescence, either independently of β-catenin, or by coactivating together with β-catenin and p300 the transcription of growth suppressive genes such as *ARF* (and perhaps other genes). The anti-tumorigenic response elicited by β-catenin may involve the accumulation of p53 by increasing ARF, in some cases, or may act independently of p53. The oncogenic activities of β-catenin are marked with black (dark) arrows, while the tumor suppressive effects are marked with red (light) arrows. (Modified from ref. 74.)

transduction of β-catenin into mouse embryo fibroblasts that harbor wild type p53 confers growth arrest and a senescence-like phenotype, and does not induce growth, as intuitively expected.[68] This is interpreted to indicate that normal cells respond to deregulated β-catenin by activating p53 that will suppress growth or induce apoptosis as part of an anticancer response to oncogenic β-catenin (Fig. 2). Only when the p53 response is eliminated, as in p53-/- or ARF-/- cells, β-catenin can manifest its oncogenic capacity by overcoming density-dependent growth inhibition and by cooperating with Ras in cell transformation.[68] Therefore, the oncogenic effects of β-catenin are more likely to occur in cells that have lost wild type p53, a common event observed at early stages of colon cancer development. In agreement with this view many tumors that harbor deregulated β-catenin have mutated p53.[69] Moreover, in a significant portion of colorectal tumors, transcriptional silencing of the ARF promoter by methylation ablates the expression of ARF.[70] Interestingly, p53-mediated growth arrest is only one facet of the p53 response to aberrant β-catenin activation. When the level of wild type p53 is elevated above a certain threshold, p53 can induce the downregulation of β-catenin by enhancing its degradation by the ubiquitin-proteasome system,[71-73] thereby antagonizing the oncogenic pressure exerted by β-catenin.

In addition to p53, recent studies have shown that the gene for promyelocytic leukemia (PML) tumor suppressor protein is a major transcriptional target of β-catenin in certain renal carcinoma cells.[74] Retrovirus mediated transduction of PML into these cells reduces dramatically their tumorigenic capacity in nude mice, in a p53-independent manner (Fig. 2 and ref. 74). The induction of PML by oncogenic Ras was demonstrated to cause a senescence-like growth arrest that is mediated by p53 and PML.[75,76] In contrast, the PML-induced tumor suppression in renal carcinoma cells does not involve p53 and thus points to a novel PML-mediated antioncogenic response triggered by β-catenin (Fig. 2 and ref. 74). An effort directed to understanding the interplay between β-catenin and PML reveled that PML can cooperate with β-catenin and the histone deacetylase p300/CBP in transcriptional activation of the *ARF* tumor suppressor gene promoter, but not of the proto-oncogene *cyclin D1* (Fig. 2 and ref. 74). Therefore, PML may exert its tumor suppressive effects in cells harboring oncogenic β-catenin by cooperating with β-catenin in the specific activation of tumor suppressor genes.

Signaling by Plakoglobin

Plakoglobin (γ-catenin), a close homologue of β-catenin, can bind to most of β-catenin's partners in AJ, in the degradation machinery (APC-GSK3β-axin) and with LEF/TCF transcription factors.[77] In addition, plakoglobin is a component of desmosomes, where it links desmosomal cadherins to intermediate filaments. Unlike with β-catenin, the role of plakoglobin in transcriptional activation is not well understood. Increased levels of plakoglobin expression compromises the degradation of β-catenin thus leading to nuclear accumulation of the endogenous β-catenin.[78-80] Consequently, LEF/TCF-dependent transactivation in plakoglobin transfected cells mostly results from indirect effects leading to elevated β-catenin that becomes engaged with LEF/TCF factors in transcription.[77] While this may occur in certain cases of vast plakoglobin overexpression, it is noteworthy that plakoglobin can activate the transcription of LEF/TCF-regulated gene promoters including *cyclin D1*, *Nr-CAM* and the synthetic promoter TOPFLASH, also in β-catenin-null ES cells.[47] This indicates that plakoglobin can cooperate with LEF/TCF factors, in the absence of β-catenin, to activate LEF/TCF responsive target genes, but the conditions under which such transcriptional capacity is exerted remain to be determined.

While both β-catenin and plakoglobin can bind LEF-1 in vitro, plakoglobin is inefficient in forming a ternary complex containing LEF1 and LEF1-binding DNA sequences.[80,81] This may result from an inhibition caused by the N and C terminal regions of plakoglobin that can fold back on the central domain containing the binding site.[80] Moreover, TCF4 can apparently bind to different regions of β-catenin and plakoglobin and the binding to plakoglobin, in vitro, can hinder the transcriptional activity of the complex.[81] It is still possible that yet unidentified cofactors involved in plakoglobin-mediated transcription in cells, that were absent in the in vitro reactions employed in these studies, prevented the assembly of active plakoglobin transcriptional complexes.

Several studies suggest that plakoglobin may have a tumor suppression function. For example, the plakoglobin locus is subject to loss of heterozygosity (LOH) in sporadic human breast and ovarian tumors,[82] and mutations in plakoglobin were only very rarely observed in cancer cells. In some transformed cells, overexpression of plakoglobin suppresses their tumorigenicity in mice.[83] Functional differences between β-catenin and plakoglobin are also indicated from knock out and transgenic animal studies. Knock out mice for β-catenin die before gastrulation,[84] while plakoglobin-null animals undergo organogenesis and die from heart failure resulting from defects in desmosomes assembly.[85,86] The expression of β-catenin or plakoglobin in transgenic mice under control of an epidermis-specific keratin promoter resulted in different effects on the regulation of hair follicle cell growth: while β-catenin induced hyperproliferation and tumor formation,[87] plakoglobin reduced cell proliferation.[88] These differences between β-catenin and plakoglobin are probably cell type specific, since plakoglobin was shown to function as an oncogene, transforming certain rat epithelial cells by inducing the *myc* gene.[89] However, since this cell line expresses endogenous β-catenin, the possibility that plakoglobin acts indirectly to affect the signaling pool of β-catenin could not be excluded.

The Reciprocal Relationship between Cadherin-Mediated Adhesion and β-Catenin Signaling

The interaction of β-catenin with cadherins and LEF/TCF family members is mediated by the central armadillo repeat domain of β-catenin, making these interactions mutually exclusive.[77] The recruitment of β-catenin into AJ, when E-cadherin expression is elevated, could decrease the signaling pool of β-catenin thereby antagonizing LEF/TCF-mediated transactivation.[79,90-93] Indeed, cadherin-dependent inhibition of growth was suggested to involve the sequestration of the signaling pool of β-catenin.[94,95] However, the loss of E-cadherin expression, at least in some cancer cells, does not result in constitutive β-catenin-LEF/TCF transcriptional activation.[96,97] This indicates that the degradation machinery of β-catenin can effectively remove excess cytosolic β-catenin resulting from the loss of E-cadherin, thereby preventing the generation of an oncogenic signal by β-catenin. Additional events, including compromised proteasomal degradation of β-catenin are probably necessary to induce β-catenin-mediated oncogenesis, even in the absence of AJ assembly.

In colorectal tumors, often defective in β-catenin degradation, the central part of the tumor displays a more differentiated phenotype that contains high levels of E-cadherin, at the cell membrane, colocalizing with β-catenin. In contrast, at the invasive front of these tumors, β-catenin is mainly found in the nuclei of dissociated tumor cells that also lost junctional E-cadherin[98] (reviewed in ref. 99). Interestingly, this loss of E-cadherin in the invasive cells is apparently reversible, and regulated by the tumor cell micro-environment, since metastatic lymph nodes of the same tumor display a central differentiated area with well developed E-cadherin and β-catenin-containing AJ.[98] The loss of E-cadherin resulting in the disruption of AJ and cell polarity, allows tumor cell metastasis, and the translocation of β-catenin into the nucleus might be required to induce the expression of β-catenin target genes that promote invasion.

The colon carcinoma cell line SW480 can mimic, in culture, many parameters of this behavior seen in colorectal cancer and their metastases. These cells harbor mutant APC resulting in impaired β-catenin degradation and constitutive activation of β-catenin-LEF/TCF signaling. Sparse cultures of SW480 cells closely resemble the invasive phenotype of colon carcinoma cells, characterized by extensive nuclear β-catenin and LEF/TCF-mediated transcription, but very low levels of E-cadherin. The reduced levels of E-cadherin result from transcriptional repression of the E-cadherin gene promoter by the E-box binding transcriptional repressor Slug, which is apparently induced by β-catenin-LEF/TCF signaling in these cells.[100] Dense cultures of SW480 cells display a more differentiated phenotype, including high levels of E-cadherin that results in the recruitment of β-catenin to AJ, thereby inhibiting β-catenin target gene activation (including Slug). Deregulated β-catenin signaling in colon cancer cells may thus repress the tumor suppressor E-cadherin, by inducing its transcriptional repressor Slug that may result in the inhibition of AJ assembly thereby contributing to invasion and metastasis.[100]

Such repression in E-cadherin expression, involving β-catenin signaling, was recently observed also during hair follicle morphogenesis, and is apparently achieved by stabilization of the signaling pool of β-catenin and increased expression of LEF1.[101] The mechanisms by which activated β-catenin signaling represses E-cadherin transcription in hair bud development are unknown yet.

Regulation of Cell-Cell Adhesion by Growth Factor Receptor Driven Phosphorylation

The phosphorylation of catenins on tyrosine residues by specific tyrosine kinases regulates their interaction with various protein partners, by increasing or decreasing their binding affinity. Tyrosine phosphorylation of catenins may have a role in regulating the sorting of these proteins to specific cellular compartments, and this phosphorylation was shown to decrease cell-cell adhesion, while dephosphorylation increases cell-cell adhesion.[1] β-catenin, plakoglobin and p120 are phosphorylated on tyrosine by protein kinases of the Src family and by transmembrane receptors tyrosine kinases (RTKs). Such phosphorylation of β-catenin promotes the disruption of AJ,[102,103] and also increases the affinity of β-catenin to the TATA binding protein (TBP), a general positive regulator of transcription.[104] The phosphorylation on a different tyrosine residue of β-catenin can inhibit its binding to α-catenin and reduces AJ assembly.[105] Expression of the v-src oncogene also leads to tyrosine phosphorylation of β-catenin, decreased AJ assembly and increased invasion.[106,107] Epidermal growth factor (EGF), hepatocyte growth factor/scatter factor (HGF/SF) and fibroblast growth factor (FGF) can all induce epithelial to mesenchymal transition (EMT) that results in dismantling of AJ and dramatic changes in cell morphology from an epithelial to a fibroblastic phenotype.[108,109] Both EGF and HGF/SF were shown to induce the phosphorylation of β-catenin and plakoglobin in cancer cells.[110] These growth factors operate via their respective receptor tyrosine kinases, often localizing at AJ, as demonstrated for the EGF[111] and HGF receptors[112] (see also Chapters 14, Day et al and 17, Boyer). When activated, they phosphorylate components of AJ, including β-catenin, thereby contributing to the disruption of AJ observed in cancer cells. In ras-transformed cells displaying poorly organized AJ this was attributed to enhanced tyrosine phosphorylation of β-catenin[102] resulting in decreased β-catenin binding to cadherin in vitro.[103] Dismantling of AJ by HGF/SF stimulation requires, in addition, activation of phosphatidylinositol 3-kinase (PI3K) and mitogen activated protein kinase (MAPK) by HGF/SF, since inhibition of these kinases prevents the disassembly of AJ.[113] AJ disruption can also be achieved via phosphorylation by casein kinase II (CKII), on serine residues, of the cadherin cytoplasmic tail that enhances its binding to β-catenin.[114] Point mutations in these cadherin tail phosphorylation sites, were shown to reduce its binding to β-catenin.[114,115]

While tyrosine kinase receptor signaling was shown to regulate cell adhesion,[116] AJ may in turn influence tyrosine kinase signaling. Reduced cell proliferation characteristic of dense cell cultures is mediated, in part, by cadherin-containing AJ that render some cells insensitive to growth factor stimulation.[117] In such cell cultures, dimerization and tyrosine kinase activation of the EGF receptor (in response to EGF), are inhibited. In addition, cadherin-dependent growth inhibition in cells grown in a three dimensional culture is mediated by inactivating tyrosine kinase signaling by increasing tyrosine phosphatase activity.[118] In such cells, the assembly of AJ also prevented the activation of the EGF receptor pathway by TGFα.

In contrast to E-cadherin, N-cadherin enhances tyrosine kinase signaling that enhances cell motility, and high levels of N-cadherin were detected in invasive tumor cell lines.[119,120] The ability of N-cadherin to promote cell motility, invasion and metastasis apparently involves the activation of FGF receptor signaling and its interaction with the extracellular domain of N-cadherin.[119-122]

References

1. Hirohashi S. Inactivation of the E-cadherin-mediated cell adhesion system in human cancers. Am J Pathol 1998; 153:333-339.
2. Ben-Ze'ev A. The dual role of cytoskeletal anchor proteins in cell adhesion and signal transduction. Ann NY Acad Sci 1999; 886:37-47.
3. Ben-Ze'ev A, Geiger B. Differential molecular interactions of β-catenin and plakoglobin in adhesion, signaling and cancer. Curr Opin Cell Biol 1998; 10:629-639.
4. Nagafuchi A. Molecular architecture of adherens junctions. Curr Opin Cell Biol 2001; 13:600-603.
5. Ben-Ze'ev A. Cytoskeletal and adhesion proteins as tumor suppressors. Curr Opin Cell Biol 1997; 9:99-108.
6. Perl A, Wilgenbus P, Dahl U et al. A causal role for E-cadherin in the transition from adenoma to carcinoma. Nature 1998; 392:190-193.
7. Conacci-Sorrell M, Zhurinsky J, Ben-Ze'ev A. The cadherin-catenin adhesion system in signaling and cancer. J Clin Invest 2002; 109:987-991.
8. Jiang WG. E-cadherin and its associated protein catenins, cancer invasion and metastasis. Br J Surg 1996; 83:437-446.

9. Papadavid E, Katsambas A. The interactions and role of epithelial cadherin and catenins in tumorigenicity. Int J Dermatol 2001; 40:254-257.
10. Vleminckx K, Vakaet L, Mareel M et al. Genetic manipulation of E-cadherin expression by epithelial tumor cells reveals an invasion suppressor role. Cell 1991; 66:107-119.
11. Ozawa M, Baribault H, Kemler R. The cytoplasmic domain of the cell adhesion molecule uvomorulin associates with three independent proteins structurally related in different species. EMBO J 1989; 8:1711-1717.
12. Nagafuchi A, Ishihara S, Tsukita S. The roles of catenins in the cadherin-mediated cell adhesion: Functional analysis of E-cadherin-α catenin fusion molecules. J Cell Biol 1994; 127:235-245.
13. Anastasiadis PZ, Reynolds AB. The p120 catenin family: Complex roles in adhesion, signaling and cancer. J Cell Sci 2000; 113:1319-1334.
14. Hajra KM, Fearon ER. Cadherin and catenin alterations in human cancer. Genes Chromosomes Cancer 2002; 34:255-268.
15. Hirano S, Kimoto N, Shimoyama Y et al. Identification of a neural α-catenin as a key regulator of cadherin function and multicellular organization. Cell 1992; 70:293-301.
16. Oda T, Kanai Y, Shimoyama Y et al. Cloning of the human α-catenin cDNA and its aberrant mRNA in a human cancer cell line. Biochem Biophys Res Commun 1993; 193:897-904.
17. Bullions L, Notterman D, Chung L et al. Expression of wild-type α-catenin protein in cells with a mutant α-catenin gene restores both growth regulation and tumor suppressor activities. Mol Cell Biol 1997; 17:4501-4508.
18. Vasioukhin V, Bauer C, Degenstein L et al. Hyperproliferation and defects in epithelial polarity upon conditional ablation of α-catenin in skin. Cell 2001; 104:605-617.
19. Oyama T, Kanai Y, Ochiai A et al. A truncated β-catenin disrupts the interaction between E-cadherin and α-catenin: A cause of loss of intercellular adhesiveness in human cancer cell lines. Cancer Res 1994; 54:6282-6287.
20. Polakis P. Wnt signaling and cancer. Genes Dev 2000; 14:1837-1851.
21. Bienz M, Clevers H. Linking colorectal cancer to Wnt signaling. Cell 2000; 103:311-320.
22. Peifer M, Polakis P. Wnt signaling in oncogenesis and embryogenesis - a look outside the nucleus. Science 2000; 287:1606-1609.
23. Behrens J, Jerchow B, Wurtele M et al. Functional interaction of an axin homolog, conductin, with β-catenin, APC, and GSK3β. Science 1998; 280:596-599.
24. Kishida S, Yamamoto H, Ikeda S et al. Axin, a negative regulator of the wnt signaling pathway, directly interacts with adenomatous polyposis coli and regulates the stabilization of β-catenin. J Biol Chem 1998; 273:10823-10826.
25. Ikeda S, Kishida S, Yamamoto H et al. Axin, a negative regulator of the Wnt signaling pathway, forms a complex with GSK-3β and β-catenin and promotes GSK-3β-dependent phosphorylation of β-catenin. EMBO J 1998; 17:1371-1384.
26. Amit S, Hatzubai A, Birman Y et al. Axin-mediated CKI phosphorylation of β-catenin at Ser 45: A molecular switch for the Wnt pathway. Genes Dev 2002; 16:1066-1076.
27. Liu C, Li Y, Semenov M et al. Control of β-catenin phosphorylation/degradation by a dual-kinase mechanism. Cell 2002; 108:837-847.
28. Yanagawa S, Matsuda Y, Lee JS et al. Casein kinase I phosphorylates the Armadillo protein and induces its degradation in Drosophila. EMBO J 2002; 21:1733-1742.
29. Polakis P. Casein kinase 1: A Wnt'er of disconnect. Curr Biol 2002; 12:R499-R501.
30. Tolwinski NS, Wehrli M, Rives A et al. Wg/Wnt signal can be transmitted through arrow/LRP5,6 and Axin independently of Zw3/Gsk3β activity. Dev Cell 2003; 4:407-418.
31. Bienz M, Clevers H. Armadillo/β-catenin signals in the nucleus—proof beyond a reasonable doubt? Nat Cell Biol 2003; 5:179-182.
32. Hecht A, Litterst C, Huber O et al. Functional characterization of multiple transactivating elements in β-catenin, some of which interact with the TATA-binding protein in vitro. J Biol Chem 1999; 274:18017-18025.
33. Hecht A, Vleminckx K, Stemmler M et al. The p300/CBP acetyltransferases function as transcriptional coactivators of β-catenin in vertebrates. EMBO J 2000; 19:1839-1850.
34. Takemaru K, Yamaguchi S, Lee YS et al. Chibby, a nuclear β-catenin-associated antagonist of the Wnt/Wingless pathway. Nature 2003; 422:905-909.
35. Wodarz A, Nusse R. Mechanisms of Wnt signaling in development. Annu Rev Cell Dev Biol 1998; 14:59-88.
36. Polakis P. Mutations in the APC gene and their implications for protein structure and function. Curr Opin Genet Dev 1995; 5:66-71.
37. Laurent-Puig P, Beroud C, Soussi T. APC gene: Database of germline and somatic mutations in human tumors and cell lines. Nucleic Acids Res 1998; 26:269-270.

38. Satoh S, Daigo Y, Furukawa Y et al. AXIN1 mutations in hepatocellular carcinomas, and growth suppression in cancer cells by virus-mediated transfer of AXIN1. Nat Genet 2000; 24:245-250.
39. Liu W, Dong X, Mai M et al. Mutations in AXIN2 cause colorectal cancer with defective mismatch repair by activating β-catenin/TCF signalling. Nat Genet 2000; 26:501.
40. Wu R, Zhai Y, Fearon ER et al. Diverse mechanisms of β-catenin deregulation in ovarian endometrioid adenocarcinomas. Cancer Res 2001; 61:8247-8255.
41. Shtutman M, Zhurinsky J, Simcha I et al. The cyclin D1 gene is a target of the β-catenin/LEF-1 pathway. Proc Natl Acad Sci USA 1999; 96:5522-5527.
42. Tetsu O, McCormick F. β-catenin regulates expression of cyclin D1 in colon carcinoma cells. Nature 1999; 398:422-426.
43. He T, Sparks A, Rago C et al. Identification of c-MYC as a target of the APC pathway. Science 1998; 281:1509-1512.
44. Pennica D, Swanson TA, Welsh JW et al. WISP genes are members of the connective tissue growth factor family that are up-regulated in wnt-1-transformed cells and aberrantly expressed in human colon tumors. Proc Natl Acad Sci USA 1998; 95:14717-14722.
45. Xu L, Corcoran RB, Welsh JW et al. WISP-1 is a Wnt-1- and β-catenin-responsive oncogene. Genes Dev 2000; 14:585-595.
46. Su F, Overholtzer M, Besser D et al. WISP-1 attenuates p53-mediated apoptosis in response to DNA damage through activation of the Akt kinase. Genes Dev 2002; 16:46-57.
47. Conacci-Sorrell ME, Ben-Yedidia T, Shtutman M et al. Nr-CAM is a target gene of the β-catenin/LEF-1 pathway in melanoma and colon cancer and its expression enhances motility and confers tumorigenesis. Genes Dev 2002; 16:2058-2072.
48. Gradl D, Kuhl M, Wedlich D. The Wnt/Wg signal transducer β-catenin controls fibronectin expression. Mol Cell Biol 1999; 19:5576-5587.
49. Hlubek F, Jung A, Kotzor N et al. Expression of the invasion factor laminin γ2 in colorectal carcinomas is regulated by β-catenin. Cancer Res 2001; 61:8089-8093.
50. Crawford H, Fingleton B, Rudolph-Owen L et al. The metalloproteinase matrilysin is a target of β-catenin transactivation in intestinal tumors. Oncogene 1999; 18:2883-2891.
51. Brabletz T, Jung A, Dag S et al. β-catenin regulates the expression of the matrix metalloproteinase-7 in human colorectal cancer. Am J Pathol 1999; 155:1033-1038.
52. Takahashi M, Tsunoda T, Seiki M et al. Identification of membrane-type matrix metalloproteinase-1 as a target of the β-catenin/Tcf4 complex in human colorectal cancers. Oncogene 2002; 21:5861-5867.
53. MacDougall JR, Matrisian LM. Contributions of tumor and stromal matrix metalloproteinases to tumor progression, invasion and metastasis. Cancer Metastasis Rev 1995; 14:351-362.
54. Zhang X, Gaspard JP, Chung DC. Regulation of vascular endothelial growth factor by the Wnt and K-ras pathways in colonic neoplasia. Cancer Res 2001; 61:6050-6054.
55. Yamada T, Takaoka A, Naishiro Y et al. Transactivation of the multidrug resistance 1 gene by T-cell factor 4/β-catenin complex in early colorectal carcinogenesis. Cancer Res 2000; 60:4761-4766.
56. Roose J, Huls G, van Beest M et al. Synergy between tumor suppressor APC and the β-catenin-Tcf4 target Tcf1. Science 1999; 285:1923-1926.
57. Hovanes K, Li T, Munguia J et al. β-catenin-sensitive isoforms of lymphoid enhancer factor-1 are selectively expressed in colon cancer. Nat Genet 2001; 28:53-57.
58. Peifer M. Developmental biology: Colon construction. Nature 2002; 420:274-275, 277.
59. Korinek V, Barker N, Moerer P et al. Depletion of epithelial stem-cell compartments in the small intestine of mice lacking Tcf-4. Nat Genet 1998; 19:379-383.
60. van de Wetering M, Sancho E, Verweij C et al. The β-catenin/TCF-4 complex imposes a crypt progenitor phenotype on colorectal cancer cells. Cell 2002; 111:241-250.
61. Batlle E, Henderson JT, Beghtel H et al. β-catenin and TCF mediate cell positioning in the intestinal epithelium by controlling the expression of EphB/ephrinB. Cell 2002; 111:251-263.
62. Reya T, Duncan AW, Ailles L et al. A role for Wnt signalling in self-renewal of haematopoietic stem cells. Nature 2003; In press PMID 12717451.
63. Willert K, Brown JD, Danenberg E et al. Wnt proteins are lipid-modified and can act as stem cell growth factors. Nature 2003; In press PMID 12717450.
64. Lowe SW. Activation of p53 by oncogenes. Endocr Relat Cancer 1999; 6:45-48.
65. Sherr CJ, Weber JD. The ARF/p53 pathway. Curr Opin Genet Dev 2000; 10:94-99.
66. Oren M, Damalas A, Gottlieb T et al. Regulation of p53: Intricate loops and delicate balances. Ann NY Acad Sci 2002; 973:374-383.
67. Damalas A, Ben-Ze'ev A, Simcha I et al. Excess β-catenin promotes accumulation of transcriptionally active p53. EMBO J 1999; 18:3054-3063.

68. Damalas A, Kahan S, Shtutman M et al. Deregulated β-catenin induces a p53- and ARF-dependent growth arrest and cooperates with Ras in transformation. EMBO J 2001; 20:4912-4922.
69. Kinzler KW, Vogelstein B. Lessons from hereditary colorectal cancer. Cell 1996; 87:159-170.
70. Esteller M, Tortola S, Toyota M et al. Hypermethylation-associated inactivation of p14(ARF) is independent of p16(INK4a) methylation and p53 mutational status. Cancer Res 2000; 60:129-133.
71. Sadot E, Geiger B, Oren M et al. Down-regulation of β-catenin by activated p53. Mol Cell Biol 2001; 21:6768-6781.
72. Liu J, Stevens J, Rote C et al. Siah-1 mediates a novel β-catenin degradation pathway linking p53 to the adenomatous polyposis coli protein. Mol Cell 2001; 7:927-936.
73. Matsuzawa S, Reed J. Siah-1, SIP, and Ebi collaborate in a novel pathway for β-catenin degradation linked to p53 responses. Mol Cell 2001; 7:915-926.
74. Shtutman M, Zhurinsky J, Oren M et al. PML is a target gene of β-catenin and plakoglobin, and coactivates β-catenin-mediated transcription. Cancer Res 2002; 62:5947-5954.
75. Ferbeyre G, de Stanchina E, Querido E et al. PML is induced by oncogenic ras and promotes premature senescence. Genes Dev 2000; 14:2015-2027.
76. Pearson M, Carbone R, Sebastiani C et al. PML regulates p53 acetylation and premature senescence induced by oncogenic Ras. Nature 2000; 406:207-210.
77. Zhurinsky J, Shtutman M, Ben-Ze'ev A. Plakoglobin and β-catenin: Protein interactions, regulation and biological roles. J Cell Sci 2000; 113:3127-3139.
78. Miller J, Moon R. Analysis of the signaling activities of localization mutants of β-catenin during axis specification in Xenopus. J Cell Biol 1997; 139:229-243.
79. Simcha I, Shtutman M, Salomon D et al. Differential nuclear translocation and transactivation potential of β-catenin and plakoglobin. J Cell Biol 1998; 141:1433-1448.
80. Zhurinsky J, Shtutman M, Ben-Ze'ev A. Differential mechanisms of LEF/TCF family-dependent transcriptional activation by β-catenin and plakoglobin. Mol Cell Biol 2000; 20:4238-4252.
81. Miravet S, Piedra J, Miro F et al. The transcriptional factor Tcf-4 contains different binding sites for β-catenin and plakoglobin. J Biol Chem 2002; 277:1884-1891.
82. Aberle H, Bierkamp C, Torchard D et al. The human plakoglobin gene localizes on chromosome 17q21 and is subjected to loss of heterozygosity in breast and ovarian cancers. Proc Natl Acad Sci USA 1995; 92:6384-6388.
83. Simcha I, Geiger B, Yehuda-Levenberg S et al. Suppression of tumorigenicity by plakoglobin: An augmenting effect of N-cadherin. J Cell Biol 1996; 133:199-209.
84. Haegel H, Larue L, Ohsugi M et al. Lack of β-catenin affects mouse development at gastrulation. Development 1995; 121:3529-3537.
85. Bierkamp C, Mclaughlin K, Schwarz H et al. Embryonic heart and skin defects in mice lacking plakoglobin. Dev Biol 1996; 180:780-785.
86. Ruiz P, Brinkmann V, Ledermann B et al. Targeted mutation of plakoglobin in mice reveals essential functions of desmosomes in the embryonic heart. J Cell Biol 1996; 135:215-225.
87. Gat U, DasGupta R, Degenstein L et al. De Novo hair follicle morphogenesis and hair tumors in mice expressing a truncated β-catenin in skin. Cell 1998; 95:605-614.
88. Charpentier E, Lavker R, Acquista E et al. Plakoglobin suppresses epithelial proliferation and hair growth in vivo. J Cell Biol 2000; 149:503-520.
89. Kolligs F, Kolligs B, Hajra K et al. γ-catenin is regulated by the APC tumor suppressor and its oncogenic activity is distinct from that of β-catenin. Genes Dev 2000; 14:1319-1331.
90. Heasman J, Crawford A, Goldstone K et al. Overexpression of cadherins and underexpression of β-catenin inhibit dorsal mesoderm induction in early Xenopus embryos. Cell 1994; 79:791-803.
91. Fagotto F, Funayama N, Gluck U et al. Binding to cadherins antagonizes the signaling activity of β-catenin during axis formation in Xenopus. J Cell Biol 1996; 132:1105-1114.
92. Sadot E, Simcha I, Shtutman M et al. Inhibition of β-catenin-mediated transactivation by cadherin derivatives. Proc Natl Acad Sci USA 1998; 95:15339-15344.
93. Orsulic S, Huber O, Aberle H et al. E-cadherin binding prevents β-catenin nuclear localization and β-catenin/LEF-1-mediated transactivation. J Cell Sci 1999; 112:1237-1245.
94. Gottardi C, Wong E, Gumbiner B. E-cadherin suppresses cellular transformation by inhibiting β-catenin signaling in an adhesion independent manner. J Cell Biol 2001; 153:1049-1060.
95. Stockinger A, Eger A, Wolf J et al. E-cadherin regulates cell growth by modulating proliferation-dependent β-catenin transcriptional activity. J Cell Biol 2001; 154:1185-1196.
96. Caca K, Kolligs F, Ji X et al. β- and γ-catenin mutations, but not E-cadherin inactivation, underlie T-cell factor/lymphoid enhancer factor transcriptional deregulation in gastric and pancreatic cancer. Cell Growth Differ 1999; 10:369-376.
97. van de Wetering M, Barker N, Harkes I et al. Mutant E-cadherin breast cancer cells do not display constitutive Wnt signaling. Cancer Res 2001; 61:278-284.

98. Brabletz T, Jung A, Reu S et al. Variable β-catenin expression in colorectal cancers indicates tumor progression driven by the tumor environment. Proc Natl Acad Sci USA 2001; 98:10356-10361.
99. Barker N, Clevers H. Tumor environment: A potent driving force in colorectal cancer? Trends Mol Med 2001; 7:535-537.
100. Conacci-Sorrell M, Simcha I, Ben-Yedidia T et al. Autoregulation of E-cadherin expression by cadherin-cadherin interactions: The roles of β-catenin signaling, Slug and ERK. Cell Biol 2003; 163:847-857.
101. Jamora C, DasGupta R, Kocieniewski P et al. Links between signal transduction, transcription and adhesion in epithelial bud development. Nature 2003; 422:317-322.
102. Kinch M, Clark G, Der C et al. Tyrosine phosphorylation regulates the adhesions of ras-transformed breast epithelia. J Cell Biol 1995; 130:461-471.
103. Roura S, Miravet S, Piedra J et al. Regulation of E-cadherin/Catenin association by tyrosine phosphorylation. J Biol Chem 1999; 274:36734-36740.
104. Piedra J, Martinez D, Castano J et al. Regulation of β-catenin structure and activity by tyrosine phosphorylation. J Biol Chem 2001; 276:20436-20443.
105. Piedra J, Miravet S, Castano J et al. p120 Catenin-associated Fer and Fyn tyrosine kinases regulate β-catenin Tyr-142 phosphorylation and β-catenin-α-catenin interaction. Mol Cell Biol 2003; 23:2287-2297.
106. Matsuyoshi N, Hamaguchi M, Taniguchi S et al. Cadherin-mediated cell-cell adhesion is perturbed by v-src tyrosine phosphorylation in metastatic fibroblasts. J Cell Biol 1992; 118:703-714.
107. Behrens J, Vakaet L, Friis R et al. Loss of epithelial differentiation and gain of invasiveness correlates with tyrosine phosphorylation of the E-cadherin/β-catenin complex in cells transformed with a temperaturesensitive v-SRC gene. J Cell Biol 1993; 120:757-766.
108. Savagner P. Leaving the neighborhood: Molecular mechanisms involved during epithelial-mesenchymal transition. BioEssays 2001; 23:912-923.
109. Hay E. An overview of epithelio-mesenchymal transformation. Acta Anat (Basel) 1995; 154:8-20.
110. Shibamoto S, Hayakawa M, Takeuchi K et al. Tyrosine phosphorylation of β-catenin and plakoglobin enhanced by hepatocyte growth factor and epidermal growth factor in human carcinoma cells. Cell Adhes Commun 1994; 1:295-305.
111. Hoschuetzky H, Aberle H, Kemler R. β-catenin mediates the interaction of the cadherin-catenin complex with epidermal growth factor receptor. J Cell Biol 1994; 127:1375-1380.
112. Crepaldi T, Pollack A, Prat M et al. Targeting of the SF/HGF receptor to the basolateral domain of polarized epithelial cells. J Cell Biol 1994; 125:313-320.
113. Potempa S, Ridley A. Activation of both MAP kinase and phosphatidylinositide 3-kinase by Ras is required for hepatocyte growth factor/scatter factor-induced adherens junction disassembly. Mol Biol Cell 1998; 9:2185-2200.
114. Lickert H, Bauer A, Kemler R et al. Casein kinase II phosphorylation of E-cadherin increases E-cadherin/β-catenin interaction and strengthens cell-cell adhesion. J Biol Chem 2000; 275:5090-5095.
115. Simcha I, Kirkpatrick C, Sadot E et al. Cadherin sequences that inhibit β-catenin signaling: A study in yeast and mammalian cells. Mol Biol Cell 2001; 12:1177-1188.
116. Daniel J, Reynolds A. Tyrosine phosphorylation and cadherin/catenin function. BioEssays 1997; 19:883-891.
117. Takahashi K, Suzuki K. Density-dependent inhibition of growth involves prevention of EGF receptor activation by E-cadherin-mediated cell-cell adhesion. Exp Cell Res 1996; 226:214-222.
118. St Croix B, Sheehan C, Rak JW et al. E-Cadherin-dependent growth suppression is mediated by the cyclin-dependent kinase inhibitor p27(KIP1). J Cell Biol 1998; 142:557-571.
119. Nieman M, Prudoff R, Johnson K et al. N-cadherin promotes motility in human breast cancer cells regardless of their E-cadherin expression. J Cell Biol 1999; 147:631-644.
120. Hazan R, Phillips G, Qiao R et al. Exogenous expression of N-cadherin in breast cancer cells induces cell migration, invasion, and metastasis. J Cell Biol 2000; 148:779-790.
121. Williams E, Williams G, Howell F et al. Identification of an N-cadherin motif that can interact with the fibroblast growth factor receptor and is required for axonal growth. J Biol Chem 2001; 276:43879-43886.
122. Suyama K, Shapiro I, Guttman M et al. A signaling pathway leading to metastasis is controlled by N-cadherin and the FGF receptor. Cancer Cell 2002; 2:301-314.

CHAPTER 13

Hepatocyte Growth Factor Regulates Transitions between Epithelial and Mesenchymal Cellular Phenotypes during Normal Development and in Disease

Regina M. Day, Angelina Felici and Donald P. Bottaro

Introduction

Transitions between epithelial and mesenchymal cell phenotypes occur as part of normal organ development and wound healing, and are also observed in cancer, fibrosis and other diseases. Although the latter processes involve dysregulated interconversion of epithelial and mesenchymal cell phenotypes, the molecular mechanisms and sequences of cellular changes often closely resemble events that occur as a part of normal development and tissue repair. While primarily epithelial to mesenchymal transition (EMT) has been implicated in cancer and fibrosis, both EMT and mesenchymal to epithelial transition (MET) occur during organogenesis and development.

Hepatocyte growth factor (HGF), upon binding to its cell surface receptor c-Met, stimulates mitogenesis, migration, and morphogenesis in a wide range of cellular targets including epithelial and endothelial cells, hematopoietic cells, neurons, melanocytes, as well as hepatocytes.[1] These pleiotropic effects are fundamentally important during development, organogenesis and tissue regeneration,[2-6] and suggest that HGF signaling may regulate EMT and MET in these contexts. HGF signaling is also strongly linked to cancer, including colon, breast, lung, thyroid, renal carcinomas, melanoma and several sarcomas, as well as glioblastoma.[6] Inherited mutations in the gene encoding c-Met found in patients with human renal papillary carcinoma cause constitutive receptor kinase activation.[7,8] Inappropriate or over-expression of normal HGF and/or c-Met resulting in autocrine or paracrine stimulation of cellular targets also contributes to cancer progression.[6] Importantly, in addition to its mitogenic activity, HGF initiates a program of cell dissociation and increased cell motility coupled with increased protease production that has been shown to promote cellular invasion through extracellular matrices, and that closely resembles tumor metastasis in vivo.[6]

Upon stimulation with HGF, the c-Met receptor kinase undergoes autophosphorylation on tyrosine (Y) residues, two of which (Y1349 and Y1356) form a multisubstrate binding site for several signal transducers including phosphatidylinositol 3-kinase (PI3K), phospholipase C (PLC)-gamma, Src, Shp-2 phosphatase, STAT3, and the adapter proteins Crk, Grb2, Shc, and Gab1 (reviewed in ref. 6), among others. Gab1 also associates with activated c-Met in other ways as described in more detail below. HGF activation of these effectors is well documented in several model systems, tissues, and in vivo, and other less well characterized receptor-proximal

Rise and Fall of Epithelial Phenotype: Concepts of Epithelial-Mesenchymal Transition, edited by Pierre Savagner.

HGF effectors have been identified in some of these contexts. Downstream HGF mitogenic signaling involves the Ras/MEK/ERK pathway, motogenic signaling involves Rac, Rho, Rho kinase, PAK and cdc42 regulation of the actin cytoskeleton, cell dissociation involves junctional proteins such as beta-catenin and E-cadherin, while cell adhesion involves focal adhesion kinase (FAK) and associated focal adhesion components.[6] Many of the latter also mediate HGF-directed cellular shape changes and, with metalloproteinases, plasminogens and their associated inhibitors, extracellular matrix invasion and branching morphogenesis. Importantly, these subcellular effectors are shared by other external signaling systems in a context-dependent manner.

While the events resulting from inappropriate HGF signaling in disease can resemble, at least in part, transitions between epithelial and mesenchymal cell types, they rarely recapitulate these developmental transitions completely. Understanding the mechanisms of HGF signaling during development and in disease has provided insight into both. Parallels drawn to EMT and MET in each case have also provided testable hypotheses about HGF signaling pathways, but evidence suggests that EMT and MET are not initiated or controlled by any single factor such as HGF. In addition, the developmental or pathophysiological context ultimately define the response to extracellular signaling molecules via intracellular effectors and responsive subcellular machinery controlling cell adhesion, motility, proliferation and gene expression.

Developmental Transitions between Epithelial and Mesenchymal Phenotypes

Developmental EMT occurs normally during gastrulation, heart formation, neural crest cell migration, palate fusion, and lens development[9-11] (see also Chapters 1-3). In vivo and experimental models for developmental EMT indicate that epithelial cells undergo a variety of changes including: actin cytoskeletal reorganization resulting in the loss of cell polarity and cell flattening; changes in intermediate filament expression from keratin to vimentin; the loss of desmosomes and adherens junctions, matrix metalloproteinase (MMP) production and increased cell mobility.[9,11] Several intracellular signaling cascades potently activated by HGF in other contexts, such the PI3K pathway, are key regulators of these events, but occur in response to TGF-beta and other extracellular signals in many of these developmental phenotypic transitions.

HGF signaling contributes to the earliest developmental EMT, which occurs during gastrulation (see also Chapter 2, Morali et al). Cells detach from the epiblast epithelium, ingress through the primitive streak, and move between the epiblast and ventral hypoblast to form mesoderm.[12] At the same time, changes occur in the classes of cadherins expressed on the surfaces of migrating epiblast cells: E-cadherin is downregulated while N-cadherin expression increases as cells migrate through the primitive streak.[13,14] Following this conversion from E- to N-cadherin expression, most epiblast cells differentiate into skeletal muscle in vitro.[15] Activin, fibroblast growth factors (FGFs) and Wnt-1 are among the cytokines present in the primitive streak in the chick embryo, although these do not appear to be sufficient to stimulate the switch in cadherin class expression or myogenesis in epiblast cell cultures.[16-23] HGF is expressed in Hensen's node during primitive streak formation in the chick embryo, and HGF added to the epiblast can induce ectopic primitive streaks.[24-29] Later in development, HGF stimulates muscle precursor cells of the dermomyotome to migrate into the limb buds, and HGF gene deletion in mice results in incomplete development of limb skeletal muscle, as well as other defects and they die in utero.[30-35] Similarly, c-Met knockout mice fail to express Pax-3, a transcription factor considered to be a marker for myogenic precursor cells, and fail to fully develop skeletal muscle in limbs.[30] In studies by DeLuca et al,[23] HGF promoted the switch from E- to N-cadherin expression in epiblast cells in vivo, although N-cadherin expression could be induced in the absence of HGF in cultured prestreak epiblast cells simply by detaching and replating them. In addition, only a subset of cultured prestreak epiblast cells responded to HGF by cadherin class switching, leading those investigators to conclude that HGF alone was not sufficient for the phenotypic changes occurring during primitive streak formation, but promoted progression along this developmental pathway.[15,23] The observations that deletion of the genes for HGF or c-Met results in embryonic lethality after gastrulation is completed indicates that HGF signaling is not required for this to occur, and reinforce the concept of functional redundancy among the many cytokines present in the primitive streak as it undergoes EMT.

Many epithelia, such as those which create airway and exocrine glands, do not arise from EMT, but from the budding and folding of preexisting epithelia through branching morphogenesis. This process requires the coordination of many of the same cellular functions involved in EMT, such as dissociation of adherens junctions, relocalization of desmosomal proteins and E-cadherin throughout the cell membrane, transient depolarization, cellular proliferation, survival, migration, invasion, differentiation, and remodeling of the extracellular matrix, except that it does not result in the loss of epithelial characteristics such as tight junctions.[36-38] Gab-1-derived signals are critical mediators of this process downstream of activated c-Met. Gab1 is a scaffold protein that contains an amino-terminal pleckstrin homology domain that binds the membrane lipid phosphatidylinositol 3,4,5-trisphosphate targeting Gab1 to the plasma membrane,[39,40] a carboxyl-terminal proline-rich domain that mediates binding to phosphorylated c-Met (MBD), as well as potential binding sites for Src homology 2 (SH2) and Src homology 3 (SH3) domains.[41-43] Gab1 association with c-Met is thought to be both indirect through Grb2 and involving Y1356,[44-46] and direct through the MBD and phosphorylated Y1349[42,46] or, as recently suggested, through a 16 residue Met-binding motif in Gab1 and the negative charge of Y1349 and the carboxyl-terminal lobe of the c-Met kinase domain.[47] Following c-Met activation, Gab1 diversifies the signal downstream of c-Met receptor through its ability to couple with multiple signal transduction pathways. c-Met/Gab1 association and phosphorylation provides binding sites for Grb2, PI3K, Shc, PLC-gamma-1, Crk-L, and Shp-2.[43] Overexpression of Gab1 in epithelial cells is sufficient to promote the adherens junction reorganization, cell dispersal, actin reorganization and branching tubulogenesis characteristic of HGF stimulation.[42] In addition, structure-function studies with chimeric c-Met receptors have revealed that disruption of c-Met/Gab1 binding impairs the ability of MDCK cells to form branching tubules upon c-Met activation,[48,49] and that Gab1 overexpression rescues c-Met-dependent tubulogenic responses in these cells.[50] In contrast to Gab1, other Gab family members (Gab2 and Gab3) are recruited to c-Met receptor in a strictly Grb2-dependent way and fail to promote a morphogenic response, suggesting that Gab1 has a unique functional role downstream of c-Met.[47] Together, these data strongly support a role for Gab1 as a critical mediator of HGF-stimulated epithelial morphogenesis (reviewed in ref. 51). Consistent with such a pivotal role, like HGF and c-Met knockout mice, Gab1-deficient embryos display defects in the migration of myogenic precursors into the limb bud and die in utero.[52]

While many epithelial organ structures are derived from existing epithelial cell sheets or tubes, the development of some organs, including those of the genitourinary system and peritoneal membranes, often involves mesenchymal to epithelial transition (MET). Some insight into the potential biological roles and molecular signaling pathways of HGF in developmental MET may be gleaned from mammalian kidney development, where HGF is among several critical growth factors and cytokines acting over an extended period during which cellular intermediates are found that express both mesenchymal and epithelial markers (for reviews of MET during kidney development see refs. 53 and 54).

The precursor of the adult mammalian kidney, metanephrogenic mesenchyme, develops into nephrons when mesenchymally derived signals induce nearby Wolffian ducts to produce ureteric buds, which then invade the mesenchyme and begin to branch. HGF is among the mesenchymal signaling molecules expressed early in this process, which include interleukin-6, fibroblast growth factors and bone morphogenic protein-7, and its impact at this stage has been largely extrapolated from observations made using tissue and organ culture systems. Along with tubulogenesis and differentiation, matrix formation and its component signaling processes, as well as precisely regulated cell proliferation and apoptosis, are essential for continued nephron maturation. Intracellular signal transduction mediating MET includes STAT3, WT-1, Pax-2, Pax-3, Pod-1, N-myc, and the beta-catenin/TCF/LEF pathways.[53,55] Again, while several of these effectors have been shown in other contexts to be activated by HGF, its precise role throughout nephrogenesis has not yet been determined. The outcome of the epithelial conversion is characterized by a change in matrix proteins and cell-cell adhesion molecules. During MET, mesenchyme-specific expression of collagens I and III and fibronectin are replaced by epithelial-specific expression of syndecan, entactin, collagen IV and laminin A_2. Cell adhesion is gradually altered, and early events include increased expression of cadherin-6 and the appearance of ZO-1, contributing to the development of tight junctions. E-cadherin upregulation occurs later and correlates with completion of the transformation.[53-55]

The roles of known intracellular mediators of HGF signaling at work early in nephrogenesis remain largely uncharacterized in that context, with a few exceptions such as formin IV.[54] Formins are implicated in the formation of actin cables in yeast, stress fibers in cultured cells and cytokinesis in many cell types.[56] Analysis of HGF driven tubulogenesis in cultured renal epithelial cells revealed that formin IV was colocalized in a submembranous band with the actin cytoskeleton in quiescent epithelial cells, but with HGF stimulation rapidly translocated into the cytosol and nucleus in an ERK-dependent manner.[57] Interestingly, the role of formin IV in most other model systems where HGF signaling has been well characterized is unclear. In light of its potential role in renal development, the generality of HGF-stimulated formin activation and its integration with other HGF activated pathways warrant further study.

The Role of HGF Signaling in Tumor-Associated EMT

Several signaling pathways are common to EMTs observed in embryonic development and tumor progression, when epithelial cells transiently or even stably lose polarity and acquire mesenchymal characteristics. Although EMT often involves a change to fibroblastoid spindle-shaped cell morphology, changes in observed epithelial and mesenchymal gene expression patterns vary widely among experimental model systems. Many important intracellular effectors of EMT and tumor progression are also downstream effectors in the HGF signaling pathway, but frequently their importance in either process has been studied outside of this context. Nonetheless, it is relevant to cite examples of this here as it is likely a connection will be found.

EMTs have been extensively characterized in breast carcinoma, where loss of epithelial polarity, disorganization of tight junctions, loss of cell E-cadherin and beta-catenin-associated adhesion complexes, cell-induced alterations in the basement membrane, and increased cell motility and migration are observed.[58-61] However, the role of EMT in carcinogenesis and metastasis is controversial. While some of the molecular events characteristic of EMT clearly contribute to malignant progression, others appear to be independent of that process.[60,62,63] For example, EpH4 murine mammary epithelial cells transfected with constitutively activated MEK1 are overtly transformed and rapidly form invasive adenocarcinomas when orthotopically implanted in syngeneic mice, but this highly invasive phenotype does not require EMT.[64] The transfected EpH4 cells display a highly metastatic phenotype characterized by disruption of cell-cell adhesion complexes, severe disorganization of the actin cytoskeleton and mislocalization of ZO-1 and beta-catenin away from cell-cell junctions.[64] However, the MEK1-transformed EpH4 cells retained expression of the epithelial specific genes E-cadherin and keratin 18, and failed to express the mesenchymal-specific genes smooth muscle actin and vimentin, suggesting that constitutive MEK1 signaling is sufficient to transform EpH mammary epithelial cells and enhance invasiveness and metastasis in vivo without complete and stable acquisition of a mesenchymal phenotype. Despite this controversy, the EMT concept has provided an important mechanistic framework for analyzing the functions of genes relevant to tumor progression. Specific changes in gene expression in breast cancer cell lines associated with EMT have been investigated using gene array analysis; these include the upregulation of mesenchymal proteins such as vimentin, TIP47 and STAT3.[59] Specific oncogenes have been implicated in the transformation of these epithelial cells to a mesenchymal phenotype. Over-expression of Ha-Ras,[65-67] c-Fos,[68,69] c-Jun,[70] mutant 70z-Cbl[71] and Crk[72] have been shown to induce EMT-like changes in epithelial cell lines. Most, if not all, of these proteins or their wild type counterparts act downstream of c-Met in the HGF signaling pathway.

HGF and its receptor c-Met have also been directly implicated in EMT during tumor progression and metastasis. Increased expression of c-Met protein, or activated forms of c-Met, have been observed in breast carcinoma cells, with highest expression levels correlating directly with invasive growth.[73] HGF is also expressed in some carcinoma cells, providing an autocrine loop for c-Met signaling; this is especially significant since only mesenchymal, and not epithelial cell types usually produce HGF. In situ hybridization analysis of RNA expression showed that both HGF and its receptor c-Met had increased expression in invasive breast carcinomas compared with normal breast epithelium.[58] The expression of HGF in these cells required the activation of c-Src, and c-Src activation coupled with STAT3 overexpression resulted in strong expression of HGF in carcinoma cells. Both c-Src and STAT3 can be found constitutively activated in breast carcinomas, thus providing a mechanism for the upregulation of HGF expression.[58,74] c-Met activating mutations are associated with hereditary and sporadic forms of papillary renal carcinoma, and in metastatic head and neck squamous-cell carcinoma, but not always in the corresponding primary tumors.[7,75] Partially activating point mutations in c-Met found in the germline of patients with type 1 papillary renal carcinoma, when expressed in liver oval cells, induce increased motility, morphological changes and matrix invasion characteristic of EMT, as well as anchorage independent growth suggestive of tumorigenicity.[76]

Although the overexpression of HGF and/or c-Met in breast carcinoma may increase the metastatic potential of these cells, the HGF pathway may not be sufficient for complete transition from epithelial to mesenchymal phenotypes. Experiments have shown that Ha-Ras-transformed mammary epithelial cells EpH4 (EpRas) transiently treated with HGF acquire a reversible spindle-shaped migratory phenotype (scattering), characterized by loss of cell polarity and increased invasiveness.[67] However, HGF treatment induced no lasting changes in epithelial or mesenchymal marker protein expression: EpRas cells continued to express the epithelial markers beta4-integrin, E-cadherin and ZO-1, and failed to express the mesenchymal marker vimentin.[67] Upon HGF withdrawal from the medium, cell polarity was reestablished, and normal localization of ZO-1 and E-cadherin was restored.[67] These results suggest that although HGF may play a role in some aspects of epithelial plasticity, such as the scattering of spindle-shaped cells, this factor alone was not sufficient for the dramatic and irreversible switch to a mesenchymal gene expression pattern observed in EpRas cells.

HGF stimulates cell dissociation and increased motility in two-dimensional cultures of Madin-Darby canine kidney (MDCK) epithelial cells, which have been used extensively as an in vitro model for HGF-induced EMT. In this system, both HGF and V12Ras initially induce centrifugal spreading of cell colonies followed by disruption of cell-cell junctions and cell scattering. These changes require activation of PI3K, MAP kinase and Rac by Ras downstream of activated Met, while a different signaling pathway not activated by Ras regulates desmosomes and tight junctions.[77] While the signaling pathways downstream of Ras that are involved in the HGF response have not been characterized in complete detail, Ras is known to activate the p42/p44 MAPK cascade, the guanine nucleotide exchange factor for the Ras-related protein, Ral-GDS, PI3K, which in turn may lead to the activation of Rac, and the novel Ras effector 1 (Nore 1)/Mst1(mammalian Ste-20 kinase) signaling pathway which promotes apoptosis.[77,78] Consistent with these findings, inhibition of MEK1 using pharmacological agents restored normal epithelial morphology and E-cadherin expression in V12ras transformed MDCK cells.[79] Taken together, these studies suggest that activation of PI3K, Rac (through PI3K), and p42/p44 MAPK are all necessary but not sufficient to mimic the effect of activated Ras downstream of HGF/c-Met, and that another signaling downstream of Ras is likely to be required to achieve HGF-induced EMT in MDCK cells.

Cell migration requires precise control, which is altered or lost when tumor cells become invasive and metastatic. Beta-catenin, an essential component of adherens junctions linking cadherins to actin microfilaments, is highly regulated during EMT.[80] In a recent study, beta-catenin was found associated with c-Met in adherens junctions and this subpopulation of beta-catenin was tyrosine phosphorylated, liberated from junctional complexes and translocated into the nucleus upon HGF treatment in primary hepatocytes cultures.[81] In other cell culture models, tyrosine phosphorylation of beta-catenin inhibited direct interaction with E-cadherin, resulting in decreased cell-cell adhesion.[80] HGF treatment also elevated cytoplasmic beta-catenin levels, which are normally minimal, and enhanced beta-catenin/LEF-dependent transcription in mammary epithelial cells and the bladder carcinoma cell line NBT II.[82,83] Artificial stabilization of cytoplasmic beta-catenin levels in NBT II cells itself induced cell migration, and enhanced the apparent potency of HGF stimulated migration.[83] Moreover, cells expressing stabilized beta-catenin also exhibit increased tumorigenic potential in nude mouse xenograft models, suggesting a critical role for beta-catenin levels in EMT and tumorigenesis.[83]

During EMT, cells dissociate from each other and from the extracellular matrix through the destabilization of focal adhesions and hemidesmosomes that anchor actin microfilaments and cytokeratins inside the cell to specific extracellular matrix proteins (e.g., fibronectin and collagens) through integrins.[11] In recent years it has become apparent that HGF/c-Met signaling cooperates with integrin signaling pathways to regulate some EMT-associated cellular responses. One example of this interaction is seen in the induction of c-Met phosphorylation by cell adhesion and integrin activation.[74,84] Activation of beta-catenin signaling during migration induced expression of the secreted, integrin-binding glycophosphoprotein osteopontin, which facilitates migration and invasion in mammary epithelial cells and has been implicated in breast cancer.[74,84,85] Interestingly, osteopontin has been reported to activate c-Met and thereby upregulate c-Met expression,[74] and to be induced by HGF treatment of the rat bladder carcinoma cell line NBT II, pointing to a role for osteopontin as an autocrine mediator of HGF-driven invasive growth.[74,83,86,87]

Important mediators of HGF-induced cell dispersion and the breakdown of epithelial junctions during EMT are the substrates SHIP-1, c-Cbl, CrkL/CrkII and Gab1, acting downstream of activated c-Met. The SH2-containing inositol 5-phosphatase (SHIP)-1, involved in inositol phospate/phosphatidylinositol metabolism, binds to Y1356 at the multifunctional docking site of c-Met in response to HGF. Overexpression of SHIP-1 in MDCK cells resulted in a morphology characteristic of EMT, involving the loss of adherens junctions and redistribution of the cortical actin, development of stress fibers and spontaneous motility in the absence of HGF.[88] While MAPK plays a role in the partial dissociation of cell-cell contacts and cell dispersion, the observed HGF-stimulated alterations of actin distribution mediated by SHIP-1 required the activation of PI3K but not MEK.[88]

One of the most highly (tyrosine) phosphorylated intracellular proteins in response to HGF stimulation is the docking protein/ubiquitin ligase c-Cbl. This proto-oncogene may amplify c-Met signaling through the recruitment of signal transducers such as Grb2, PI3K, Src, and Crk, and may regulate signal duration by stimulating the ubiquitination and internalization of c-Met.[89] Naturally occurring c-Cbl mutants devoid of ubiquitin ligase activity, such as 70z-Cbl, have been found in preB cell lymphomas. Expression of 70z-Cbl in MDCK cells resulted in the loss of cell-cell contacts, cell spreading, initiation of cell dispersal and alterations in cell morphology characteristic of a mesenchymal fibroblastic phenotype.[90] Overexpression of the Crk family members CrkL and CrkII in MDCK cells or in the well-differentiated breast cancer epithelial cells T47D caused a similar change in morphology, breakdown of adherens junctions, spreading of epithelial colonies and the formation of lamellipodia in the absence of HGF,[72] suggesting that CrkL and CrkII may be the downstream targets of c-Cbl activation leading to EMT. In addition, cytoskeletal rearrangements and cell spreading stimulated by either HGF treatment or Crk overexpression could be blocked by expression of a dominant negative mutant of the small GTP binding protein Rac1,[72] suggesting that Rac1 is an essential mediator of CrkII-induced lamellipodia formation and cell spreading.

Finally, Gab-1, a ubiquitous multisubstrate docking protein acting downstream activated c-Met, plays a major positive role in the PI3K and MAPK pathways that are critical to HGF-induced EMT.[91,92] Predominantly localized at sites of cell-cell contact in epithelial cells, Gab-1 controls multiple HGF responses involved in EMT, including remodeling of epithelial junctions, cell proliferation, migration and invasion.[39,43,50,52,93-99] HGF stimulates the rapid phosphorylation of Gab1 on specific tyrosine residues which in turn recruit several SH2 domain-containing proteins that initiate Crk, PI3K and SHP-2 signaling cascades. In addition to mediating PI3K activation, Gab1 is also a downstream target of PI3K: it binds specifically to the PI3K product phosphatidylinositol 3,4,5-trisphosphate through its PH domain and is thereby recruited to the plasma membrane in the vicinity of c-Met. This positive feedback loop was found to be critical for Gab1 phosphorylation and HGF-induced epithelial morphogenesis.[40,50,100] The observations that Gab1 knockout embryo cells displayed impaired HGF-stimulated MAPK activation[99] and that the disruption of Shp-2 binding sites on Gab1 decreased Ras/MAPK activation following HGF treatment,[43,101] suggest that Gab1 activates the MAPK pathway primarily through SHP-2.

Occurrence of EMT in Tissue Fibrosis and Its Inhibition by HGF

In fibrotic tissue, myofibroblasts have been traditionally thought to evolve exclusively via the trans-differentiation of fibroblasts normally present in the tissue.[102,103] However, recent studies have provided substantial evidence for EMT as an important process in the genesis myofibroblasts.[103,104] Local EMT has been shown during renal fibrogenesis using unilateral ureteral obstruction (UUO) to produce experimental renal fibrosis in one kidney of genetically engineered (gamma-GT.Cre x R26R) mice.[104] These mice acquired irreversible expression of beta-galactosidase (beta-gal) specifically in mature renal tubule epithelial cells. Within 10 days of UUO, animals displayed fibrosis in the treated kidney without systemically compromised renal function. Renal tubule epithelial cells in the treated kidney, identified by beta-Gal expression, acquired expression of fibroblast specific protein 1 (FSP1), indicating a shift from epithelial to mesenchymal gene expression patterns.[104] No beta-gal positive/FSP1 positive cells were found in glomeruli from the contralateral kidneys.[104] Thus, in this model of renal fibrosis, EMT clearly contributes to myofibroblast ontogenesis. Moreover, in biopsies of renal tubular epithelial cells from a variety of renal diseases, in situ hybridization revealed the expression of markers for cellular proliferation (PCNA, Mib-1) together with markers for mesenchymal transformation (vimentin, alpha-smooth muscle actin and the extracellular matrix proteins collagens I and III), whereas specific epithelial cell markers (ZO-1 and cytokeratins) were lost.[105]

Renal fibrosis-associated EMT, like breast carcinoma EMT, involves the loss of apical-basal polarity, loss of intercellular tight junctions via decreased expression of E-cadherin, acquisition of spindle-shaped mesenchymal morphology, formation of actin stress fibers, extracellular matrix turnover, and increased cell migration and invasion.[103] This phenotypic transition has also been reported in human glomerulonephritis with progressive tubulointerstitial fibrosis.[103,106] Transdifferentiated cells begin expressing alpha-smooth muscle actin (a marker for myofibroblasts), but in some cases maintain expression of epithelial cytokeratins, allowing clear identification of intermediately transformed cells.[106] The inducer for EMT in renal fibrosis has been postulated to be TGF-beta, although interleukin-1, FGF-2 and angiotensin II may also participate.[103,106] TGF-beta stimulates the upregulation of connective tissue growth factor (CTGF), which in turn contributes to the increased deposition of collagen I and fibronectin.[107] EMT in fibrosis is negatively regulated by HGF.[108] HGF abrogates TGF-beta-induced EMT both phenotypically and morphologically in renal tubulointerstitial fibrosis in experimentally obstructed kidneys,[108] and suppresses alpha-smooth muscle actin expression as well as TGF-beta induced CTGF upregulation.[107-109]

Summary

Transitions between epithelial and mesenchymal cell phenotypes occur throughout normal development and in disease, although the degree to which pathological EMT recapitulates normal development varies in extent and molecular mechanism in a context dependent manner. HGF acts in concert with other cytokines, matrix components, and environmental cues to provide direction to these multistep transitions, often by modulating shared intracellular signaling pathways. While the basic routes of HGF signaling gleaned from experimental model systems operate consistently in cells transiting between epithelial and mesenchymal phenotypes, local cues ultimately define the precise subset of molecular events and can significantly affect the biological outcome of HGF signaling as it occurs in EMT. For example, HGF signaling appears to antagonize fibrotic EMT, but typically promotes oncogenic EMT. This divergent biological outcome may result from mutations in cancer cells, differences in the extracellular milieu in the two diseases, or both. Further study of HGF signaling in these contexts may reveal important interactions with other pathways that explain the divergent biological responses, and may also identify critical molecular targets and events for the development of novel therapeutic strategies to combat fibrosis and cancer.

References

1. Michalopoulos GK, DeFrances MC. Liver regeneration. Science 1997; 276:60-66.
2. Birchmeier C, Gherardi E. Developmental roles of HGF/SF and its receptor, the c-Met tyrosine kinase. Trends Cell Biol 1998; 8:404-410.
3. Streit AC, Stern CD. Competence for neural induction: HGF/SF, HGF1/MSP and the c-Met receptor. Ciba Found Symp 1997; 212:155-165.
4. Birchmeier W, Brinkmann V, Niemann C et al. Role of HGF/SF and c-Met in morphogenesis and metastasis of epithelial cells. Ciba Found Symp 1997; 212:230-240.
5. Funakoshi H, Nakamura T. Hepatocyte growth factor: From diagnosis to clinical applications. Clin Chim Acta 2003; 327:1-23.
6. Zhang YW, Vande Woude GF. HGF/SF-Met signaling in the control of branching morphogenesis and invasion. J Cell Biochem 2003; 88:408-417.
7. Schmidt L, Duh FM, Chen F et al. Germline and somatic mutations in the tyrosine kinase domain of the MET proto-oncogene in papillary renal carcinomas. Nat Genet 1997; 16:68-73.
8. Jeffers M, Schmidt L, Nakaigawa N et al. Activating mutations for the met tyrosine kinase receptor in human cancer. Proc Natl Acad Sci USA 1997; 94:11445-11450.
9. Kang P, Svoboda KK. PI-3 kinase activity is required for epithelial-mesenchymal transformation during palate fusion. Dev Dyn 2002; 225:316-321.
10. Zuk A, Hay ED. Expression of beta 1 integrins changes during transformation of avian lens epithelium to mesenchyme in collagen gels. Dev Dyn 1994; 201:378-393.
11. Savagner P. Leaving the neighborhood: Molecular mechanisms involved during epithelial-mesenchymal transition. Bioessays 2001; 23:912-923.
12. Bellars R. The primitive streak. Anat Embryo 1986; 174:1-14.
13. Edelman GM, Gallin WJ, Delouvee A et al. Early epochal maps of two different cell adhesion molecules. Proc Natl Acad Sci USA 1983; 80:4384-4388.
14. Hatta K, Takeichi M. Expression of N-cadherin adhesion molecules associated with early morphogenetic events in chick development. Nature 1986; 320:447-449.
15. George-Weinstein M, Gerhart J, Blitz J et al. N-cadherin promotes the commitment and differentiation of skeletal muscle precursor cells. Dev Biol 1997; 185:14-24.
16. Mitrani E, Gruenbaum Y, Shohat H et al. Fibroblast growth factor during mesoderm induction in the early chick embryo. Development 1990; 109:387-393.
17. Mitrani E, Ziv T, Shimoni Y et al. Activin can induce the formation of axial structures and is expressed in the hypoblast of the chick. Cell 1990; 63:495-501.
18. Cooke J, Wong A. Growth-factor-related proteins that are inducers in early amphibian development may mediate similar steps in amniote (bird) embryogenesis. Development 1991; 111:197-212.
19. Stern CD, Yu RT, Kakizuka A et al. Activin and its receptors during gastrulation and the later phases of mesoderm development in the chick embryo. Dev Biol 1995; 172:192-205.
20. Kisbert A, Ortner H, Cooke J et al. The chick Brachyury gene: Developmental expression pattern and response to axial induction by localized activin. Dev Biol 1995; 168:406-415.
21. George-Weinstein M, Gerhart J, Reed R et al. Intrinsic and extrinsic regulation of the development of myogenic precursors in the chick embryo. Basic Appl Myology 1996; 6:417-430.

22. George-Weinstein M, Dare J, Gerhart J et al. Unexpected effects of Wnt-1 on the differentiation and adhesion of myogenic cells. Basic Appl Myology 1998; 8:303-314.
23. DeLuca SM, Gerhart J, Cochran E et al. Hepatocyte growth factor/scatter factor promotes a switch from E- to N-cadherin in chick embryo epiblast cells. Exp Cell Res 1999; 251:3-15.
24. Stern CD, Ireland GW, Herrick SE et al. Epithelial scatter factor and development of the chick embryonic axis. Development 1990; 110:1271-1284.
25. Ireland GW, Stern CD, Stoker M. Human MRC-5 cells induce a secondary primitive streak when grafted into chick embryos. J Anat 1987; 152:223-224.
26. Stern CD, Ireland GW. HGF/SF: A neural inducing molecule in vertebrate embryos? EXS 1993; 65:369-380.
27. Thery C, Sharpe MJ, Batley SJ et al. Expression of HGF/SF, HGF1/MSP, and c-met suggests new functions during early chick development. Dev Gen 1995; 17:90-101.
28. Streit A, Stern CD, Thiery C et al. A role for HGF/SF in neural induction and its expression in Hensen's node during gastrulation. Development 1995; 121:813-824.
29. Streit A, Sockanathan S, Perez L et al. Preventing the loss of competence for neural induction: HGF/SF, L5 and Sox-2. Development 1997; 124:1191-1202.
30. Bladt F, Riethmacher D, Isenmann S et al. Essential role for the c-met receptor in the migration of myogenic precursor cells into the limb bud. Nature 1995; 376:768-771.
31. Brand-Saberi B, Muller TS, Wilting J et al. SF/HGF induces emigration of myogenic cells at interlimb level in vivo. Dev Biol 1996; 179:303-308.
32. Yang X, Vogan K, Gruss P et al. Expression of the met receptor tyrosine kinase in muscle progenitor cells in somites and limbs is absent in Splotch mice. Development 1996; 122:2163-2171.
33. Uehara Y, Minowa O, Mori C et al. Placental defect and embryonic lethality in mice lacking hepatocyte growth factor/scatter factor. Nature 1995; 373:702-705.
34. Brand-Sabieri B, Gamel AJ, Krenn V et al. N-cadherin is involved in myoblast migration and muscle differentiation in the avian limb bud. Dev Biol 1996; 178:160-173.
35. Schmidt C, Bladt F, Goedecke S et al. Scatter factor/hepatocyte growth factor is essential for liver development. Nature 1995; 373:699-702.
36. Pollack AL, Runyan RB, Mostov KE. Morphogenetic mechanisms of epithelial tubulogenesis: MDCK cell polarity is transiently rearranged without loss of cell-cell contact during scatter factor/hepatocyte growth factor-induced tubulogenesis. Dev Biol 1998; 204:64-79.
37. Metzger RJ, Krasnow MA. Genetic control of branching morphogenesis. Science 1999; 284:1635-1639.
38. Samakovlis C, Manning G, Steneberg P et al. Genetic control of epithelial tube fusion during *Drosophila* tracheal development. Development 1996; 122:3531-3536.
39. Maroun CR, Moscatello DK, Naujokas MA et al. A conserved inositol phospholipid binding site within the pleckstrin homology domain of the Gab1 docking protein is required for epithelial morphogenesis. J Biol Chem 1999; 274:31719-31726.
40. Rodrigues GA, Falasca M, Zhang Z et al. A novel positive feedback loop mediated by the docking protein Gab1 and phosphatidylinositol 3-kinase in epidermal growth factor receptor signaling. Mol Cell Biol 2000; 20:1448-1459.
41. Holgado-Madruga M, Emlet DR, Moscatello DK et al. A Grb2-associated docking protein in EGF- and insulin-receptor signaling. Nature 1996; 379:560-564.
42. Weidner KM, Di Cesare S, Sachs M et al. Interaction between Gab1 and the c-Met receptor tyrosine kinase is responsible for epithelial morphogenesis. Nature 1996; 384:173-176.
43. Schaeper U, Gehring NH, Fuchs KP et al. Coupling of Gab1 to c-Met, Grb2, and Shp2 mediates biological responses. J Cell Biol 2000; 149:1419-1432.
44. Bardelli A, Longati P, Gramaglia D et al. Gab1 coupling to the HGF/Met receptor multifunctional docking site requires binding of Grb2 and correlates with the transforming potential. Oncogene 1997; 15:3103-3111.
45. Fixman ED, Holgado-Madruga M, Nguyen L et al. Efficient cellular transformation by the Met oncoprotein requires a functional Grb2 binding site and correlates with phosphorylation of the Grb2-associated proteins, Cbl and Gab1. J Biol Chem 1997; 272:20167-20172.
46. Nguyen L, Holgado-Madruga M, Maroun C et al. Association of the multisubstrate docking protein Gab1 with the hepatocyte growth factor receptor requires a functional Grb2 binding site involving tyrosine 1356. J Biol Chem 1997; 272:20811-20819.
47. Lock LS, Frigault MM, Saucier C et al. Grb2-independent recruitment of Gab1 requires the C-terminal lobe and structural integrity of the Met receptor kinase domain. J Biol Chem 2003; 278:30083-30090.
48. Zhu H, Naujokas MA, Fixman ED et al. Tyrosine 1356 in the carboxyl-terminal tail of the HGF/SF receptor is essential for the transduction of signals for cell motility and morphogenesis. J Biol Chem 1994; 269:29943-29948.

49. Fournier TM, Kamikura D, Teng K et al. Branching tubulogenesis but not scatter of madin-darby canine kidney cells requires a functional Grb2 binding site in the Met receptor tyrosine kinase. J Biol Chem 1996; 271:22211-22217.
50. Maroun CR, Holgado-Madruga M, Royal I et al. The Gab1 PH domain is required for localization of Gab1 at sites of cell-cell contact and epithelial morphogenesis downstream from the met receptor tyrosine kinase. Mol Cell Biol 1999; 19:1784-1799.
51. Furge KA, Zhang YW, Vande Woude GF. Met receptor tyrosine kinase: Enhanced signaling through adapter proteins. Oncogene 2000; 19:5582-5589.
52. Sachs M, Brohmann H, Zechner D et al. Essential role of Gab1 for signaling by the c-Met receptor in vivo. J Cell Biol 2000; 150:1375-1384.
53. Barasch J. Genes and proteins involved in mesenchymal to epithelial transition. Curr Opin Nephrol Hypertens 2001; 10:429-436.
54. Davies JA, Fisher CE. Genes and proteins in renal development. Exp Nephrol 2002; 10:102-113.
55. Wiggan O, Fadel MP, Hamel PA. Pax3 induces cell aggregation and regulates phenotypic mesenchymal-epithelial interconversion. J Cell Sci 2002; 115:517-529.
56. Evangelista M, Zigmond S, Boone C. Formins: Signaling effectors for assembly and polarization of actin filaments. J Cell Sci 2003; 116:2603-2611.
57. O'Rourke DA, Liu ZX, Sellin L et al. Hepatocyte growth factor induces MAPK-dependent formin IV translocation in renal epithelial cells. J Am Soc Nephrol 2000; 11:2212-2221.
58. Elliott BE, Hung WL, Boag AH et al. The role of hepatocyte growth factor (scatter factor) in epithelial-mesenchymal transition and breast cancer. Can J Physiol Pharmacol 2002; 80:91-102.
59. Mellick AS, Day CJ, Weinstein SR et al. Differential gene expression in breast cancer cell lines and stroma-tumor differences in microdissected breast cancer biopsies revealed by display array analysis. Int J Cancer 2002; 100:172-80.
60. Boyer B, Valles AM, Thiery JP. Model systems of epithelium-mesenchyme transitions. Acta Anat 1996; 156:227-239.
61. Birchmeier C, Birchmeier W, Brand-Saberi B. Epithelial-mesenchymal transitions in cancer progression. Acta Anat 1996; 156:217-26.
62. Cui W, Fowlis DJ, Bryson S et al. TGFbeta1 inhibits the formation of benign skin tumors, but enhances progression to invasive spindle carcinomas in transgenic mice. Cell 1996; 86:531-542.
63. Bakin AV, Tomlinson AK, Bhowmick NA. Phosphatidylinositol 3-kinase function is required for transforming growth factor ß-mediated epithelial to mesenchymal transition and cell migration. J Biol Chem 2000; 275:36803-36810.
64. Pinkas J, Leder P. MEK1 signaling mediates transformation and metastasis of EpH4 mammary epithelial cells independent of an epithelial to mesenchymal transition. Cancer Res 2002; 62:4781-4790.
65. Oft M, Peli J, Rudaz C. TGF-beta1 and Ha-Ras collaborate in modulating the phenotypic plasticity and invasiveness of epithelial tumor cells. Genes Dev 1996; 10:2462-2477.
66. Oft M, Heider KH, Beug H. TGF-beta signaling is necessary for carcinoma cell invasiveness and metastasis. Curr Biol 1998; 8:1243-1252.
67. Janda E, Lehmann K, Killisch I et al. Ras and TGF-beta cooperatively regulate epithelial cell plasticity and metastasis: Dissection of Ras signaling pathways. J Cell Biol 2002; 156:299-313.
68. Reichmann E, Schwarz H, Deiner EM et al. Activation of an inducible c-FosER fusion protein causes loss of epithelial polarity and triggers epithelial-fibroblastoid cell conversion. Cell 1992; 7:1103-1116.
69. Oft M, Akhurst RJ, Balmain A. Metastasis is driven by sequential elevation of H-ras and Smad2 levels. Nat Cell Biol 2002; 4:487-94.
70. Fialka I, Schwarz H, Reichmann E et al. The estrogen-dependent c-Jun ER protein causes a reversible loss of mammary epithelial cell polarity involving a destabilization of adherens junctions. J Cell Biol 1996; 132:1115-32.
71. Fournier TM, Lamorte L, Maroun CR et al. Cbl-transforming variants trigger a cascade of molecular alterations that lead to epithelial mesenchymal conversion. Mol Biol Cell 2000; 11:3397-410.
72. Lamorte L, Royal I, Naujokas M et al. Crk adapter proteins promote an epithelial-mesenchymal-like transition and are required for HGF-mediated cell spreading and breakdown of epithelial adherens junctions. Mol Biol Cell 2002; 13:1449-61.
73. Jin L, Fuchs A, Schnitt SJ et al. Expression of scatter factor and c-met receptor in benign and malignant breast tissue. Cancer 1997; 79:749-60.
74. Tuck AB, Elliott BE, Hota C. Integrin-dependent migration of human mammary epithelial cells involves activation of the hepatocyte growth factor receptor (Met). J Cell Biochem 2000; 78:465-75.
75. Di Renzo MF, Olivero M, Martone T et al. Somatic mutations of the MET oncogene are selected during metastatic spread of human HNSC carcinomas. Oncogene 2000; 19:1547-55.

76. Giordano S, Maffe A, Williams TA et al. Different point mutations in the met oncogene elicit distinct biological properties. FASEB J 2000; 14:399-406.
77. Potempa S, Ridley AJ. Activation of both MAP kinase and phosphatidylinositide 3-kinase by Ras is required for hepatocyte growth factor/scatter factor-induced adherens junction disassembly. Mol Biol Cell 1998; 9:2185-200.
78. Yart A, Mayeux P, Raynal P. Gab1, SHP-2 and Other Novel Regulators of Ras: Targets for Anticancer Drug Discovery? Curr Cancer Drug Targets 2003; 3:177-92.
79. Chen Y, Lu Q, Schneeberger EE. Restoration of tight junction structure and barrier function by down-regulation of the mitogen-activated protein kinase pathway in ras-transformed Madin-Darby canine kidney cells. Mol Biol Cell 2000; 11:849-862.
80. Balsamo J, Arregui C, Leung T et al. The nonreceptor protein tyrosine phosphatase PTP1B binds to the cytoplasmic domain of N-cadherin and regulates the cadherin-actin linkage. J Cell Biol 1998; 143:523-32.
81. Monga SP, Mars WM, Pediaditakis P et al. Hepatocyte growth factor induces Wnt-independent nuclear translocation of beta-catenin after Met-beta-catenin dissociation in hepatocytes. Cancer Res 2002; 62:2064-71.
82. Papkoff J, Aikawa M. WNT-1 and HGF regulate GSK3 beta activity and beta-catenin signaling in mammary epithelial cells. Biochem Biophys Res Commun 1998; 247:851-8.
83. Muller T, Bain G, Wang X et al. Regulation of epithelial cell migration and tumor formation by beta-catenin signaling. Exp Cell Res 2002; 280:119-33.
84. Tuck AB, Hota C, Wilson SM et al. Osteopontin-induced migration of human mammary epithelial cells involves activation of EGF receptor and multiple signal transduction pathways. Oncogene 2003; 22:1198-205.
85. Tuck AB, Arsenault DM, O'Malley FP et al. Osteopontin induces increased invasiveness and plasminogen activator expression of human mammary epithelial cells. Oncogene 1999; 18:4237-46.
86. Nagai S, Hashimoto S, Yamashita T et al. Comprehensive gene expression profile of human activated T(h)1- and T(h)2-polarized cells. Int Immunol 2000; 13:367-76.
87. Medico E, Gentile A, Lo Celso C. Osteopontin is an autocrine mediator of hepatocyte growth factor-induced invasive growth. Cancer Res 2001; 61:5861-8.
88. Mancini A, Koch A, Wilms R et al. The SH2-containing inositol 5-phosphatase (SHIP)-1 is implicated in the control of cell-cell junction and induces dissociation and dispersion of MDCK cells. Oncogene 2002; 21:1477-84.
89. Petrelli A, Gilestro GF, Lanzardo S et al. The endophilin-CIN85-Cbl complex mediates ligand-dependent downregulation of c-Met. Nature 2002; 416:187-90.
90. Fournier TM, Lamorte L, Maroun CR et al. Cbl-transforming variants trigger a cascade of molecular alterations that lead to epithelial mesenchymal conversion. Mol Biol Cell 2000; 11:3397-410.
91. Lock LS, Maroun CR, Naujokas MA et al. Distinct recruitment and function of Gab1 and Gab2 in Met receptor-mediated epithelial morphogenesis. Mol Biol Cell 2002; 13:2132-46.
92. Maroun CR, Naujokas MA, Park M. Membrane targeting of Grb2-associated binder-1 (Gab1) scaffolding protein through Src myristoylation sequence substitutes for Gab1 pleckstrin homology domain and switches an epidermal growth factor response to an invasive morphogenic program. Mol Biol Cell 2003; 14:1691-708.
93. Holgado-Madruga M, Moscatello DK, Emlet DR et al. Grb2-associated binder-1 mediates phosphatidylinositol 3-kinase activation and the promotion of cell survival by nerve growth factor. Proc Natl Acad Sci USA 1997; 94:12419-24.
94. Korhonen JM, Said FA, Wong AJ et al. Gab1 mediates neurite outgrowth, DNA synthesis, and survival in PC12 cells. J Biol Chem 1999; 274:37307-14.
95. Kamikura DM, Khoury H, Maroun C et al. Enhanced transformation by a plasma membrane-associated met oncoprotein: Activation of a phosphoinositide 3'-kinase-dependent autocrine loop involving hyaluronic acid and CD44. Mol Cell Biol 2000; 20:3482-96.
96. Lamorte L, Kamikura DM, Park M. A switch from p130Cas/Crk to Gab1/Crk signaling correlates with anchorage independent growth and JNK activation in cells transformed by the Met receptor oncoprotein. Oncogene 2000; 19:5973-81.
97. Garcia-Guzman M, Dolfi F, Zeh K et al. Met-induced JNK activation is mediated by the adapter protein Crk and correlates with the Gab1-Crk signaling complex formation. Oncogene 1999; 18:7775-86.
98. Gual P, Giordano S, Williams TA et al. Sustained recruitment of phospholipase C-gamma to Gab1 is required for HGF-induced branching tubulogenesis. Oncogene 2000; 19:1509-18.
99. Itoh M, Yoshida Y, Nishida K et al. Role of Gab1 in heart, placenta, and skin development and growth factor- and cytokine-induced extracellular signal-regulated kinase mitogen-activated protein kinase activation. Mol Cell Biol 2000; 20:3695-704.

100. Isakoff SJ, Cardozo T, Andreev J et al. Identification and analysis of PH domain-containing targets of phosphatidylinositol 3-kinase using a novel in vivo assay in yeast. EMBO J 1998; 17:5374-87.
101. Maroun CR, Naujokas MA, Holgado-Madruga M et al. The tyrosine phosphatase SHP-2 is required for sustained activation of extracellular signal-regulated kinase and epithelial morphogenesis downstream from the met receptor tyrosine kinase. Mol Cell Biol 2000; 20:8513-25.
102. Cooper JA. Pulmonary fibrosis: Pathways are slowly coming into light. Am J Respir Cell Mol Biol 2000; 22:520-523.
103. Lan HY. Tubular epithelial-myofibroblast transdifferentiation mechanisms in proximal tubule cells. Curr Opin Nephrol Hypertens 2003; 12:25-9.
104. Iwano M, Plieth D, Danoff TM et al. Evidence that fibroblasts derive from epithelium during tissue fibrosis. J Clin Invest 2002; 110:341-50.
105. Rastaldi MP, Ferrario F, Giardino L et al. Epithelial-mesenchymal transition of tubular epithelial cells in human renal biopsies. Kidney Int 2002; 62:137-46.
106. Roberts IS, Burrows C, Shanks JH et al. Interstitial myofibroblasts: Predictors of progression in membranous nephropathy. J Clin Pathol 1997; 50:123-127.
107. Inoue T, Okada H, Kobayashi T et al. Hepatocyte growth factor counteracts transforming growth factor-beta1, through attenuation of connective tissue growth factor induction, and prevents renal fibrogenesis in 5/6 nephrectomized mice. FASEB J 2003; 17:268-70.
108. Liu Y. 2002. Hepatocyte growth factor and the kidney. Curr Opin Nephrol Hypertens 2002; 11:23-30.
109. Yang J, Dai C, Liu Y. Hepatocyte growth factor gene therapy and angiotensin II blockade synergistically attenuate renal interstitial fibrosis in mice. J Am Soc Nephrol 2002; 13:2464-77.

CHAPTER 14

The Role of Insulin-Like Growth Factors in the Epithelial to Mesenchymal Transition

Sylvia Julien-Grille, Robert Moore, Laurence Denat, Olivier G. Morali, Véronique Delmas, Alfonso Bellacosa and Lionel Larue

Abstract

IGFs (insulin-like growth factors) are peptides known to stimulate a wide range of actions on different tissues. Indeed, IGFs can stimulate anabolism, acute metabolic effects as well as enhancing more chronic effects such as cell proliferation and differentiation together with protecting cells from apoptosis. Recently, it was shown that IGFs induce an epithelial to mesenchymal transition (EMT), a crucial morphogenic event during development and transformation. Here, the cellular and molecular aspects of IGF-induced EMT are reviewed. Major signaling pathways downstream of IGFs are described in order to introduce molecules that are believed to convey the EMT signal. The roles and targets of these molecules are analysed. The importance of IGFs in cellular events which when dysregulated lead to neoplasia is discussed in this review.

Introduction

The insulin-like growth factor (IGF) family contains two highly similar, single-chain ligands of ~7.5 kDa termed IGF-I and IGF-II. In most species, the major forms of IGF-I and IGF-II are molecules of 70 and 67 amino acids, respectively. However, there are additional isoforms of IGF-I and IGF-II which arise from differential promoter usage, variable transcription initiation sites, differential RNA splicing and alternative polyadenylation site usage.

Both mature proteins have 4 domains: B (N-terminal), C, A and D (C-terminal). This organization is similar to that of pro-insulin which also contains domains B, C and A. The amino acid similarity between insulin (Ins) and IGFs is restricted to the A and B domains but nevertheless insulin, IGF-I and IGF-II are believed to have evolved from a single common ancestor. The primary sequences of both IGF-I and IGF-II have been highly conserved in mammals (for review see ref. 1).

IGFs are ligands which bind to the insulin receptor (IR), the IGF-I receptor (IGF-1R) and the IGF-II receptor (IGF-2R). IR and IGF-1R both are receptor protein tyrosine kinases (RPTK) with intrinsic tyrosine kinase activity. Their overall structure resembles that of the archetypal epidermal growth factor receptor. The mature IR and IGF-1R are large transmembrane glycoproteins (~300-350kDa) composed of two α and two β subunits. The α and β subunits are proteolytically cleaved from a single common proreceptor and covalently linked into dimers by disulphide bonds. The two α subunits are held together by at least two disulphide bonds whereas only a single -S-S- bond is responsible for the α-β linkage. These dimers then form a functional $\alpha_2\beta_2$ heterotetrameric complex. The putative ligand binding pockets are located in the cysteine rich portion of the extracellular α subunits. About one third of the β subunit is extracellular

Rise and Fall of Epithelial Phenotype: Concepts of Epithelial-Mesenchymal Transition, edited by Pierre Savagner.

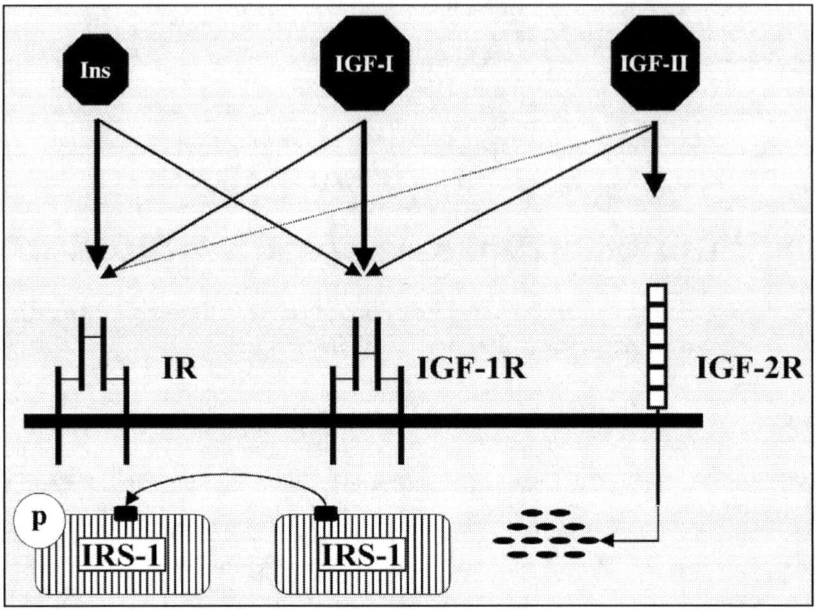

Figure 1. Affinity of interactions of various insulin ligands with their receptors. Ins (Insulin) and IGF (insulin-like growth factors) peptides have different affinities for insulin receptor (IR), IGF-1R and IGF-2R. Thickness of the arrow represents the affinity of a ligand for a receptor. IGF-I has a higher affinity for IGF-1R than for InsR. Binding of the ligands to InsR or IGF-1R initiates signalling pathways such as IRS-1 (insulin receptor substrate 1) pathway. Binding of IGF-II to IGF-2R leads to the degradation of this ligand.

and the remainder spans the membrane and contains tyrosine kinase domain, ATP and substrate binding pockets and autophosphorylation sites. In sharp contrast, IGF-2R is a single protein of ~300kDa, devoid of any apparent intrinsic catalytic activity. The same molecule has also been identified as a cation independent receptor for mannose 6-phosphate, being involved in the transport of enzymes linked with phosphorylated mannose to the lysosome. It is believed that IGF-2R is a multifunctional protein binding both IGF-II and proteins tagged with a phosphorylated mannose recognition marker (for review see ref. 1). It is unclear, however, whether any signal transduction pathway relevant to Ins or IGFs is activated either upon ligand binding at the cell surface or upon ligand-induced receptor internalization.

Any study of Ins, IGF-I and IGF-II ligands and their receptor is made infinitely more complicated by the fact that both the receptors and ligands are promiscuous; practically all of the three receptors bind heterologous ligands, but in a concentration-dependent manner. For example, IGF-1R binds these ligands with the following relative affinities: IGF-I > IGF-II >> Insulin; and IGF-2R: IGF-II> IGF-I, but Ins does not bind at physiological concentrations to IGF-1R or IGF-2R (Fig. 1). Obviously, each receptor has the highest affinity for its cognate ligand but typically a 100 to 1000 times lower affinity for any of the others. The most crucial point to note is that the affinities of either IGF-I or IGF-II for the IGF-1R are so favourable that in mammals many of the biological effects attributed to IGF-I and IGF-II are believed to be mediated exclusively via IGF-1R.

The biological effects of IGF ligands (but not Ins) are further modulated by a family of structurally related, secreted proteins termed IGF binding proteins (IGFBP). They are present ubiquitously in the circulation and in the extracellular space. IGFBP bind to the IGFs with affinities comparable to those for the IGFRs. IGFBP regulate IGFs bioavaibility by (i) protecting them from degradation, (ii) facilitating their transport and (iii) promoting or inhibiting IGF binding to their receptors.

IGFs are produced during both prenatal and postnatal phases where they exert a general anabolic or growth promoting effect. During this time, endocrine IGFs are mainly synthesized by hepatocytes. In human, the serum concentrations of IGF-I and IGF-II increase with age. At 30 weeks of gestation, the serum levels of IGF-I and IGF-II are 20 ng/mL and 100 ng/mL, respectively; at birth, they are 100 ng/mL and 300 ng/mL, respectively; and after birth, they are 200 ng/mL and 700 ng/mL, respectively. IGF-1R is broadly expressed in various tissues during embryonic development. Postnatally, in addition, local production of IGFs begins and continues in numerous cell types including the ovary, testis, nervous system, bones, skeletal muscles. When locally synthesized IGFs have paracrine and autocrine functions.[1]

IGFs are believed to be important for growth and development and their role has been examined genetically in the mouse. Somatic undergrowth is a typical characteristic of the mutant mice. If the IGF-II gene is disrupted, for example, birth weight is 61.5% of the weight of wild-type mice. When IGF-1R gene is disrupted, embryo birth weight is 46% the weight of wild-type littermates. Mutation of both the IGF-II and IGF-1R genes gives an embryo birth weight which is 34% of that of wild-type littermates. The comparison of these phenotypes suggests that, in vivo, IGF-II signals through another receptor which was identified as IR.

The powerful mitogenic effects of IGFs on many cell types partially explain these phenotypes. Indeed, IGFs induce cell growth and cell division in vivo and in vitro: (1) IGF-I stimulates DNA synthesis within granulosa and granulosa-luteal cells; (2) IGF-I mediates proliferative effects of oestradiol during the endometrial proliferative phase; (3) IGFs stimulate spermatogonial DNA synthesis and Sertoli and Leydig cell division; (4) IGFs regulate proliferation of prostate epithelial cells and of bladder urothelial and smooth cells; (5) IGFs promote axonal growth and partial regeneration even in adult animals; (6) IGFs promote astrocyte proliferation after physical injury in central nervous system; (7) IGFs are key regulators of bone formation by stimulating osteoblast proliferation; (8) IGFs induce proliferation of skeletal muscle cells.[1] IGFs also regulate cell number via regulation of apoptosis. In most cases, IGFs are anti-apoptotic and can therefore be classified as survival factors. The overall mitogenic effect of IGFs suggests that abnormal activation of the IGF pathway could be a crucial event in the formation of many cancers. Indeed, IGF signaling is dysregulated in certain pancreatic cancers.

Moreover, IGF participates in physiological invasion mechanisms occurring during development, such as trophoblast invasion of the endometrium during implantation.[1]

Surprisingly, IGF promotes the acquisition or maintenance of the differentiated status of certain cell types: (1) IGF stimulates differentiation of skeletal muscle cells and Leydig cells; (2) IGF promotes oestradiol production by granulosa and granulosa-luteal cells. IGF-I also synergises with FSH (follicle stimulating hormone) and hCG (human chorionic gonadotropin) by increasing oestradiol and progesterone production; (3) IGF modulates androgen production by Sertoli and Leydig cells; (4) IGF may also be important in neuronal differentiation, and promotes myelination in central and peripheral nervous systems.[1] Therefore, IGF stimulates both proliferation and differentiation, commonly accepted as being two mutually exclusive cellular events.

This chapter will describe a recently identified biological effect of IGF peptides namely their role in the epithelial to mesenchymal transition (EMT). First, the cellular and molecular events induced by IGFs and associated with EMT will be described. As the induction of EMT by IGFs must proceed through the activation of IGF-1R, the main signaling cascades induced by IGF-1R will be described. Finally, we will explain how the molecular events of this EMT are initiated in the light of the known signaling pathways downstream from IGFRs.

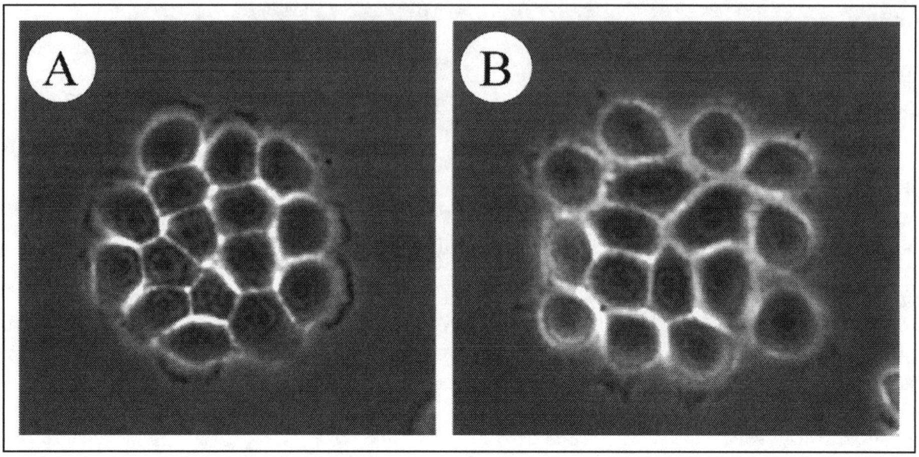

Figure 2. Phase contrast microscopy of NBT-II cells. The NBT-II cells were treated (B) or not (A) with 100 ng/mL of IGF-II for 1 hour. Note that the cells located at the periphery of the colony spread and loose the cell-cell contacts.

IGFs and Regulation of Cell-Cell Adhesion: Cellular and Molecular Aspects

The effects of IGF-I and IGF-II on cell-cell adhesion have been studied using in vitro models including the epithelial cell line NBT-II (nara bladder tumor II).[2] The NBT-II cell line is derived from a chemically induced rat bladder carcinoma. This cell line possesses a typical epithelial cell morphology with extensive cell-cell contacts and has also been used to study the epithelial to mesenchymal transition induced by various factors.[3-5] Importantly, cell-cell adhesion between NBT-II cells is mainly mediated by E-cadherin.

IGFs Induce EMT

In the absence of IGFs peptides and Ins, NBT-II cells have the classical morphology of epithelial cells: they are polarized and tightly attached to one another. Upon either IGF-I or IGF-II treatment (subsequently referred to as IGF), NBT-II cells lose their cell-cell contacts, flatten and spread (Fig. 2). Other epithelial cells, such as MCF7 and MDCK cells, or embryonic stem (ES) cells undergo the same morphological changes upon IGF treatment but the mesenchymal NIH-3T3 cells do not. Importantly, these morphological modifications appear very quickly after IGF addition (typically within one hour of application) and are not associated with cell division.

The visible loss of cell-cell contacts is associated with various molecular events: (i) the rapid internalisation of E-cadherin and desmoplakin leading to the disruption of junctional complexes such as adherens junctions, desmosomes and certainly gap junctions; and (ii) after 4 days of treatment with IGF the neoexpression of the mesenchymal-specific marker vimentin. Therefore, IGF induces a series of cellular and molecular modifications characteristic of standard EMT. This transition is reversible. After removing IGF from the medium, NBT-II cells revert to an epithelial morphology within 24 hours, and E-cadherin is relocalized at cell-cell contacts.[3]

A basic EMT can be defined as the transition between tightly attached and polarized epithelial cells to a set of loosely attached and nonpolarized mesenchymal cells. In addition to the characteristics cited above, a full EMT is associated with cellular motility. IGF is able to induce the migration of certain epithelial cells such as MCF7[6] and melanoma cells[7] but not all (e.g., NBT-II cells).[3] epithelial cells. In this respect, IGFs induce a basic or a full EMT according to the cell type.

In conclusion, IGF affects cell-cell adhesion. The cadherin/catenin complex is certainly one of the most important complexes involved in cell-cell adhesion. Consequently, the signaling link between IGF-1R and the cadherin/catenin complex has been studied in further detail.

IGF-1R Interacts Indirectly with E-Cadherin and β-Catenin

The cadherin cell adhesion molecules bind to one another homophilically at the cell surface in a Ca^{2+}-dependent manner. By binding cells opposing each other they physically adhere cells to each other forming large networks or even tissues. The cytoplasmic domain of cadherin binds to β-catenin, which in turn binds to α-catenin, which connects the cadherin/catenin complex to the actin-based cytoskeleton. In a large majority of epithelial cells, IGF-1R and E-cadherin are coexpressed. They can be coimmunoprecipitated and so constitute a membranous complex.[3,8] According to the results of immunoprecipitations, the interaction of IGF-1R with E-cadherin, β-catenin or α-catenin does not appear to dissociate cadherins from catenins.[3,8] Presumably, a IGF-1R/E-cadherin/β-catenin/α-catenin supra-molecular complex is present at the cell surface of various cells.

The cytoplasmic domain of E-cadherin is sufficient to interact with the cytoplasmic domain of IGF-1Rβ subunit.[3] However, according to protein-protein interaction assays, IGF-1R does not interact directly either with E-cadherin or β-catenin.[3] The nature of the molecules linking members of the complex together has yet to be determined.

IGFs Redistribute Proteins of Adherens Junctions

In epithelial cells non exposed to IGF, E-cadherin and β-catenin are mainly concentrated at cell-cell contacts and IGF-1R is present both at cell-cell contacts and in the cytoplasm. Exposure to IGF induces the redistribution of E-cadherin and IGF-1R from the membrane to the cytoplasm. After IGF treatment, E-cadherin is concentrated in a ring around the nucleus. The constant cycling of E-cadherin between the cytoplasm and the cell membrane has been demonstrated.[9,10] Large amounts of IGF could affect this equilibrium: E-cadherin could be internalised faster than it is readdressed to the membrane.

These results obtained in vitro are consistent with a correlation of expression of IGF-II, IGF-1R and E-cadherin at the time of gastrulation (Fig. 3). At this time, IGF-II is mainly expressed by mesenchymal cells (mesoderm), E-cadherin is mainly absent in murine mesodermal cells[11] and IGF-1R is located at the membrane of epithelial cells (ectoderm and endoderm) and in the cytoplasm for mesodermal cells.

Moreover, the degradation, albeit minor, of E-cadherin has been reported as being associated with IGF treatment[3] and this is supported by the partial colocalization of a subset of E-cadherin molecules with LAMP1 (lysosomal associated protein), an endosomal and lysosomal marker. Thus, E-cadherin may be degradated in LAMP1-positive organelles.

In conclusion, IGF induces rapid internalization of E-cadherin, associated with both slight degradation and sequestration in vesicles located around the nucleus.[3] The reversibility of the EMT can now be explained by a simple hypothesis: the stored E-cadherin is rapidly readdressed to the membrane once IGF is removed from the medium.

Distribution of β-catenin is also affected by IGF. β-catenin is important for cell-cell adhesion and also in signal transduction via the Wnt signaling pathway. In the absence of Wnt, β-catenin is part of a complex containing GSK-3β (glycogen synthase kinase-3β), APC (adenomatous polyposis coli), and axin. GSK-3β is a serine-threonine kinase which phosphorylates β-catenin. β-catenin is then ubiquitinated and degradated by proteasomes.[12,13] Upon binding of Wnt to its cell surface receptor, termed Frizzled, the serine-threonine kinase Dishevelled (Dsh) is activated.[14-16] Dsh then inhibits the catalytic activity of GSK-3β by phosphorylation. Unphosphorylated and hence undegradable β-catenin accumulates in the cytoplasm and is translocated with TCF/LEF (T cell factor/lymphoid enhancer factor) factors—DNA bending proteins—into the nucleus (for review see ref. 17). β-catenin/TCF complex can induce or repress the expression of a variety of target genes such as cyclin D1, c-Myc, T-brachyury, c-Jun, Fra-1, matrix-metalloprotease-7 (MMP-7), fibronectin, cyclo-oxygenase-2, m-Mitf,

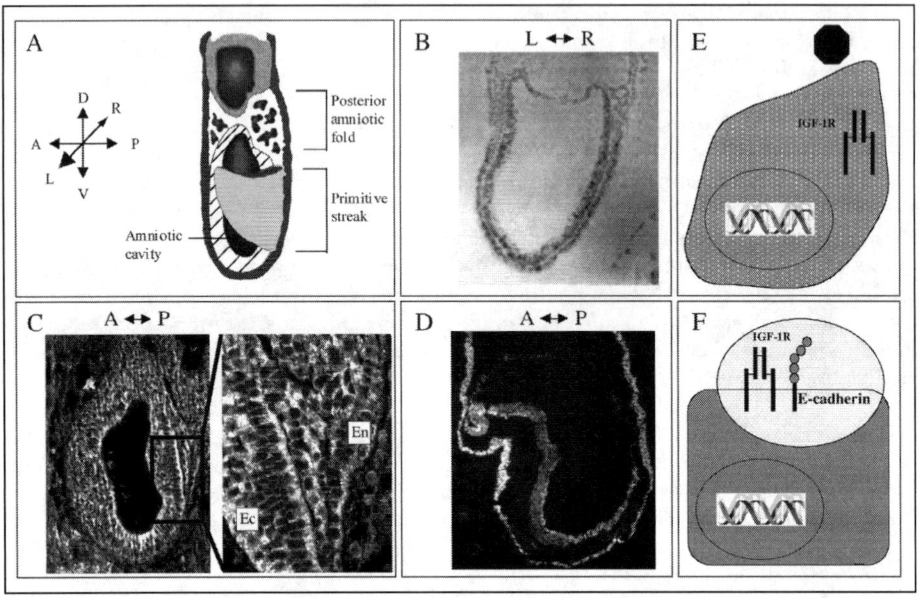

Figure 3. Localization of IGF-II, IGF-1R and E-cadherin in a mouse gastrulating embryo. A) Schematic transversal section of a murine gastrulating embryo. The extra-embryonic tissues are located on the upper part of the scheme and the embryonic tissues on the bottom part. The endoderm (dark) is surrounding the mesoderm (grey) and the ectoderm (striated). The amniotic cavity is surrounded by the epithelial cells of the ectoderm. D means dorsal; V= ventral; A= anterior; P= posterior; R= right; and L= left [136] B) In situ hybridization of mouse embryo with a specific probe for IGF-II mRNA. Note that the signal is mainly located in the mesoderm. The section is longitudinal.[137] C) Immunofluorescence showing the localization of IGF-1R at the cell-cell contact in ectodermal and endodermal cells and in the cytoplasm in the mesodermal cells. The image located on the right of the panel is an enlargement of the image located on the left. D) Immunofluorescence showing the localization of E-cadherin in the mouse embryo.[11] E) Schematic localization of IGF-II and IGF-1R in mesodermal cells. F) Schematic localization of IGF-1R and E-cadherin in ectodermal cells. Ec= ectoderm; En= endoderm. The mesoderm is located between the ectoderm and endoderm.

EphB2 and EphB3 receptors, and ephrin-B1.[18-27] Some of these genes, such as cyclin D1, are expressed ubiquitously and some, such as m-Mitf in the melanocyte lineage, are cell specific.[28,29] IGF redistributes β-catenin from the cell membrane to the nucleus, and induces the translocation of TCF3 from the cytoplasm to the nucleus.[3] None of the genes known to be activated/ repressed by β-catenin is, alone, able to induce an EMT. As not all β-catenin target genes have been identified, some could be involved in EMT or/and the classical IGF signaling pathway may induce expression of genes involved in EMT.

In summary, IGFs induce a reversible EMT which is based on the reduction of the cell-cell adhesive properties of E-cadherin and the redistribution of the proteins associated with it. Indeed, IGF (i) rapidly delocalizes E-cadherin from cell-cell contacts, (ii) disrupts the interaction of E-cadherin with β-catenin, by phosphorylation, (iii) induces a limited degradation of E-cadherin, (iv) activates genes via β-catenin/TCF and (v) obviously induces the IGF signaling pathway.

Signaling Pathways Activated by IGF-1R

IGF-1R is a RPTK (for reviews see refs. 30-32) and IGF binding leads to the subsequent activation of multiple signaling pathways, which are described below. This signaling multiplicity could explain the diversity of biological effects elicited by IGFs.

Activation of IGF-1R by IGF

IGFs bind to the extracellular domain of IGF-1R and initiate a conformational change in the quiescent receptor that is transmitted to the intracellular β subunit. The tyrosine kinase activity of IGF-1R is thereby activated and results in trans-autophosphorylation of the β-subunit on three tyrosines within the catalytic core: tyrosine residues 1161, 1165 and 1166 according to the numbering of the IGF-1R precursor, tyrosine residues 1131, 1135 and 1136 in α+β IGF-1R protein or tyrosine residues 421, 425 and 426 in the IGF-1R β-subunit, see GenBank accession number P08069. In the following sections, we will refer to the IGF-1R β-subunit to position specific amino acids. This autophosphorylation enhances the capability of the kinase to phosphorylate other substrates. These three phosphotyrosines participate in the activation of cytoplasmic signaling pathways.

Intracellular Signaling Pathways

The three phosphotyrosines located in the IGF-1R β-subunit at positions 421, 425 and 426 participate in the activation of cytoplasmic signaling pathways by two different processes:

i. the tyrosine-phosphorylated IGF-1R acts only as a platform for the recruitment of specific cytoplasmic proteins, such as Grb (growth factor receptor-bound).
ii. the tyrosine-phosphorylated IGF-1R recruits and phosphorylates various particular cytoplasmic proteins, such as PI-3K (phosphoInositol 3-kinase), Shc (SH2 containing proto-oncogene), Crk (chicken tumor virus CT10 regulator of kinase) and IRS (insulin receptor substrate). The recruitment and tyrosine phosphorylation of these proteins by IGF-1R either leads to the activation of the enzymatic activity of PI-3K or creates new docking sites for the recruitment of cytoplasmic molecules interacting with Shc, Crk or IRS.

Thus, five cytoplasmic protein families interact with IGF-1R and propagate the signal from the exterior to the cytoplasm of the cells:

1. Grb is a large family of adaptor proteins containing SH2 (Src homology 2) and SH3 domains. Some of its members, including various Grb10 isoforms, interact via their SH2 domain with a tyrosine-phosphorylated form of IGF-1R. Tyrosines 425 and 426 of β-subunit are implicated in the binding of Grb10 isoforms.[33] When Grb10 isoforms bind to IGF-1R in a similar manner as to IR, it will involve tyrosine residues from the catalytic domain (implicating tyrosines 425 and 426 in the IGF-1R β-subunit).[34] The role of Grb10 in IGF signaling is controversial. As a function of the cell type, the overexpression of Grb10 can inhibit or increase the mitogenic effects of IGF.[35,36] Nevertheless, recent data suggest that Grb10 inhibits substrate phosphorylation by active IGF-1R.[37]
2. Class I PI-3K is a heterodimer composed of a regulatory subunit, p85, and a tyrosine kinase catalytic subunit, p110. Different classes of PI-3K were characterized. In this chapter, we will focus on class I PI-3K and will just refer to PI-3K. Both subunits contain SH2 domains. The YXXM motif (implicating amino-acids (AA) 633-636) of IGF-1R β-subunit is necessary to interact with SH2 domains of p85.[1] p85 is phosphorylated by activated IGF-1R. This phosphorylation induces a conformational change of p85 that drives the activation of the catalytic subunit, p110. The PI-3K pathway can lead to activation of proteins containing a PH (Pleckstrin homology) domain such as PDK1 (phosphoinositide dependent kinase) and Akt (Fig. 4A1,A2).
3. SHC is also an adaptor protein that contains SH2 domains and also numerous tyrosine residues which can be phosphorylated by IGF-1R. The NPXY motif (AA237-240) of the β-subunit of IGF-1R is necessary for the interaction of IGF-1R with SH2 domains of SHC.[38] The tyrosine-phosphorylated SHC mainly binds to the Grb2/Sos complex which drives the activation of the MAPK pathway (Fig. 4B1, B2).[39-41]
4. The Crk adaptor protein family includes three members (Crk-I, Crk-II and Crk-L) containing both SH2 and SH3 domains. Tyrosine phosphorylation of Crk-II is rapid following its interaction with IGF-1R. The tyrosines 233, 240 and 540 of the β-subunit of IGF-1R are required for the interaction with SH2 domains of Crk-II. The tyrosine-phosphorylated

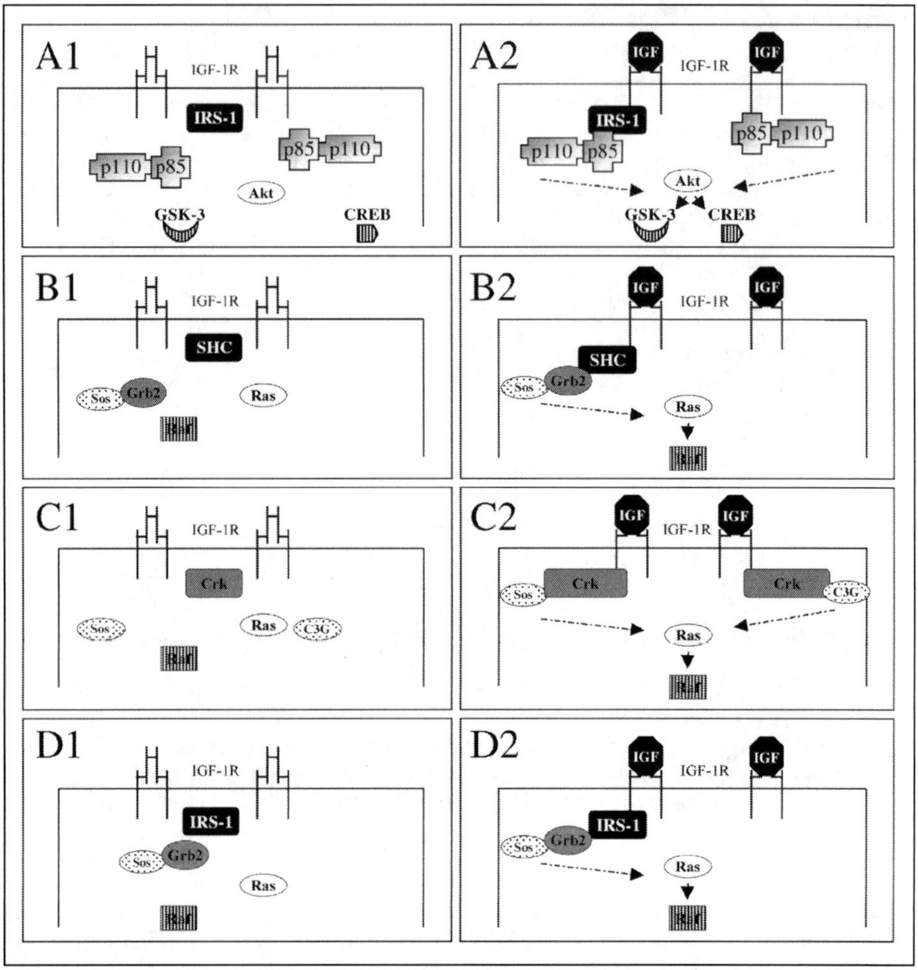

Figure 4. Signaling pathways activated by IGF-1R. Five major cytoplasmic protein families interact with IGF-1R (insulin growth factor-I receptor): Grb (growth factor receptor-bound), PI-3K (phosphoinositides-3kinase), SHC (SH2 containing proto-oncogene), Crk (chicken tumor virus CT10 regulator of kinase) and IRS (insulin receptor substrate) families. This figure presents schematically various signaling pathways initiated by PI-3K (A1,A2), SHC (B1,B2), Crk (C1,C2), Grb2 (B1,B2,D1,D2) and IRS-1 (A1,A2,D1,D2). On the left panels (1), the cell is not induced, and on the right (2) the cell is induced by IGF-I (insulin growth factor-I). A1) PI-3K is an inactive heterodimer before induction. This protein is composed of the p110 regulatory subunit and of the p85 catalytic subunit. A2) After IGF induction PI-3K is activated after its direct interaction with IGF-1R or via IRS-1. Active PI-3K produces D3-phosphorylated phosphoinositides. These molecules participate in the activation of several types of proteins such as the serine-threonine kinase AKT. This specific enzyme is able to repress or to activate various targets. For instance, the activity of GSK-3 is repressed and the activity of the transcription factor CREB is activated. B1,B2) SHC can be involved as IRS-1 in the induction of the MAPK signaling pathway (see D). C1) Before induction, Crk, Sos and C3G are cytoplasmic and inactive. C2) After induction, Crk interacts with IGF-1R and recruits Sos and C3G to the membrane to activate them to induce the MAPK pathway. D1) Grb2 and Sos (Son of sevenless) interact and are localized in the cytoplasm. Sos is inactive. D2) Grb2 interacts with IRS-1. The recruitment of Grb2 by IRS1 activates Sos. This leads to the activation of the MAPK pathway going through the activation of Ras and Raf.

Crk-II in turn interacts with two GEF (guanine nucleotide-exchange factor) proteins, C3G and Sos[42] and these factors drive the activation of the MAPK pathway (Fig. 4C1, C2).[43,44]

5. Members of the IRS adaptor protein family contain a PTB (phosphotyrosine binding) domain, a C-terminal region rich in tyrosine residues and a PH domain, but no SH2 domain. IRS-1 and IRS-2 are rapidly phosphorylated by activated IGF-1R. The NPXY motif (AA237-240) of the β-subunit of IGF-1R is necessary for the interaction with PTB domain of IRS-1 and IRS-2.[45] IRS-1 and IRS-2 are phosphorylated on various tyrosines by activated IGF-1R. The various phosphorylated residues on IRS are the docking sites for (i) p85 (ii) the three adaptor proteins Grb2, Crk-II and Nck (non catalytic region of tyrosine kinase), (iii) SHP-2 (Src homology 2 domain-containing protein tyrosine phosphatase, also known as Syp or PTP1D), which is a tyrosine phosphatase[46,47] and (iv) Fyn and Csk (C-terminal Src kinase), which are Src family kinases (SFK).[48]

The formation of the IRS-1/Grb2 complex propagates the signal induced initially by the IGF ligand through Sos (Son of sevenless). Before IGF induction, Sos and Grb2 interact and are localized in the cytoplasm, where Sos is inactive. The recruitment, mediated by IRS-1, of Grb2, and therefore Sos, at the membrane activates Sos. Sos substrates, such as Ras, are also present at the membrane. Sos is a GEF protein activating the MAPK pathway (mitogen activated protein kinase (Fig. 4D1, D2). Thus, the MAPK pathway is induced but will be described later. As mentioned above, Crk proteins also activate the MAPK pathway.

Nck- and Csk-activated signaling pathways lead to cytoskeletal reorganization. Csk activates FAK (focal adhesion kinase) increasing the turnover of focal contacts.[49] Nck targets members of the Pak ($P21^{rac/cdc42}$-activated serine/threonine kinase) family to the plasma membrane. Then, Pak interacts with the GTP-bound Rac/Cdc42 and is activated. Activated Pak1 phosphorylates MLCK (myosin light chain kinase) promoting actin polymerisation, a crucial event during physiological or pathological migration of cells.[50]

Independently, it was shown that the protein Src is associated with the IGF signaling pathway.[51] Currently, however, the molecular link between IGF-1R and Src is unclear.

In summary, IGFs cause: (i) the reorganization of cytoskeleton through Csk-FAK and NCK-PAK; (ii) the activation of MAPK pathway through Grb2-Sos-Ras; and (iii) the activation of pathways downstream from PI-3K, such as the AKT (or PKB) pathway. The MAPK, AKT and Src pathways will be described in more detail below.

MAPK Signaling Pathway

MAPK pathways can be described as a successive activation of three kinase families (i) MAPKKK (MAP kinase kinase kinase) phosphorylates and activates MAPKK, (ii) MAPKK then activates by phosphorylation MAPK and (iii) MAPK phosphorylates numerous proteins driving the biological effects of the pathway. Cytokines, stress and growth factors activate the MAPK pathway.

IGF-activation of the MAPK pathway is initiated by the activation of Ras. Ras is a GTP-binding protein kinase that alternates between an active state when bound to GTP and an inactive state when bound to GDP. Ras activation is accelerated by GEFs which catalyse the GDP/GTP exchange. Downstream from IGF, Sos and C3G activate Ras. One of the main targets of Ras is the serine/threonine kinase Raf, which is a MAPKKK. In mammals, three Raf proteins have been isolated: A-Raf, B-Raf and C-Raf (or Raf-1). The affinity of Raf for Ras-GTP is greater than its affinity for Ras-GDP. Activated Ras recruits Raf to the plasma membrane where it is phosphorylated and activated. Full activation of A-Raf and C-Raf requires the phosphorylation of two residues, serine 338 and tyrosine 341. Ras phosphorylates the serine 338 whereas Src phosphorylates the tyrosine 341.[52] The full activation of B-Raf requires only the phosphorylation of the serine 338.

Activated Raf phosphorylates MAPKKs, termed MEK1 and MEK2 (Map/Erk kinase 1 and 2), in the growth factor-activated MAPK pathway. This phosphorylation occurs on the tyrosine residue contained in the TXY motifs.

Activated threonine/tyrosine kinases MEKs phosphorylate MAPKs named ERK1 and 2 (extracellular signal-regulated kinases 1 and 2) in the growth factor-activated MAPK pathway. Serine/threonine kinases ERKs phosphorylate target proteins on serine or threonine residues of PX(T/S)P motifs. Some transcription factors are phosphorylated by ERKs.[53]

Moderate activation of MAPK pathway induces cell division. Indeed, active ERKs enhance the expression of Cyclin D1 and inhibit the expression of p27. Various findings suggest that Shc mediates most of mitogenic effects of IGFs. Strong activation of the MAPK pathway induces cell division arrest and, in some cell types, cell differentiation. These biological effects explain the important role of the MAPK pathway during development. Interestingly, the MAPK pathway is implicated in the loss of E-cadherin dependent adherens junctions upon HGF/SF induction and Src activation.[54]

PI-3K and AKT Signaling Pathway

PI-3K are heterodimers composed of a regulatory subunit, p85, and a catalytic subunit, p110. p85 binds directly to the tyrosine phosphorylated forms of IGF-1R and IRS-1. These interactions trigger tyrosine phosphorylation of p85. This phosphorylation induces a conformational change of p85 that activates the catalytic subunit, p110. Moreover, Ras-GTP can directly activate p110.[55,56]

Active PI-3K phosphorylates PtdIns (phosphatidylinositol), PtdIns(4)P and Ptd(4,5)P on the 3'-OH group of the inositol ring producing Ptd(3)P, PtdIns(3,4)P_2 and PtdIns(3,4,5)P3 (subsequently referred to as D3-phosphorylated phosphoinositides) respectively. (see Fig. 5). amphipathic PtdIns(3,4)P_2 and PtdIns(3,4,5)P3 molecules bind to proteins containing a PH domain. Serine/threonine kinases, such as AKT (= RAC = PKB) and PDK1 (phosphatidylinositol dependent kinase 1), are relocalized to cell membrane upon their binding to these phosphorylated PtdIns. AKT is then appropriately localized to be phosphorylated on threonine 308 by PDK1 and on serine 473 by PDK2. More controversial is the identity of PDK2, the kinase(s) responsible for Ser-473/474 phosphorylation.[57] AKT phosphorylated on T308 and S473 is fully active. It phosphorylates threonine or serine residues of RXRXX(T/S) motifs in numerous molecules. AKT regulates the activity of several transcription factors: CREB (cAMP responsive element binding protein), members of forkhead family and Ets-2. Phosphorylation of CREB by AKT stimulates CREB-dependent transcription.[58] FKHR (forkhead in rhabdomyosarcome) and FKHRL1 (forkhead in rhabdomyosarcome like 1) are phosphorylated and inhibited by AKT.[59,60] Ets-2 is phosphorylated by JNK-2 in cells in which AKT is also activated, leading to the activation of Ets-2-dependent transcription.[61]

AKT promotes cell cycle progression, cell survival, and tumour cell invasion.[62] Interestingly, AKT phosphorylates and inhibits GSK-3β thereby, presumably, linking IGF and Wnt pathways.

Moreover, Ras-mediated reorganisation of the actin cytoskeleton and cell migration depend on PI-3K. Indeed, some membrane lipid targets of PI-3K regulate: (i) the activity and structure of proteins that bind to actin; (ii) the GTPase Rac thereby inducing the formation of membrane folds.

Src Signaling Pathway

Src is the prototype SFK protein. It contains: (1) an amino-terminal myristylation sequence, (2) a single U specific region, (3) a SH2 domain, (4) a SH3 domain, (5) a tyrosine kinase domain that contains tyrosine-416, (6) a carboxy-terminal domain that contains tyrosine-537 (for review see ref. 63). Autophosphorylation of Src on tyrosine-416 is required for optimal activity of the enzyme.

The activity of Src depends on its phosphorylation and two main intramolecular interactions, involving the SH2 and SH3 domains. Indeed, Src is inactive (i) when the carboxy-terminal part of Src interacts by appropriate folding of the protein with its own SH2 domain and (ii) possibly when the catalytic domain interacts with the SH3 domain. The interaction between the carboxy-terminal part and the SH2 domain is only possible when the tyrosine-537 is

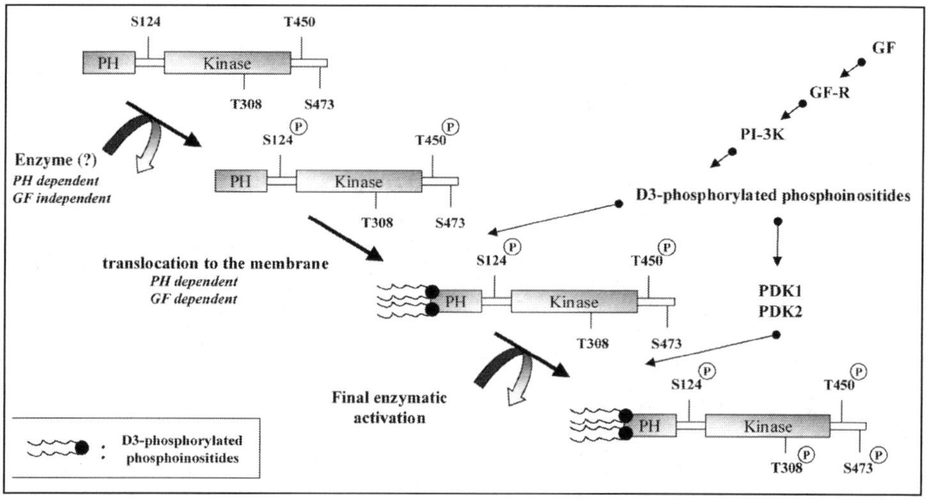

Figure 5. Model of AKT activation. AKT activation follows three major steps. During the first step, AKT acquires a precise conformation and becomes phosphorylated on serine 124 (S124) and on threonine 450 (T450) residues. This step is dependent on the plekstrin homology (PH) domain of AKT but not on growth factors (GF). The second step is initiated by the activation of growth factor receptors (GF-R) by their respective ligands. Active growth factor receptors then activate phosphoinositides 3-kinase (PI-3K). This enzyme synthetize D3-phosphorylated phosphoinositides. The hydrophilic part (black dots) interacts directly with the PH domain of AKT whereas the hydrophobic part (black broken line) interacts with lipidic bilayer of the membrane. Therefore, the D3-phosphorylated phosphoinositides/AKT complex is located at the membrane. During the third step, AKT is phosphorylated by two kinases. Phosphoinositides dependent kinase 1 (PDK1) is a kinase localized at the membrane that phosphorylates AKT on residue threonine 308 (T308). Phosphoinositides dependent kinase 2 (PDK2) phosphorylates AKT on residue serine 473 (S473). This final form of AKT is fully active and can phosphorylate target proteins.

phosphorylated. Inactive Src is found in the perinuclear region of the cell and is most likely associated with endosomal membranes.[64] The activation of Src induces SH3-dependent association with actin and peripheral targeting.[65,66] Src can be activated in several ways. First, dephosphorylation of tyrosine-537 by PTP-α, PTP1, SHP-1, or SHP-2 phosphatases activates Src. Also, Src-interacting partners can compete with intramolecular interactions between Src domains: PDGFR (platelet-derived growth factor receptor) and FAK can bind to the SH2 or SH3 domains of Src and, thereby, activate it (for review see ref. 63).

Activation of Src by IGF-I has been shown in neuroblastoma and in 3T3-L1 preadipocyte cells.[51,67] Interestingly, inactivation of Src by IGF-I has also been shown in NIH-3T3 cells.[68] Therefore, the activation/repression of Src by IGF-I is cell type-specific. The mechanisms by which IGF-I activates/represses Src still remain unknown.

Src participates in mitogenic signaling of IGF-I.[51] More precisely, Src participates in the activation of the MAPK pathway that is downstream from IGF. As concerns EMT, Src regulates cell-cell adhesion:

1. Src reduces the conductance of small molecules through gap junctions. The phosphorylation of connexin 43 has been implicated in this regulation.[69-71]
2. Src is involved in EGF-induced disruption of desmosomes.[72,73]
3. Src induces the loss of adherens junctions[74,75] and this is associated with the tyrosine phosphorylation of both β-catenin and p-120^{ctn} and the redistribution of E-cadherin to the cytosol.

In conclusion, IGF is implicated in cell differentiation, division and survival, and also in the regulation of cell-cell adhesion and in cytoskeletal remodelling. It is now well established that

IGFs activate MAPK, FAK, Pak and PI-3K pathways[50,76-80] and may activate Src.[51,67] A series of apparently unrelated observations could link these molecular pathways to adhesion and cytoskeletal cellular events:

1. MAPK is involved in the disruption of E-cadherin-dependent adherens junctions upon HGF/SF induction.[54]
2. FAK is involved in the turn-over of focal contacts.[81]
3. Pak stimulates actin polymerisation, a crucial molecular event during cell migration.[50]
4. PI-3K is involved in Ras-induced actin reorganization and cell migration. PI-3K also activates AKT, which inhibits GSK-3β, a regulator of β-catenin degradation.

The mechanism by which IGFs induce an EMT is still poorly understood and involves at least the disruption of cell-cell adhesion via the disorganisation of components present in adherens junctions and desmosomes. The next paragraph describes various evidence identifying the molecules involved in the induction of EMT by IGF.

Pathways Activated by IGF and Implicated in the Induction of the EMT

IGF induces an EMT associated with various molecular events: (i) the redistribution of E-cadherin, β-catenin and desmoplakin from plasma membrane to cytoplasm; (ii) the cytoplasmic sequestration and slight degradation of E-cadherin and the translocation of β-catenin to the nucleus; (iii) the neoexpression of mesenchymal markers. How do these molecular events are linked to the signaling pathways induced by IGF?

Effectors Implicated in Redistribution of Proteins

E-cadherin and β-catenin localize at the surface of epithelial cells. After IGF induction, they are internalized, and no longer colocalized.

Internalization of E-Cadherin and β-Catenin

IGF-1R forms a complex with E-cadherin and β-catenin. Following ligand binding, IGF-1R aggregates in coated pits and the aggregates are quickly internalised. Subsequently, IGF-1R is recycled to the cell surface. The mechanisms of internalisation of E-cadherin and β-catenin are not yet known, but various models can be suggested:

i. E-cadherin and β-catenin may be internalised passively, following the sorting of IGF-1R to which they are bound.
ii. Recently, it was shown that HGF (hepatocyte growth factor) induces endocytosis of E-cadherin after tyrosine-phosphorylation and ubiquitination of this protein.[82] A similar mechanism is plausible for IGF. E-cadherin becomes phosphorylated on tyrosines 755 and 756 (numbered according to the murine E-cadherin sequence) upon HGF induction. This form of E-cadherin can interact with Hakai.[82] Hakai is a E3 ubiquitin ligase that induces E2-dependent ubiquitination of E-cadherin in vivo and in vitro. Ubiquitination of membrane proteins triggers their internalisation and targeting to lysosomes for degradation. The mechanism responsible for this internalisation has not been determined. However, membrane proteins that have to be endocytosed are classically recognized by AP (adaptor proteins). When membrane-docked AP is bound to the protein destined to be internalised it can recruit clathrin. Polymerisation of clathrin induces the formation of a endocytotic vesicle (for reviews see refs. 83,84). Therefore, internalisation of phosphorylated and ubiquitinated E-cadherin could be driven by the interaction of E-cadherin with proteins of endocytotic machinery.
iii. Moreover, IGFs activate AKT.[80] This serine-threonine kinase activates Rab proteins. The Rab proteins regulate the rate of vesicle fusion involved in protein trafficking.[85] Interestingly, the expression of a constitutively active form of AKT mimics IGF induction by inducing an EMT.[86] This transition is associated with the internalisation of E-cadherin and β-catenin. In conclusion, AKT may be an important effector of IGF during EMT and is a candidate for the stimulation of endocytosis.

Regulation of E-Cadherin/β-Catenin Interaction

After IGF-induction, the localization of IGF-1R, E-cadherin and β-catenin changes dramatically; IGF-1R is recycled, E-cadherin is trapped in organelles surrounding the nucleus and β-catenin is concentrated within the nucleus. The molecular mechanisms responsible for the partition of these three proteins are not known. After their internalisation, E-cadherin and β-catenin no longer colocalize. Post-translational modifications, in particular phosphorylation, are involved in the regulation of protein-protein interactions, and this type of mechanism may be involved in the association/dissociation of cadherin/catenins.

As mentioned earlier, E-cadherin can be tyrosine-phosphorylated. The tyrosine-phosphorylation of E-cadherin decreases its binding to p120^{ctn} and increases its binding to Hakai.[82] The residues implicated are tyrosines 755 and 756, numbered according to the murine E-cadherin sequence. IGF can activate the Src and Fyn tyrosine kinases.[51,67,87] Src was first implicated in regulating adherens junctions at the end of the 1980s.[88] Also, Src and Yes are abundant at adherens junctions of epithelial cells[89] and Src induces tyrosine phosphorylation of E-cadherin.[82] So, Src could participate in the regulation of the interaction between E-cadherin and its partners. This leads to various possibilities. One of them is that Src induces the tyrosine-phosphorylation of E-cadherin after IGF induction, and as a result the interaction of the phosphorylated form of E-cadherin with p120 is inhibited and the interaction with Hakai is favoured. The interaction of E-cadherin with Hakai could lead to the internalisation of this new complex. In addition to activating tyrosine kinases, IGF can also activate SHP-2 tyrosine phosphatase. SHP2, PTPμ and PTP1B, can interact with cadherins in vivo, suggesting that cadherins were previously tyrosine-phosphorylated.[90] The apparent discrepancy of the kinase and phosphatase inductor effect of IGF could be better understood when it was shown that SHP-2 phosphatase can activate Src by tyrosine-537 dephosphorylation. The tyrosine-phosphorylation status of E-cadherin seems tightly regulated and IGF could be involved in this process. Consequently, IGF may contribute to the maintenance/formation of the adherens junctions.

Tyrosine-phosphorylation of other classical cadherins, including N- and VE-cadherin, has been described.[91,92] It appears that this process induces cellular events typically associated with the particular cadherin involved.[82,93,94] Increased confluence of endothelial cells in culture is accompanied by a decrease in the tyrosine phosphorylation of VE-cadherin.[92] VE-PTP (vascular endothelial-protein tyrosine phosphatase), an endothelial receptor-type phosphatase, is a transmembrane binding partner of VE-cadherin that associates through an extracellular domain and reduces the tyrosine phosphorylation of VE-cadherin.[95]

Moreover, E-cadherin can be serine-phosphorylated.[96] The serine-phosphorylation of E-cadherin increases the affinity of its binding to β-catenin. The serine residues implicated map to a region from amino-acid 833 to 862 (numbered according to the murine E-cadherin sequence). This region of E-cadherin is also necessary for the interaction with β-catenin. Of the eight serines in this region, Ser-853 and Ser-855 seem to be particularly important[96] and are phosphorylated by Casein kinase II.[96] Another serine, Ser-849, is phosphorylated by GSK-3β. The phosphorylation status of these serine residues in E-cadherin upon IGF induction has, unfortunately, not been determined by phosphopeptide mapping. As E-cadherin does not bind to β-catenin after IGF induction, it seems likely that the serine phosphorylation of E-cadherin is abolished.

In contrast, tyrosine-phosphorylation of β-catenin decreases its affinity for E-cadherin.[97-99] Tyrosine-654 of β-catenin seems to be the main residue involved in the modulation of β-catenin binding to E-cadherin.[97] Src phosphorylates β-catenin on tyrosine 654[97] and IGFs activate Src in certain cell types (see above). Thus, Src could act downstream from IGF to tyrosine-phosphorylate β-catenin and to disrupt the interaction between molecules involved in cell-cell adhesion.

β-catenin is not only involved in cell-cell adhesion when associated to cadherins but is also involved in cell signaling when associated to APC, axin and GSK-3β, and in regulation of gene transcription when associated to TCF/LEF proteins. The status of β-catenin phosphorylation is critical for its various protein-protein interactions.[100,101] To elucidate the molecular details of the interactions of β-catenin with its partners, the crystal structure of the central core region of β-catenin has been solved.[102] Armadillo repeats of the core region of β-catenin interact with E-cadherin, APC and many other proteins. The three dimensional structure of this region reveals a superhelix that forms a long positively-charged groove. This structure suggests that this segment can interact with acidic, negatively charged regions. Indeed, the region of E-cadherin which binds to β-catenin is highly acidic. One can imagine that addition of a phosphate group to E-cadherin reinforces the acidic properties of the β-catenin-binding domain. In contrast, addition of phosphate groups to β-catenin would diminish the positive nature of the groove and thereby perturb interaction with E-cadherin.[103,104]

Effectors Implicated in the Stability of E-Cadherin and β-Catenin Proteins

IGF induces a slight degradation of the extracellular part of E-cadherin.[3] This may occur in lysosomes as E-cadherin is partially colocalized with LAMP1, an endosome-lysosome marker.[105] The full process of E-cadherin degradation is still unknown although a parallel between IGF and HGF may exists. In the presence of HGF, the tyrosine phosphorylated form of E-cadherin becomes ubiquitinated by Hakai and consequently targeted to lysosomes.[82] The level of expression and the degradation of E-cadherin has not been studied after long periods of IGF induction. Stable expression of active AKT is associated with a low level of E-cadherin protein, mainly due to repression of the E-cadherin promoter.[86] The repression of E-cadherin expression is certainly required to "lock" the cells in a mesenchymal state. During embryonic development this process must be rapid and strictly timed, but during transformation it can occur more gradually.

IGF and Wnt pathways both result in nuclear translocation of β-catenin and transcriptional regulation of gene expression.[3,17] The IGF pathway leads to an increase of the β-catenin half-life: IGF induces tyrosine-phosphorylation of β-catenin and doubles the half-life of this protein from 3 to 6 hours[98,106] and AKT serine phosphorylates and inhibits GSK-3β.[107] So, AKT may participate in the stabilisation of β-catenin upon IGF induction. Interestingly, the expression of active AKT results in a reduction in the amount of β-catenin protein without any modification of the transcriptional level of the gene.[86] Therefore, there is an apparent discrepancy between the effect of IGF and AKT. This discrepancy can be explained by the following reasons:

i. The timing of induction is certainly very important. The effect of IGF on the amount of cytoplasmic β-catenin was studied after 17 h of IGF induction whereas studies on the effect of AKT have been performed on cells constitutively expressing an active form of AKT. The long-term regulation of the amount of β-catenin could be different from the short-term regulation.

ii. AKT-induced downregulation of β-catenin may be consecutive to the downregulation of E-cadherin. Indeed, binding to E-cadherin stabilizes β-catenin in both Drosophila and during mouse development.[108,109]

iii. β-catenin can be located in three distinct areas of the cell: plasma membrane, cytoplasm and nucleus. In epithelial cells, which do not express an exogenous active form of AKT, the vast majority of β-catenin is present at the plasma membrane. β-catenin is mainly present in the cytoplasm of cells expressing an active exogenous form of AKT.[86] Even though the total (plasma membrane + cytoplasm + nucleus) amount of β-catenin is lower in cells expressing exogenous AKT than in parental cells, quantitatively the amount of β-catenin in the cytoplasm is certainly higher and much lower at the plasma membrane in comparison.

iv. The mechanism of GSK-3β regulation revealed by recent crystallographic and biochemical studies is another important issue.[110,111] Wnt and IGF/AKT signaling pathways affect two distinct pools of GSK-3β that in turn target different substrates. AKT can potentiate the Wnt pathway but AKT signaling alone cannot initiate the Wnt pathway.[112-114] Consequently, AKT signaling is not the only pathway induced by IGF that is implicated in the regulation of the level of β-catenin. Finally, even though two pools of GSK-3β exist, one associated to IGF and the other to the Wnt, the barrier between the two may be not total.

Effectors Implicated in Transcriptional Regulations

During EMT, there is a massive shift of gene expression. We will focus this paragraph on the regulation of the E-cadherin gene and the role of β-catenin as a transcription factor.

Activation of Genes by the β-Catenin Transcription Factor

The role of β-catenin/TCF as a bipartite transcription factor was first shown in 1996.[115-117] Nevertheless, the induction of the β-catenin/TCF pathway by IGF is still controversial: Playford et al[98] reported that IGF alone cannot induce β-catenin/TCF-dependent transcriptional activation; in contrast, two groups[3,106] have reported that it could. This obvious discrepancy may be explained by the nature of the experiments and that the conclusions cannot be generalized. Indeed, the different conditions of induction may explain the conflicting findings:

i. Playford and colleagues (2000)[98] induced cells for no more than 6 hours with 0.65×10^{-8} M of IGF-I;
ii. Morali and colleagues (2001)[3] and Desbois-Mouthon and colleagues (2001)[106] induced their cellular systems for 16 hours with 1.3×10^{-8} M of IGF-II or 17 hours with 10^{-8} M of IGF-I, respectively. The activation of β-catenin/TCF by IGF may possibly require a period of greater than 6 hours.

The target genes of the β-catenin/TCF complex isolated from different systems include Cyclin D1, c-Myc, T-Brachyury, c-Jun, Fra-1, matrix-metalloprotease-7 (MMP-7), Fibronectin, Cyclo-oxygenase-2, m-Mitf, EphB2 and EphB3 receptors and ephrin-B1.[18-27]

T-Brachyury belongs to T-box genes family. In mouse, zebrafish and frogs, T-box genes are expressed in mesoderm or mesoderm precursors around the time of gastrulation. In Xenopus laevis, Brat induces mesodermal markers such as XmyoD, *Xwnt8*, goosecoid and XFKH-1.[118-122] MMP-7 (matrilysin), a metalloproteinase, degrades extracellular matrix components and facilitates cell migration/invasion.[123,124] Thus, these three proteins could be related to EMT, but none of them is involved in EMT per se.

The expression of Myc and Cyclin D1 genes has been determined after IGF-II treatment, in epithelial and ES cells. In epithelial cells, Myc, but not cyclin D1 expression, increased; in ES cells, the opposite was observed.[3,125] Thus such induction seems to be cell specific. This leads to complexity, but also to cell specificity, of each combination of molecular pathways. Obviously, the combination of a series of molecular events (effectors, signaling pathways, transcription factors) will lead to the appropriate associated cellular event, in our case EMT. IGF alone cannot recapitulate EMT completely or prefectly; IGF needs to be associated with other factors to induce the molecular events correctly (such as the full downregulation of E-cadherin expression) and to assure cell specificity. This complexity leads to the specificity observed during development but not during oncogenesis.

Regulation of *E-Cadherin* Expression

Downregulation of E-cadherin seems to be intimately involved in physiological and pathological EMT. In most cases, the downregulation of E-cadherin occurs by transcriptional repression. This repression can be due to methylation of the promoter and/or binding of repressors to the promoter (for review see ref. 126). Snail, SIP1, E47 and Ets-1 transcription factors have all recently been independently identified as strong repressors of the E-cadherin promoter.[127-131]

The direct effect of IGFs on the activity of the E-cadherin promoter has not been studied. However, in stable transfectant cells, it was shown that active AKT represses the E-cadherin promoter activity[86] and Snail mRNA is more abundant.[86] The connection between AKT and Snail may be due to the presence of putative binding sites for CREB, members of ForkHead family and Ets-2, in the Snail promoter, because all these transcription factors are direct targets of AKT. So, AKT could indirectly repress E-cadherin promoter via Snail. This does not exclude a possible role for SIP1, E47, Ets or other transcription factors associated or potentially associated with AKT.

In conclusion, the major roles of IGF peptides are associated with cell proliferation and differentiation and as anti-apoptotic proteins. Recent work has lead to a variety of new insights into the biological effects of IGF factors. In particular, it is now clear that IGF is involved in EMT, a crucial cellular mechanism. Our understanding of the molecular mechanisms involved in the interaction between IGF and the IGF-1R/cadherin/catenin system is not clearly yet complete. The in vivo relevance of IGF to correct cell-cell adherens regulation and to EMT also require further elucidation.[132]

Mitogenicity, apoptosis and cell-cell interactions must be strictly regulated to maintain tissue homeostasis. Indeed, dysregulation of IGF signaling can lead to cancer, and IGF-1R is overexpressed in many types of tumour.[133-135] Previously, the presence of this receptor was mainly considered in the light of cell proliferation. We must now consider the role of IGF in EMT. This step forward may well open new possibilities for cancer therapeutics and for gene therapy.

Acknowledgments

This work was supported by grants from Ligue Nationale contre le Cancer, ARC, Gefluc and Fondation de France. We are grateful to Silvia Martinozzi and Patrick Pla for their comments on the manuscript. We thank Pennsylvanie de la Corbeille de Venus for providing additionnal information and Cesfo for providing various supplies.

References

1. Ron G, Rosenfeld CTR. The IGF system. Totowa, New Jersey: Humana Press, 1999.
2. Toyoshima K, Ito N, Hiasa Y et al. Tissue culture of urinary bladder tumor induced in a rat by N-butyl-N-(-4-hydroxybutyl)nitrosamine: Establishment of cell line, Nara Bladder Tumor II. J Natl Cancer Inst 1971; 47(5):979-985.
3. Morali OG, Delmas V, Moore R et al. IGF-II induces rapid beta-catenin relocation to the nucleus during epithelium to mesenchyme transition. Oncogene 2001; 20(36):4942-4950.
4. Valles AM, Boyer B, Badet J et al. Acidic fibroblast growth factor is a modulator of epithelial plasticity in a rat bladder carcinoma cell line. Proc Natl Acad Sci USA 1990; 87(3):1124-1128.
5. Gavrilovic J, Moens G, Thiery JP et al. Expression of transfected transforming growth factor alpha induces a motile fibroblast-like phenotype with extracellular matrix-degrading potential in a rat bladder carcinoma cell line. Cell Regul 1990; 1(13):1003-1014.
6. Guvakova MA, Adams JC, Boettiger D. Functional role of alpha-actinin, PI 3-kinase and MEK1/2 in insulin-like growth factor I receptor kinase regulated motility of human breast carcinoma cells. J Cell Sci 2002; 115(Pt 21):4149-4165.
7. Li Y, Bhargava MM, Joseph A et al. Goldberg ID. Effect of hepatocyte growth factor/scatter factor and other growth factors on motility and morphology of nontumorigenic and tumor cells. In Vitro Cell Dev Biol Anim 1994; 30A(2):105-110.
8. Guvakova MA, Surmacz E. Overexpressed IGF-I receptors reduce estrogen growth requirements, enhance survival, and promote E-cadherin-mediated cell-cell adhesion in human breast cancer cells. Exp Cell Res 1997; 231(1):149-162.
9. Bauer A, Lickert H, Kemler R et al. Modification of the E-cadherin-catenin complex in mitotic Madin-Darby canine kidney epithelial cells. J Biol Chem 1998; 273(43):28314-28321.
10. Le TL, Yap AS, Stow JL. Recycling of E-cadherin: A potential mechanism for regulating cadherin dynamics. J Cell Biol 1999; 146(1):219-232.
11. Butz S, Larue L. Expression of catenins during mouse embryonic development and in adult tissues. Cell Adhes Commun 1995; 3(4):337-352.
12. Yost C, Torres M, Miller JR et al. The axis-inducing activity, stability, and subcellular distribution of beta-catenin is regulated in Xenopus embryos by glycogen synthase kinase 3. Genes Dev 1996; 10(12):1443-1454.
13. Aberle H, Bauer A, Stappert J et al. Beta-catenin is a target for the ubiquitin-proteasome pathway. Embo J 1997; 16(13):3797-3804.
14. Axelrod JD, Miller JR, Shulman JM et al. Differential recruitment of Dishevelled provides signaling specificity in the planar cell polarity and Wingless signaling pathways. Genes Dev 1998; 12(16):2610-2622.
15. Yanagawa S, van Leeuwen F, Wodarz A et al. The dishevelled protein is modified by wingless signaling in Drosophila. Genes Dev 1995; 9(9):1087-1097.
16. Karasawa T, Yokokura H, Kitajewski J et al. Frizzled-9 is activated by Wnt-2 and functions in Wnt/beta -catenin signaling. J Biol Chem 2002; 277(40):37479-37486.

17. Novak A, Dedhar S. Signaling through beta-catenin and Lef/Tcf. Cell Mol Life Sci 1999; 56(5-6):523-537.
18. Shtutman M, Zhurinsky J, Simcha I et al. The cyclin D1 gene is a target of the beta-catenin/LEF-1 pathway. Proc Natl Acad Sci USA 1999; 96(10):5522-5527.
19. He TC, Sparks AB, Rago C et al. Identification of c-MYC as a target of the APC pathway. Science 1998; 281(5382):1509-1512.
20. Mann B, Gelos M, Siedow A et al. Target genes of beta-catenin-T cell-factor/lymphoid-enhancer-factor signaling in human colorectal carcinomas. Proc Natl Acad Sci USA 1999; 96(4):1603-1608.
21. Arnold SJ, Stappert J, Bauer A et al. Brachyury is a target gene of the Wnt/beta-catenin signaling pathway. Mech Dev 2000;91(1-2):249-258.
22. Brabletz T, Jung A, Dag S et al. Beta-catenin regulates the expression of the matrix metalloproteinase-7 in human colorectal cancer. Am J Pathol 1999; 155(4):1033-1038.
23. Crawford HC, Fingleton BM, Rudolph-Owen LA et al. The metalloproteinase matrilysin is a target of beta-catenin transactivation in intestinal tumors. Oncogene 1999; 18(18):2883-2891.
24. Gradl D, Kuhl M, Wedlich D. The Wnt/Wg signal transducer beta-catenin controls fibronectin expression. Mol Cell Biol 1999; 19(8):5576-5587.
25. Howe LR, Subbaramaiah K, Chung WJ et al. Transcriptional activation of cyclooxygenase-2 in Wnt-1-transformed mouse mammary epithelial cells. Cancer Res 1999; 59(7):1572-1577.
26. Takeda K, Yasumoto K, Takada R et al. Induction of melanocyte-specific microphthalmia-associated transcription factor by Wnt-3a. J Biol Chem 2000; 275(19):14013-14016.
27. Batlle E, Henderson JT, Beghtel H et al. Beta-catenin and TCF mediate cell positioning in the intestinal epithelium by controlling the expression of EphB/ephrinB. Cell 2002; 111(2):251-263.
28. Fuse N, Yasumoto K, Takeda K et al. Molecular cloning of cDNA encoding a novel microphthalmia-associated transcription factor isoform with a distinct amino-terminus. J Biochem (Tokyo) 1999; 126(6):1043-1051.
29. Amae S, Fuse N, Yasumoto K et al. Identification of a novel isoform of microphthalmia-associated transcription factor that is enriched in retinal pigment epithelium. Biochem Biophys Res Commun 1998; 247(3):710-715.
30. Schlessinger J. Cell signaling by receptor tyrosine kinases. Cell 2000; 103(2):211-225.
31. Favelyukis S, Till JH, Hubbard SR et al. Structure and autoregulation of the insulin-like growth factor 1 receptor kinase. Nat Struct Biol 2001; 8(12):1058-1063.
32. Ullrich A, Schlessinger J. Signal transduction by receptors with tyrosine kinase activity. Cell 1990; 61(2):203-212.
33. Dong LQ, Farris S, Christal J et al. Site-directed mutagenesis and yeast two-hybrid studies of the insulin and insulin-like growth factor-1 receptors: The Src homology-2 domain-containing protein hGrb10 binds to the autophosphorylated tyrosine residues in the kinase domain of the insulin receptor. Mol Endocrinol Nov 1997; 11(12):1757-1765.
34. Dong LQ, Farris S, Christal J et al. Site-directed mutagenesis and yeast two-hybrid studies of the insulin and insulin-like growth factor-1 receptors: The Src homology-2 domain-containing protein hGrb10 binds to the autophosphorylated tyrosine residues in the kinase domain of the insulin receptor. Mol Endocrinol 1997; 11(12):1757-1765.
35. Morrione A, Valentinis B, Resnicoff M et al. The role of mGrb10alpha in insulin-like growth factor I-mediated growth. J Biol Chem 1997; 272(42):26382-26387.
36. Wang J, Dai H, Yousaf N et al. Grb10, a positive, stimulatory signaling adapter in platelet-derived growth factor BB-, insulin-like growth factor I-, and insulin-mediated mitogenesis. Mol Cell Biol 1999; 19(9):6217-6228.
37. Stein EG, Gustafson TA, Hubbard SR. The BPS domain of Grb10 inhibits the catalytic activity of the insulin and IGF1 receptors. FEBS Lett 2001; 493(2-3):106-111.
38. Dey BR, Frick K, Lopaczynski W et al. Evidence for the direct interaction of the insulin-like growth factor I receptor with IRS-1, Shc, and Grb10. Mol Endocrinol 1996; 10(6):631-641.
39. Sasaoka T, Rose DW, Jhun BH et al. Evidence for a functional role of Shc proteins in mitogenic signaling induced by insulin, insulin-like growth factor-1, and epidermal growth factor. J Biol Chem 1994; 269(18):13689-13694.
40. Rozakis-Adcock M, Fernley R, Wade J et al. The SH2 and SH3 domains of mammalian Grb2 couple the EGF receptor to the Ras activator mSos1. Nature 1993; 363(6424):83-85.
41. Lowenstein EJ, Daly RJ, Batzer AG et al. The SH2 and SH3 domain-containing protein GRB2 links receptor tyrosine kinases to ras signaling. Cell 1992; 70(3):431-442.
42. Matsuda M, Hashimoto Y, Muroya K et al. CRK protein binds to two guanine nucleotide-releasing proteins for the Ras family and modulates nerve growth factor-induced activation of Ras in PC12 cells. Mol Cell Biol 1994; 14(8):5495-5500.

43. Ohba Y, Mochizuki N, Yamashita S et al. Regulatory proteins of R-Ras, TC21/R-Ras2, and M-Ras/R-Ras3. J Biol Chem 2000; 275(26):20020-20026.
44. Gotoh T, Niino Y, Tokuda M et al. Activation of R-Ras by Ras-guanine nucleotide-releasing factor. J Biol Chem 1997; 272(30):18602-18607.
45. Craparo A, O'Neill TJ, Gustafson TA. NonSH2 domains within insulin receptor substrate-1 and SHC mediate their phosphotyrosine-dependent interaction with the NPEY motif of the insulin-like growth factor I receptor. J Biol Chem 1995; 270(26):15639-15643.
46. Kuhne MR, Pawson T, Lienhard GE et al. The insulin receptor substrate 1 associates with the SH2-containing phosphotyrosine phosphatase Syp. J Biol Chem 1993; 268(16):11479-11481.
47. Qu CK. The SHP-2 tyrosine phosphatase: Signaling mechanisms and biological functions. Cell Res 2000; 10(4):279-288.
48. Courtneidge SA, Fumagalli S, Koegl M et al. The Src family of protein tyrosine kinases: Regulation and functions. Dev Suppl 1993:57-64.
49. Tobe K, Sabe H, Yamamoto T et al. Csk enhances insulin-stimulated dephosphorylation of focal adhesion proteins. Mol Cell Biol 1996; 16(9):4765-4772.
50. Li W, Fan J, Woodley DT. Nck/Dock: An adapter between cell surface receptors and the actin cytoskeleton. Oncogene 2001; 20(44):6403-6417.
51. Boney CM, Sekimoto H, Gruppuso PA et al. Src family tyrosine kinases participate in insulin-like growth factor I mitogenic signaling in 3T3-L1 cells. Cell Growth Differ 2001; 12(7):379-386.
52. Mason CS, Springer CJ, Cooper RG et al. Serine and tyrosine phosphorylations cooperate in Raf-1, but not B-Raf activation. Embo J 1999; 18(8):2137-2148.
53. Treisman R. Journey to the surface of the cell: Fos regulation and the SRE. Embo J 1995; 14(20):4905-4913.
54. Herrera R. Modulation of hepatocyte growth factor-induced scattering of HT29 colon carcinoma cells. Involvement of the MAPK pathway. J Cell Sci 1998; 111(Pt 8):1039-1049.
55. Kodaki T, Woscholski R, Hallberg B et al. The activation of phosphatidylinositol 3-kinase by Ras. Curr Biol 1994;4(9):798-806.
56. Rodriguez-Viciana P, Warne PH, Vanhaesebroeck B et al. Activation of phosphoinositide 3-kinase by interaction with Ras and by point mutation. Embo J 1996; 15(10):2442-2451.
57. Chan TO, Tsichlis PN. PDK2: A complex tail in one Akt. Sci STKE 2001; 2001(66):E1.
58. Du K, Montminy M. CREB is a regulatory target for the protein kinase Akt/PKB. J Biol Chem 1998; 273(49):32377-32379.
59. Brunet A, Bonni A, Zigmond MJ et al. Akt promotes cell survival by phosphorylating and inhibiting a Forkhead transcription factor. Cell 1999; 96(6):857-868.
60. Tang ED, Nunez G, Barr FG et al. Negative regulation of the forkhead transcription factor FKHR by Akt. J Biol Chem 1999; 274(24):16741-16746.
61. Smith JL, Schaffner AE, Hofmeister JK et al. Ets-2 is a target for an akt (Protein kinase B)/jun N-terminal kinase signaling pathway in macrophages of motheaten-viable mutant mice. Mol Cell Biol 2000; 20(21):8026-8034.
62. Testa JR, Bellacosa A. AKT plays a central role in tumorigenesis. Proc Natl Acad Sci USA 2001; 98(20):10983-10985.
63. Frame MC. Src in cancer: Deregulation and consequences for cell behaviour. Biochim Biophys Acta 2002; 1602(2):114-130.
64. Kaplan KB, Swedlow JR, Varmus HE et al. Association of p60c-src with endosomal membranes in mammalian fibroblasts. J Cell Biol 1992; 118(2):321-333.
65. Fincham VJ, Unlu M, Brunton VG et al. Translocation of Src kinase to the cell periphery is mediated by the actin cytoskeleton under the control of the Rho family of small G proteins. J Cell Biol 1996; 135(6 Pt 1):1551-1564.
66. Fincham VJ, Brunton VG, Frame MC. The SH3 domain directs acto-myosin-dependent targeting of v-Src to focal adhesions via phosphatidylinositol 3-kinase. Mol Cell Biol 2000; 20(17):6518-6536.
67. Bence-Hanulec KK, Marshall J, Blair LA. Potentiation of neuronal L calcium channels by IGF-1 requires phosphorylation of the alpha1 subunit on a specific tyrosine residue. Neuron 2000; 27(1):121-131.
68. Arbet-Engels C, TartareDeckert S, Eckhart W. C-terminal Src kinase associates with ligand-stimulated insulin-like growth factor-I receptor. J Biol Chem 1999; 274(9):5422-5428.
69. Azarnia R, Reddy S, Kmiecik TE et al. The cellular src gene product regulates junctional cell-to-cell communication. Science 1988; 239(4838):398-401.
70. Toyofuku T, Yabuki M, Otsu K et al. Functional role of c-Src in gap junctions of the cardiomyopathic heart. Circ Res 1999; 85(8):672-681.
71. Toyofuku T, Akamatsu Y, Zhang H et al. C-Src regulates the interaction between connexin-43 and ZO-1 in cardiac myocytes. J Biol Chem 2001; 276(3):1780-1788.

72. Rodier JM, Valles AM, Denoyelle M et al. pp60c-src is a positive regulator of growth factor-induced cell scattering in a rat bladder carcinoma cell line. J Cell Biol 1995; 131(3):761-773.
73. Boyer B, Roche S, Denoyelle M et al. Src and Ras are involved in separate pathways in epithelial cell scattering. Embo J 1997; 16(19):5904-5913.
74. Volberg T, Geiger B, Dror R et al. Modulation of intercellular adherens-type junctions and tyrosine phosphorylation of their components in RSV-transformed cultured chick lens cells. Cell Regul 1991; 2(2):105-120.
75. Gomez S, del Mont Llosas M, Verdu J et al. Independent regulation of adherens and tight junctions by tyrosine phosphorylation in Caco2 cells. Biochim Biophys Acta 1999; 1452(2):121-132.
76. Hansson A, Thoren M. Activation of MAP kinase in Swiss 3T3 fibroblasts by insulin-like growth factor-I. Growth Regul 1995; 5(2):92-100.
77. Casamassima A, Rozengurt E. Insulin-like growth factor I stimulates tyrosine phosphorylation of p130(Cas), focal adhesion kinase, and paxillin. Role of phosphatidylinositol 3'-kinase and formation of a p130(Cas).Crk complex. J Biol Chem 1998; 273(40):26149-26156.
78. Lee CH, Li W, Nishimura R et al. Nck associates with the SH2 domain-docking protein IRS-1 in insulin-stimulated cells. Proc Natl Acad Sci USA 1993; 90(24):11713-11717.
79. Myers Jr MG, Backer JM, Sun XJ et al. IRS-1 activates phosphatidylinositol 3'-kinase by associating with src homology 2 domains of p85. Proc Natl Acad Sci USA 1992; 89(21):10350-10354.
80. Alessi DR, Andjelkovic M, Caudwell B et al. Mechanism of activation of protein kinase B by insulin and IGF-1. Embo J 1996; 15(23):6541-6551.
81. Parsons JT, Martin KH, Slack JK et al. Focal adhesion kinase: A regulator of focal adhesion dynamics and cell movement. Oncogene 2000; 19(49):5606-5613.
82. Fujita Y, Krause G, Scheffner M et al. Hakai, a c-Cbl-like protein, ubiquitinates and induces endocytosis of the E-cadherin complex. Nat Cell Biol 2002; 4(3):222-231.
83. Kirchhausen T. Clathrin adaptors really adapt. Cell 2002; 109(4):413-416.
84. Pearse BM, Smith CJ, Owen DJ. Clathrin coat construction in endocytosis. Curr Opin Struct Biol 2000; 10(2):220-228.
85. Barbieri MA, Kohn AD, Roth RA et al. Protein kinase B/akt and rab5 mediate Ras activation of endocytosis. J Biol Chem 1998; 273(31):19367-19370.
86. Grille SJ, Bellacosa A, Upson J et al. The protein kinase akt induces epithelial mesenchymal transition and promotes enhanced motility and invasiveness of squamous cell carcinoma lines. Cancer Res 2003; 63(9):2172-2178.
87. Sun XJ, Pons S, Asano T et al. The Fyn tyrosine kinase binds Irs-1 and forms a distinct signaling complex during insulin stimulation. J Biol Chem 1996; 271(18):10583-10587.
88. Warren SL, Nelson WJ. Nonmitogenic morphoregulatory action of pp60v-src on multicellular epithelial structures. Mol Cell Biol 1987; 7(4):1326-1337.
89. Tsukita S, Oishi K, Akiyama T et al. Specific proto-oncogenic tyrosine kinases of src family are enriched in cell-to-cell adherens junctions where the level of tyrosine phosphorylation is elevated. J Cell Biol 1991; 113(4):867-879.
90. Ukropec JA, Hollinger MK, Salva SM et al. SHP2 association with VE-cadherin complexes in human endothelial cells is regulated by thrombin. J Biol Chem 2000; 275(8):5983-5986.
91. Lee MM, Fink BD, Grunwald GB. Evidence that tyrosine phosphorylation regulates N-cadherin turnover during retinal development. Dev Genet 1997; 20(3):224-234.
92. Lampugnani MG, Corada M, Andriopoulou P et al. Cell confluence regulates tyrosine phosphorylation of adherens junction components in endothelial cells. J Cell Sci 1997; 110(Pt 17):2065-2077.
93. Daniel JM, Reynolds AB. Tyrosine phosphorylation and cadherin/catenin function. Bioessays 1997; 19(10):883-891.
94. Esser S, Lampugnani MG, Corada M et al. Vascular endothelial growth factor induces VE-cadherin tyrosine phosphorylation in endothelial cells. J Cell Sci 1998; 111(Pt 13):1853-1865.
95. Nawroth R, Poell G, Ranft A et al. VE-PTP and VE-cadherin ectodomains interact to facilitate regulation of phosphorylation and cell contacts. Embo J 2002; 21(18):4885-4895.
96. Lickert H, Bauer A, Kemler R et al. Casein kinase II phosphorylation of E-cadherin increases E-cadherin/beta-catenin interaction and strengthens cell-cell adhesion. J Biol Chem 2000; 275(7):5090-5095.
97. Roura S, Miravet S, Piedra J et al. Regulation of E-cadherin/Catenin association by tyrosine phosphorylation. J Biol Chem 1999; 274(51):36734-36740.
98. Playford MP, Bicknell D, Bodmer WF et al. Insulin-like growth factor 1 regulates the location, stability, and transcriptional activity of beta-catenin. Proc Natl Acad Sci USA 2000; 97(22):12103-12108.

99. Hu P, O'Keefe EJ, Rubenstein DS. Tyrosine phosphorylation of human keratinocyte beta-catenin and plakoglobin reversibly regulates their binding to E-cadherin and alpha-catenin. J Invest Dermatol 2001; 117(5):1059-1067.
100. Rubinfeld B, Albert I, Porfiri E et al. Binding of GSK3beta to the APC-beta-catenin complex and regulation of complex assembly. Science 1996; 272(5264):1023-1026.
101. Hecht A, Kemler R. Curbing the nuclear activities of beta-catenin. Control over Wnt target gene expression. EMBO Rep 2000; 1(1):24-28.
102. Huber AH, Nelson WJ, Weis WI. Three-dimensional structure of the armadillo repeat region of beta-catenin. Cell 1997; 90(5):871-882.
103. Huber AH, Weis WI. The structure of the beta-catenin/E-cadherin complex and the molecular basis of diverse ligand recognition by beta-catenin. Cell 2001; 105(3):391-402.
104. Graham TA, Weaver C, Mao F et al. Crystal structure of a beta-catenin/Tcf complex. Cell 2000; 103(6):885-896.
105. Chen JW, Pan W, D'Souza MP et al. Lysosome-associated membrane proteins: Characterization of LAMP-1 of macrophage P388 and mouse embryo 3T3 cultured cells. Arch Biochem Biophys 1985; 239(2):574-586.
106. Desbois-Mouthon C, Cadoret A, Blivet-Van Eggelpoel MJ et al. Insulin and IGF-1 stimulate the beta-catenin pathway through two signalling cascades involving GSK-3beta inhibition and Ras activation. Oncogene 2001; 20(2):252-259.
107. Cross DA, Alessi DR, Cohen P et al. Inhibition of glycogen synthase kinase-3 by insulin mediated by protein kinase B. Nature 1995; 378(6559):785-789.
108. Grevengoed EE, Loureiro JJ, Jesse TL et al. Abelson kinase regulates epithelial morphogenesis in Drosophila. J Cell Biol 2001; 155(7):1185-1198.
109. Larue L, Antos C, Butz S et al. A role for cadherins in tissue formation. Development 1996; 122(10):3185-3194.
110. Frame S, Cohen P, Biondi RM. A common phosphate binding site explains the unique substrate specificity of GSK3 and its inactivation by phosphorylation. Mol Cell 2001; 7(6):1321-1327.
111. Dajani R, Fraser E, Roe SM et al. Crystal structure of glycogen synthase kinase 3 beta: Structural basis for phosphate-primed substrate specificity and autoinhibition. Cell 2001; 105(6):721-732.
112. Fukumoto S, Hsieh CM, Maemura K et al. Akt participation in the Wnt signaling pathway through Dishevelled. J Biol Chem 2001; 276(20):17479-17483.
113. Chen RH, Ding WV, McCormick F. Wnt signaling to beta-catenin involves two interactive components. Glycogen synthase kinase-3beta inhibition and activation of protein kinase C. J Biol Chem 2000; 275(23):17894-17899.
114. Ding VW, Chen RH, McCormick F. Differential regulation of glycogen synthase kinase 3beta by insulin and Wnt signaling. J Biol Chem 2000; 275(42):32475-32481.
115. Molenaar M, van de Wetering M, Oosterwegel M et al. XTcf-3 transcription factor mediates beta-catenin-induced axis formation in Xenopus embryos. Cell 1996; 86(3):391-399.
116. Behrens J, von Kries JP, Kuhl M et al. Functional interaction of beta-catenin with the transcription factor LEF-1. Nature 1996; 382(6592):638-642.
117. Huber O, Korn R, McLaughlin J et al. Nuclear localization of beta-catenin by interaction with transcription factor LEF-1. Mech Dev 1996; 59(1):3-10.
118. Horb ME, Thomsen GH. A vegetally localized T-box transcription factor in Xenopus eggs specifies mesoderm and endoderm and is essential for embryonic mesoderm formation. Development 1997; 124(9):1689-1698.
119. Papaioannou VE, Silver LM. The T-box gene family. Bioessays 1998; 20(1):9-19.
120. Jacobs-Cohen RJ, Spiegelman M, Bennett D. Abnormalities of cells and extracellular matrix of T/T embryos. Differentiation 1983; 25(1):48-55.
121. Wilson V, Manson L, Skarnes WC et al. The T gene is necessary for normal mesodermal morphogenetic cell movements during gastrulation. Development 1995; 121(3):877-886.
122. Rashbass P, Wilson V, Rosen B et al. Alterations in gene expression during mesoderm formation and axial patterning in Brachyury (T) embryos. Int J Dev Biol 1994; 38(1):35-44.
123. von Bredow DC, Nagle RB, Bowden GT et al. Degradation of fibronectin fibrils by matrilysin and characterization of the degradation products. Exp Cell Res 1995; 221(1):83-91.
124. Cuvelier A, Kuntz C, Sesboue R et al. Metalloproteinases in the extracellular matrix: Structure and activity. Rev Mal Respir 1997; 14(1):1-10.
125. Delmas V et al. unpublished.
126. Thiery JP. Epithelial-mesenchymal transitions in tumour progression. Nat Rev Cancer 2002; 2(6):442-454.
127. Cano A, Perez-Moreno MA, Rodrigo I et al. The transcription factor snail controls epithelial-mesenchymal transitions by repressing E-cadherin expression. Nat Cell Biol 2000; 2(2):76-83.

128. Batlle E, Sancho E, Franci C et al. The transcription factor snail is a repressor of E-cadherin gene expression in epithelial tumour cells. Nat Cell Biol 2000; 2(2):84-89.
129. Comijn J, Berx G, Vermassen P et al. The two-handed E box binding zinc finger protein SIP1 downregulates E-cadherin and induces invasion. Mol Cell 2001; 7(6):1267-1278.
130. Perez-Moreno MA, Locascio A, Rodrigo I et al. A new role for E12/E47 in the repression of E-cadherin expression and epithelial-mesenchymal transitions. J Biol Chem 2001; 276(29):27424-27431.
131. Rodrigo I, Cato AC, Cano A. Regulation of E-cadherin gene expression during tumor progression: The role of a new Ets-binding site and the E-pal element. Exp Cell Res 1999; 248(2):358-371.
132. Morali OG, Jouneau A, McLaughlin KJ et al. IGF-II promotes mesoderm formation. Dev Biol 2000; 227(1):133-145.
133. Sartor O, Cooper MR, Khleif SN et al. Suramin decreases circulating levels of insulin-like growth factor-I. Am J Med 1994; 96(4):390.
134. Sell C, Dumenil G, Deveaud C et al. Effect of a null mutation of the insulin-like growth factor I receptor gene on growth and transformation of mouse embryo fibroblasts. Mol Cell Biol 1994; 14(6):3604-3612.
135. Xie SP, Pirianov G, Colston KW. Vitamin D analogues suppress IGF-I signalling and promote apoptosis in breast cancer cells. Eur J Cancer 1999; 35(12):1717-1723.
136. Hogan B, Beddington R, Costantini F et al. Manipulating the mouse embryo : A laboratory manual. Cold Spring Harbor Laboratory Press, 1994.
137. Lee JE, Pintar J, Efstratiadis A. Pattern of the insulin-like growth factor II gene expression during early mouse embryogenesis. Development 1990; 110(1):151-159.

Chapter 15

TGFβ-Dependent Epithelial-Mesenchymal Transition

Marie-Luce Vignais and Patrick Fafet

Summary

The transforming growth factor β (TGFβ) is involved in a whole range of biological functions, from cell growth to cell differentiation and apoptosis. The role of TGFβ in epithelial-mesenchymal-transitions (EMTs) has been shown for both embryonic development and tumorigenesis. All three TGFβ mammalian isoforms—TGFβ1, TGFβ2 and TGFβ3—can regulate EMTs, with distinct outcomes depending on the tissue and on the state of cell differentiation. This diversity in the TGFβ response relies on a complex network of signals starting with different sets of TGFβ receptors and subsequently involving distinct TGFβ-dependent pathways. The purpose of this review is to recapitulate the current knowledge on the various signaling pathways—including the Smads, Ras, p38MAPK, RhoA and PI3K—which, upon activation by TGFβ can together give rise to TGFβ-induced EMT phenotypes.

Introduction

The transforming growth factor β (TGFβ) is implicated in a wide variety of biological functions ranging from growth control to cell differentiation, immune response and apoptosis.[1-9] The transforming growth factor beta (TGFβ) polypeptide superfamily includes over 40 related members belonging to the TGFβ, activin, nodal and BMP (bone morphogenetic) families which are conserved throughout species, including vertebrates, insects and nematodes.

Three forms of TGFβ are known in mammals, the TGFβ1, TGFβ2 and TGFβ3 which are homodimeric proteins of 25kDa and are 65 to 80% identical to one another. Although these three isoforms demonstrate overlapping properties, they also display specific features. This is illustrated for instance by the distinct roles of the 3 TGFβs in embryonic cardiac differentiation (see below) and, more generally, by the specific phenotypes of the murine knock-outs such as the defective haematopoiesis in TGFβ1 knock-outs, the abnormal lung development and cleft palate observed in TGFβ3 null mice or in the reduced apoptosis of Tgfbeta2(-/-) Tgfbeta3(-/-) double-deficient mice.[10-14] Supporting the notion that TGFβs play distinct roles during development, the three isoforms TGFβ1, TGFβ2 and TGFβ3 are differentially expressed during embryogenesis.[15] More generally, the TGFβs can demonstrate antithetic physiological effects which will depend upon both the cell type, its state of differentiation and the cellular microenvironment.

The present review focuses on the role the three mammalian TGFβs, TGFβ1, TGFβ2 and TGFβ3, in the epithelial-mesenchymal-transitions (EMTs) which occur in both embryonic development and cancer progression and on the signaling pathways which are involved in these regulations.

Rise and Fall of Epithelial Phenotype: Concepts of Epithelial-Mesenchymal Transition, edited by Pierre Savagner.

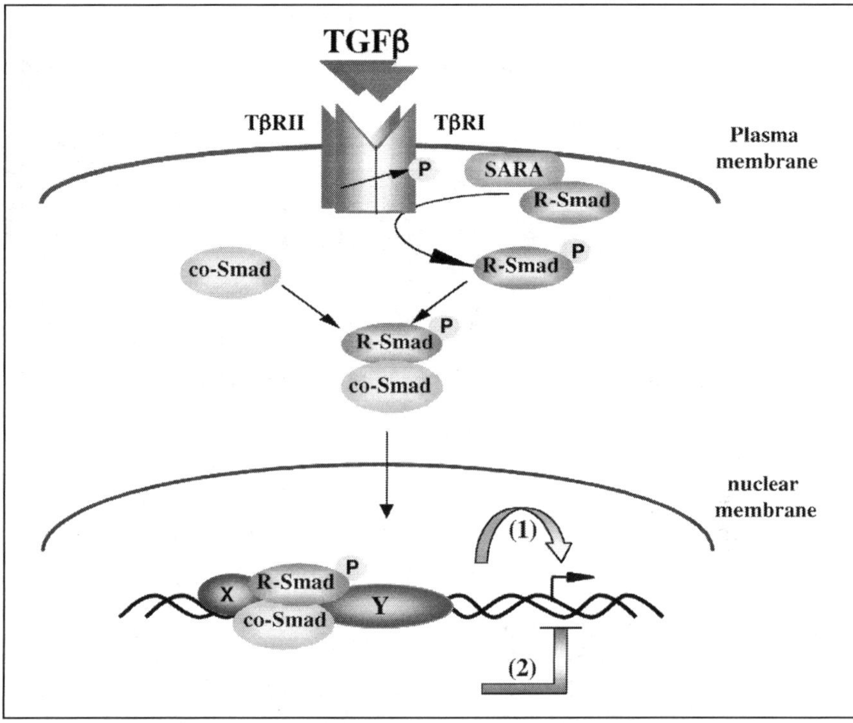

Figure 1. The Smad signaling pathway. TGFβ binding to type II and type I receptors (TβRII and TβRI) leads to receptor oligomerization and activation. Recruitment of the R-Smads to the receptor complex involves the membrane bound protein SARA (Smad anchor for receptor activation). Upon phosphorylation by the serine/threonine kinase activity of the receptor, R-Smads can associate with the coSmad Smad4 and translocate to the nucleus. Together with interacting transcription factors (X and Y), Smad can bind specific gene promoters and either activate (1) or repress (2) transcription of the target genes, depending on the biological context.

TGFβ Signaling

The effects of TGFβ are triggered by the activation of heteromeric complexes of TGFβ transmembrane type II (TβRII) and type I (TβRI) receptors which leads to the phosphorylation of downstream targets. Whereas TβRII binds TGFβ1 and TGFβ3 with high affinity, its affinity for TGFβ2 is much poorer. Therefore, TGFβ2 signaling also depends upon type III receptors (TβRIII) which enable its binding to TβRII and TβRI. Ligand binding to TβRII results in activation of TβRI and downstream signaling. At least two types of TβRI are capable of interacting with TβRII. The ALK5 (activin-like receptor kinase 5) is considered as the prototypical TGFβ receptor whereas ALK2 can interact with the type II receptor for either TGFβ, activin or BMP (bone morphogenetic protein). Because these various receptors can be differentially expressed in the different cells of a given tissue, one gets a first hint as to how cells within the same microenvironment can respond differently to local TGFβ concentrations.

Several signaling pathways have been identified downstream of the TGFβ receptors, among which the Smad pathway is the best characterized (Fig. 1). The R-Smads (TGFβ-Responsive Smads), Smad2 and Smad3, are cytoplasmic proteins which are activated at the cell membrane by the TGFβ receptors. They associate with the coSmad Smad4 and translocate to the nucleus where they bind target genes and regulate transcription.[16-18] Aside from the thoroughly characterized Smad signaling pathway, the transforming growth factor β (TGFβ) signaling can lead

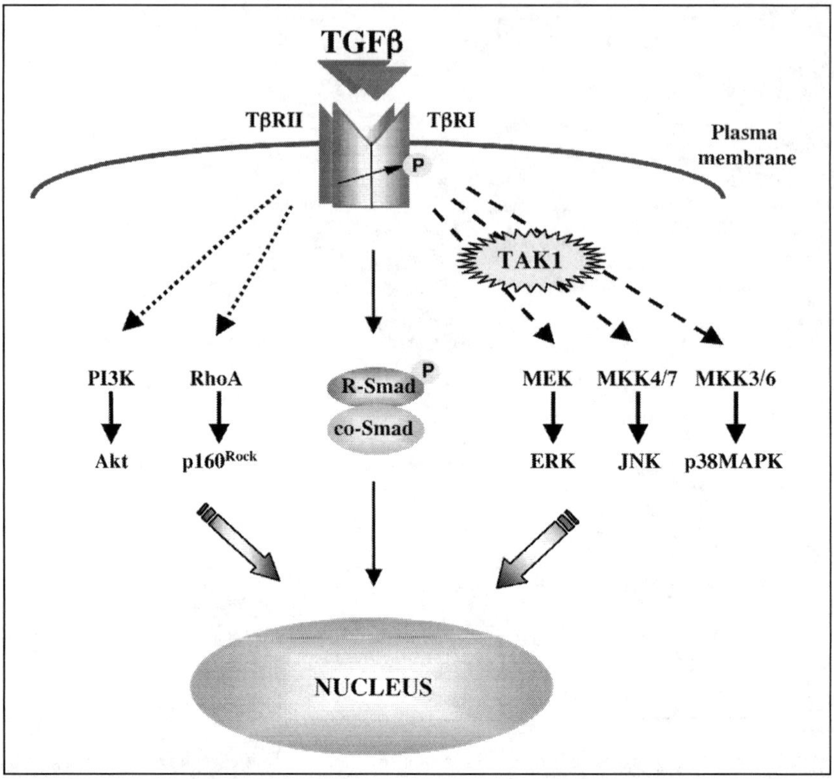

Figure 2. The TGFβ-dependent signaling pathways. TGFβ can induce Smad-independent signaling pathways. These include the three MAP kinase pathways leading to the activation of ERK, JNK and p38MAPK. TGFβ has also been shown to activate the phosphatidylinositol 3-kinase (PI3K) and the protein kinase Akt as well as RhoA and p160^{ROCK}. However, the mechanisms of activation of these Smad-independent signaling pathways remain to be precisely characterized as does their connection to the TGFβ receptor. The exact positioning of TAK1 (TGFβ-activated kinase 1) and its role for the activation of the MAPK pathways remain to be established.

to the activation of ERK (extracellular signal-regulated kinase), JNK (Jun N-terminal kinase) and p38MAPK (p38 mitogen-activated protein kinase).[4] Although some of this signaling involves the kinase TAK1 (TGFβ-activated kinase 1),[19-21] the mechanisms of these signaling pathways as well as their connection to the TGFβ receptors remain to be more precisely defined. Alternatively, activation of PI3K/Akt (phosphatidylinositol-3 kinase) as well as of the Rho GTPases have also been reported[22,23] (Fig. 2).

Role of TGFβ in Tumorigenesis and EMT

TGFβ plays an ambiguous role as it is both a tumor suppressor and a tumor promoter.[8,24,25] TGFβ was named transforming growth factorβ for its capacity to induce, together with TGFα, a transformed phenotype in normal rat kidney fibroblasts.[26] Although TGFβ was originally purified from human placenta and platelets,[27,28] most cells secrete it. Secretion of TGFβ is increased in many tumor cells which respond to TGFβ by enhanced invasiveness. This process, which contributes to metastasis formation, is due to an increased mesenchymal-transdifferentiation in epithelial cells.[29] TGFβ was first shown to induce epithelial-mesenchymal-transitions (EMTs) in immortalized mammary epithelial NMuMG cells.[30] This differentiation from epithelial to fibroblastic phenotype was observed within 16 hours after addition of TGFβ1 and was accompanied by a decreased expression of the epithelial markers E-cadherin, ZO-1, and desmoplakin I and II, an increased expression of mesenchymal fibronectin markers and by a reorganization of actin stress fibers. In a different study, mice with TGFβ1 expression targeted to keratinocytes and exposed to carcinogenesis treatment showed an enhanced malignant progression.[31] It took several years for what appears now as a fundamental TGFβ-dependent physiological effect to come to the front line. In the past few years, there has been a wealth of data which further emphasized the role of TGFβ in regulating the epithelial to mesenchymal transition (see reviews 7,9,24, see also Chapter 1, Hay and Chapter 2, Morali et al). The main features of the epithelial-mesenchymal transition (EMT) induced by TGFβ are a switch from mainly cytokeratin to mainly vimentin intermediate filaments, the formation of actin stress fibers and an enhancement of cell migration. Markers of TGFβ-induced EMT also include elevated expression of N-cadherin and delocalization of E-cadherin from cell junctions.[32] The TGFβ-dependent EMTs can be fully reversible as shown in the initial experiments on NMuMG epithelial cells where removal of TGFβ1 restored the epithelial phenotype within two days.[30] The reversibility of the TGFβ-dependent EMT was also demonstrated, in the dedifferentiated mesenchymal mouse colon carcinoma cells (CT26), by the overexpression of a dominant-negative type II TβR which induced mesenchymal-to-epithelial transition and inhibited in vitro invasiveness.[33] Metastasis formation in these cells was completely abolished. Besides, reexpression of the wild-type TβRII restored the invasiveness of human colon carcinoma cells (hnPCC) which otherwise harbor a nonfunctional receptor and are noninvasive in vitro.[33]

Role of the Three TGFβ Isoforms and of Their Receptors in the EMT Process

Outside from their role in tumor EMTs, the TGFβs are also involved in epithelial-mesenchymal transitions which occur during embryogenesis as, for instance, during palatal or cardiac development.[11,34-36] The three TGFβ isoforms, TGFβ1, TGFβ2 and TGFβ3, show different spatial and temporal expression patterns in the embryo and their different physiological roles are acutely demonstrated by their distinct functions during the embryonic heart EMT. Whereas TGFβ2 and TGFβ3 are necessary respectively for chick endothelial cell separation and transformation, both growth factors have an opposite effect on chick epicardial cells as they inhibit their FGF-dependent EMT.[37,38] The complexity further increases with apparently distinct requirements for TGFβ family members depending on the species, the mouse differing from the chick by the fact that only TGFβ2 is expressed in the endocardial cushions and plays a functional role in the endocardial cushion EMT[39] (see also Chapter 4, Runyan et al).

The combinatorial diversity seen for TGFβ family members is also observed for their receptor subunits. Aside from the two well characterized TβRI and TβRII cell surface receptors, TGFβ signaling can also depend on TβRIII which is known to enhance TGFβ2 signaling by presenting the TGFβ ligand to the type II receptor. Work by Runyan (see Chapter 4) showed that, during the EMT which occurs in chick heart, an antibody-mediated block of TβRIII activity inhibited the TGFβ2-dependent endothelial cell-cell separation while, interestingly, antibody inhibition of TβRII had no effect[40] suggesting distinct receptor requirements for TGFβ-induced EMT which still need to be fully established.

Both ALK2 and ALK5 TβRI receptors are expressed in the chick atrioventricular cushion. Antibody-mediated receptor inactivation showed that ALK2, but not ALK5, was necessary for the TGFβ1-induced EMT. On the other hand, only ALK5 was able to mediate the TGFβ1-dependent regulation of PAI-1 transcription,[41] which shows that activation of different type I receptor subunits can lead to distinct biological outcomes. The ALK2/TSK-7L receptor was also shown to mediate TGFβ-dependent transdifferentiation in mouse NMuMG mammary epithelial tumor cells as the overexpression of a kinase-dead ALK2 receptor prevented them from acquiring a fibroblastoid morphology in response to TGFβ1.[30] Interestingly though, the EMT markers E-cadherin, ZO-1 or fibronectin still responded to TGFβ regulation and the cells also exhibited TGFβ-dependent growth inhibiton.[30] In the same cells, neither activin A nor bone morphogenetic protein-7 (BMP-7), which nonetheless utilize the same receptor subunit, were able to induce the morphological transformations, i.e., reorganization of the actin cytoskeleton or down regulation of E-cadherin and β-catenin,[42] which suggests that activation of ALK2 is necessary but not sufficient to induce EMT in NMuMG epithelial cells. In addition, Piek et al showed that overexpression of an activated ALK5 receptor could induce transdifferentiation within 24 hours of infection, as shown by the remodeling of the actin cytoskeleton.[42] These independent observations could appear difficult to reconcile. However, they also raise the interesting possibility of divergent TGFβ signaling emerging from distinct TGFβ subunits and which can target distinct sets of EMT markers and which altogether can generate a full mesenchymal phenotype.

Mechanisms and Signaling Pathways Involved in TGFβ-Dependent EMTs

The molecular mechanisms of TGFβ-induced EMTs are complex and presumably tissue-dependent. The existence of intricate regulatory networks could explain why the inhibition of a given signaling pathway sometimes shows little effect on the TGFβ-dependent EMT while it is found central by other experimental approaches. Overall, several signaling pathways emerge as being repeatedly found in the TGFβ-induced EMT process. Outside from the Smads, these pathways include signaling leading to the activation of Ras, p38MAPK, Rho or PI3K. The involvement of the various pathways and the crosstalks taking place are discussed below.

Role of the Smads

There have been somehow conflicting results concerning the role of the Smads in the TGFβ-induced EMTs. This ambiguity might be the consequence of concentration threshold levels needed to be passed to achieve transdifferentiation. Involvement of the Smads in the EMT process was shown for example by experiments performed in the mouse mammary epithelial cells (NMuMG) where Smad2 and Smad3 were overexpressed together with a constitutively active TβRI.[42] It was also suggested that in order to achieve invasive properties, the Smads require cooperation with other pathways such as that of Ras,[43] see also Chapter 16: Boyer. Using squamous mouse carcinomas initiated by activating mutations in the *Hras1* gene, Balmain and collaborators showed that activation of Smad2 in a background of mutant H-Ras induced an EMT characterized by a transformation from a differentiated state to a motile invasive stage. Overexpression of the mutant H-Ras by itself induced neither the changes in cell shape nor the expression of the microfilaments characteristic of the EMT.[44]

On the other hand, the role of the Smads was questioned by experiments where the overexpression of anti-Smads, like Smad7 or a Smad3 dominant negative mutant, blocked cell cycle progression of NMuMG cells and, still, had no affect on EMT markers like the E- and N-cadherin relocalization or the actin cytoskeleton remodeling.[32] This suggested that TGFβ-induced EMTs can be achieved independently of the Smads or, alternatively, that they require cellular levels of the Smads unable to achieve cell cycle regulation. In any case, the TGFβ-dependent signalings which regulate cell motility and proliferation appear to use distinct pathways.

Role of Smad-Independent Signaling Pathways

p38 MAP Kinase

The p38MAPK is involved in the signaling of various TGFβ family members. However, the molecular mechanisms which enable p38MAPK activation are not fully determined. Arteaga and collaborators provided evidence that the MKK3/6-p38MAPK-ATF2 pathway is rapidly activated in response to TGFβ and that it is involved in TGFβ-mediated EMT of NMuMG mouse mammary epithelial cells.[45] The kinase activities of both TβRI and TβRII receptors were necessary for p38MAPK activation and EMT to occur. Alternatively, overexpression of an activated ALK5 receptor also resulted in EMT which was concomitant with phosphorylation of MKK3/6 and p38MAPK.[45] The role of p38 for TGFβ-induced EMT in NMuMG cells was further substantiated by experiments by Zhang and collaborators who, in addition, could establish a dichotomy in the TGFβ signalings leading on one hand to EMT and on the other hand to TGFβ-dependent cell growth control.[46] These results, together with the findings of Bhowmick et al[47] suggested that p38MAPK activity is required, though not sufficient, to induce TGFβ-mediated EMT.

Rho and PI3K

Signaling through RhoA or the PI3-kinase/Akt pathway has also been proposed for TGFβ-dependent EMTs,[23,32] see also Chapter 17, Nakagawa et al. TGFβ1 rapidly activates RhoA and p160^{Rock} in NMuMG mammary epithelial cells.[32] This activation correlated with two markers of the TGFβ-dependent EMT, i.e., delocalization of E-cadherin from cell junctions and expression of N-cadherin at cell margins. Neither of these were affected by altering either the Rac1 or JNK^{APF} pathways. Interestingly, NMuMG cell proliferation remained sensitive to TGFβ during this EMT transition.[32] At least two components of the TGFβ-induced EMT are regulated by the RhoA pathway, which lead to actin cytoskeleton remodeling and disassembly of the adherens junctions. This was shown in NMuMG cells that overexpressed a kinase dead mutant of p160^{ROCK} unable to interact with Rho, which still underwent E-cadherin delocalization from the adherens junctions in response to TGFβ but failed to acquire stress fibers and a prototypical fibroblastoid morphology.

The phosphatidylinositide 3-kinase (PI3K)/Akt pathway was shown to contribute to EMT-related tight junctions disruption. This response correlated with the delocalization of E-cadherin, ZO-1 and integrin β_1 from the cell junctions while cells acquired a spindle morphology.[23] By administrating to MMTV-PyVmT transgenic mice a soluble Fc:TGFβ type II receptor fusion protein containing the extracellular domain of the TβRII, Arteaga and collaborators established a correlation between the observed reduction of tumor cell motility, intravasion and metastasis formation and the inhibition of Akt activity.[48] Such a role for the Akt protein kinase in EMT processes was recently confirmed by the expression of a constitutively active Akt in carcinoma lines which led to characteristic EMT modifications like down-regulation of the epithelial markers desmoplakin, E-cadherin and β-catenin and up-regulation of the fibroblastic marker vimentin.[49]

Cross-Talk between Signaling Pathways

TGFβ-dependent EMTs are either inhibited or enhanced by various signaling pathways. Cytokines which induce these pathways can be produced in an autocrine manner. Alternatively, they can be secreted by stromal cells and constitute therefore active mediators of the dialogue which often takes place between cells undergoing EMT and cells from the stroma. Such is the case, for instance, for colon carcinoma whose TGFβ1-dependent EMT is accelerated by the tumor necrosis factor-α (TNF-α) produced by the surrounding activated macrophages.[50] Members of the TGFβ-superfamily themselves can also regulate the TGFβ-dependent EMT process. Their role as well as the particular case of Ras are discussed below.

Role of Members of the TGFβ Superfamily

Members of the bone morphogenetic protein (BMP) family can have either enhancing or inhibiting properties on the TGFβ-induced EMTs, as illustrated by the myocardial BMP2 which acts in synergy with TGFβ3 for the initiation of EMT whereas BMP7 counteracts the TGFβ1-induced EMT in renal tubular epithelial cells.[51,52]

The Case of Ras

The relationships between TGFβ and Ras are complex because of the intermingling between the two signaling pathways. TGFβ can on its own activate the Ras/MAPK pathway. In addition, the Ras/MAPK pathway can be activated by other cytokines, such as EGF, and regulate various steps of Smad signaling (see for a review ref. 21). Several reports indicate a synergy between TGFβ and Ras/MAPK pathways in promoting EMT. EGF-dependent activation of Ras/MAPK was shown to stimulate TGFβ-induced EMT.[53] In nontransformed mammary epithelial cells, expression of a mutant Ras was shown to have two effects on the TGFβ response : it prevented growth inhibition by TGFβ1 and, in addition, promoted EMT.[33,43,54] Importantly, the Ras-transformed fibroblastoid cells could secrete TGF-beta themselves, leading to autocrine amplification of the phenomenon. In kidney MDCK epithelial cells, sustained activation of Raf induced cell invasiveness which depended on Raf-induced secretion of TGFβ.[55]

Conclusion and Perspectives

A number of studies suggest that distinct TGFβ signaling pathways regulate growth inhibition and epithelial-mesenchymal-transition. This raises the fascinating possibility that antagonists specific for the EMT pathway might be generated, that do not affect the control of cell proliferation. Under such circumstances, one could envision drugs for cancer therapy which would provide a block to metastasis spreading, while keeping the powerful capacity of TGFβ to control cell proliferation. This should emphasize the current efforts in developing TGFβ signaling inhibitors.[25]

Acknowledgements

We are grateful to J.C. Dantonel for critical reading of the manuscript.

References

1. Moses HL, Serra R. Regulation of differentiation by TGF-beta. Curr Opin Genet Dev 1996; 6:581-6.
2. Capdevila J, Belmonte JC. Extracellular modulation of the Hedgehog, Wnt and TGF-beta signalling pathways during embryonic development. Curr Opin Genet Dev 1999; 9:427-33.
3. Massague J, Blain SW, Lo RS. TGFbeta signaling in growth control, cancer, and heritable disorders. Cell 2000; 103:295-309.
4. Derynck R, Akhurst RJ, Balmain A. TGF-beta signaling in tumor suppression and cancer progression. Nat Genet 2001; 29:117-29.
5. Hill CS. TGF-beta signalling pathways in early Xenopus development. Curr Opin Genet Dev 2001; 11:533-40.
6. Gorelik L, Flavell RA. Transforming growth factor-beta in T-cell biology. Nat Rev Immunol. 2002; 2:46-53.
7. Moustakas A, Pardali K, Gaal A et al. Mechanisms of TGF-beta signaling in regulation of cell growth and differentiation. Immunol Lett 2002; 82:85-91.
8. Wakefield LM, Roberts AB. TGF-beta signaling: Positive and negative effects on tumorigenesis. Curr Opin Genet Dev 2002; 12:22-9.
9. Govinden R, Bhoola KD Genealogy, expression, and cellular function of transforming growth factor-beta. Pharmacol Ther 2003; 98:257-65.
10. Dickson MC, Martin JS, Cousins FM et al. Defective haematopoiesis and vasculogenesis in transforming growth factor-beta 1 knock out mice. Development 1995; 121:1845-54.
11. Kaartinen V, Voncken JW, Shuler C et al. Abnormal lung development and cleft palate in mice lacking TGF-beta 3 indicates defects of epithelial-mesenchymal interaction. Nat Genet 1995; 11:415-21.
12. Sanford LP, Ormsby I, Gittenberger-de Groot AC et al. TGFbeta2 knockout mice have multiple developmental defects that are nonoverlapping with other TGFbeta knockout phenotypes. Development 1997; 124:2659-70.
13. Taya Y, O'Kane S, Ferguson MW. Pathogenesis of cleft palate in TGF-beta3 knockout mice. Development 1999; 126:3869-79.
14. Dunker N, Krieglstein K. Reduced programmed cell death in the retina and defects in lens and cornea of Tgfbeta2(-/-) Tgfbeta3(-/-) double-deficient mice. Cell Tissue Res 2003.

15. Schmid P, Cox D, Bilbe G et al. Differential expression of TGF beta 1, beta 2 and beta 3 genes during mouse embryogenesis. Development 1991; 111:117-30.
16. Wrana JL. Regulation of Smad activity. Cell 2000;100:189-92.
17. Moustakas A, Souchelnytskyi S, Heldin CH. Smad regulation in TGF-beta signal transduction. J Cell Sci 2001; 114:4359-69.
18. Shi Y, Massague J. Mechanisms of tgf-Beta signaling from cell membrane to the nucleus. Cell 2003; 113:685-700.
19. Yamaguchi K, Nagai S, Ninomiya-Tsuji J et al. XIAP, a cellular member of the inhibitor of apoptosis protein family, links the receptors to TAB1-TAK1 in the BMP signaling pathway. EMBO J 1999; 18:179-87.
20. Behrens J. Cross-regulation of the Wnt signalling pathway: A role of MAP kinases. J Cell Sci 2000; 113(Pt 6):911-9.
21. Massague J. How cells read TGF-beta signals. Nat Rev Mol Cell Biol 2000; 1:169-78.
22. Hanafusa H, Ninomiya-Tsuji J, Masuyama N et al. Involvement of the p38 mitogen-activated protein kinase pathway in transforming growth factor-beta-induced gene expression. J Biol Chem 1999; 274:27161-7.
23. Bakin AV, Tomlinson AK, Bhowmick NA et al. Phosphatidylinositol 3-kinase function is required for transforming growth factor beta-mediated epithelial to mesenchymal transition and cell migration. J Biol Chem 2000; 275:36803-10.
24. Akhurst RJ, Derynck R. TGF-beta signaling in cancer—a double-edged sword. Trends Cell Biol 2001; 11:S44-51.
25. Dumont N, Arteaga CL. Targeting the TGFbeta signaling network in human neoplasia. Cancer Cell 2003; 3:531-6.
26. De Larco JE, Todaro GJ. Growth factors from murine sarcoma virus-transformed cells. Proc Natl Acad Sci USA 1978; 75:4001-5.
27. Frolik CA, Dart LL, Meyers CA et al. Purification and initial characterization of a type beta transforming growth factor from human placenta. Proc Natl Acad Sci USA 1983; 80:3676-80.
28. Assoian RK, Komoriya A, Meyers CA et al. Transforming growth factor-beta in human platelets. Identification of a major storage site, purification, and characterization. J Biol Chem 1983; 258:7155-60.
29. Dumont N, Arteaga CL. Transforming growth factor-beta and breast cancer: Tumor promoting effects of transforming growth factor-beta. Breast Cancer Res 2000; 2:125-32.
30. Miettinen PJ, Ebner R, Lopez AR et al. TGF-beta induced transdifferentiation of mammary epithelial cells to mesenchymal cells: Involvement of type I receptors. J Cell Biol 1994; 127:2021-36.
31. Cui W, Fowlis DJ, Bryson S et al. TGFbetal inhibits the formation of benign skin tumors, but enhances progression to invasive spindle carcinomas in transgenic mice. Cell 1996; 86:531-42.
32. Bhowmick NA, Ghiassi M, Bakin A et al. Transforming growth factor-betal mediates epithelial to mesenchymal transdifferentiation through a RhoA-dependent mechanism. Mol Biol Cell 2001; 12:27-36.
33. Oft M, Heider KH, Beug H. TGFbeta signaling is necessary for carcinoma cell invasiveness and metastasis. Curr Biol 1998; 8:1243-52.
34. Potts JD, Runyan RB. Epithelial-mesenchymal cell transformation in the embryonic heart can be mediated, in part, by transforming growth factor beta. Dev Biol 1989; 134:392-401.
35. Proetzel G, Pawlowski SA, Wiles MV et al. Transforming growth factor-beta 3 is required for secondary palate fusion. Nat Genet 1995; 11:409-14.
36. Brown CB, Boyer AS, Runyan RB et al. Requirement of type III TGF-beta receptor for endocardial cell transformation in the heart. Science 1999; 283:2080-2.
37. Boyer AS, Ayerinskas, II, Vincent EB et al. TGFbeta2 and TGFbeta3 have separate and sequential activities during epithelial-mesenchymal cell transformation in the embryonic heart. Dev Biol 1999; 208:530-45.
38. Morabito CJ, Dettman RW, Kattan J et al. Positive and negative regulation of epicardial-mesenchymal transformation during avian heart development. Dev Biol 2001; 234:204-15.
39. Camenisch TD, Molin DG, Person A et al. Temporal and distinct TGFbeta ligand requirements during mouse and avian endocardial cushion morphogenesis. Dev Biol 2002; 248:170-81.
40. Boyer AS, Runyan RB. TGFbeta Type III and TGFbeta Type II receptors have distinct activities during epithelial-mesenchymal cell transformation in the embryonic heart. Dev Dyn 2001; 221:454-9.
41. Lai YT, Beason KB, Brames GP et al. Activin receptor-like kinase 2 can mediate atrioventricular cushion transformation. Dev Biol 2000; 222:1-11.
42. Piek E, Moustakas A, Kurisaki A et al. TGF-(beta) type I receptor/ALK-5 and Smad proteins mediate epithelial to mesenchymal transdifferentiation in NMuMG breast epithelial cells. J Cell Sci 1999; 112(Pt 24):4557-68.

43. Oft M, Peli J, Rudaz C et al. TGF-beta1 and Ha-Ras collaborate in modulating the phenotypic plasticity and invasiveness of epithelial tumor cells. Genes Dev 1996; 10:2462-77.
44. Oft M, Akhurst RJ, Balmain A. Metastasis is driven by sequential elevation of H-ras and Smad2 levels. Nat Cell Biol 2002; 4:487-94.
45. Bakin AV, Rinehart C, Tomlinson AK et al. p38 mitogen-activated protein kinase is required for TGFbeta-mediated fibroblastic transdifferentiation and cell migration. J Cell Sci 2002; 115:3193-206.
46. Yu L, Hebert MC, Zhang YE. TGF-beta receptor-activated p38 MAP kinase mediates Smad-independent TGF-beta responses. EMBO J 2002; 21:3749-59.
47. Bhowmick NA, Zent R, Ghiassi M et al. Integrin beta 1 signaling is necessary for transforming growth factor-beta activation of p38MAPK and epithelial plasticity. J Biol Chem 2001; 276:46707-13.
48. Muraoka RS, Dumont N, Ritter CA et al. Blockade of TGF-beta inhibits mammary tumor cell viability, migration, and metastases. J Clin Invest 2002; 109:1551-9.
49. Grille SJ, Bellacosa A, Upson J et al. The protein kinase Akt induces epithelial mesenchymal transition and promotes enhanced motility and invasiveness of squamous cell carcinoma lines. Cancer Res 2003; 63:2172-8.
50. Bates RC, Mercurio AM. Tumor necrosis factor-alpha stimulates the epithelial-to-mesenchymal transition of human colonic organoids. Mol Biol Cell 2003; 14:1790-800.
51. Nakajima Y, Yamagishi T, Hokari S et al. Mechanisms involved in valvuloseptal endocardial cushion formation in early cardiogenesis: Roles of transforming growth factor (TGF)-beta and bone morphogenetic protein (BMP). Anat Rec 2000; 258:119-27.
52. Zeisberg M, Hanai J, Sugimoto H et al. BMP-7 counteracts TGF-beta1-induced epithelial-to-mesenchymal transition and reverses chronic renal injury. Nat Med 2003; 9:964-8.
53. Grande M, Franzen A, Karlsson JO et al. Transforming growth factor-beta and epidermal growth factor synergistically stimulate epithelial to mesenchymal transition (EMT) through a MEK-dependent mechanism in primary cultured pig thyrocytes. J Cell Sci 2002; 115:4227-36.
54. Janda E, Lehmann K, Killisch I et al. Ras and TGF[beta] cooperatively regulate epithelial cell plasticity and metastasis: Dissection of Ras signaling pathways. J Cell Biol 2002; 156:299-313.
55. Lehmann K, Janda E, Pierreux CE et al. Raf induces TGFbeta production while blocking its apoptotic but not invasive responses: A mechanism leading to increased malignancy in epithelial cells. Genes Dev 2000; 14:2610-22.

Chapter 16

The Ras and Src Signaling Cascades Involved in Epithelial Cell Scattering

Brigitte Boyer

Summary

This contribution is to discuss key elements of the intracellular signaling cascades that bring about the cellular transformation referred to as epithelium-to-mesenchyme transition (EMT). EMT can be defined as the events that allows epithelial cells to dissociate from the tissue from which they originate and to migrate freely. EMT is therefore thought to play a fundamental role during the early steps of invasion and metastasis of carcinoma cells. Among biological agents which have been identified as inducers of EMT are a number of cytokines and extracellular matrix macromolecules. The coordinated changes in cell morphology, associated with the induction of cell motility and the disruption of intercellular junctions, are the consequence of signaling cascades emanating from the activation of the receptors specific for the inducing molecules and leading to changes in gene expression. Two of the transduction cascades that play a crucial role in the generation of EMT, namely the Src kinase family and the Ras signal, are discussed extensively with respect to their contribution of scattering signals.

Introduction

EMT is a process during which cells loose their epithelial characteristics and acquire a fibroblastic-like phenotype (Fig. 1). In other words, cells loose the intercellular contacts (adherens junctions, desmosomes, tight junctions) that seal the epithelial sheet, start to scatter and to move by crawling on extracellular matrix components (Fig. 2). Thus, EMT can be seen as a complex process that integrates disruption of intercellular junctions and cell migration. This process participates in several morphogenetic processes during embryonic development, including gastrulation, neural crest cell emigration and somite formation, and it is also likely to play a key role during the metastasis of carcinoma cells. In vitro models for EMT have contributed to demonstrate that cell dissociation and motility are induced by stimuli provided by either growth factors or extracellular matrix components. Activation of transmembrane receptors to these extracellular ligands initiates cascades of events that lead ultimately to cytoskeleton reorganization and gene transcription. Understanding how these cellular changes are orchestrated during EMT requires characterization of the signal targets engendered by the activated receptors. For the seek of clarity, this article is concentrated on the signals emanating from receptors binding to soluble growth factors.

The Signaling Pathways Leading to EMT

Besides the TGFβ-induced signaling cascade, which has been demonstrated to participate in a number of EMTs[1,2] and the Wnt signaling whose role in EMT remains elusive,[3,4] the vast

Rise and Fall of Epithelial Phenotype: Concepts of Epithelial-Mesenchymal Transition, edited by Pierre Savagner.

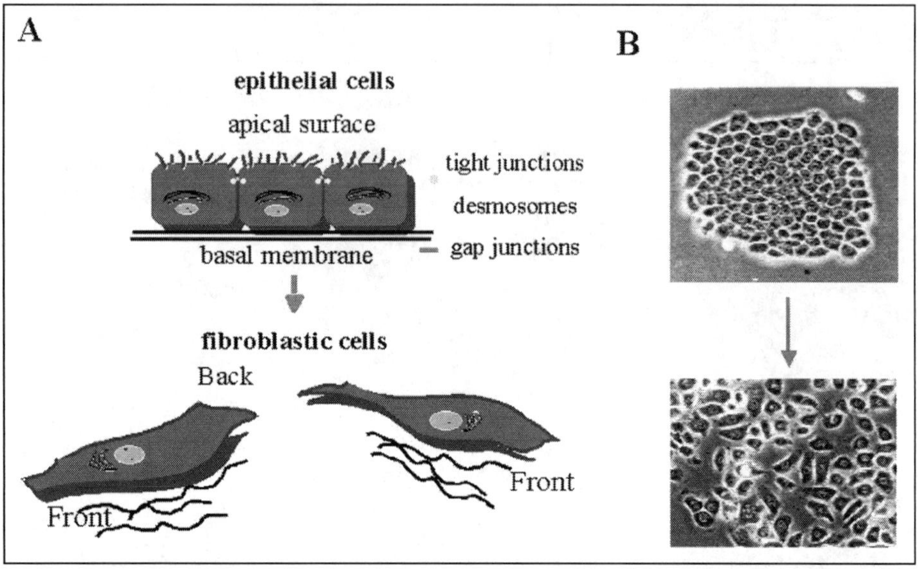

Figure 1. A) Schematic representation of EMT. Epithelial cells laying on a basement membrane that isolates the epithelial tissue from the neighbouring connective tisssue, undergo an EMT event. The intercellular junctions disappear, the morphology of the cells is transformed, the apicobasolateral polarity is replaced by a front end-back end polarity that reflects the direction of cell movement. Cells begin to crawl on extracellular matrix molecules deposited around the cell surface. B) The NBT-II cell line. Under standard conditions of culture, the rat bladder carcinoma cell line NBT-II assumes an epithelial morphology. If growth factors binding to RTK are added into the culture medium, NBT-II cells are transformed into elongated, motile fibroblast-like cells.

majority of growth factors endowed with a cell scattering activity bind to receptors with ligand-inducible intrinsic kinase activity (RTK). Activation of RTK by their cognate ligands generates cascades of cytoplasmic events initiated by the autophosphorylation of specific tyrosines on the activated receptors. With respect to growth factor-induced epithelial cell scattering, a major question concerns the specific transduction pathway involved in EMT, as opposed

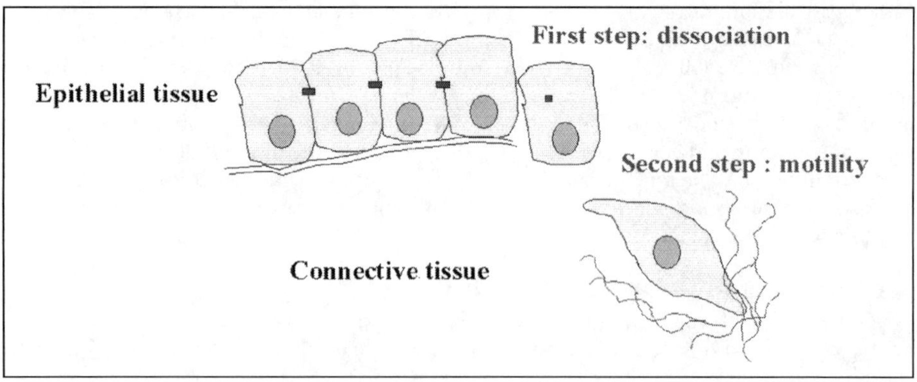

Figure 2. EMT is constituted of cell dissociation followed by motility of individualized cells. The dissociated epithelial cells have to break the basement membrane in order to be able to migrate freely in the connective tissue.

to other functions attributed to tyrosine kinase receptors (growth, survival, differentiation...). In several cases, it has been clearly demonstrated that the biological function depends on specific motifs on the activated receptor. For instance, only one of five C-terminal tyrosine residues of neu/erbB2 is responsible for transmitting the mitogenic signal.[5] Ligand-dependent activation of RTK with morphogenetic activity has also been studied : For example, the C-terminal domain of let-23, an EGF receptor homolog of *C. elegans*, can be subdivided into elements with different cell-type specific functions.[6] However, defining the specific function of domains in the cytoplasmic portion of growth factor-receptors is not always possible. The c-met receptor, for instance, comprises a short C-terminal sequence that serves as a docking site for several cytoplasmic effectors. Deciphering which phosphorylated tyrosine of the docking site is selectively implicated in cell scattering or tubulogenesis has led to conflicting results.[7-10] Another approach commonly used to identify signaling molecules involved in various cell functions consists of overexpressing wild-type or mutant (transdominant-negative or constitutively activated) forms of signaling molecules and analyzing the biological consequences of the overexpression. Using this strategy, we demonstrated that Src tyrosine kinase is a positive regulator of growth-factor induced cell scattering, as opposed to cell proliferation.[11]

Src Involvement in EMT

Src belongs to a family of cytoplasmic tyrosine kinases. These enzymes have a pivotal role in the regulation of a variety of biological functions which are associated with changes in cell morphology, including malignant transformation,[12] epithelial plasticity,[13] and modulation of intercellular adhesion.[14] In addition, the Src family is required during mitogenesis induced by EGF, platelet-derived growth factor and colony-stimulating factor-1.[15,16]

The Src family is composed of nine members, all of which contain one SH2 and one SH3 domain. In this family, Src, Fyn and Yes are ubiquitously expressed while the other members have a more restricted pattern of expression. The specificity of action of each individual member is not clear, as knock-out experiments have pointed to the possible redundancy of function among the Src family during mouse.[17] We demonstrated that Src family kinases are required during in vitro cell scattering, probably by phosphorylating substrates important for signaling.[18] The kinase activity of Src family kinases is also necessary during the gastrulation of Xenopus embryo, and particularly during the phase of convergence-extension movements of the ectodermal sheet, a process that requires epithelial cell motility.[19] Fyn and Src share redundant functions during cell scattering, both in vitro and in vivo, during Xenopus gastrulation.[18,19] Since the SH2 and SH3 domains, together with the kinase domain, are required for cell scattering, these data suggest that the SH2 and SH3 domains of Src and Fyn, which are highly homologous, bind to the same target proteins.

The mechanisms whereby Src kinases exert their functions are still unclear (Fig. 3). Three hypotheses have emerged: Src may phosphorylate directly specific substrates which are mainly cytoskeletal-based components, and molecules localized in cell-cell and cell-substrate adhesion sites,[20-23] the tyrosine phosphorylation of which could in turn alter the cellular architecture or intervene in the cellular machinery needed to switch epithelial cells to a fibroblastic phenotype. Src can also induce the transcription of specific genes, such as Myc.[24] Finally, Src activity could interact with other signaling pathways: For example, activated Src can bind Shc, an early element of the Ras cascade, leading to the activation of this pathway.[25,26] The three possible modes of action of Src are not mutually exclusive and might depend on the cellular context. In most cases, the kinase activity of Src is required in order to exert its functions. However, the possibility still exists that Src functions may rely only on its SH2- and SH3- based protein-protein interactions. For example, over-expression of the noncatalytic domain of Src is sufficient to induce assembly of integrin-based focal contacts, which are necessary elements of the cell motility machinery.[27]

During NBT-II epithelial cell scattering, Src does not act upstream of the Ras/MAPK pathway nor induce transcriptional activity but rather phosphorylates cytoskeleton-linked proteins.[18] However, recent data obtained in *Drosophila* suggested that Src may act upstream of Jun Kinase

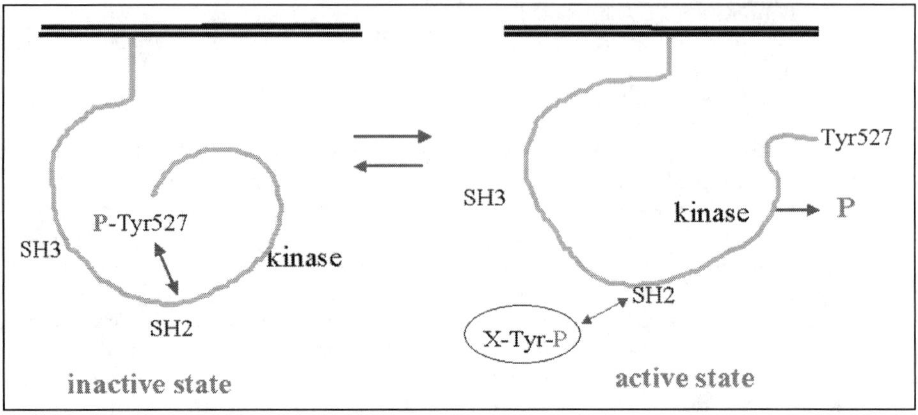

Figure 3. Conformational changes of Src kinase. The cytoplasmic Src kinase is anchored in the plasma membrane through its N-terminal myristylated domain. When the regulatory C-terminal tyrosine residue (Y527 in the chicken sequence) is phosphorylated, it interacts intramolecularly with the SH2 domain of Src, blocking the kinase domain in a close conformation and thereby preventing it from phosphorylating specific substrates. The dephosphorylation of the C-terminal tyrosine or a strong interation of Src SH2 domain with a tyrosine-phosphorylated protein disrupts the intramolecular bond. Src is then in an open conformation compatible with its enzymatic activity.

and Djun in epidermal closure, a process that implies epithelial remodeling.[28] The apparent discrepancy between this report and our results may be due to the fact that epithelial cell shape changes during epidermal closure are not comparable to EMT processes. Alternatively, since Jun Kinase could only partially rescue the phenotype caused by Src loss-of-function mutations, it may indicate that activation of Jun kinase is an accessory pathway that could not be identified in our experimental system.

Src is localized in discrete sites on the plasma membrane in epithelial cells. It is enriched in E-cadherin-associated adherens junctions and in integrin-based cell-substrate contacts. This subcellular localization is thought to reflect the major sites of its activity during cell scattering. In that respect, it is noteworthy that elevated Src activity disrupts E-cadherin-associated intercellular junctions. Moreover, this effect is dependent on integrin-regulated focal adhesion kinase (FAK) phosphorylation on Src-specific sites.[27] Thus, one major role of Src could be to drive adhesion changes associated with loss of intercellular contacts and formation of integrin-based cell-matrix contacts, that are thought to play a major role in cell scattering.

The Ras Pathway

Beside Src kinases, we demonstrated a role for Ras signaling pathway in EGF-induced epithelial cell scattering, corroborating the implication of Ras pathway in Scatter Factor-induced signaling.[29-31] Accordingly, the SH2-SH3 domain-containing adaptor protein Grb2 that connects activated RTK to the Ras pathway participates iÇ·HGF-induced cell scattering.[32] The small GTP-binding protein Ras has multiple effector molecules, each of which defines a pathway with specific functions (Figs. 4, 5). Ras is known to participate in several signaling pathways leading to activation of the phosphoinositide-3-kinase (PI3K),[33-35] RalGDS and Raf/MAPK phosphorylation cascades.[36] In addition, Ras has been shown to activate the small GTP-binding protein Rac.[37]

The Raf-MEK-MAPK Pathway

Among the different Ras effectors, one the best characterized is the serine/threonine kinase Raf that initiates a series of phosphorylations resulting in the activation of MAP Kinase (MAPK). The regulation by MAPK has been implicated in a wide variety of biological processes,

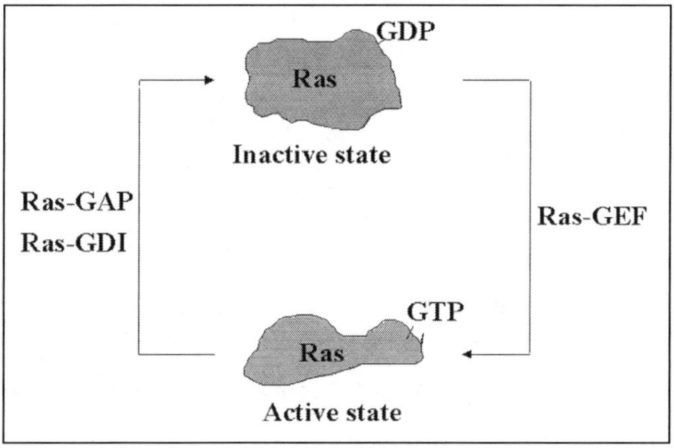

Figure 4. The cycle of Ras activity. Ras belongs to a family of monomeric G proteins able to bind GDP and GTP nucleotides and endowed with an intrinsic GTPase activity. When Ras binds to GDP, it is in an inactif state while binding to GTP correlates with an active state. Ras-GEF (guanine exchange factors) activate Ras while Ras-GAP (GTPase activating protein) and Ras-GDI (GDP dissociation inhibitor) inactivate Ras.

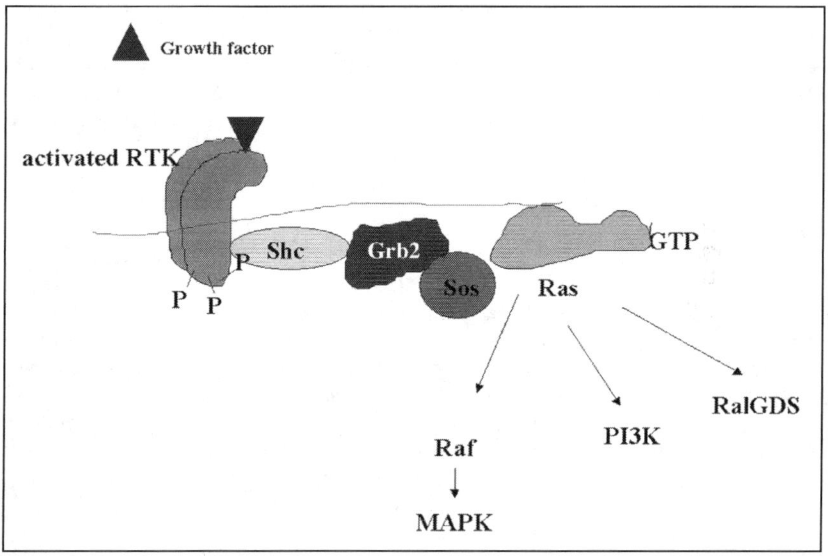

Figure 5. The Ras effectors. Following activation of RTK, Ras is activated by a cascade of molecular events. The adaptor molecule Shc binds to a phosphorylated tyrosine of RTK cytoplasmic tail. Consequently, the Grb2-Sos complex that binds constitutively to Shc is brought into the vicinity of Ras, anchored to the plasma membrane. Sos, a Ras-GEF, can then activate Ras, which in turn can activate a series of effectors.

including cell proliferation, integrin-mediated cell adhesion, secretion, neuronal cell differentiation, oocyte maturation, and activation of B-and T-cells. A role of the MAPK pathway has also been demonstrated in HGF/SF-induced scattering[38-40] and keratinocyte cell dispersion depends, at least in part, upon the sustained activation of MAPK.[41] Interestingly, a prolonged nuclear retention of activated MAPK is necessary for HGF-induced cell motility of MDCK cells, while EGF, which elicits only a brief nuclear accumulation of MAPK fails to induce cell motility.[42] In EGF-stimulated NBT-II cells, overexpression of active MEK1, which stimulates the activity of MAPK, induces cell dissociation but not cell motility,[43] suggesting that the level of MAPK activity may govern its role during cell scattering. A EGF-induced brief nuclear accumulation of the active enzyme would induce cell dissociation while its HGF-induced sustained nuclear retention would elicit cell motility. Activation of MAPK may well serve to induce the transcription of specific genes. Accordingly, the Jun/Fos complex of transcription factors, known to be activated by the Ras/MAPK pathway, is able to rescue the blockade of EGF-induced cell scattering induced by a transdominant-negative mutant of Ras.[18] MAPK may also act in cell scattering by activating transcription factors other than the Jun/fos complex, such as Slug, a transcription factor involved in cell scattering events, that has been demonstrated to be downstream of Ras.[44] Alternatively, MAPK could phosphorylate cytoplasmic substrates that remain yet to be identified. P38, another member of the MAPK family, has also been implicated in cell motility stimulated by platelet-derived growth factor and it is activated via a Ras-dependent pathway.[45]

The PI3K Pathway

Apart from MAPK, Ras is also likely to activate other effector molecules involved in EMT. The phosphatidylinositol 3-kinase (PI3K) may well serve such a role, as its function in epithelial cell scattering has been demonstrated in several examples of substrate- or growth factor-induced cell migration.[39,46,47] PI3K mediates the Ras control of the cytoskeleton[35] and is responsible for actin rearrangement induced by activated FAK and by Rac1 and Cdc42.[48] Furthermore, the increased motility observed in response to HGF treatment of the human breast epithelial cell line 184BJ is blocked by the PI3K inhibitor wortmannin[49] and carcinoma invasion promoted by $\alpha6\beta4$ integrin involves PI3K activity.[50] However, as other studies failed to confirm the requirement of PI3K in cell motility, its function may depend on the experimental system.[51] Nevertheless, when needed, PI3K is likely to be involved in cell motility by activating the small GTPase Rac of the Rho family,[52-54] whose role in cell migration and invasiveness is well established.[55-58]

The Rho Family

As far as EMT is concerned, Rac is required downstream of HGF/SF and Ras since a dominant-negative Rac1 inhibits SF- or Ras-induced MDCK cell scattering,[31,39] although activated Rac1 alone does not induce this response.[31,59] Rac may be necessary in the initiation of EMT because it stimulates the p21(cdc42/rac) activated kinase Pak1 which could in turn induce the phosphorylation of myosin light chain, thus linking the Rac-induced pathway to proteins affecting directly cell movement.[60,61] In contrast to its positive role in cell motility, Rac has also been shown unambiguously to strengthen cell-cell adhesion, thereby preventing tumour cell invasiveness.[59,62] Moreover, activities of both Rac and Rho, another small GTPase of the Rho family, are required for the formation of adherens junctions.[62-64] The apparent discrepancy between the two antagonistic roles attributed to Rac may be solved when considering that the type of response following Rac activation depends on the cell substrate. On substrates permissive for locomotion and under conditions preventing the formation of adherens junctions, expression of active Rac promotes a motile behavior whereas on substrates impeding cell motility Rac-dependent cell-cell adhesion is favoured.[53] Like Rac, Rho has complex functions in cell scattering, probably related to its attributed role in the assembly of focal contacts and actin stress fibers in fibroblastic cells.[65] Rho plays a positive role in Colony-Stimulating Factor-1 induced macrophage translocation[57] and in the migration and metastatic properties of human hepatocellular carcinomas.[66] The mechanism whereby Rho regulates cytokinesis and cell motility is partly accounted by its ability to stimulate the phosphorylation of myosin light chain[67] and of adducin, an actin-binding protein.[68] In contrast to its positive role in cell motility, Rho activity may antagonize epithelial cell scattering signals,[31] by reinforcing cell-cell adhesion sites.

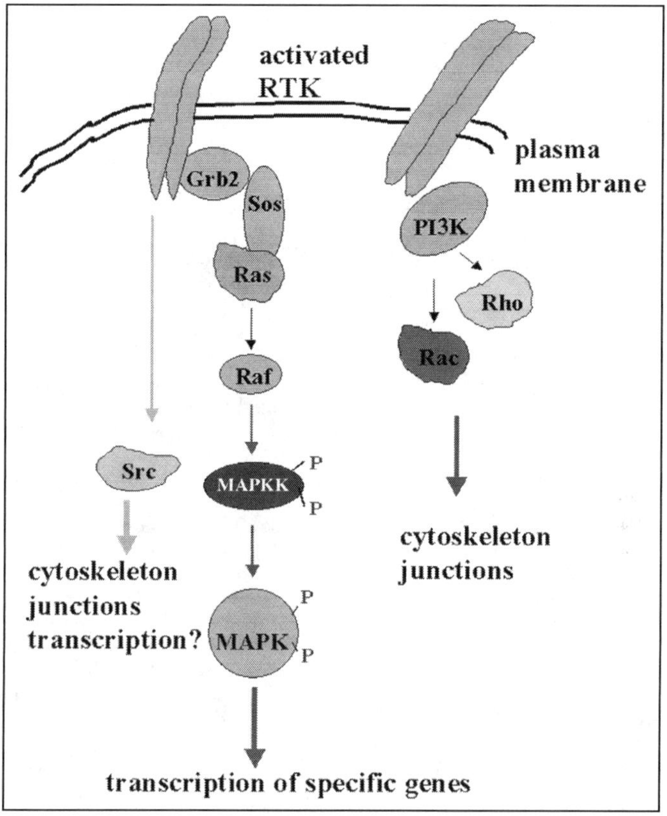

Figure 6. Schematic view of the signaling pathways involved in EMT. Growth factor binding to RTK elicits a series of intracellular signals. The Src pathway is mainly involved in EMT due to the ability of Src kinase to directly phosphorylate cellular components of the epithelial architecture. The Ras pathway implicates several Ras effectors, among which MAPK and Rac.

Conclusion

EMTs are complex processes involving simultaneous changes in the cell biology. The cytoskeleton is altered, leading to changes of cell morphology, the intercellular epithelial junctions disappear and cells start moving. These coordinated changes are controlled by signaling molecules that appear to have specific functions with respect to the several aspects of EMT changes (Fig. 6). During EMT, the activated signaling molecules are assembled within a signaling network whose role is to orchestrate the changes at the cellular level.

Although EMTs are induced by a wide variety of molecules (growth factors, extracellular matrix components) that stimulate signaling pathways shared by all types of eukaryotic cells, the epithelial phenotype is maintained under most circonstances and EMT appears as an occasional, tightly regulated process. Up to now, we do not understand the molecular control of EMT events but answering the question will be a challenging issue in this area of research for the upcoming years.

References

1. Sun D, Vanderburg CR, Odierna GS et al. TGFbeta3 promotes transformation of chicken palate medial edge epithelium to mesenchyme in vitro. Development 1998; 125:95-105.
2. Viebahn C. Epithelio-mesenchymal trasnformation during formation of the mesoderm in the mammalian embryo. Acta Anatomica 1995; 154:79-97.
3. Hay ED, Zuk A. Transformations between epithelium and mesenchyme: Normal, pathological, and experimentally induced. Am J Kidney Dis 1995; 26:678-690.
4. Kim K, Lu Z, Hay ED. Direct evidence for a role of beta-catenin/LEF-1 signaling pathway in induction of EMT. Cell Biol Int 2002; 26:463-476.
5. Ben-Levy R, paterson HF, Marshall CJ et al. A single autophosphorylation site confers oncogenicity to the Neu/ErbB-2 receptor and enables coupling to the MAP kinase pathway. EMBO J 1994; 13:3302-3311.
6. Aroian RV, Lesa GM, Sternberg PW. Mutations in the Caenorhabditis elegans let-23 EGFR-like gene define elements important for cell-type specificity and function. EMBO J 1994; 13:360-366.
7. Ponzetto C, Bardelli A, Zhen Z et al. A multifunctional docking site mediates signaling and transformation by the hepatocyte growth factor/scatter factor receptor family. Cell 1994; 77:261-271.
8. Tulasne D, Paumelle R, Weidner KM et al. The multisubstrate docking site of the MET receptor is dispensable for MET-mediated RAS signaling and cell scattering. Mol Biol Cell 1999; 10:551-565.
9. Weidner KM, Sachs M, Birchmeier W. Mutation of juxtamembrane tyrosine residue 1001 suppresses loss of function mutations of the met receptor in epithelial cells. Proc Natl Acad Sci USA 1995; 92:2597-2601.
10. Zhu H, Naujokas MA, Fixman ED et al. Tyrosine 1356 in the carboxy-terminal tail of the HGF/SF receptor is essential for the transduction of signals for cell motility and morphogenesis. J Biol Chem 1994; 269:29943-29948.
11. Rodier J-M, Vallés AM, Denoyelle M et al. $pp60^{c\text{-}src}$ is a positive regulator of growth factor-induced cell scattering in a rat bladder carcinoma cell line. J Cell Biol 1995; 131:761-773.
12. Parsons JT, Weber MJ. Genetics of src: Structure and functional organization of a tyrosine protein kinase. Current Topics in Microbiology and Immunology 1989; 147:79-127.
13. Warren SL, Nelson WJ. Nonmitogenic morphoregulatory action of $pp60^{v\text{-}src}$ on multicellular epithelial structures. Mol Cell Biol 1987; 7:1326-1337.
14. Volberg T, Geiger B, Dror R et al. Modulation of intercellular adherens-type junctions and tyrosine phosphorylation of their components in RSV-transformed cultured chick lens cells. Cell Regulation 1991; 2:105-120.
15. Luttrell DK, Luttrell LM, Parsons SJ. Augmented mitogenic responsiveness to epidermal growth factor in murine fibroblasts that overexpress $pp60^{c\text{-}src}$. Mol Cell Biol 1988; 8:497-501.
16. Roche S, Koegl M, Barone MV, Roussel MF et al. DNA synthesis induced by some but not all growth factors requires Src family protein tyrosine kinases. Mol Cell Biol 1995; 15.
17. Thomas SM, Soriano P, Imamoto A. Specific and redundant roles of Src and Fyn in organizing the cytoskeleton. Nature 1995; 376:267-271.
18. Boyer B, Roche S, Denoyelle M et al. Src and Ras are involved in separate pathways in epithelial cell scattering. EMBO J 1997; 196:5904-5913.
19. Denoyelle M, Vallés AM, Lentz D et al. Src tyrosine kinases regulate gastrulation movements in Xenopus embryo. Differentiation 2001; 69:38-48.
20. Clark EA, Brugge JS. Redistribution of activated $pp60^{c\text{-}src}$ to integrin-dependent cytoskeletal complexes in thrombin-stimulated platelets. Mol Cell Biol 1993; 13:1863-1871.
21. Kellie S, Horvath AR, Elmore MA. Cytoskeletal targets for oncogenic tyrosine kinases. J Cell Sci 1991; 99:207-211.
22. Okamura H, Resh MD. p80/85 cortactin associates with the Src SH2 domain and colocalizes with v-Src in transformed cells. J Biol Chem 1995; 270:26613-26618.
23. Wu H, Reynolds AB, Kanner SB et al. Identification of a novel cytoskeleton-associated $pp60^{src}$ substrate. Mol Cell Biol 1991; 11:5113-5124.
24. Barone MV, Courtneidge SA. Myc but not Fos rescue of PDGF signaling block caused by kinase-inactive Src. Nature 1995; 378:509-512.
25. McGlade J, Cheng A, Pelicci G et al. Shc proteins are phosphorylated and regulated by the v-Src and v-Fps protein-tyrosine kinases. Proc Natl Acad Sci USA 1992; 89:8869-8873.
26. Nagai K, Takata M, Yamamura H et al. Tyrosine phosphorylation of Shc is mediated through Lyn and Syk in B cell receptor signaling. J Biol Chem 1995; 270:6824-6829.
27. Avizienyte E, Wyke AW, Jones RJ et al. Src-induced de-regulation of E-cadherin in colon cancer cells requires integrin signaling. Nature Cell Biol 2002.
28. Tateno M, Nishida Y, Adachi-Yamada T. Regulation of JNK by Src during Drosophila development. Science 2000; 287:324-327.

29. Graziani A, Gramaglia D, dalla Zonca P et al. Hepatocyte growth factor/scatter factor stimulates the ras-guanine nucleotide exchanger. J Biol Chem 1993; 268:9165-9168.
30. Pelicci G, Giordano S, Zhen Z et al. The motogenic and mitogenic responses to HGF are amplified by the Shc adaptor protein. Oncogene 1995; 10:1631-1638.
31. Ridley AJ, Comoglio PM, Hall A. Regulation of scatter factor/hepatocyte growth factor responses by Ras, Rac, and Rho in MDCK cells. Mol Cell Biol 1995; 15:1110-1122.
32. Gay B, Suarez S, Weber C et al. Effect of potent and selective inhibitors of the Grb2 SH2 domain on cell motility. J Biol Chem 1999; 274:23311-23315.
33. Kotani K, Yonezawa K, Hara K et al. Involvement of phosphoinositide 3-kinase in insulin- or IGF-1-induced membrane ruffling. EMBO J 1994; 13:2313-21.
34. Rodriguez-Viciana P, Warne PH, Dhand R et al. Phosphatidylinositol-3-OH kinase as a direct target of Ras. Nature 1994; 370:527-32.
35. Rodriguez-Viciana P, Warne PH, Khwaja A et al. Role of phosphoinositide 3-OH kinase in cell transformation and control of the actin cytoskeleton by Ras. Cell 1997; 89:457-67.
36. Trelsman R. Regulation of transcription by MAP kinase cascade. Curr Opin Cell Biol 1996; 8:205-215.
37. Van Aelst L, D'Souza-Schorey C. Rho GTPases and signaling networks. Genes Dev 1997; 11:2295-322.
38. Herrera R. Modulation of hepatocyte growth factor-induced scattering of HT29 colon carcinoma cells. J Cell Sci 1998; 111:1039-1049.
39. Potempka S, Ridley AJ. Activation of both MAPK and phosphatidylinositide 3-kinase by Ras is required for hepatocyte growth factor/scatter factor-induced adherens junction disassembly. Mol Biol Cell 1998; 9:2185-2195.
40. Tanimura S, Chatani Y, Hoshino R et al. Activation of the 41/43kDa mitogen-activated protein kinase signaling is required for hepatocyte growth factor-induced cell scattering. Oncogene 1998; 17:57-65.
41. McCawley LJ, Li S, Wattenberg EV et al. Sustained activation of the mitogen-activated protein kinase pathway. A mechanism underlying receptor tyrosine kinase specificity for matrix metalloproteinase-9 induction and cell migration. J Biol Chem 1999; 274:4347-53.
42. Tanimura S, Nomura K, Ozaki K et al. Prolonged nuclear retention of activated Extracellular signal-Regulated Kinase 1/2 is required for Hepatocyte Growth Factor-induced cell motility. J Biol Chem 2002; 277:28256-28264.
43. Edme N, Downward J, Thiery JP et al. Ras induces NBT-II epithelial cell scattering through the coordinate activities of Rac and MAPK pathways. J Cell Sci 2002; 115:2591-2601.
44. Savagner P, Yamada KM, Thiery JP. The zinc-finger protein slug causes desmosome dissociation, an initial and necessary step for growth factor-induced epithelial-mesenchymal transition. J Cell Biol 1997; 137:1403-19.
45. Matsumoto T, Yokote K, Tamura K et al. Platelet-derived growth factor activates p38 mitogen-activated protein kinase through a Ras-dependent pathway that is important for actin reorganization and cell migration. J Biol Chem 1999; 274:13954-13960.
46. Pilkington MF, Sims SM, Dixon SJ Wortmannin inhibits spreading and chemotaxis of rat osteoclasts in vitro. J Bone Miner Res 1998; 13:688-694.
47. Reiske HR, Kao SC, Cary LA et al. Requirement of phosphatidylinositol 3-kinase in focal adhesion kinase-promoted cell migration. J Biol Chem 1999; 274:12361-12366.
48. Keely PJ, Westwick JK, Whitehead IP et al. Cdc42 and Rac1 induce integrin-mediated cell motility and invasiveness through PI(3)K. Nature 1997; 390:632-6.
49. Day RM, Cioce V, Breckenridge D et al. Differential signaling by alternative HGF isoforms through c-Met: Activation of both MAP kinase and PI 3-kinase pathways is insufficient for mitogenesis. Oncogene 1999; 18:3399-406.
50. Shaw LM, Rabinovitz I, Wang HH et al. Activation of phosphoinositide 3-OH kinase by the alpha6beta4 integrin promotes carcinoma invasion. Cell 1997; 91:949-60.
51. Li E, Stupack D, Klemke R et al. Adenovirus endocytosis via alpha(v) integrins requires phosphoinositide-3-OH kinase. J Virol 1998; 72:2055-2061.
52. Hooshmand-Rad R, Claesson-Welsh L, Wennstrom S et al. Involvement of phosphatidylinositide 3'-kinase and Rac in platelet-derived growth factor-induced actin reorganization and chemotaxis. Exp Cell Res 1997; 234:434-441.
53. Sander EE, van Delft S, ten Klooster JP et al. Matrix dependent Tiam1/Rac signaling in epithelial cells promotes either cell-cell adhesion or cell migration and is regulated by phosphatidylinositol 3-kinase. J Cell Biol 1998; 143:1385-1398.
54. Stephens L, Hawkins PT, Eguinoa A et al. A heterotrimeric GTPase-regulated isoform of PI3K and the regulation of its potential effectors. Philosophical Transactions of the Royal Society of London - Series B: Biological Sciences 1996; 351:211-215.

55. Arber S, Barbayannis FA, Hanser H et al. Regulation of actin dynamics through phosphorylation of cofilin by LIM-kinase. Nature 1998; 393:805-809.
56. Gimond C, van der Flier A, van Delft S et al. Induction of cell scattering by expression of β1 integrins in β1-deficient epithelial cells requires activation of members of the Rho family of GTPases and downregulation of cadherin and catenin function. J Cell Biol 1999; 147:1325-1340.
57. Jones GE, Allen WE, Ridley AJ. The Rho GTPases in macrophage motility and chemotaxis. Cell Adhes Commun 1998; 6:237-245.
58. Keely PJ, Westwick JK, Whitehead IP et al. Cdc42 and Rac1 induce integrin-mediated cell motility and invasiveness through PI(3)K. Nature 1997; 390:632-636.
59. Hordijk PL, ten Klooster JP, van der Kammer RA et al. Inhibition of invasion of epithelial cells by Tiam1-Rac signaling. Science 1997; 21:1464-1466.
60. Knaus UG, Bokoch GM. The p21Rac/Cdc42-activated kinases (PAKs). International J Biochem Cell Biol 1998; 30:857-862.
61. Sells M, Boyd J, Chernoff J. p21-activated kinase 1 (Pak1) regulates cell motility in mammalian fibroblasts. J Cell Biol 1999; 145:837-849.
62. Takaishi K, Sasaki T, Kotani H et al. Regulation of cell-cell adhesion by rac and rho small G proteins in MDCK cells. J Cell Biol 1997; 139:1047-1059.
63. Braga VM, Del Maschio A, Machesky L et al. Regulation of cadherin function by Rho and Rac: Modulation by junction maturation and cellular context. Mol Biol Cell 1999; 10:9-22.
64. Braga VM, Machesky LM, Hall A et al. The small GTPases Rho and Rac are required for the establishment of cadherin-dependent cell-cell contacts. J Cell Biol 1997; 137:1421-1431.
65. Ridley AJ, Hall A. The small GTP-binding protein rho regulates the assembly of focal adhesions and actin stress fibers in response to growth factors. Cell 1992; 70:389-399.
66. Genda T, Sakamoto M, Ichida T et al. Cell motility mediated by Rho and Rho-associated protein kinase plays a critical role in intrahepatic metastasis of human hepatocellular carcinoma. Hepatology 1999; 30:1027-1036.
67. Amano M, Ito M, Kimura K et al. Phosphorylation and activation of myosin by Rho-associated kinase (Rho-kinase). J Biol Chem 1996; 271:20246-20249.
68. Fukata M, Kuroda S, Nakagawa M et al. Cdc42 and Rac1 regulate the interaction of IQGAP1 with beta-catenin. J Biol Chem 1999; 274:26044-50.

CHAPTER 17

Regulation of E-Cadherin-Mediated Cell-Cell Adhesion by Rho Family GTPases

Masato Nakagawa, Nanae Izumi and Kozo Kaibuchi

Summary

Cell-cell adhesions are rearranged dynamically during tissue development and tumor metastasis. Recently, Rho-family GTPases, including RhoA, Rac1, and Cdc42, have emerged as key regulators of cadherin-mediated cell-cell adhesion. Following the identification and characterization of regulators and effectors of Rho GTPases, signal transduction pathways from cadherin to Rho GTPases and, in turn, from Rho GTPases to cadherin are beginning to be clarified.

Introduction

Although cell-cell adhesion may at first seem static, dynamic rearrangements of cell-cell adhesion underlie a diverse range of physiological processes, including induction of morphological changes during tissue development (such as the compaction of the eight-cell embryo), cell scattering, wound healing, and synaptogenesis (Fig. 1) see also Chapter 8, Arnoux et al.[1-4] Such rearrangements are also involved in tumor metastasis.

Cadherins comprise a major group of cell-cell adhesion molecules that mediate intercellular adhesion by Ca^{2+}-dependent homophilic interactions.[1-4] By forming homodimers, cadherins can cluster through a zipper-like mechanism, while their intracellular domain is anchored to the actin cytoskeleton through cytoplasmic proteins, α-catenin and β-catenin (Fig. 2a). The anchoring of cadherins to the actin cytoskeleton and the clustering of cadherins are necessary for the development of strong and rigid adhesions.[5] Indeed, in primary cultures of α-catenin-null keratinocytes, adhesive contacts are not sealed because of a lack of linkage of epithelial-cadherin (E-cadherin) to the actin cytoskeleton.[6] Loss of α-catenin expression also occurs in malignant carcinomas, which show a scattered-cell phenotype.[7,8]

Cell-cell adhesions appear to be rearranged constantly through remodeling of the cadherin-catenin complex.[9] Recently, Rho family GTPases—including RhoA, Rac1, and Cdc42—have emerged as important regulators of this process.

Cadherin-Mediated Cell-Cell Adhesions

More than 80 members of the cadherin superfamily have been identified in the human genome, including classic cadherins, Fat-like cadherins, and seven-pass transmembrane cadherins.[4,10] All members of the cadherin superfamily possess extracellular cadherin (EC) domains (or cadherin repeats) typically organized in tandem repeats, which mediate calcium-dependent homophilic interactions (Fig. 2b). Cadherin adhesive activity is regulated by extracellular calcium ions and cytoplasmic signaling events. Calcium binding to linker regions between EC domains allows cadherin molecules to arrange in a rigid and organized

Rise and Fall of Epithelial Phenotype: Concepts of Epithelial-Mesenchymal Transition, edited by Pierre Savagner.

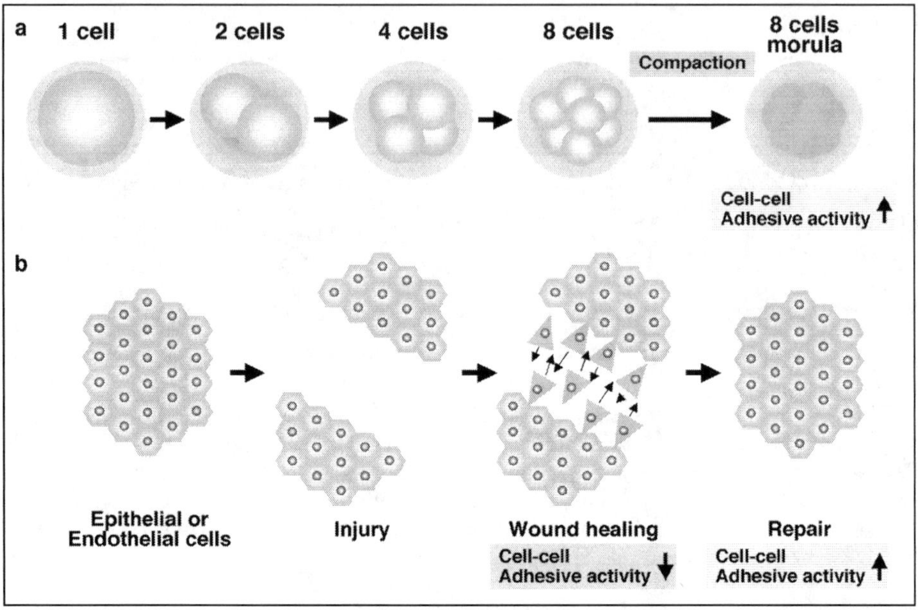

Figure 1. Dynamic rearrangements of cell-cell adhesion. a) Compaction occurs between the 8-cell and 16-cell stages. The embryo develops from a collection of loosely adherent cells into a tightly compacted form. This morphogenic process involves the rapid activation of E-cadherin at the cell surface without marked change of its expression. b) When a sheet of epithelial or endothelial cells is wounded, the edge cells are dissociated from neighboring cells and undergo migration to heal the wound. In this process, cell-cell adhesion is transiently perturbed in migrating areas, enabling dynamic migration.

structure that is resistant to proteolysis.[11] Under these conditions, cadherin can form a cis-dimer and a trans-dimer, developing cadherin clustering (a zipper-like structure), whereas removal of calcium ions by EGTA leads to disordered cadherin structure and dysfunction of cadherin-mediated cell-cell adhesion. The cytoplasmic domain of classic cadherin can be divided into two portions: (1) the carboxy-terminal "distal β-catenin-binding domain" (DβD); and (2) the juxtamembrane domain (JMD), which is the p120 catenin-binding (p120ctn-binding) site thought to regulate clustering of cadherins (Fig. 2b).[1-4]

Beta-catenin, a member of the armadillo-repeat containing proteins, was originally identified as a protein that interacts with DβD of cadherin and acts as a bridge connecting cadherin to the actin cytoskeleton through α-catenin. The linkage between cadherin and actin cytoskeleton through β-catenin and α-catenin contributes to stable cell-cell adhesion.[5]

Unlike β-catenin, p120ctn, another member of the armadillo-repeat containing proteins, does not bind α-catenin; it is thought to play a pivotal role in cadherin clustering by acting on JMD of the cadherin cytoplasmic domain.[12] Recent studies suggest that p120ctn acts as a switch, both inhibiting and promoting cadherin function, depending on circumstances. In Colo 205 carcinoma cells, E-cadherin function is defective despite an intact E-cadherin-catenin complex.[13] Expression of p120ctn mutant, lacking the amino-terminal domain, restores normal E-cadherin function in a dominant negative manner.[13] The data suggest that an aberrant constitutive signaling through p120ctn can severely inhibit E-cadherin-mediated cell-cell adhesion under some unusual circumstances. Alternatively, other reports suggest that p120ctn has positive effects on cadherin clustering under other circumstances.[12] Thus, p120ctn may have several biological effects on cadherin activity acting on JMD, Rho GTPases (described below), or other unknown factors. The various functions of p120ctn probably are tightly regulated by posttranslational modifications (such as phosphorylation) and its subcellular

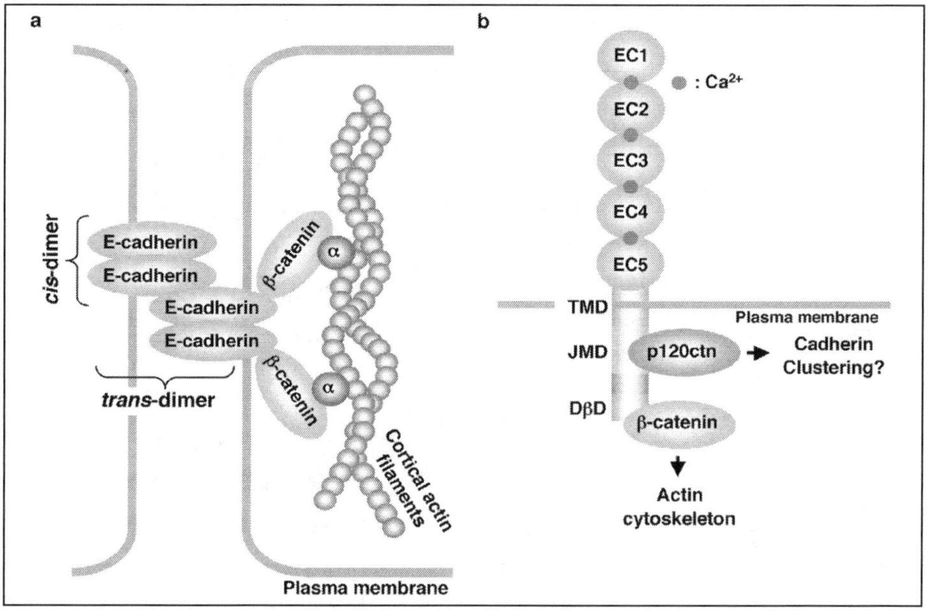

Figure 2. Cadherin-mediated cell-cell adhesion. a) Cadherin, a calcium-dependent adhesion molecule, is linked to bundles of actin filaments through β-catenin and α-catenin (α). Cadherins can form cis- and trans-dimers, thereby forming rigid adhesions. b) The E-cadherin molecule is composed of repeated EC domains (or cadherin domains) that mediate calcium-dependent homophilic interactions; a transmembrane domain (TMD); a juxtamembrane domain (JMD); and a distal β-catenin-binding domain (DβD).

localization. Because there are also a large number of splice variants, different isoforms may affect cadherin function by recruiting distinct binding partners to the cadherin complex.[14] Further studies, especially identification of p120ctn-interacting proteins, are necessary to elucidate how p120ctn regulates cadherin activity and the signaling pathways that are involved.

Rho-Family GTPases

Rho-family GTPases are involved in GDP/GTP-binding and GTPase activities. They cycle between a GTP-bound active state and a GDP-bound inactive state (Fig. 3). The nucleotide state of Rho GTPases is controlled by three classes of key regulators: (1) guanine nucleotide exchange factors (GEFs), which promote the exchange of GDP for GTP; (2) Rho GDP dissociation inhibitors (Rho GDIs), which interact with GDP-bound Rho GTPases and inhibit the exchange of GDP for GTP; and (3) GTPase-activating proteins (GAPs), which enhance the intrinsic GTPase activities of Rho GTPases. These regulators ensure that the activation and inactivation of Rho GTPases are tightly regulated spatio-temporally to generate specific and localized effects.[15,16] The modes of action of Rho-family GTPases have been elucidated by the identification and characterization of specific effectors of these GTPases.[15,17,18] Such effector molecules interact with Rho GTPases only in their GTP-bound state. We discuss below those effectors that function in the regulation of cadherin-mediated cell-cell adhesion.

Localization of Rho GTPases

Biochemical and immunofluorescence studies in swiss3T3 fibroblasts and NIE-115 neuronal cells indicate that Rho GTPases probably shuttle between the cytosol and specific membrane sites following extracellular stimulation.[19,20] Green fluorescent protein (GFP) and time-lapse microscopy allow us to monitor the dynamic relocalization of these molecules. Where, then, is the localization of Rho GTPases in epithelial cells? Several groups have shown that

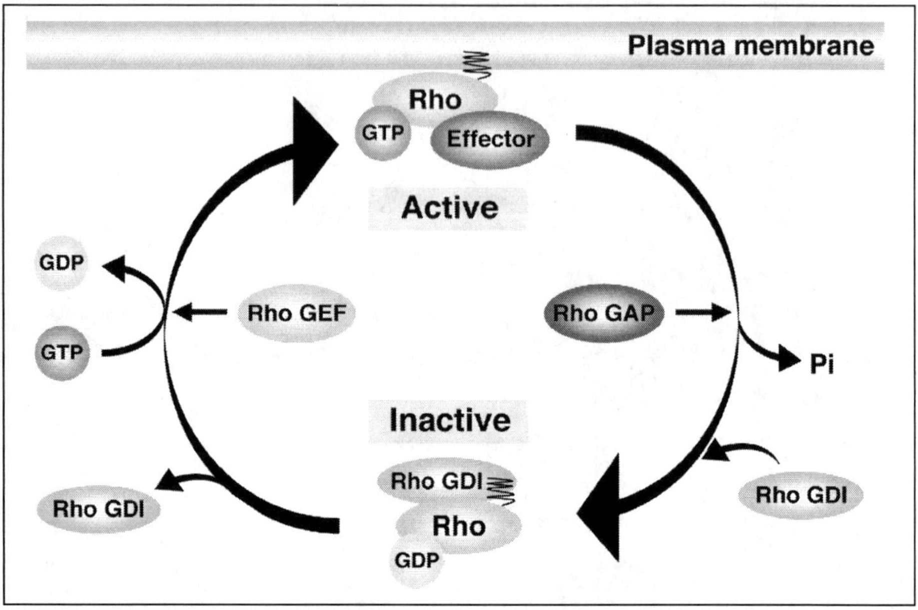

Figure 3. Regulation of Rho-family GTPases. In resting cells, Rho exists mostly in the GDP-bound form (GDP·Rho) and in complexes with Rho GDI in the cytosol. On stimulation with extracellular signals, Rho is likely to be dissociated from Rho GDI and targeted to specific membranes by its carboxy-terminal prenyl group. At the membrane, specific GEFs for Rho are activated and GDP·Rho is then converted to GTP·Rho. GTP·Rho interacts with its specific effectors and exerts its functions. GAPs enhance the GTPase activity of Rho and reconvert Rho to its inactive GDP-bound form. Rho GDI can then form a complex with GDP·Rho and extract it from the membrane back into the cytosol.

Rac1 and Cdc42 are localized at sites of cell-cell contact.[21-24] Rac1 seems to colocalize with E-cadherin in Mardin-Darby canine kidney II (MDCKII) epithelial cells at sites of cell-cell contact, and is translocated to the cytosol during disruption of E-cadherin-mediated cell-cell adhesion by Ca^{2+} chelation.[24] Furthermore, engagement of E-cadherin to the apical membrane by anti-E-cadherin antibodies causes the recruitment of GFP-Rac1. Colocalization of Rac1 with E-cadherin is also observed in E-cadherin-positive vesicles in human keratinocytes, although the nature of these vesicles is somewhat mysterious.[25] In addition to its localization at regions of cell-cell contact, Cdc42 is also found at the Golgi apparatus, where it is thought to regulate the establishment of epithelial cell polarity during secretory and endocytic transport.[26] Unlike Rac1 and Cdc42, RhoA is mostly localized in the cytosol.[21]

It is possible that some GEFs or GAPs are recruited to cadherin-based adhesion sites through PDZ domain-mediated protein-protein interactions. In fact, it has been reported that PDZ-GEF1 (a GEF for Rap1 GTPase, a member of the Ras-family GTPases) is recruited to β-catenin at sites of cell-cell contact through interaction with KIAA0705, which is a PDZ domain-containing scaffolding protein.[27] Beta-catenin has a PDZ-binding consensus sequence at its carboxy-terminal end (-D-T-D-L) and is able to interact with several PDZ domain-containing scaffolding proteins, such as membrane-associated guanylate kinase with inverted orientation (MAGI-1),[28] synaptic scaffolding molecule (S-SCAM),[29] KIAA0705,[27] and Lin7.[30] As several GEFs for Rho GTPases have a PDZ domain, such as Tiam1,[31] STEF[32] (GEFs for Rac1), PDZ-RhoGEF,[33] and leukemia-associated Rho guanine nucleotide exchange factor (LARG)[34,35] (GEFs for RhoA), this might be a mechanism by which Rho GTPases are activated at sites of cell-cell contact.

Rho GTPases Regulate Cadherin-Mediated Cell-Cell Adhesion

Several lines of evidence suggest that Rho-family GTPases are required for cadherin-mediated cell-cell adhesion. Braga et al first reported that microinjection of dominant-negative Rac1 ($Rac1^{N17}$) and C3 botulinum toxin, which inactivates RhoA, into keratinocytes inhibits the accumulation of cadherin at sites of cell-cell contact upon calcium-induced intercellular adhesion.[36] The effects of Rac1 and RhoA on the localization of cadherin depend on the cell type and the junctional maturity (that is, time elapsed after induction of adhesion).[37] Takaishi et al showed that overexpression of $Rac1^{V12}$ in MDCKII epithelial cells induces the accumulation of E-cadherin, β-catenin, and actin filaments at sites of cell-cell contact, whereas overexpression of $Rac1^{N17}$ reduces their accumulation.[21] Moreover, it was shown that Cdc42, as well as Rac1, are required for E-cadherin-mediated cell-cell adhesion in MDCKII cells.[22,38] Tiam1, one of the GEFs for Rac1, is localized at cell-cell contacts and inhibits hepatocyte growth factor-induced (HGF-induced) cell scattering in MDCKII cells, probably by increasing E-cadherin-mediated cell-cell adhesion.[39]

The above studies suggest that Rac1, RhoA, and Cdc42 regulate cadherin-mediated cell-cell adhesion. However, until recently, there was no direct evidence to determine whether Rho-family GTPases regulate cadherin-mediated cell-cell adhesion by acting on the cadherin-catenin complex or by acting on the actin cytoskeleton and other components. Using a cell dissociation assay, a quantitative assay for E-cadherin activity, Fukata et al showed that Rac1 and Cdc42 regulate E-cadherin activity through the cadherin-catenin complex and that RhoA affects E-cadherin-mediated adhesive activity presumably through the actin cytoskeleton or other components, but not the cadherin-catenin complex.[40]

The best-studied effector through which Rac1 and Cdc42 can regulate E-cadherin activity is IQGAP1,[40-42] whereas the RhoA effector that participates in cadherin-mediated adhesion remains unknown. IQGAP1 is also known as an actin cross-linking protein.[43,44] IQGAP1 is localized at sites of cell-cell contact, and it negatively regulates E-cadherin-mediated cell-cell adhesion by interacting with β-catenin, causing the dissociation of α-catenin from the cadherin-catenin complex.[41] Activated Rac1 and Cdc42 positively regulate cadherin-mediated cell-cell adhesion by inhibiting the interaction of IQGAP1 with β-catenin.[40] Based on these studies, we propose that E-cadherin exists in a dynamic equilibrium between the E-cadherin–β-catenin–α-catenin complex and the E-cadherin–β-catenin–IQGAP1 complex at sites of cell-cell contact (Fig. 4). The ratio between the two complexes could be a determinant of the strength of E-cadherin-mediated cell-cell adhesion. When the amounts of activated Rac1 and Cdc42 increase, Rac1 and Cdc42 directly bind IQGAP1, thereby inhibiting the interaction of IQGAP1 with β-catenin. Under these conditions, the ratio of E-cadherin–β-catenin–α-catenin complex to E-cadherin–β-catenin–IQGAP1 complex is high. This state confers strong adhesive activity. In contrast, when the amounts of inactivated Rac1 and Cdc42 increase, IQGAP1 is freed from Rac1 and Cdc42 and interacts with β-catenin to dissociate α-catenin from the cadherin-catenin complex. In this case, the ratio of E-cadherin–β-catenin–IQGAP1 complex to E-cadherin–β-catenin–α-catenin complex is high. This results in weak adhesion and cell-cell dissociation.

Our recent experiments showed that knockdown of IQGAP1 by RNA interference (RNAi) decreases the accumulation of E-cadherin, β-catenin, and actin filaments at the sites of cell-cell contact and decreases the Rac1 activity in MDCKII cells. Moreover, RNAi of Rac1 decreased the accumulation of E-cadherin, β-catenin, and actin filaments at the cell-cell contacts.[45] It has been reported that IQGAP1 has an actin cross-linking activity,[43,44] GTP-bound Rac1/Cdc42 enhances this activity,[43,44] and IQGAP1 increases the levels of active, GTP-bound Cdc42.[46] Thus, IQGAP1 complexed with GTP-bound Rac1/Cdc42 positively regulates cadherin-mediated cell-cell adhesion through stabilization of actin filaments at the cell-cell contacts (Fig. 4).

However, it remains to be clarified in which physiological processes this Rac1/Cdc42/IQGAP1 system is involved. Recently we showed that this system functions in cell-cell

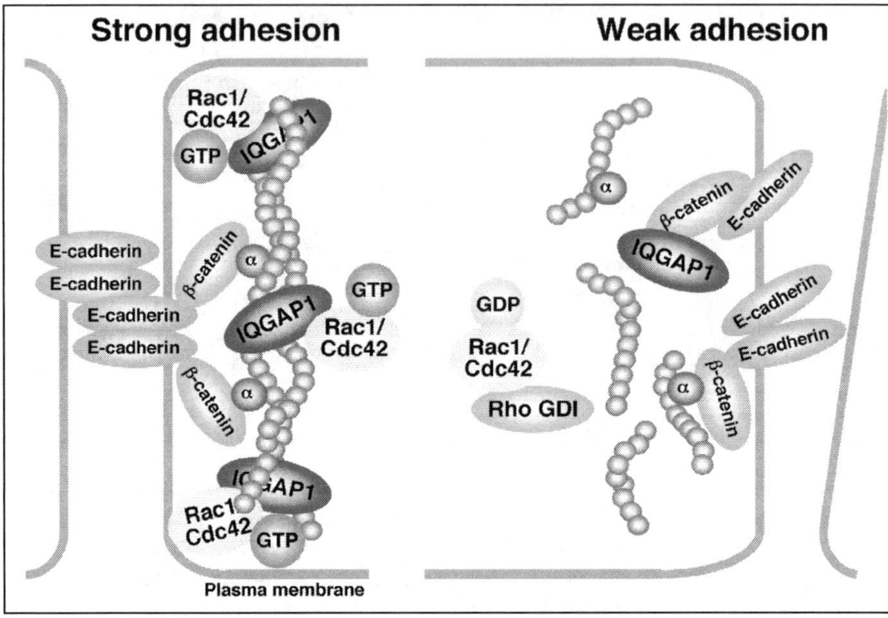

Figure 4. Role of Rac1/Cdc42 and IQGAP1 in the regulation of E-cadherin-mediated cell-cell adhesion. When Rac1/Cdc42 is in an active, GTP-bound form, IQGAP1 stabilizes the cortical actin filaments at the sites of cell-cell contact as an actin cross-linking protein. IQGAP1 maintains the amount of GTP·Rac1/Cdc42 at sites of cell-cell contact, leading to stable adhesion (left). By contrast, when Rac1/Cdc42 is inactivated, IQGAP1 is freed from Rac1 and interacts with β-catenin to dissociate α-catenin from the cadherin-catenin complex. This results in weak adhesion and cell-cell dissociation (right). Thus, IQGAP1 behaves as a both positive and negative regulator downstream of Rac1/Cdc42.

dissociation during HGF- or TPA- (12-O-tetradecanoylphorbol-13-acetate) induced cell scattering,[47] which is thought to be a model system for epithelial-mesenchymal transition and dispersal of cancer cells. Time-lapse analyses using GFP-tagged α-catenin showed that α-catenin disappears from cell-cell contacts before the cells dissociate during cell scattering. $Rac1^{V12}$, $Cdc42^{V12}$, and the dominant negative mutant of IQGAP1 (carboxy-terminal fragment of IQGAP1), which interacts with endogenous IQGAP1 and delocalizes IQGAP1 from sites of cell-cell contact, inhibit the disappearance of α-catenin from sites of cell-cell contact induced by HGF or TPA. Furthermore, on stimulation with HGF or TPA, the level of GTP-bound Rac1 and the proportion of Rac1 forming a complex with IQGAP1 decrease, and IQGAP is increasingly found in a complex with β-catenin. In support of this, α-catenin-deficient mouse teratocarcinoma F9 cells display a scattered phenotype under conditions in which parental or α-catenin-reexpressing cells form compact colonies,[48] indicating that loss of α-catenin results in loss of cell-cell adhesion and scattered phenotype. IQGAP1 dissociates α-catenin from the cadherin-catenin complex and dominant negative IQGAP1 inhibits disappearance of α-catenin from sites of cell-cell contact during cell scattering; these results indicate that the Rac1/Cdc42/IQGAP1 system is involved in cell-cell dissociation induced by HGF or TPA through α-catenin relocalization.

IQGAP1-knockout mice exhibit gastric hyperplasia and dysplasia,[49] suggesting that IQGAP1 plays an important role in reorganizing the gastric epithelium. However, IQGAP1-knockout mice do not display any developmental defects, which may be due to the functional redundancy with IQGAP2[50] and IQGAP3 (Fukata et al unpublished data). Sugimoto et al reported that IQGAP1 is upregulated by gene amplification in some diffuse types of gastric cancer.[51]

Matsumoto et al reported that IQGAP1 is highly expressed in patients with moderate and/or sever atopic dermatitis.[52] Furthermore, there seems to be a correlation between dysfunction of E-cadherin-mediated adhesion in gastric tumors, increased membrane-localization of IQGAP1, and decreased membrane-localization of α-catenin,[53] indicating that the Rac1/Cdc42/IQGAP1 system might also be involved in tumor progression. Presslauer et al found that autoantibodies in patients with autoimmune bullous skin diseases recognize IQGAP1 as an antigen.[54] IQGAP1 seems to be involved in several human diseases. Thus, IQGAP1 has an important physiological role in human cells through regulation of cadherin-mediated cell-cell adhesion.

Recently, another Ras family member, Rap1, has been implicated in cadherin-mediated adhesion. In Drosophila, analysis of Rap1 mutants during wing development shows adherence junction misplacement and the dispersal of clonal cells that would otherwise stay in a coherent group.[55] However, the effects of Rap1 on cadherin-mediated cell-cell adhesion seem specific for the differentiation process during wing development, as it is not essential for the appropriate apical localization of junctional proteins. Although it has been reported that Rap1 is implicated in integrin-mediated adhesion,[56] Rap1 may be also required for cadherin-mediated cell-cell adhesion in mammalian cells.

Cadherin-Mediated Cell-Cell Adhesion Affects Rho GTPase Activity

The establishment of cadherin-mediated cell-cell adhesions requires changes in the activities of Rho GTPases, but how do the activities of Rho GTPases change during reorganization of cadherin-mediated cell-cell adhesion? The levels of GTP-bound forms of Rho GTPases can be measured by affinity precipitation.[57] Like stimulation by growth factors, the engagement of integrins with fibronectin induces the activation of Rho-family GTPases.[58] Analogies with integrin-mediated cell-substratum adhesion lead us to speculate that E-cadherin might also function as a receptor that sends signals to Rho-family GTPases. Recent observations showed that E-cadherin-mediated cell-cell interactions result in rapid activation of Cdc42 in MCF-7 epithelial cells.[59] We have also shown that E-cadherin-mediated cell-cell contact induces the activation of Rac1 in MDCKII cells and that basal levels of active Rac1 are still maintained even when cells become confluent.[24] Kovacs et al also showed that Rac1 activation is cadherin-dependent.[60]

So how does cadherin-mediated cell-cell adhesion modulate the activity of the Rho-family GTPases? It has been reported that E-cadherin-mediated cell-cell adhesion stimulates phosphatidylinositol 3-kinase (PI3-kinase) activity in MDCKII cells.[61] Moreover, PI3-kinase was shown to interact with E-cadherin[61] and β-catenin.[62] Because PI3-kinase is thought to function upstream of Rac1,[60,63] these observations indicate the possible involvement of PI3-kinase in E-cadherin-dependent Rac1 activation. Indeed, wortmannin, an inhibitor of PI3-kinase, inhibits E-cadherin-induced Rac1 activation but does not affect the localization of Rac1 or E-cadherin.[24] The activation of Rac1 through PI3-kinase by E-cadherin-mediated cell-cell adhesion seems to require at least two steps: (1) Rac1 recruitment to sites of cell-cell contact; and (2) Rac1 activation by a certain GEF (Fig. 5).

Considering that Tiam1 is localized at sites of cell-cell contact and functions downstream of PI3-kinase, Tiam1 could be a candidate for a Rac1 regulator downstream of E-cadherin. In support of this, it was shown that CD44, which binds hyaluronic acid and mediates cell-cell or cell-substratum adhesion, directly interacts with Tiam1, and binding of hyaluronic acid to CD44 stimulates Tiam1-catalyzed Rac1 activation.[64] Therefore, it might be worthwhile examining whether Tiam1 is involved in E-cadherin-induced Rac1 activation.

The activation of Rac1 and Cdc42 can be inhibited by treatment of cells with a neutralizing antibody against E-cadherin. In contrast, RhoA activity is not affected by the induction of cell-cell adhesion,[24] and it is markedly reduced as cells become confluent.[65] Recently, it was reported that RhoA activity is inhibited by cadherin engagement via p190RhoGAP.[66] These observations are consistent with the proposal that Rac1/Cdc42 and RhoA exert mutually antagonistic effects at some point during contact initiation, maturation, and reorganization, as in

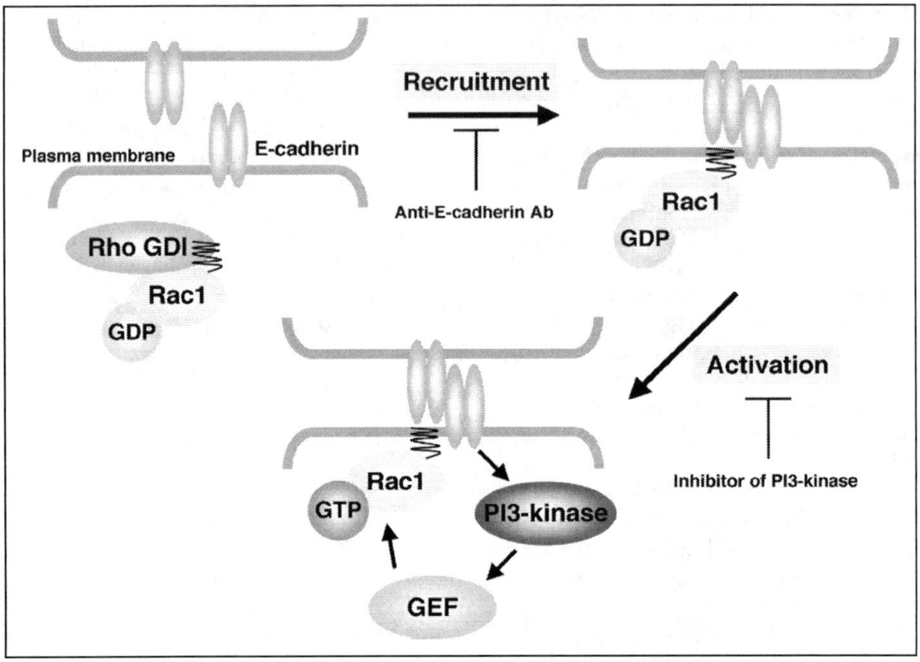

Figure 5. Mode of activation of Rac1 in the formation of E-cadherin-mediated cell-cell adhesion. Before establishment of E-cadherin-mediated cell-cell adhesion, GDP·Rac1 is sequestered in the cytosol by Rho GDI. When cadherin-mediated homophilic interactions occur, GDP·Rac1 is dissociated from Rho GDI by an unknown mechanism and is targeted to the plasma membrane. GDP·Rac1 is converted to GTP·Rac1 through the action of a GEF, such as Tiam1, downstream of PI3-kinase. Activated Rac1 then positively regulates E-cadherin-mediated cell-cell adhesion.

the case of formation of focal adhesion[67] and neurite extension.[68] It has also been shown that Tiam1 or Rac1^{V12} is able to downregulate RhoA activity and revert the mesenchymal phenotype of Ras^{V12}-transformed MDCKII cells to an epithelial phenotype.[69,70] Anjaruwee et al reported the redox-dependent downregulation of Rho by Rac via p190RhoGAP.[71] Thus, E-cadherin-mediated adhesion activates Rac1 and Cdc42 (outside-in signaling), and active Rac1 downregulates RhoA. As described above, Rac1^{V12} promotes the accumulation of E-cadherin at sites of cell-cell contact (inside-out signaling). Thus, this indicates that bidirectional signaling pathways must exist, that is, there is a positive feedback loop between E-cadherin and the Rho-family GTPases that contributes to the maintenance of junctions. Despite much progress, further studies are necessary to understand the spatio-temporal dynamics of Rho GTPase activation during reorganization of cell-cell adhesion. Whereas the existing pull-down assays enable us to measure only total amounts of GTP-bound Rho GTPases in whole cells, a fluorescent resonance energy transfer-based (FRET-based) method, such as FLAIR (fluorescence activation indicator for Rho proteins),[72] may be useful for analyzing the cellular distribution of active Rho GTPases in real time.

Conclusions and Perspectives

Numerous attempts have been made to clarify the mechanisms by which Rho-family GTPases regulate cadherin-mediated cell-cell adhesion. The discovery of specific effectors of Rho GTPases and the development of methods for detecting GTP-bound Rho GTPases contribute to our understanding of the bidirectional signaling pathways between Rho GTPases and cadherin. We know that formation of cadherin-mediated cell-cell adhesion activates Rac1 and Cdc42. In turn, activated Rac1 and Cdc42 stabilize cadherin-catenin complex acting on their effector, leading to rigid cell-cell adhesion. Although, RhoA seems to be inactivated by activation of Rac1, it is not clear how RhoA inactivation is involved in this process. In addition, it will be important to determine how cadherin-mediated cell-cell adhesion transduces signals to Rho-GTPases and how Rho GTPases cross talk each other. Recent reports have shown that Rho GTPases regulate microtubule dynamics see also Chapter 19, Gauthier-Rouviere et al.[73-77] Furthermore, p190RhoGEF, which is localized along microtubules, has been identified.[78] Others reported that cadherin-mediated cell-cell adhesion regulates microtubule dynamics[79] and cytoplasmic dynein may tether microtubules at the adherence junction by interacting with β-catenin.[80] Thus, Rho GTPases may regulate cadherin-based adhesion through reorganizing microtubules as well as the actin cytoskeleton.

Because most studies have been performed under cultured epithelial cell levels, it will be crucial to examine whether our knowledge is actually applicable to physiological processes in vivo. Indeed, several recent reports have addressed the process of compaction of the eight-cell embryo,[81] gastrulation,[82] and synaptogenesis.[83,84] It was also shown that catenins and N-cadherin, which mediates synaptogenesis, are localized in the surrounding region of the active zone, the so-called synaptic adherence junction.[10] Interestingly, still life (SIF), a Tiam1 homolog of Drosophila, is localized in this region.[85] Therefore, it raises the possibility that regulation of cadherin adhesion by the Rho family operates in the synaptogenesis and synaptic plasticity in neuronal cells as well as epithelial cells. Genetic loss-of-function mutants in Rho GTPases, their regulators, and effectors will shed light on this issue. Moreover, RNAi using siRNA in mammalian cells is a powerful tool for identifying the functions of Rho family on cadherin-mediated cell-cell adhesion.

Finally, it is also conceivable that the misregulation of cadherin-mediated cell-cell adhesion by Rho GTPases may be a cause of tumor metastasis, which very often leads to death. Therefore, it is critical to determine whether the described molecular mechanism is also involved in tumor metastasis. The Rac1/Cdc42/IQGAP1 system seems to be a good candidate for the above process. If so, modulation of the signaling pathway between cadherin and Rho GTPases may improve the treatment for tumor metastasis.

Acknowledgements

The authors thank M. Fukata, T. Watanabe, J. Noritake, and the members of our laboratory for preparing the manuscript.

The original work by the authors was supported by grants-in-aid for scientific research from the Ministry of Education, Science, and Culture of Japan, by grants from Novartis Foundation for the Promotion of Science and by grants from The Nitto Foundation.

References

1. Takeichi M. Morphogenetic roles of classic cadherins. Curr Opin Cell Biol 1995; 7(5):619-627.
2. Adams CL, Nelson WJ. Cytomechanics of cadherin-mediated cell-cell adhesion. Curr Opin Cell Biol 1998; 10(5):572-577.
3. Gumbiner BM. Regulation of cadherin adhesive activity. J Cell Biol 2000; 148(3):399-404.
4. Tepass U, Truong K, Godt D et al. Cadherins in embryonic and neural morphogenesis. Nat Rev Mol Cell Biol 2000; 1(2):91-100.
5. Tsukita S, Tsukita S, Nagafuchi A et al. Molecular linkage between cadherins and actin filaments in cell-cell adherens junctions. Curr Opin Cell Biol 1992; 4(5):834-839.
6. Vasioukhin V, Bauer C, Yin M et al. Directed actin polymerization is the driving force for epithelial cell-cell adhesion. Cell 2000; 100(2):209-219.
7. Ochiai A, Akimoto S, Shimoyama Y et al. Frequent loss of alpha catenin expression in scirrhous carcinomas with scattered cell growth. Jpn J Cancer Res 1994; 85(3):266-273.
8. Watabe M, Nagafuchi A, Tsukita S et al. Induction of polarized cell-cell association and retardation of growth by activation of the E-cadherin-catenin adhesion system in a dispersed carcinoma line. J Cell Biol 1994; 127(1):247-256.
9. Sako Y, Nagafuchi A, Tsukita S et al. Cytoplasmic regulation of the movement of E-cadherin on the free cell surface as studied by optical tweezers and single particle tracking: Corralling and tethering by the membrane skeleton. J Cell Biol 1998; 140(5):1227-1240.

10. Yagi T, Takeichi M. Cadherin superfamily genes: Functions, genomic organization, and neurologic diversity Genes Dev 2000; 14(10):1169-1180.
11. Steinberg MS, McNutt PM. Cadherins and their connections: Adhesion junctions have broader functions. Curr Opin Cell Biol 1999; 11(5):554-560.
12. Yap AS, Niessen CM, Gumbiner BM. The juxtamembrane region of the cadherin cytoplasmic tail supports lateral clustering, adhesive strengthening, and interaction with p120ctn. J Cell Biol 1998; 141(3):779-789.
13. Aono S, Nakagawa S, Reynolds AB et al. p120(ctn) acts as an inhibitory regulator of cadherin function in colon carcinoma cells. J Cell Biol 1999; 145(3):551-562.
14. Anastasiadis PZ, Moon SY, Thoreson MA et al. Inhibition of RhoA by p120 catenin. Nat. Cell Biol 2000; 2(9):637-644.
15. Kaibuchi K, Kuroda S, Amano M. Regulation of the cytoskeleton and cell adhesion by the Rho family GTPases in mammalian cells. Annu Rev Biochem 1999; 68:459-486.
16. Gulli MP, Peter M. Temporal and spatial regulation of Rho-type guanine-nucleotide exchange factors: The yeast perspective. Genes Dev 2001; 15(4):365-379.
17. Van Aelst L, D'Souza-Schorey C. Rho GTPases and signaling networks. Genes Dev 1997; 11(18):2295-2322.
18. Hall A. Rho GTPases and the actin cytoskeleton. Science 1998; 279(5350):509-514.
19. Fleming IN, Elliott CM, Exton JH. Differential translocation of Rho family GTPases by lysophosphatidic acid, endothelin-1, and platelet-derived growth factor. J Biol Chem 1996; 271(51):33067-33073.
20. Kranenburg O, Poland M, Gebbink M et al. Dissociation of LPA-induced cytoskeletal contraction from stress fiber formation by differential localization of RhoA. J Cell Sci 1997; 110(19):2417-2427.
21. Takaishi K, Sasaki T, Kotani H et al. Regulation of cell-cell adhesion by Rac and Rho small G proteins in MDCK cells. J Cell Biol 1997; 139(4):1047-1059.
22. Kuroda S, Fukata M, Fujii K et al. Regulation of cell-cell adhesion of MDCK cells by Cdc42 and Rac1 small GTPases. Biochem Biophys Res Commun 1997; 240(2):430-435.
23. Jou TS, Nelson WJ. Effects of regulated expression of mutant RhoA and Rac1 small GTPases on the development of epithelial (MDCK) cell polarity. J Cell Biol 1998; 142(1):85-100.
24. Nakagawa M, Fukata M, Yamaga M et al. Recruitment and activation of Rac1 by the formation of E-cadherin-mediated cell-cell adhesion sites. J Cell Sci 2001; 114(10):1829-1838.
25. Akhtar N, Hotchin NA. RAC1 regulates adherens junctions through endocytosis of E-cadherin. Mol Biol Cell 2001; 12(4):847-862.
26. Kroschewski R, Hall A, Mellman I. Cdc42 controls secretory and endocytic transport to the basolateral plasma membrane of MDCK cells. Nat Cell Biol 1999; 1(1):8-13.
27. Kawajiri A, Itoh N, Fukata M et al. Identification of a novel beta-catenin-interacting protein. Biochem Biophys Res Commun 2000; 273(2):712-717.
28. Dobrosotskaya IY, James GL. MAGI-1 interacts with beta-catenin and is associated with cell-cell adhesion structures. Biochem Biophys Res Commun 2000; 270(3):903-909.
29. Ide N, Hata Y, Deguchi M et al. Interaction of S-SCAM with neural plakophilin-related Armadillo-repeat protein/delta-catenin. Biochem Biophys Res Commun 1999; 256(3):456-461.
30. Perego C, Vanoni C, Massari S et al. Mammalian LIN-7 PDZ proteins associate with beta-catenin at the cell-cell junctions of epithelia and neurons. EMBO J 2000; 19(15):3978-3989.
31. Michiels F, Habets GG, Stam JC et al. A role for Rac in Tiam1-induced membrane ruffling and invasion. Nature 1995; 375(6529):338-340.
32. Hoshino M, Sone M, Fukata M et al. Identification of the stef gene that encodes a novel guanine nucleotide exchange factor specific for Rac1. J Biol Chem 1999; 274(25):17837-17844.
33. Fukuhara S, Murga C, Zohar M et al. A novel PDZ domain containing guanine nucleotide exchange factor links heterotrimeric G proteins to Rho. J Biol Chem 1999; 274(9):5868-5879.
34. Kourlas PJ, Strout MP, Becknell B et al. Identification of a gene at 11q23 encoding a guanine nucleotide exchange factor: Evidence for its fusion with MLL in acute myeloid leukemia. Proc Natl Acad Sci USA 2000; 97(5):2145-2150.
35. Taya S, Inagaki N, Sengiku H et al. Direct interaction of insulin-like growth factor-1 receptor with leukemia-associated RhoGEF. J Cell Biol 2001; 155(5):809-820.
36. Braga V, Machesky LM, Hall A et al. The small GTPases Rho and Rac are required for the establishment of cadherin-dependent cell-cell contacts. J Cell Biol 1997; 137(6):1421-1431.
37. Braga VM, Del Maschio A, Machesky L et al. Regulation of cadherin function by Rho and Rac: Modulation by junction maturation and cellular context. Mol Biol Cell 1999; 10(1):9-22.
38. Kodama A, Takaishi K, Nakano K et al. Involvement of Cdc42 small G protein in cell-cell adhesion, migration and morphology of MDCK cells. Oncogene 1999; 18(27):3996-4006.
39. Hordijk PL, ten Klooster JP, van der Kammen RA et al. Inhibition of invasion of epithelial cells by Tiam1-Rac signaling. Science 1997; 278(5342):1464-1466.

40. Fukata M, Kuroda S, Nakagawa M et al. Cdc42 and Rac1 regulate the interaction of IQGAP1 with β-catenin. J Biol Chem 1999; 274(37):26044-26050.
41. Kuroda S, Fukata M, Nakagawa M et al. Role of IQGAP1, a target of the small GTPases Cdc42 and Rac1, in regulation of E-cadherin-mediated cell-cell adhesion. Science 1998; 281(5378):832-835.
42. Li Z, Kim SH, Higgins JM et al. IQGAP1 and calmodulin modulate E-cadherin function. J Biol Chem 1999; 274(53):37885-37892.
43. Bashour AM, Fullerton AT, Hart MJ et al. IQGAP1, a Rac- and Cdc42-binding protein, directly binds and cross-links microfilaments. J Cell Biol Jun 30 1997; 137(7):1555-1566.
44. Fukata M, Kuroda S, Fujii K et al. Regulation of cross-linking of actin filament by IQGAP1, a target for Cdc42. J Biol Chem 1997; 272(47):29579-29583.
45. Noritake J, Fukuta M, Sato K et al. Positive role of IQGAP1, an effector of Rac1, in actin-meshwork formation at sites of cell-cell contact. Mol Biol Cell 2004; 15:1065-1076.
46. Swart-Mataraza JM, Li Z, Sacks DB. IQGAP1 is a component of Cdc42 signaling to the cytoskeleton. J Biol Chem 2002; 277(27):24753-24763.
47. Fukata M, Nakagawa M, Itoh N et al. Involvement of IQGAP1, an effector of Rac1 and Cdc42 GTPases, in cell-cell dissociation during cell scattering. Mol Cell Biol 2001; 21(6):2165-2183.
48. Maeno Y, Moroi S, Nagashima H et al. Alpha-catenin-deficient F9 cells differentiate into signet ring cells. Am J Pathol 1999; 154(5):1323-1328.
49. Li S, Wang Q, Chakladar A et al. Gastric hyperplasia in mice lacking the putative Cdc42 effector IQGAP1. Mol Cell Biol 2000; 20(2):697-701.
50. Brill S, Li S, Lyman CW et al. The Ras GTPase-activating-protein-related human protein IQGAP2 harbors a potential actin binding domain and interacts with calmodulin and Rho family GTPases. Mol Cell Biol 1996; 16(9):4869-4878.
51. Sugimoto N, Imoto I, Fukuda Y et al. IQGAP1, a negative regulator of cell-cell adhesion, is upregulated by gene amplification at 15q26 in gastric cancer cell lines HSC39 and 40A. J Hum Genet 2001; 46(1):21-25.
52. Matsumoto Y, Oshida T, Obayashi I et al. Identification of highly expressed genes in peripheral blood T cells from patients with atopic dermatitis. Int Arch Allergy Immunol 2002; 129(4):327-340.
53. Takemoto H, Doki Y, Shiozaki H et al. Localization of IQGAP1 is inversely correlated with intercellular adhesion mediated by E-cadherin in gastric cancers. Int J Cancer 2001; 91(6):783-788.
54. Presslauer S, Hinterhuber G, Cauza K et al. RasGAP-like protein IQGAP1 is expressed by human keratinocytes and recognized by autoantibodies in association with bullous skin disease. J Invest Dermatol 2003; 120(3):365-371.
55. Knox AL, Brown NH. Rap1 GTPase regulation of adherens junction positioning and cell adhesion. Science 2002; 295(5558):1285-1288.
56. Bos JL, de Rooij J, Reedquist KA. Rap1 signalling: Adhering to new models. Nat Rev Mol Cell Biol 2001; 2(5):369-377.
57. Braga VM. Cell-cell adhesion and signalling. Curr Opin Cell Biol 2002; 14(5):546-556.
58. Price LS, Leng J, Schwartz MA et al. Activation of Rac and Cdc42 by integrins mediates cell spreading. Mol Biol Cell 1998; 9(7):1863-1871.
59. Kim SH, Li Z, Sacks DB. E-cadherin-mediated cell-cell attachment activates Cdc42. J Biol Chem 2000; 275(47):36999-37005.
60. Kovacs EM, Ali RG, McCormack AJ et al. E-cadherin homophilic ligation directly signals through Rac and phosphatidylinositol 3-kinase to regulate adhesive contacts. J Biol Chem 2002; 277(8):6708-6718.
61. Pece S, Chiariello M, Murga C et al. Activation of the protein kinase Akt/PKB by the formation of E-cadherin-mediated cell-cell junctions. Evidence for the association of phosphatidylinositol 3-kinase with the E-cadherin adhesion complex. J Biol Chem 1999; 274(27):19347-19351.
62. Espada J, Perez-Moreno M, Braga VM et al. H-Ras activation promotes cytoplasmic accumulation and phosphoinositide 3-OH kinase association of beta-catenin in epidermal keratinocytes. J Cell Biol 1999; 146(5):967-980.
63. Kotani K, Yonezawa K, Hara K et al. Involvement of phosphoinositide 3-kinase in insulin- or IGF-1-induced membrane ruffling. EMBO J 1994; 13(10):2313-2321.
64. Bourguignon LY, Zhu H, Shao L et al. CD44 interaction with tiam1 promotes Rac1 signaling and hyaluronic acid-mediated breast tumor cell migration. J Biol Chem 2000; 275(3):1829-1838.
65. Noren NK, Niessen CM, Gumbiner BM et al. Cadherin engagement regulates Rho family GTPases. J Biol Chem 2001; 16:33305-33308.
66. Noren NK, Arthur WT, Burridge K. Cadherin engagement inhibits RhoA via p190RhoGAP. J Biol Chem 2003; in press.
67. Rottner K, Hall A, Small JV. Interplay between Rac and Rho in the control of substrate contact dynamics. Curr Biol 1999; 9(12):640-648.

68. Kozma R, Sarner S, Ahmed S et al. Rho family GTPases and neuronal growth cone remodelling: Relationship between increased complexity induced by Cdc42s, Rac1, and acetylcholine and collapse induced by RhoA and lysophosphatidic acid. Mol Cell Biol 1997; 17(3):1201-1211.
69. Sander EE, ten Klooster JP, van Delft S et al. Rac downregulates Rho activity: Reciprocal balance between both GTPases determines cellular morphology and migratory behavior. J Cell Biol 1999; 147(5):1009-1022.
70. Zondag GC, Evers EE, ten Klooster JP et al. Oncogenic Ras downregulates Rac activity, which leads to increased Rho activity and epithelial-mesenchymal transition. J Cell Biol 2000; 149(4):775-782.
71. Nimnual AS, Taylor LJ, Bar-Sagi D. Redox-dependent downregulation of Rho by Rac. Nat Cell Biol 2003; in press.
72. Kraynov VS, Chamberlain C, Bokoch GM et al. Localized Rac activation dynamics visualized in living cells. Science 2000; 290(5490):333-337.
73. Cook TA, Nagasaki T, Gundersen GG. Rho guanosine triphosphatase mediates the selective stabilization of microtubules induced by lysophosphatidic acid. J Cell Biol 1998; 141(1):175-185.
74. Ishizaki T, Morishima Y, Okamoto M et al. Coordination of microtubules and the actin cytoskeleton by the Rho effector mDia1. Nat Cell Biol 2001; 3(1):8-14.
75. Palazzo AF, Cook TA, Alberts AS et al. mDia mediates Rho-regulated formation and orientation of stable microtubules. Nat Cell Biol 2001; 3(8):723-729.
76. Palazzo AF, Joseph HL, Chen YJ et al. Cdc42, dynein, and dynactin regulate MTOC reorientation independent of Rho-regulated microtubule stabilization. Curr Biol 2001; 11(19):1536-1541.
77. Etienne-Manneville S, Hall A. Integrin-mediated activation of Cdc42 controls cell polarity in migrating astrocytes through PKC. Cell 2001; 106:489-498.
78. van Horck FP, Ahmadian MR, Haeusler LC et al. Characterization of p190RhoGEF, a RhoA-specific guanine nucleotide exchange factor that interacts with microtubules. J Biol Chem 2001; 276(7):4948-4956.
79. Chausovsky A, Bershadsky AD, Borisy GG. Cadherin-mediated regulation of microtubule dynamics. Nat Cell Biol 2000; 2(11):797-804.
80. Ligon LA, Karki S, Tokito M et al. Dynein binds to beta-catenin and may tether microtubules at adherens junctions. Nat Cell Biol 2001; 3(10):913-917.
81. Clayton L, Hall A, Johnson MH. A role for Rho-like GTPases in the polarisation of mouse eight-cell blastomeres. Dev Biol 1999; 205(2):322-331.
82. Sugihara K, Nakatsuji N, Nakamura K et al. Rac1 is required for the formation of three germ layers during gastrulation. Oncogene 1998; 17(26):3427-3433.
83. Luo L. Rho GTPases in neuronal morphogenesis. Nat Rev Neurosci 2000; 1(3):173-180.
84. Sone M, Hoshino M, Suzuki E et al. Still life, a protein in synaptic terminals of Drosophila homologous to GDP-GTP exchangers Science 1997; 275(5299):543-547.
85. Sone M, Suzuki E, Hoshino M et al. Synaptic development is controlled in the periactive zones of Drosophila synapses. Development 2000; 127(19):4157-4168.

Chapter 18

Wnt Signaling Networks and Embryonic Patterning

Michael W. Klymkowsky

Abstract

Wnt signaling lies at the heart of metazoan development. The Wnt pathway bifurcates a number of times and regulates cell polarity, migration, adhesion and gene expression. I review the complexity of this network with a focus on the role of SOX proteins in its regulation.

Introduction

There is a growing consensus that the metazoans are monophyletic, descended from urmetazoans that lived ~1.2 Gya.[1] While it seems likely that many inter- and intracellular signaling pathways were in place in this eumetaozan ancestor, there is much genomic sequence gazing and phylogenetic reconstruction remaining before we have a clear view of the metazoan Adam/Eve. Nevertheless, it is clear that a small number of conserved signaling pathways have been used repeatedly to establish body axes and tissue types. These include the Wnts, the receptor serine/threonine protein kinases, the receptor tyrosine kinases, hedgehogs and Notch/Delta.[2] I will focus on the Wnt signaling system as a key player from the earliest stages of metazoan evolution (Fig. 1).

Wnt signaling typically occurs over short ranges (juxtacrine), involves components of the cadherin cell-cell adhesion system, and can directly modulate gene expression through the LEF/TCF-subfamily of sequence-specific HMG-box DNA binding proteins. In part their autocrine/juxtacrine activity is due to the fact that Wnts are lipid-modified.[3] Molecular studies of the cnidarian *Hydra vulgaris* reveals the presence of Wnts and a number of downstream components of the Wnt signaling pathway.[4-6] Whether Wnt signaling or Wnt signaling components are present in the more basal Porifera (sponges) or the protozoan sister group of the metazoans, the Choanoflagellates,[7-10] is not yet known. Cadherin-like polypeptides have been identified in the unicellular and colonial Choanoflagellates *Monosiga brevicollis* and *Proterospongia*[11] and catenin-like polypeptides are present in the cellular slime mold *Dictyostelium discoideum*.[12] Wnt signaling has been implicated in a large number of morphogenic events during development as well as the maintenance of stem cells in their undifferentiated state.[13-15] Wnt signaling plays a key role in the regulation of epithelial-mesenchymal transition during neural crest formation.[16]

Rise and Fall of Epithelial Phenotype: Concepts of Epithelial-Mesenchymal Transition, edited by Pierre Savagner.

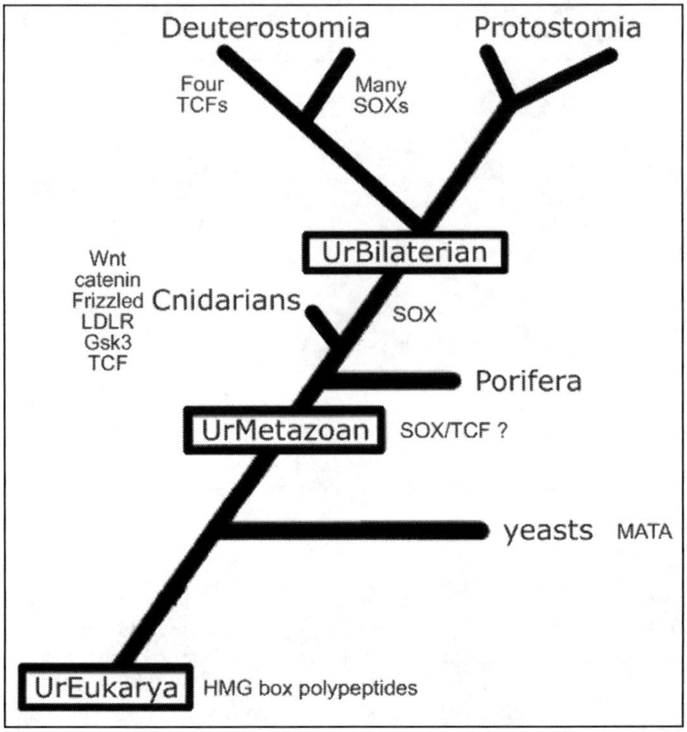

Figure 1. This tree, adopted from reference (1), indicates the appearance of various components of the Wnt/SOX system. Sequence-specific HMG box DNA-binding proteins (MATA) are present in the yeasts and other unicellular eukaryotes. A β-catenin-like protein is present in the cellular slime mold *Dicytostelium*, while cadherin-like proteins have been found in Choanoflagellates. Parts of a canonical Wnt signaling system is present in the simplest diplobastic metazoan the Cnidarian Hydra. Deutrostomia have multiple LEF/TCF polypeptides and multiple SOX proteins, while Protostomia appear to have a single TCF-like polypeptide and a smaller number of SOXs.

Wnts and Their Receptors

Wnts are secreted glycoproteins that act at short range to regulate cellular behavior.[17] In the human there are 19 known Wnts.* The downstream effects of Wnt signaling are complex and can affect cellular morphology, polarity and motility as well as gene expression through parallel, complementary and interacting pathways.[1,14,18,19] The origin of this diversity begins with the membrane-bound Wnt receptors. Wnts bind to the extracellular domains of frizzled polypeptides. These are seven-pass transmembrane proteins[20] that associate with the low-density lipoprotein receptor-related LRP5 and LPR6 polypeptides to form one type of Wnt receptor.[21,22] The mouse LPR5 gene plays a critical role in cholesterol metabolism and, together with Wnts 3a and 5, regulates glucose-induced insulin secretion.[23] LPR5 interacts directly with the downstream Wnt-signaling polypeptide Axin.[24] Ten frizzleds have been identified in human[2] and different frizzleds bind to differ Wnts with different affinities to produce different down-stream outcomes.

* Roel Nusse's lab maintains the excellent Wnt Gene Homepage where you can usually find the latest information of all components of the Wnt signaling pathway. http:// http://www.stanford.edu/~rnusse/wntwindow.html

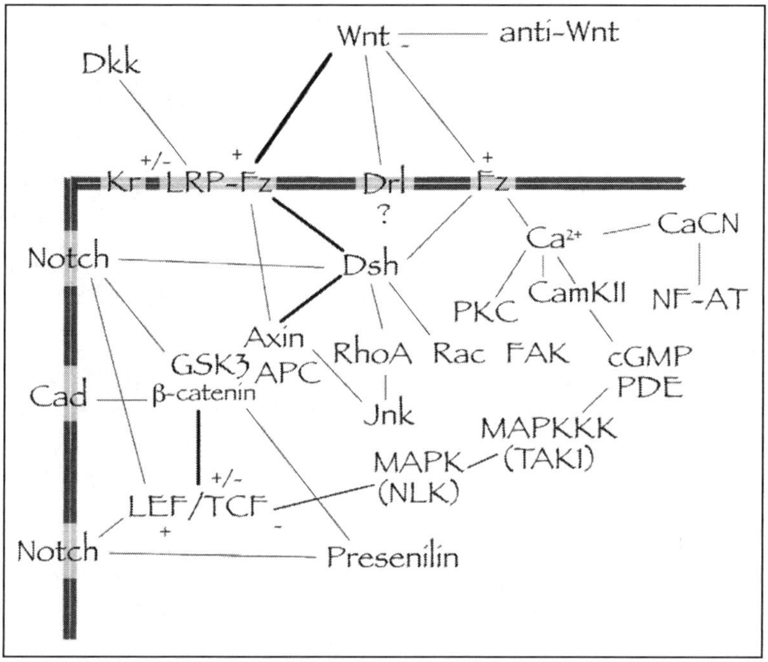

Figure 2. Wnt pathways. This is a simplified diagram of some of the interactions described in the text, but illustrates how the complexity of Wnt signaling emerges from the net-like nature of 'downstream' pathways. For example, Dsh is involved in both the regulation of β-catenin proteolysis in the canonical pathway and the regulation of Rho and Rac in the planar cell polarity (PCP) pathway. Even though Ca^{2+} mediated Wnt signaling does not involve Dsh, it can modulate the catenin-mediated pathway through effects on MAPK that modulate LEF/TCF DNA-binding. For abbreviations, see text.

The true complexity of Wnt receptors has yet to emerge completely. For example consider the secreted Wnt regulators Dickkopf-1 and 2 (Dkk1/Dkk2). Dkk1 binds to LRPs and blocks Wnt binding whereas Dkk2 acts as a Wnt signal enhancer.[25] Similarly, the secreted Wise polypeptide binds to LPR6 and can act as either a Wnt agonist or antagonist.[26] The effects of the Dkks on Wnt signaling are modulated by the Kremen-1 and -2 polypeptides. These are membrane proteins that mediate LRP endocytosis in response to Dkk binding.[27,28] In addition to the Dkks, a number of other secreted Wnt antagonists have been identified. Many are structurally related to the Wnt-binding extracellular domain of the frizzleds.[29]

A previously unexpected function of Wnts, mediated by the *Drosophila* Wnt5 homolog and the integral membrane protein Derailed (Drl), has recently been found to play an important role in neuronal guidance.[30] While Drl is a member of the RYK receptor protein tyrosine kinase-like family of proteins, it appears to mediate signaling in a kinase-independent fashion.[31,32] The extracellular domain of Drl contains a Wnt inhibitory factor (WIF) domain[33] through which it interacts with Wnt5 to regulate neuronal outgrowth.

Downstream of Frizzled, a Networking Nightmare

Multiple downstream Wnt signaling pathways have been identified. These include the catenin pathway, the Ca^{2+} "tissue separation" pathway and the "planar cell polarity" (PCP) pathway.[14] Defining the specific roles of specific pathways in specific processes is complicated by the fact that multiple Wnts, Wnt antagonists and agonists, Wnt receptor components and down-stream signaling components are often coexpressed. Moreover there is significant and apparently integral cross-talk between these signaling pathways (Fig. 2).[34] Generating a meaningful map of the interactions between these components, and the outcomes of a particular stimulus, will require new bioinformatic approaches.

Consider the PCP pathway, which acts through the cytoplasmic polypeptide Dishevelled (Dsh).[35,36] Dsh is also an integral component in the catenin-pathway (more below) and acts as a negative regulator of Notch signaling.[37] Upon Wnt-Wnt receptor interactions, Dsh forms distinct interactions the small GTPases Rho and Rac and regulates cell polarity and movement.[38] Activation of the PCP pathway activates Jun-N-terminal kinase (Jnk)[35] which initiates its own 'downstream' signaling cascade.[39] Jnk can also be activated by the catenin-pathway component Axin.[40,41] The polypeptide Dapper can associate with Dsh and inhibit both catenin and JNK mediated signaling[42] whereas the polypeptide Frodo interacts with, and facilitates Dsh signaling.[43] Whether Dapper or Frodo activities are themselves regulated in unknown.[44]

Analyzing Interactions

In the interpretation of any experimentally manipulated system, the detailed cellular context is critical. In 'genetic organisms' such as *Drosophila melanogaster* and *Caenorhabditis elegans*, where much elegant work on Wnt signaling has been done, mutations are expressed either globally or regionally as mosaics, continuously or induced at specific developmental stages or by specific treatments, e.g., via heat shock or tissue-specific promoters. Yet, phenotypic analysis is largely impressionistic and nonquantitative. This gives license to the human penchant for linearity in casual networks. Moreover, experimental interpretations are generally based upon incomplete and sometimes misleading assumptions as to the nature of the mutations used and how they interact with other cellular components.[45,46] A "dominant negative" mutation for one function may be neomorphic or hypermorphic for another.[47] Some of these issues, and their relation to Wnt signaling, are raised by the work of Brennan et al.[48]

Experimental studies of *Xenopus laevis* have provided an essential complement to genetic studies in the development of our understanding of Wnt signaling, beginning with the observation that ectopic expression of mouse int-1 induced axis duplication.[50] and continuing with the discovery that β-catenin and its paralog plakoglobin/γ-catenin (see below) produce a similar effect[51-53] and localize to nuclei.[53,54] In *X. laevis* and cultured cell systems, experimental manipulations typically involve the nonphysiological expression of 'wild type', mutated and chimeric polypeptides or the anti-sense RNA or modified DNA oligonucleotide (morpholino) mediated down regulation of translation (see refs. 55-57).

The classical studies on the vertebrate organizer revealed that a remarkably diverse set of stimuli produced similar outcomes, i.e., the induction of a second neural axis.[58,59] Within such systems, the effects of distinct perturbations appear to be channeled by the regulatory landscape, with its signaling cascades, cross-talk and feedback loops, into stereotyped outcomes. Without a clear understanding of the network dynamics in play, the interpretation of any particular experimental manipulation can be problematic. In the absence of a robust network model for Wnt signaling and its cellular and developmental context, we are often left with idiosyncratic and disconnected observations. Recently, Loose and Patient[49] have begun to construct such a network model for *Xenopus*.

The Canonical Pathway

The best-characterized Wnt-activated signaling pathway, often referred to as the canonical pathway, involves the cadherin-binding polypeptide β-catenin (see also Chapter 12, Conacci-Sorell and Ben-Ze'ev). β-catenin orthologs are found throughout the metazoans (Fig. 3).[60] A β-catenin-like polypeptide, aardvark, has been identified in the slime mold *Dictyostelium*. Like β-catenin, aardvark is associated with sites of cell-cell interaction.[12,61] Vertebrates contain as second, paralogous polypeptide plakoglobin.[62,63] *C. elegans* contain three distinct β-catenin-like polypeptides known as Bar-1, Wrm-1 and HMP-2.[64] These are highly divergent from one another and other β-catenins and appear to be specialized for specific tasks, while retaining some functional redundancy.

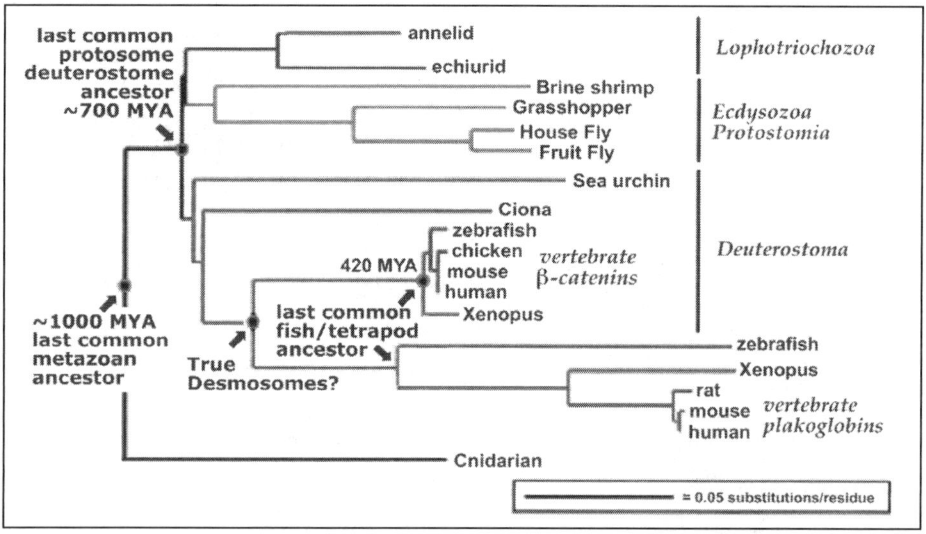

Figure 3. Catenin evolution, adapted from reference.[60] β-catenin like polypeptides are present throughout the metazoans with the possible exception of the sponges, and were presumably present in the last common 'urmetazoan' ancestor. The lophotriochozoa and ecdyozoa appear to have a single β-catenin like polypeptide. A single β-catenin-like polypeptide has been identified in echinoderms and sea squirts, members of the deuterostoma. In the vertebrate lineage, there are two paralogous polypeptides, the highly conserved β-catenins and the more divergent plakoglobins (γ-catenins). This duplication may be related to the appearance of true, intermediate filament-associated desmosomes. Intermediate filaments themselves appear to pre-date the appearance of plakoglobin.

Both β-catenin and plakoglobin bind to the cytoplasmic tails of type I cadherins and to the actin filament anchoring protein α-catenin.[65-67] Plakoglobin plays a distinct role at the desmosome (reviewed in Chapter 10, Getsios et al). The binding of plakoglobin to the cytoplasmic tail of desmosomal cadherins, desmocollins and desmogleins, blocks plakoglobin's interaction with α-catenin and facilitates interactions with intermediate filament networks via desmoplakins, plakophilins and other desmosome associated polypeptides.[62,68,69] β-catenin associates with desmosomal cadherins only in the absence of plakoglobin.[70] While the appearance of plakoglobin may reflect the appearance of vertebrate-type desmosomes, intermediate filaments are more ancient; they are present in many invertebrates, including *C. elegans*,[71,72] and annelids[73] where they have a number of essential functions.[74]

Since the origin of the vertebrates, the rate of evolutionary change in the plakoglobin has been greater than in β-catenin (Fig. 3).[60] This suggests that plakoglobin is under less severe structural constraint than is β-catenin. Nevertheless, where it has been examined, plakoglobin has been found to interact with the same polypeptides as does β-catenin.[63] There are, however, clear differences between the two polypeptides, in terms of the phenotypes associated with their absence, their relative affinities for desmosomal cadherins, and their cytoplasmic stability.[75-77] These differences relate to their distinct roles in adhesion, epithelial-mesenchymal transition and Wnt signaling. Nevertheless, recent evidence indicates that plakoglobin can replace all of the functions of β-catenin in the mouse epidermis.[78]

A particularly interesting feature of β-catenin, and presumably plakoglobin as well, is its ability to move into and out of the nucleus independently of the standard nuclear localization receptors.[79,80] This property may be related to the structural similarities between β-catenin and importin/karyopherin nuclear transport polypeptides (Fig. 4A).[81] The signature structural feature of β-catenin is a series of 12 repeats first recognized by Riggleman et al,[82] in the *Drosophila*

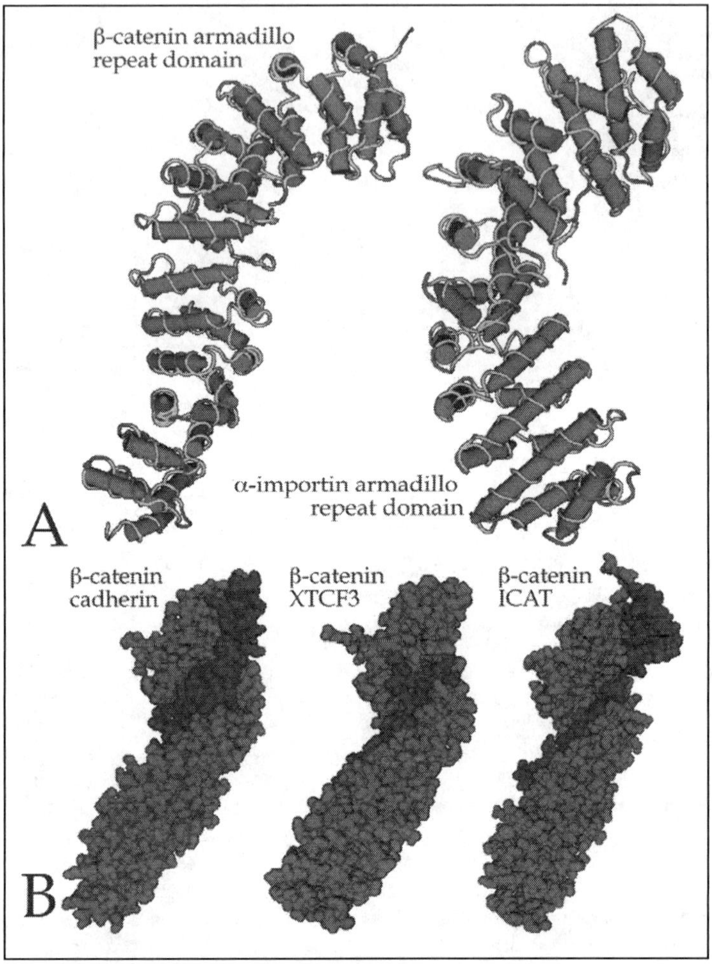

Figure 4. Catenin structure. A) The β-catenin-like proteins are characterized by a series of 12 repeated structural "armadillo" motifs, each composed of three α-helical segments. This same structural motif is found within the nuclear transport proteins α-importin and karyopherin. B) The armadillo region mediates interactions with a number of binding partners, including the C-terminal cytoplasmic tail of cadherin, the N-terminal region of LEF/TCF, the APC protein (not shown) and the small inhibitory polypeptide ICAT. Structures were downloaded from the NCBI web site and images were displayed using the Cn3D viewer.

β-catenin homolog armadillo. It is through this repeat domain that β-catenin interacts with the cytoplasmic domain of cadherins, transcription and regulatory factors, and the adenomatous polyposis coli proteins (Fig. 4B). Whether this is an example of molecular convergence or common ancestry between β-catenin and the importins is an interesting question.

While vertebrate and invertebrate β-catenins perform similar functions, they are not totally interchangeable. Neither vertebrate β-catenin nor plakoglobin can rescue the signaling defects associated with null mutations in *Drosophila* armadillo, although both can rescue the associated adhesion defects.[83] In the mouse, the defects associated with the absence of plakoglobin are partially ameliorated, but not completely rescued by the presence of β-catenin and visa versa.[70,78]

Dishevelled

Wnt signaling pathways fall into two groups, those that involve the cytoplasmic polypeptide Dishevelled (Dsh) and those that do not (Fig. 2). Dsh is directly involved in both the canonical, catenin and planar cell polarity (PCP) pathways, but not in Wnt-signaling via altered intracellular [Ca^{2+}] and PKC activity[14,84] which appear to be mediated by cyclic GMP.[85] That said, the Wnt-Ca^{2+} pathway can modulate the output of the canonical catenin-pathway via the activation of MAPK cascades[86] and Wnt/Dsh may be involved in the regulation of cell motility via regulation of focal adhesion kinase (FAK).[14]

Dsh polypeptides have been identified throughout the metazoans and are characterized by the presence of DIX, PDZ and DEP domains. The DIX domain mediates dimerization. The PDZ domain has been implicated in membrane receptor clustering. The DEP domain is found in a number of regulators of G-protein signaling (RGS) polypeptides.[84] In mouse and human there are three Dsh polypeptides, Dvl-1, Dvl-2 and Dvl-3. Dsh interacts with a number of intracellular structures, including actin fibers, vesicular membranes.[87] and ephrin receptors.[88] Exactly how Wnt binding to frizzled 'activates' Dsh remains unclear.

Catenin Degradation

In the 'resting' state, β-catenin not associated with a cadherin cytoplasmic tail, or some other polypeptide, is unstable. A degradosome formed by Axin1 or its paralog Axin2/Axil/Conductin, Adenomatous Polyposis Coli (APC) and glycogen synthase kinase-3 (GSK3) associates with β-catenin, leading to its phosphorylated, ubiquination and degradation in proteosomes.[89] Canonical pathway activation of Dsh suppresses the degradation of 'free' β-catenin, leading to its nuclear accumulation.[79,80,90] This in turn positively regulate Axin2 expression,[91] which should destabilizes cytoplasmic Catenins.

Plakoglobin, like β-catenin, binds to APC.[92,93] and Axin.[94,95]but its cytoplasmic form is generally more stable than β-catenin.[77] While the degradation of β-catenin depends quite sensitively on GSK3-mediated phosphorylation sites within its N-terminal domain, plakoglobin's stability is relatively unaffected by such mutations, at least in some mammalian cell test system.[77] In *Xenopus* embryos deletion of the N-terminal domain does stabilize plakoglobin.[96] Plakoglobin and β-catenin form an interacting network in which cytosolic plakoglobin acts as a β-catenin buffer, modulating the rate of β-catenin turnover and binding to other polypeptides. While β-catenin acts exclusively as a proto-oncogene, plakoglobin can act both as an oncogene and a tumor suppressor depending on the cellular/tissue context.[97-100]

In addition to its role in the turnover of β-catenin, APC is involved in a number of other cellular processes. It appears act as a nuclear export factor.[101-103] and is associated with nucleoli, microtubules, kinetochores, and cortical actin networks.[104-108] APC-microtubule interaction appear to be regulated by GSK3β.[109] APC moves along microtubules, and accumulates at their "+" ends in a process that may involve direct interactions with kinesins.[110] Cui et al[111] suggest that β-catenin's localization and turnover depends upon kinesin/APC interactions.

The degradation of β-catenin can also be mediated by interactions with presenilin-1/γ-secretase in an Axin-independent pathway.[112,113] and by other degradative systems.[114] Presenilin appears to be an integral component of adherence and synaptic junctions and mediates the proteolytic processing of cadherins.[115-117] Loss of presenilin has been associated with increased cytoplasmic β-catenin levels in skin.[118] Other components of the catenin-degradosome have alternative functions as well. Axin forms a complex with MEKK1, activated JNK[40] and the LRP5 Wnt coreceptor polypeptide.[24] GSK3 regulates a number of target molecules, including eukaryotic initiation factor 2B, and cyclin D1, c-myc, the Hedgehog signaling effecter Cubitus interruptus, and Notch (see Wnt Web Site).[119-121] In turn, GSK3's activity is regulated by a number of 'upstream' pathways, including the insulin, growth factor activated MAPK and PKA pathways.[122] In the canonical Wnt pathway, the assembly of the catenin degradosome regulates GSK3/β-catenin interactions, while the activity of 'free' GSK3 is regulated by priming kinases.[122] GSK3 activity is itself regulated by phosphorylation and by the small GSK3 binding proteins, GBP/FRAT.[123] GSK3 is involved in the nuclear export of a number of transcription factors [124,125]

β-Catenin and Transcription Factor Interactions

β-catenin interacts with the cytoplasmic domain of E-cadherin through its armadillo-repeat domain.[126] This same region of β-catenin interacts with APC[127] and transcription factors of the LEF/TCF[128,129] and SOX families (see below). The N-terminal domains of the LEF/TCF polypeptides bind in a mutually exclusive manner to the cadherin/APC-binding region of β-catenin. The interaction between β-catenin and LEF/TCF polypeptides and other binding partners are inhibited by the small and widely expressed ~9kDa polypeptide ICAT[130-132] (Fig. 4B). Duplin[133] binds to β-catenin and inhibits is ability to interact with TCFs. ICAT has also been found to block the interaction of β-catenin with mammalian androgen receptors[134] and its over-expression can induce G_2 arrest and apoptosis in cells carrying mutations in β-catenin, APC or Axin.[135] Whether the interactions between ICAT, Duplin and β-catenin stabilize β-catenin or are themselves regulated has not yet been reported.

Catenins and LEF/TCFs

The LEF/TCF proteins were initially identified as factors regulating the expression of the T-cell receptor.[136-138] They are members of a larger group of sequence specific HMG-box containing DNA binding proteins (Fig. 5). The HMG box is a highly charged, ~80 amino acid polypeptide motif that recognizes the minor groove of the DNA helix and induces a 90-120° bend when bound to 'naked' DNA.[139,140] This bending is due to the insertion of nonpolar amino acid side chains between base pairs.[141] DNA bending suggests that LEF/TCFs act as architectural factors, producing specific three dimensional structures that modulate gene expression. By comparing known LEF/TCF binding sites in *Xenopus* (Table 1), we can identify a consensus LEF/TCF DNA binding site as 5'$^A/_T$ $^A/_T$ $^C/_G$ AA $^A/_T$ G 3'. Quantitative studies suggest that optimal DNA binding sites are larger, from 10-12 base pairs in length.[142]

The initial discovery of β-catenin's and plakoglobin's interactions with LEF/TCF proteins was made through yeast two-hybrid screens and confirmed by various *in vitro* and *in vivo* assays.[143-145] Early on it was obvious that different TCFs had distinct activities in different systems.[146] In mammals, there are four known LEF/TCFs: TCF1, LEF1, TCF3 and TCF4. Their OMIM designations are TCF7 (189908), LEF1 (153245), TCF7-like 1 (604652), TCF7-like 2 (602228). A number of differentially spliced transcripts have been described, and in some cases the major expressed form lacks the N-terminal β-catenin interaction domain.[147] Orthologs of TCF1, LEF, TCF3 and TCF4 have been identified in *Xenopus laevis* and *Dario rerio*, whereas *Drosophila melangoaster* (dTCF/pangolin) and *C. elegans* (POP-1) each appear to possess only a single member of the family.[3]

In addition to binding to β-catenin and plakoglobin, LEF/TCF polypeptides bind to transcriptional corepressors of the Groucho/TLE (transducin-like enhancer of split) family[148] as well as a number of other polypeptides, such as ALY,[149] the pair-like homeodomain contain polypeptide Alx4[150] and SMADs.[151,152] LEF1 has also been reported to bind to the Notch intracellular domain, which acts as a coactivator.[153]

It is commonly assumed that β-catenin acts as an coactivating accessory factor when associated with DNA-bound LEF/TCF polypeptides. That the truth is rather more complex is suggested by a number of observations. The most compelling is that in the early *C. elegans* embryo Wnt signaling leads to the disappearance of the LEF/TCF-like POP-1 polypeptide from embryonic nuclei.[19,154,155] This observation indicates that activated β-catenin can regulate the intracellular distribution of LEF/TCFs. A similar conclusion was reached in extensive analyses of the mode of action of membrane-anchored forms of plakoglobin[94,156] and armadillo.[157] In the only published examination of the interactions between β-catenin, LEF1 and chromatin, Tutter et al[158] found that LEF1's N-terminal β-catenin-binding domain blocks LEF1 binding to chromatin but not to naked DNA. The binding of β-catenin to LEF1 obviates LEF1's self-inhibition. Catenin-LEF complex binding to assembled chromatin required the p300

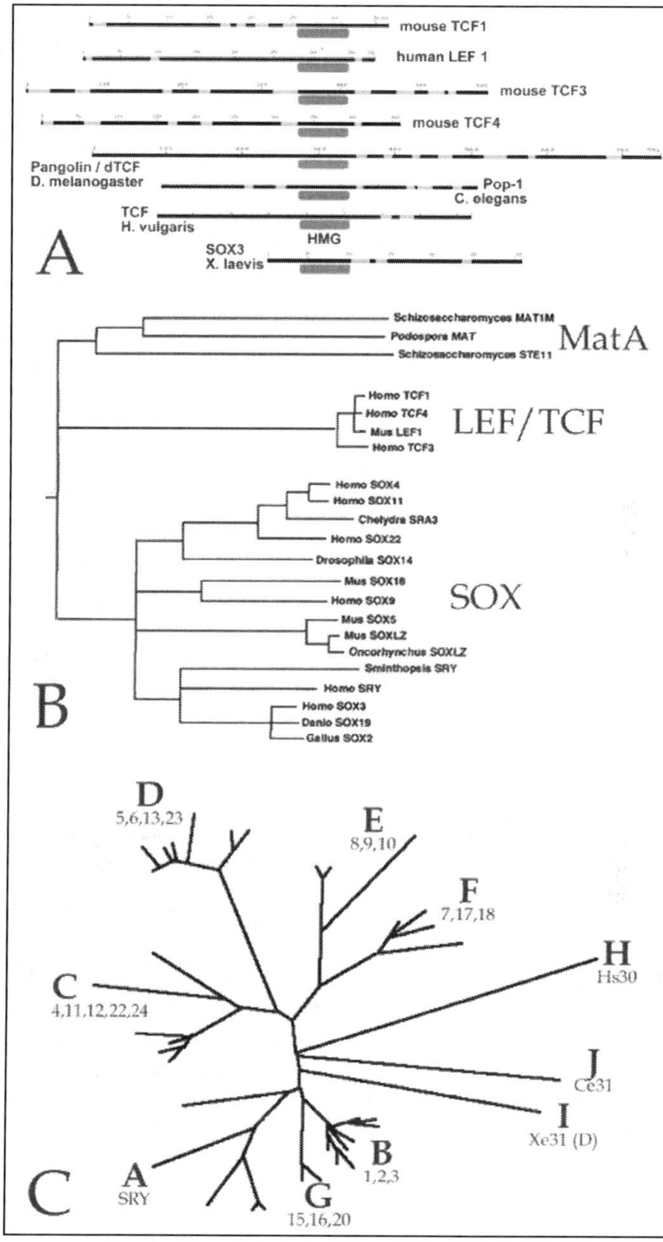

Figure 5. The sequence-specific HMG box DNA-binding proteins. A) The HMG boxes of the vertebrate LEF/TCF polypeptides TCF1, LEF1, TCF3 and TCF4 are aligned together those of Pangolin from *D. melanogaster*, POP-1 from *C. elegans* and TCF from *H. vulgaris*. Images are taken from the BLAST web site and compared to the *Xenopus* SOX3 polypeptide. B) A sequence comparison tree shows the clustering of the MATA, LEF/TCF and SOX polypeptides into distinct groups. Modified from reference (157). C) A complete sequence comparison of the entire SOX family, showing the divisions into 10 subgroups (A-J). The numerical designations of some of the SOX proteins in each group are indicated. The branch lengths reflect the degree of divergence within each subgroup (some subgroups have only a single member). The B group appears to be most highly conserved. Modified from reference 157.

Table 1. Identified TCF/SOX sites in Xenopus genes are compared. The direction to the transcription start site is indicated with an arrow. "SOX" indicates whether the identified TCF site is also a SOX site.

Gene I Site			SOX	References
Xnr5	"A"	←ATCAAAC	N[1]	174,207
	"B"	ATGAAAG→	N	
XSiamois	"0"	← TTCAAAG	_	168
	"1"	ATCAAAG→	_	
	"3"	ATCAAAG→	_	
XTwin	site 1	← TACAAAG	_	200
	site 2	ATCAAAG→	_	
	site 3	← AACAAAG	_	
Xnr3	WRE 2A	ATCAAAG→	_	201[2]
	WRE 2B	AACAAAG→	_	
XBra	TCF Site I	←ATCAATG	N	202
	TCF Site II	ATCAAAG→	N	
Bra[3]	TCF Site II	←AACAATG	_	203
	TCF Site I	TTCAAAG→	N	
Xmyf-5		TTCAAG →	_	204
XFibronectin		←TTCAAAG	N	205
Xengrailed	1	TACAAAG→ (?)	N	195
	2	←ATCAAAG	N	
	3	AACAAAG→ (?)	N	
XSlug distal		TACAAAG→	N	206
proximal		←TACAAAG (?)	N	
Concensus	$^{A}/_{T}$ $^{A}/_{T}$ $^{C}/_{G}$AA$^{A}/_{T}$$^{C}/_{G}$			

[1] Adjacent to two SOX3 binding sites (see ref. 174); [2] The WRE 1 of Xnr3 appears to be indirect; [3] mouse Brachyury

coactivator protein and active chromatin remodeling. Whether the N-terminal domains of other TCF polypeptides acts in a similar self-inhibiting manner remains to be determined.

Snider et al[159] found that I-mfa and a related *Xenopus* protein XIC bind XTCF3 and block its interactions with DNA. I-mfa and related proteins have also been found to interact with Axin[160] and regulate Wnt and c-jun/Jnk signaling pathways. MAP kinase activity can act antagonistically to modulate catenin/TCF DNA-binding.[18] The MAP kinase kinase kinase TAK1 activates the MAP kinase NLK (Nemo-like kinase or NLK), which then phosphorylates the β-catenin/TCF complex, a modification that inhibits complex binding to DNA. How these 'alternative' regulators of LEF/TCF activity interact remains poorly understood.

Xenopus β-Catenin and LEF/TCFs

The initial studies of β-catenin-LEF/TCF interactions were carried out in *X. laevis*, where sperm-entry initiates and biases the direction of a cortical rotation that in turn specifies the future dorsal-ventral axis.[76,161] This rotation leads to a number of molecular asymmetries, most notably higher β-catenin activity on the pro-dorsal compared to the pro-ventral side of the embryo.[162,163] In early embryos, the ventral expression of factors that 'activate' β-catenin or plakoglobin[51,53,54] or that are 'downstream' of β-catenin lead to the formation of a second dorsal axis.[76] Dorsal over-expression of cadherins, which sequester β-catenin, or the depletion

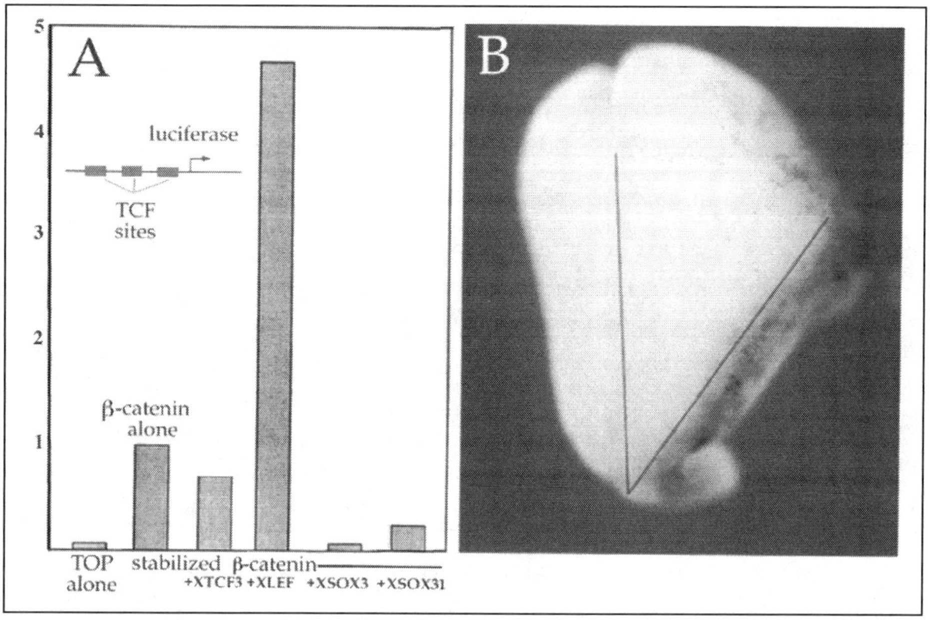

Figure 6. Wnt/SOX interactions. Different assays examine different aspects of the Wnt/SOX interaction. A) In cultured mammalian cells, the TOPFLASH reporter is commonly used to assay β-catenin/TCF mediated gene regulatory activity. In this assay system, a mutationally stabilized form of β-catenin activates the reporter (the level is set arbitrarily to 1). Coexpression of β-catenin and XLEF1 leads to a synergistic activation, coexpression of β-catenin and XTCF3 suppresses activation. Coexpression of XSOX3 or XSOXD (XSOX31) repress activation. B) In vivo dorsal expression of exogenous XSOX3 suppresses dorsal axis duplication induced by exogenous β-catenin[153] and ventralize embryos[165,171] whereas XSOX31 has little if any effect. A neurula state embryo, stained for exogenous XSOX31, is shown. The endogenous and β-catenin-induced axes are indicated by the lines.

of β-catenin leads to the ventralization of the embryo.[52] In this system, ventral expression of mouse LEF1 with or without its β-catenin binding domain can induce a second axis, indicating that LEF1 acts positively on catenin-regulated genes.[143] *Xenopus* LEF1 behaves in a similar manner.[164] In contrast, the dorsal expression of a wild type[165] or a mutated form of *Xenopus* TCF3, unable to bind to β-catenin, leads to embryo ventralization.[56,144,164]

The functional nature of the catenin-LEF/TCF complex has been studied extensively in various cultured cells using various reporters. The most widely used is the TOPFLASH reporter, which contains three LEF/TCF consensus binding sites upstream of a minimal *c-fos* promoter driving firefly luciferase expression (Fig. 6). Molenaar et al[144] found that an HA-epitope tagged form of XTCF3 activated this reporter. This activation may be an artifact of using an epitope-tagged XTCF3 polypeptide; when a different epitope is used, XTCF3 represses β-catenin's ability to activity the reporter[94] (Fig. 6). XTCF3 normally acts negatively on target promoters within early *Xenopus*[56,146] and zebrafish[166] embryos. This conclusion been confirmed in studies of the *Siamois* promoter, a homeobox gene associated with dorsal axis formation in *Xenopus.*[167] Brannon et al[168] found that the deletion of three XTCF3 binding sites a minimal *Siamois* promoter had little effect on dorsal expression but derepressed the ventral expression of the construct. In this light, the ventralizing effects of I-mfa/XIC expression[159] are somewhat confusing, since inhibition of XTCF3 binding to DNA should lead to dorsalization. The situation is further complicated, however, but the fact that both XTCF1 and XTCF4 are also supplied maternally in Xenopous.[56,169]

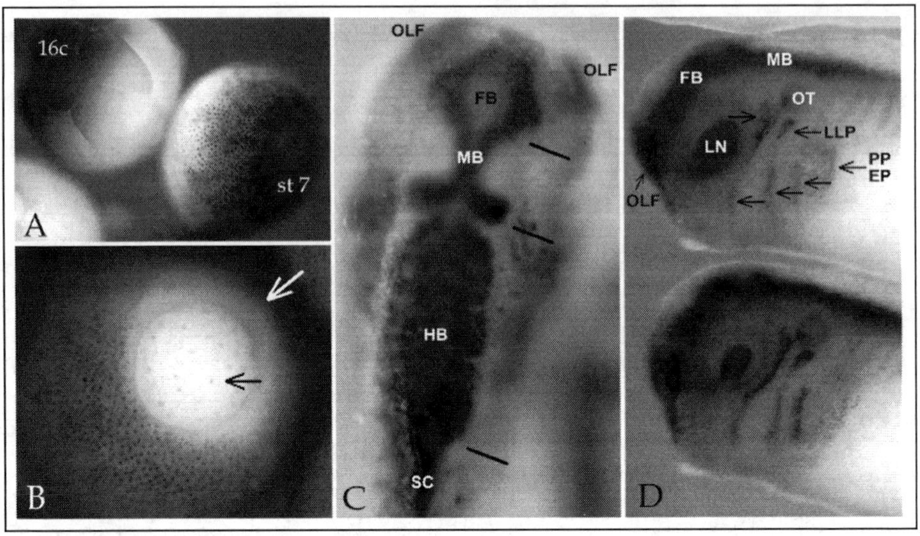

Figure 7. SOX3 distribution in *Xenopus*. Embryos were stained using the antiXSOX3 antibody described in reference 174. In the early cleavage stages (A – "16c"), XSOX3 staining is primarily cytoplasmic and concentrated to the animal hemisphere. In the later blastula (A – "st. 7") staining is nuclear except in those cell that are in the midst of mitosis. B) In gastrula stage embryos, XSOX3 staining can be clearly seen in the nuclei of the vegetal endodermal cells of the yolk plug (black arrow). XSOX3 appears to disappear from the region surrounding the blastopore (white arrow) prior to involution. In later tadpole stage embryos (C and D), XSOX3 staining is largely restricted to the nervous system and epidermal placodes. In panel C boundaries between forebrain ("FB"), midbrain ("MB"), hindbrain ("HB") and spinal cord ("SC") are marked by blank lines. The olfactory placodes ("OLF") are also stained. In panel D, XSOX3 expression can be seen in the lens ("LN"), the otic vesicle ("OT"), the lateral line placodes (upper arrows – "LLP") and in the pharyngeal pouches ("PP"), olfactory ("OLF") and epibranchial placodes ("EP").

SOXs as Modulators of the Catenin Wnt Pathway

The SOXs are a large subgroup of the sequence specific, HMG box DNA binding proteins (Fig. 5).[170] They are grouped together based on a greater than 50% identity of their HMG boxes to the HMG box of the mammalian Sex-Related on the Y (SRY) polypeptide.[170] Related MatA polypeptides are found in the yeasts and other eukaryotes but SOXs appear to be unique to the metazoans.[171] Their consensus binding site is often given as 5' $^{A}/_{T}$ $^{A}/_{T}$CAAAG 3'[172] but their optimal binding, like that of the LEF/TCF proteins, appears to involve 10-12 base pair sequences.[142,173] *Xenopus* SOXs 3 and 17β recognize a core ATTGTT motif and not to common LEF/TCF concensus sequences (see Table 1).[165,174] Moreover, the number of SOX polypeptides that have been studied in detail remains small and efficient DNA interactions may require cooperative binding with partner proteins.[175,176]

Outside of the HMG-box, the level of heterogeneity between SOXs increases dramatically.[170,177] Based on sequence analyses, the SOX family is currently divided into 10 groups A-J (Fig. 5B,C). The B-group is further divided into two sub-groups, the activating B_1 sub-group and the repressive B_2 subgroups,[178] although recent studies in the early *Xenopus* embryo indicates that the B_1 SOX, XSOX3 acts as a repressor on some target genes.[174] Based upon phylogenetic analyses it appears that the B group SOXs are the most evolutionarily conserved, i.e., most slowly changing, group of SOX proteins (Fig. 5). In this light is worth noting that B-group SOX3 protein is maternally expressed in zebrafish[208] and *Xenopus* (Fig. 7)[179] and that SOX2 and SOX3 are expressed very early in the mouse.[180] These early SOX proteins appear to play an essential role in the patterning of the early embryo.[174,180,208]

While the HMG box is commonly thought of exclusively as a DNA binding domain, it has also been found to act as a protein-protein interaction domain.[181] In a number of cases, SOX proteins act via synergistic interactions with other transcription factors. For example there are interactions between specific POU-type homeodomain proteins and specific SOXs.[171,182] There is also evidence that SOX11 interacts with the NLK MAPK kinase in a manner analogous to NLK-TCF interaction.[183]

SOX-Catenin Interactions

The first evidence that SOX proteins could modulate Wnt-catenin-mediated gene expression was the observation by Zorn et al.[165] XSOX17α and β and XSOX3 inhibited β-catenin-dependent axis duplication in *X. laevis*. A number of other SOX polypeptides have been tested for their ability to inhibit catenin-mediated gene expression. What emerges from this analysis is that β-catenin inhibitory activity of a SOX varies dramatically depending upon the system examined. For example both XSOX2 and XSOXD, renamed XSOX31,[170] inhibit the β-catenin-induced activation of the TOPFLASH reporter in cultured mammalian cells, whereas XSOX2 but not XSOXD block β-catenin induced axis duplication in the *Xenopus* embryo (Fig. 6).[174] Not all SOXs interfere with β-catenin-mediate gene regulation. Human SOX7 blocks β-catenin mediated gene expression,[184] while XSOX5 and XSOX11 appear inactive in this assay (unpubl. obs).

The most unexpected finding from the work of Zorn et al[165] was that XSOX17β and XSOX3 bound to β-catenin in GST-pull down and coimmunoprecipitation experiments. There is no obvious sequence similarly between the β-catenin binding domain of LEF/TCF polypeptides, which is located at the N-terminus of these polypeptides, and the β-catenin binding domain of the SOX17β polypeptides, which is located in the C-terminus and appears to include a portion of the HMG box and NLS sequence. Moreover, outside of the HMG box, there is no obvious sequence similarity between XSOX3 and XSOX17β. It is possible that the structural similarity between β-catenin and nuclear localization signal receptor proteins is at least in part responsible for these interactions.

We have found that mutations that eliminate the binding of XSOX3 to simple target DNA sequences, but leave binding to β-catenin intact, leave the polypeptide's ability to inhibit the TOPFLASH reporter in cultured mammalian cells intact. A number of these mutations, however, abolish the ability to inhibit embryonic axis formation.[174] This leads to the conclusion that the ability of XSOX3 and other SOX proteins, e.g., human SOX7,[184] to inhibit the TOPFLASH reporter in cultured mammalian cells is due to competition with LEF/TCFs for β-catenin. In the *Xenopus* embryo, however, different mechanisms are involved; these appears to depends upon SOX's ability to bind specific DNA target sequences.[174]

Generally, β-catenin is thought to be the limiting factor in the canonical Wnt signaling pathway. In immunoprecipitation experiments with anti-XSOX3 and anti-XTCF3 antibodies, we have consistently found that the levels of endogenous XSOX3 and XTCF3 that coprecipitate with endogenous β-catenin are similar and extremely low. The bulk of the soluble β-catenin is presumably associated with other polypeptides, such as ICAT, and so is unavailable for interactions with SOX/TCF proteins.

SOX Modulation of Catenin-Signaling

Some cells express multiple LEF/TCFs, SOXs and regulatory polypeptides. For example an RT-PCR analysis of 293T cells revealed the presence of all four of the human LEF/TCF RNAs.[94] The different SOXs are distributed in dynamic and complex patterns. Consider early *Xenopus* development. XSOX3 (Fig. 7),[179,185] XSOX7,[186] and XSOX11 (originally designated XLS13A and B)[187] are supplied as maternal RNAs. At the onset of neural plate differentiation, these proteins are joined by XSOX2,[188] XSOXD,[189] XSOX11[183] and presumably other SOXs as well. As the neural crest differentiates, XSOX5,[190] XSOX9[57] and XSOX10[191,192] are expressed and required for neural crest differentiation. How these SOX proteins interact with Wnt signaling systems associated with neural crest[16,193,194] and neural plate[195-198] differentiation and patterning, not to mention later developmental events, remains to be studied systematically. It is already clear, however, that manipulation of SOXs levels leads to profound changes in embryonic patterning.[57,188,189] Given the complex and dynamic nature of Wnt signaling,[199] the competition of SOXs and LEF/TCFs for β-catenin binding, as well as the possibility that SOXs and LEF/TCFs may directly compete for binding to a subset of LEF/TCF regulatory sequences, the task of generating an intelligible map of Wnt/SOX interactions should keep students of the subject busy for some time to come.

Acknowledgements

I would like to offer a preemptive apology to all of those whose work has either escaped my attention, which I have failed to appreciate, or was otherwise been unable to incorporate into this review. I thank Gerard Schlosser for a tutorial in *Xenopus* neuroanatomy. I thank all of the current members of the my lab, Tamara Basta, Alex Brown, Shana Fawcett, Tom Gardner, Jon Keeney, Laura Minna-Choi, Evan Trudeau, Neha Valeja and Chi Zhang for their contributions and inspiration. This work has been supported by grants from the NIH (GM54001) and March of Dimes Birth Defects Foundation.

References

1. Muller WE, Bohm M, Grebenjuk VA et al. Conservation of the positions of metazoan introns from sponges to humans. Gene 2002; 295(2):299-309.
2. Pires-daSilva A, Sommer RJ. The evolution of signalling pathways in animal development. Nat Rev Genet 2003; 4(1):39-49.
3. Willert K, Brown JD, Danenberg E et al. Wnt proteins are lipid-modified and can act as stem cell growth factors. Nature 2003; 423(6938):448-52.
4. Hobmayer E, Hatta M, Fischer R et al. Identification of a Hydra homologue of the beta- catenin/ plakoglobin/armadillo gene family. Gene 1996; 172(1):155-9.
5. Hobmayer B, Rentzsch F, Kuhn K et al. WNT signalling molecules act in axis formation in the diploblastic metazoan Hydra. Nature 2000; 407(6801):186-9.
6. Steele RE. Developmental signaling in Hydra: What does it take to build a "simple" animal? Dev Biol 2002; 248(2):199-219.
7. King N, Carroll SB. A receptor tyrosine kinase from choanoflagellates: Molecular insights into early animal evolution. Proc Natl Acad Sci USA 2001; 98(26):15032-7.
8. Snell EA, Furlong RF, Holland PW. Hsp70 sequences indicate that choanoflagellates are closely related to animals. Curr Biol 2001; 11(12):967-70.
9. Lang BF, O'Kelly C, Nerad T et al. The closest unicellular relatives of animals. Curr Biol 2002; 12(20):1773-8.
10. Rokas A, King N, Finnerty J et al. Conflicting phylogenetic signals at the base of the metazoan tree. Evol Dev 2003; 5(4):346-59.
11. King N, Hittinger CT, Carroll SB. Evolution of key cell signaling and adhesion protein families predated the origin of animals. Science 2003; 301:361-163.
12. Grimson MJ, Coates JC, Reynolds JP et al. Adherens junctions and beta-catenin-mediated cell signalling in a nonmetazoan organism. Nature 2000; 408(6813):727-31.
13. Hendriks B, Reichmann E. Wnt signaling: A complex issue. Biol Res 2002; 35(2):277-86.
14. Pandur P, Maurus D, Kuhl M. Increasingly complex: New players enter the Wnt signaling network. Bioessays 2002; 24(10):881-4.
15. Reya T, Duncan AW, Ailles L et al. A role for Wnt signalling in self-renewal of haematopoietic stem cells. Nature 2003; 423(6938):409-14.
16. Wu J, Saint-Jeannet JP, Klein PS. Wnt-frizzled signaling in neural crest formation. Trends Neurosci 2003; 26(1):40-5.
17. Miller JR. The Wnts. Genome Biol 2002; 3(1):REVIEWS3001.
18. Behrens J. Cross-regulation of the Wnt signalling pathway: A role of MAP kinases. J Cell Sci 2000; 113(Pt 6):911-9.
19. Korswagen HC. Canonical and noncanonical Wnt signaling pathways in Caenorhabditis elegans: Variations on a common signaling theme. Bioessays 2002; 24(9):801-10.
20. Malbon CC, Wang H, Moon RT. Wnt signaling and heterotrimeric G-proteins: Strange bedfellows or a classic romance? Biochem Biophys Res Commun 2001; 287(3):589-93.
21. Tamai K, Semenov M, Kato Y et al. LDL-receptor-related proteins in Wnt signal transduction. Nature 2000; 407(6803):530-5.

22. Wehrli M, Dougan ST, Caldwell K et al. arrow encodes an LDL-receptor-related protein essential for wingless signalling. Nature 2000; 407(6803):527-30.
23. Fujino T, Asaba H, Kang MJ. et al. Low-density lipoprotein receptor-related protein 5 (LRP5) is essential for normal cholesterol metabolism and glucose-induced insulin secretion. Proc Natl Acad Sci USA 2003; 100(1):229-34.
24. Mao J, Wang J, Liu B et al. Low-density lipoprotein receptor-related protein-5 binds to Axin and regulates the canonical Wnt signaling pathway. Mol Cell 2001; 7(4):801-9.
25. Wu W, Glinka A, Delius H et al. Mutual antagonism between dickkopf1 and dickkopf2 regulates Wnt/beta-catenin signalling. Curr Biol 2000; 10(24):1611-4.
26. Itasaki N, Jones CM, Mercurio S et al. Wise, a context-dependent activator and inhibitor of Wnt signalling. Development 2003; 130(18):4295-305.
27. Mao B, Wu W, Davidson G et al. Kremen proteins are Dickkopf receptors that regulate Wnt/beta-catenin signalling. Nature 2002; 417(6889):664-7.
28. Mao B, Niehrs C. Kremen2 modulates Dickkopf2 activity during Wnt/lRP6 signaling. Gene 2003; 302(1-2):179-83.
29. Jones SE, Jomary C. Secreted Frizzled-related proteins: Searching for relationships and patterns. Bioessays 2002; 24(9):811-20.
30. Yoshikawa S, McKinnon RD, Kokel M et al. Wnt-mediated axon guidance via the Drosophila Derailed receptor. Nature 2003; 422(6932):583-8.
31. Halford MM, Stacker SA. Revelations of the RYK receptor. Bioessays 2001; 23(1):34-45.
32. Yoshikawa S, Bonkowsky JL, Kokel M et al. The derailed guidance receptor does not require kinase activity in vivo. J Neurosci 2001; 21(1):RC119.
33. Patthy L. The WIF module. Trends Biochem Sci 2000; 25(1):12-3.
34. Medina A, Reintsch W, Steinbeisser H. Xenopus frizzled 7 can act in canonical and noncanonical Wnt signaling pathways: Implications on early patterning and morphogenesis. Mech Dev 2000; 92(2):227-37.
35. Boutros M, Paricio N, Strutt DI et al. Dishevelled activates JNK and discriminates between JNK pathways in planar polarity and wingless signaling. Cell 1998; 94(1):109-18.
36. Boutros M, Mlodzik M. Dishevelled: At the crossroads of divergent intracellular signaling pathways. Mech Dev 1999; 83(1-2):27-37.
37. Axelrod JD, Matsuno K, Artavanis TS et al. Interaction between Wingless and Notch signaling pathways mediated by dishevelled. Science 1996; 271(5257):1826-32.
38. Habas R, Dawid IB, He X. Coactivation of Rac and Rho by Wnt/Frizzled signaling is required for vertebrate gastrulation. Genes Dev 2003; 17(2):295-309.
39. Weston CR, Davis RJ. The JNK signal transduction pathway. Curr Opin Genet Dev 2002; 12(1):14-21.
40. Zhang Y, Neo SY, Wang X et al. Axin forms a complex with MEKK1 and activates c-Jun NH(2)-terminal kinase/stress-activated protein kinase through domains distinct from Wnt signaling. J Biol Chem 1999; 274(49):35247-54.
41. Zhang Y, Neo SY, Han J et al. Dimerization choices control the ability of axin and dishevelled to activate c-Jun N-terminal kinase/stress-activated protein kinase. J Biol Chem 2000; 275(32):25008-14.
42. Cheyette BN, Waxman JS, Miller JR et al. Dapper, a Dishevelled-associated antagonist of beta-catenin and JNK signaling, is required for notochord formation. Dev Cell 200; 2(4):449-61.
43. Gloy J, Hikasa H, Sokol SY. Frodo interacts with Dishevelled to transduce Wnt signals. Nat Cell Biol 2002; 4(5):351-7.
44. Katoh M. Identification and characterization of human DAPPER1 and DAPPER2 genes in silico. Int J Oncol 2003; 22(4):907-13.
45. Thieffry D, Romero D. The modularity of biological regulatory networks. Biosystems 1999; 50(1):49-59.
46. Jong de, Modeling H. Modeling and simulation of genetic regulatory systems: A literature review. J Comput Biol 2002; 9(1):67-103.
47. Muller HJ. Further studies on the nature and causes of gene mutations. Sixth Int Cong Genet 1932; 1:213-255.
48. Brennan K, Klein T, Wilder E et al. Wingless modulates the effects of dominant negative notch molecules in the developing wing of Drosophila. Dev Biol 1999; 216(1):210-29.
49. Loose M, Patient R. A genetic regulatory network for Xenopus mesendodermal formation. Devel Biol 2004; 271:467-478.
50. McMahon AP, Moon RT. Ectopic expression of the proto-oncogene int-1 in Xenopus embryos leads to duplication of the embryonic axis. Cell 1989; 58:1075-1084.

51. McCrea PD, Brieher WM, Gumbiner BM. Induction of a secondary body axis in Xenopus by antibodies to β-catenin. J Cell Biol 1993; 123(2):477-84.
52. Heasman J, Crawford A, Goldstone K et al. Overexpression of cadherins and underexpression of β-catenin inhibit dorsal mesoderm induction in early Xenopus embryos. Cell 1994; 79:791-803.
53. Funayama N, Fagotto F, McCrea P et al. Embryonic axis induction by the armadillo repeat domain of beta- catenin: Evidence for intracellular signaling. J Cell Biol 1995; 128(5):959-68.
54. Karnovsky A, Klymkowsky MW. Anterior axis duplication in Xenopus induced by the over-expression of the cadherin-binding protein plakoglobin. Proc Natl Acad Sci USA 1995; 92(10):4522-6.
55. Carl TF, Dufton C, Hanken J et al. Inhibition of neural crest migration in Xenopus using antisense slug RNA. Dev Biol 1999; 213(1):101-115.
56. Houston DW, Kofron M, Resnik E et al. Repression of organizer genes in dorsal and ventral Xenopus cells mediated by maternal XTcf3. Development 2002; 129(17):4015-25.
57. Spokony RF, Aoki Y, Saint-Germain N et al. The transcription factor Sox9 is required for cranial neural crest development in Xenopus. Development 2002; 129(2):421-32.
58. Grunz H. Developmental biology of amphibians after Hans spemann in germany. Int J Dev Biol 2001; 45(1 Spec No):39-50.
59. Holtfreter JF. A new look at Spemann's organizer. Dev Biol (N Y 1985) 1988; 5:127-50.
60. Schneider SQ, Finnerty JR, Martindale MQ. Protein evolution: Structurefunction relationships of the oncogene beta-catenin in the evolution of multicellular animals. J Exp Zool 2003; 295B(1):25-44.
61. Coates JC, Grimson MJ, Williams RS et al. Loss of the beta-catenin homologue aardvark causes ectopic stalk formation in Dictyostelium. Mech Dev 2002; 116(1-2):117-27.
62. Gelderloos J, Witcher L, Cowin P et al. Plakoglobin: The other ARM of vertebrates. In: Cowin P, Klymkowsky MW, editors. Cytoskeletal-membrane interactions and signal transduction. Austin, Texas: Landes Bioscience, 1997:13-30.
63. Zhurinsky J, Shtutman M, Ben-Ze'ev A. Plakoglobin and beta-catenin: Protein interactions, regulation and biological roles. J Cell Sci 2000; 113(Pt 18):3127-39.
64. Natarajan L, Witwer NE, Eisenmann DM. The divergent Caenorhabditis elegans beta-catenin proteins BAR-1, WRM-1 and HMP-2 make distinct protein interactions but retain functional redundancy in vivo. Genetics 2001; 159(1):159-72.
65. Steinberg MS, McNutt PM. Cadherins and their connections: Adhesion junctions have broader functions. Curr Opin Cell Biol 1999; 11(5):554-60.
66. Pokutta S, Weis WI. The cytoplasmic face of cell contact sites. Curr Opin Struct Biol 2002; 12(2):255-62.
67. Perez-Martin J, Espinosa M. A genetic system to study the in vivo role of transcriptional regulators in Escherichia coli. Gene 1992; 116(1):75-80.
68. Kowalczyk AP, Bornslaeger EA, Norvell SM et al. Desmosomes: Intercellular adhesive junctions specialized for attachment of intermediate filaments. Int Rev Cytol 1999; 185(237):237-302.
69. Garrod DR, Merritt AJ, Nie Z. Desmosomal adhesion: Structural basis, molecular mechanism and regulation (Review). Mol Membr Biol 2002; 19(2):81-94.
70. Bierkamp C, Schwarz H, Huber O et al. Desmosomal localization of beta-catenin in the skin of plakoglobin null-mutant mice [In Process Citation]. Development 1999; 126(2):371-81.
71. Bartnik E, Weber K. Intermediate filaments in the giant muscle cells of the nematode Ascaris lumbricoides; abundance and three-dimensional complexity of arrangements. Eur J Cell Biol 1987; 45:291-301.
72. Bartnik E, Weber K. Widespread occurrence of intermediate filaments in invertebrates: Common principles and diversion. Eur J Cell Bio 1989; 50:17-33.
73. Johansen KM, Johansen J. Filarin, a novel invertebrate intermediate filament protein present in axons and perikarya of developing and mature leech neurons. J Neurobiol 1995; 27(2):227-39.
74. Karabinos A, Schmidt H, Harborth J et al. Essential roles for four cytoplasmic intermediate filament proteins in Caenorhabditis elegans development. Proc Natl Acad Sci USA 2001; 98(14):7863-8.
75. Kofron M, Spagnuolo A, Klymkowsky M et al. The roles of maternal alpha-catenin and plakoglobin in the early Xenopus embryo. Development 1997; 124(8):1553-60.
76. St Amand AL, Klymkowsky MW. Cadherins and catenins, Wnts and SOXs: Embryonic patterning in Xenopus. Int Rev Cytol 2001; 203:291-355.
77. Williams BO, Barish GD, Klymkowsky MW et al. A comparative evaluation of beta-catenin and plakoglobin signaling activity. Oncogene 2000; 19(50):5720-8.
78. Posthaus H, Williamson L, Baumann D et al. Beta-Catenin is not required for proliferation and differentiation of epidermal mouse keratinocytes. J Cell Sci 2002; 115(Pt 23):4587-95.
79. Fagotto F, Gluck U, Gumbiner BM. Nuclear localization signal-independent and importin/karyopherin- independent nuclear import of beta-catenin. Curr Biol 1998; 8(4):181-90.

80. Yokoya F, Imamoto N, Tachibana T et al. Beta-catenin can be transported into the nucleus in a Ran-unassisted manner. Mol Biol Cell 1999; 10(4):1119-31.
81. Lee SJ, Imamoto N, Sakai H et al. The adoption of a twisted structure of importin-beta is essential for the protein-protein interaction required for nuclear transport. J Mol Biol 2000; 302(1):251-64.
82. Riggleman B, Wieschaus E, Schedl P. Molecular analysis of the armadillo locus: Uniformly distributed transcripts and a protein with novel internal repeats are associated with a Drosophila segment polarity gene. Genes Dev 1989; 3(1):96-113.
83. White P, Aberle H, Vincent JP. Signaling and adhesion activities of mammalian beta-catenin and plakoglobin in drosophila. J Cell Biol 1998; 140(1):183-95.
84. Wharton KA, Jr. Runnin' with the Dvl: Proteins that associate with Dsh/Dvl and their significance to Wnt signal transduction. Dev Biol 2003; 253(1):1-17.
85. Ahumada A, Slusarski DC, Liu X et al. Signaling of rat Frizzled-2 through phosphodiesterase and cyclic GMP. Science 2002; 298(5600):2006-10.
86. Ishitani T, Kishida S, Hyodo-Miura J et al. The TAK1-NLK mitogen-activated protein kinase cascade functions in the Wnt-5a/Ca(2+) pathway to antagonize Wnt/beta-catenin signaling. Mol Cell Biol 2003; 23(1):131-9.
87. Capelluto DG, Kutateladze TG, Habas R et al. The DIX domain targets dishevelled to actin stress fibres and vesicular membranes. Nature 2002; 419(6908):726-9.
88. Tanaka M, Kamo T, Ota S et al. Association of Dishevelled with Eph tyrosine kinase receptor and ephrin mediates cell repulsion. Embo J 2003; 22(4):847-58.
89. Aberle H, Bauer A, Stappert J et al. β-catenin is a target for the ubiquitin-proteasome pathway. Embo J 1997; 16(13):3797-804.
90. Wiechens N, Fagotto F. CRM1- and Ran-independent nuclear export of beta-catenin. Curr Biol 2001; 11(1):18-27.
91. Leung JY, Kolligs FT, Wu R et al. Activation of AXIN2 expression by beta-catenin-T cell factor. A feedback repressor pathway regulating Wnt signaling. J Biol Chem 2002; 277(24):21657-65.
92. Hulsken J, Birchmeier W, Behrens J. E-cadherin and APC compete for the interaction with beta-catenin and the cytoskeleton. J Cell Biol 1994:2061-9.
93. Rubinfeld B, Souza B, Albert I et al. The APC protein and E-cadherin form similar but independent complexes with alpha-catenin, beta-catenin, and plakoglobin. J Biol Chem 1995; 270(10):5549-55.
94. Klymkowsky MW, Williams BO, Barish GD et al. Membrane-anchored plakoglobins have multiple mechanisms of action in Wnt signaling. Mol Biol Cell 1999; 10(10):3151-3169.
95. Kodama S, Ikeda S, Asahara T et al. Axin directly interacts with plakoglobin and regulates its stability. J Biol Chem 1999; 274(39):27682-8.
96. Rubenstein A, Merriam J, Klymkowsky MW. Localizing the adhesive and signaling functions of plakoglobin. Dev Genet 1997; 20(2):91-102.
97. Simcha I, Geiger B, Yehuda-Levenberg S et al. Suppression of tumorigenicity by plakoglobin: An augmenting effect of N-cadherin. J Cell Biol 1996; 133:199-209.
98. Kolligs FT, Kolligs B, Hajra KM et al. Gamma-catenin is regulated by the APC tumor suppressor and its oncogenic activity is distinct from that of beta-catenin. Genes Dev 2000; 14(11):1319-31.
99. Shtutman M, Zhurinsky J, Oren M et al. PML is a target gene of beta-catenin and plakoglobin, and coactivates beta-catenin-mediated transcription. Cancer Res 2002; 62(20):5947-54.
100. Winn RA, Bremnes RM, Bemis L et al. Gamma-Catenin expression is reduced or absent in a subset of human lung cancers and reexpression inhibits transformed cell growth. Oncogene 2002; 21(49):7497-506.
101. Rosin-Arbesfeld R, Townsley F, Bienz M. The APC tumour suppressor has a nuclear export function. Nature 2000; 406(6799):1009-12.
102. Rosin-Arbesfeld R, Cliffe A, Brabletz T et al. Nuclear export of the APC tumour suppressor controls beta-catenin function in transcription. Embo J 2003; 22(5):1101-13.
103. Neufeld KL, Nix DA, Bogerd H et al. Adenomatous polyposis coli protein contains two nuclear export signals and shuttles between the nucleus and cytoplasm. Proc Natl Acad Sci USA 2000; 97(22):12085-90.
104. Neufeld KL, White RL. Nuclear and cytoplasmic localizations of the adenomatous polyposis coli protein. Proc Natl Acad Sci USA 1997; 94(7):3034-9.
105. Fodde R, Kuipers J, Rosenberg C et al. Mutations in the APC tumour suppressor gene cause chromosomal instability. Nat Cell Biol 2001; 3(4):433-8.
106. McCartney BM, McEwen DG, Grevengoed E et al. Drosophila APC2 and Armadillo participate in tethering mitotic spindles to cortical actin. Nat Cell Biol 2001; 3(10):933-8.

107. Kaplan KB, Burds AA, Swedlow JR et al. A role for the Adenomatous Polyposis Coli protein in chromosome segregation. Nat Cell Biol 2001; 3(4):429-32.
108. Townsley FM, Bienz M. Actin-dependent membrane association of a Drosophila epithelial APC protein and its effect on junctional Armadillo. Curr Biol 2000; 10(21):1339-48.
109. Zumbrunn J, Kinoshita K, Hyman AA et al. Binding of the adenomatous polyposis coli protein to microtubules increases microtubule stability and is regulated by GSK3 beta phosphorylation. Curr Biol 2001; 11(1):44-9.
110. Jimbo T, Kawasaki Y, Koyama R et al. Identification of a link between the tumour suppressor APC and the kinesin superfamily. Nat Cell Biol 2002; 4(4):323-7.
111. Cui H, Dong M, Sadhu DN et al. Suppression of kinesin expression disrupts adenomatous polyposis coli (APC) localization and affects beta-catenin turnover in young adult mouse colon (YAMC) epithelial cells. Exp Cell Res 2002; 280(1):12-23.
112. Kang D, Soriano S, Xia X et al. Presenilin couples the paired phosphorylation of beta-catenin independent of Axin. Implications for beta-Catenin Activation in Tumorigenesis. Cell 2002; 110(6):751.
113. Kang DE, Soriano S, Frosch MP et al. Presenilin 1 facilitates the constitutive turnover of beta-catenin: Differential activity of Alzheimer's disease-linked PS1 mutants in the beta-catenin-signaling pathway. J Neurosci 1999; 19(11):4229-37.
114. Mak BC, Takemaru K, Kenerson HL et al. The tuberin-hamartin complex negatively regulates Beta -catenin signaling activity. J Biol Chem 2003; 278(8):5947-51.
115. Marambaud P, Shioi J, Serban G et al. A presenilin-1/gamma-secretase cleavage releases the E-cadherin intracellular domain and regulates disassembly of adherens junctions. Embo J 2002; 21(8):1948-56.
116. Baki L, Marambaud P, Efthimiopoulos S et al. Presenilin-1 binds cytoplasmic epithelial cadherin, inhibits cadherin/p120 association, and regulates stability and function of the cadherin/catenin adhesion complex. Proc Natl Acad Sci USA 2001; 98(5):2381-6.
117. Georgakopoulos A, Marambaud P, Efthimiopoulos S et al. Presenilin-1 forms complexes with the cadherin/catenin cell-cell adhesion system and is recruited to intercellular and synaptic contacts. Mol Cell 1999; 4(6):893-902.
118. Xia X, Qian S, Soriano S et al. Loss of presenilin 1 is associated with enhanced beta-catenin signaling and skin tumorigenesis. Proc Natl Acad Sci USA 2001; 98(19):10863-8.
119. Price MA, Kalderon D. Proteolysis of the Hedgehog signaling effector Cubitus interruptus requires phosphorylation by Glycogen Synthase Kinase 3 and Casein Kinase 1. Cell 2002; 108(6):823-35.
120. Jia J, Amanai K, Wang G et al. Shaggy/GSK3 antagonizes Hedgehog signalling by regulating Cubitus interruptus. Nature 2002; 416(6880):548-52.
121. Foltz DR, Santiago MC, Berechid BE et al. Glycogen synthase kinase-3beta modulates notch signaling and stability. Curr Biol 2002; 12(12):1006-11.
122. Cohen P, Frame S. The renaissance of GSK3. Nat Rev Mol Cell Biol 2001; 2(10):769-76.
123. Thomas GM, Frame S, Goedert M et al. A GSK3-binding peptide from FRAT1 selectively inhibits the GSK3-catalysed phosphorylation of axin and beta-catenin. FEBS Lett 1999; 458(2):247-51.
124. Franca-Koh J, Yeo M, Fraser E et al. Dale TC. The regulation of glycogen synthase kinase-3 nuclear export by Frat/GBP. J Biol Chem 2002; 277(46):43844-8.
125. Beals CR, Sheridan CM, Turck CW et al. Nuclear export of NF-ATc enhanced by glycogen synthase kinase-3. Science 1997; 275(5308):1930-4.
126. Huber AH, Weis WI. The structure of the beta-catenin/E-cadherin complex and the molecular basis of diverse ligand recognition by beta-catenin. Cell 2001; 105(3):391-402.
127. Eklof Spink K, Fridman SG, Weis WI. Molecular mechanisms of beta-catenin recognition by adenomatous polyposis coli revealed by the structure of an APC-beta-catenin complex. Embo J 2001; 20(22):6203-12.
128. Graham TA, Weaver C, Mao F et al. Crystal structure of a beta-catenin/Tcf complex. Cell 2000; 103(6):885-96.
129. Poy F, Lepourcelet M, Shivdasani RA et al. Structure of a human Tcf4-beta-catenin complex. Nat Struct Biol 2001; 8(12):1053-7.
130. Tago K, Nakamura T, Nishita M et al. Inhibition of Wnt signaling by ICAT, a novel beta-catenin-interacting protein. Genes Dev 2000; 14(14):1741-9.
131. Graham TA, Clements WK, Kimelman D et al. The crystal structure of the beta-catenin/ICAT complex reveals the inhibitory mechanism of ICAT. Mol Cell 2002; 10(3):563-71.
132. Daniels DL, Weis WI. ICAT inhibits beta-catenin binding to Tcf/Lef-family transcription factors and the general coactivator p300 using independent structural modules. Mol Cell 2002; 10(3):573-84.

133. Kobayashi M, Kishida S, Fukui A et al. Nuclear localization of Duplin, a beta-catenin-binding protein, is essential for its inhibitory activity on the Wnt signaling pathway. J Biol Chem 2002; 277(8):5816-22.
134. Song LN, Herrell R, Byers S et al. Beta-catenin binds to the activation function 2 region of the androgen receptor and modulates the effects of the N-terminal domain and TIF2 on ligand-dependent transcription. Mol Cell Biol 2003; 23(5):1674-87.
135. Sekiya T, Nakamura T, Kazuki Y et al. Overexpression of Icat induces G(2) arrest and cell death in tumor cell mutants for adenomatous polyposis coli, beta-catenin, or Axin. Cancer Res 2002; 62(11):3322-6.
136. Waterman ML, Fischer WH, Jones KA. A thymus-specific member of the HMG protein family regulates the human T cell receptor C alpha enhancer. Genes Dev 1991; 5(4):656-69.
137. Oosterwegel MA, van de Wetering ML, Holstege FC et al. TCF-1, a T cell-specific transcription factor of the HMG box family, interacts with sequence motifs in the TCR beta and TCR delta enhancers. Int Immunol 1991; 3(11):1189-92.
138. Travis A, Amsterdam A, Belanger C et al. LEF-1, a gene encoding a lymphoid-specific protein with an HMG domain, regulates T-cell receptor alpha enhancer function. Genes Dev 1991; 5(5):880-94.
139. Giese K, Cox J, Grosschedl R. The HMG domain of lymphoid enhancer factor 1 bends DNA and facilitates assembly of functional nucleoprotein structures. Cell 1992; 69(1):185-95.
140. Bewley CA, Gronenborn AM, Clore GM. Minor groove-binding architectural proteins: Structure, function, and DNA recognition. Annu Rev Biophys Biomol Struct 1998; 27(105):105-31.
141. Love JJ, Li X, Case DA et al. Structural basis for DNA bending by the architectural transcription factor LEF-1. Nature 1995; 376:791-5.
142. van Beest M, Dooijes D, van De Wetering M et al. Sequence-specific high mobility group box factors recognize 10-12-base pair minor groove motifs. J Biol Chem 2000; 275(35):27266-73.
143. Behrens J, von Kries JP, Kuhl M et al. Functional interaction of β-catenin with the transcription factor LEF-1. Nature 1996; 382:638-642.
144. Molenaar M, van de Wetering M, Oosterwegel M et al. XTcf-3 transcription factor mediates β-catenin-induced axis formation in Xenopus embryos. Cell 1996; 86:391-399.
145. Huber O, Korn R, McLaughlin J et al. Nuclear localization of beta-catenin by interaction with transcription factor LEF-1. Mech Dev 1996; 59(1):3-10.
146. Klymkowsky MW. Minireviews, minidogmas and mythinformation. BioEssays 1997; 20:537-539.
147. Roose J, Huls G, van Beest M et al. Synergy between tumor suppressor APC and the beta-catenin-tcf4 target tcf1. Science 1999; 285(5435):1923-6.
148. Brantjes H, Roose J, van De Wetering M et al. All Tcf HMG box transcription factors interact with Groucho-related corepressors. Nucleic Acids Res 2001; 29(7):1410-9.
149. Bruhn L, Munnerlyn A, Grosschedl R. ALY, a context-dependent coactivator of LEF-1 and AML-1, is required for TCRalpha enhancer function. Genes Dev 1997; 11(5):640-53.
150. Boras K, Hamel PA. Alx4 binding to LEF-1 regulates N-CAM promoter activity. J Biol Chem 2002; 277(2):1120-7.
151. Nishita M, Hashimoto MK, Ogata S et al. Interaction between Wnt and TGF-beta signalling pathways during formation of Spemann's organizer. Nature 2000; 403(6771):781-5.
152. Labbe E, Letamendia A, Attisano L. Association of Smads with lymphoid enhancer binding factor 1/T cell-specific factor mediates cooperative signaling by the transforming growth factor-beta and wnt pathways. Proc Natl Acad Sci USA 2000; 97(15):8358-63.
153. Ross DA, Kadesch T. The notch intracellular domain can function as a coactivator for LEF-1. Mol Cell Biol 2001; 21(22):7537-44.
154. Rocheleau CE, Downs WD, Lin R et al. Wnt signaling and an APC-related gene specify endoderm in early C. elegans embryos. Cell 1997; 90(4):707-16.
155. Thorpe CJ, Schlesinger A, Carter JC et al. Wnt signaling polarizes an early C. elegans blastomere to distinguish endoderm from mesoderm [In Process Citation]. Cell 1997; 90(4):695-705.
156. Merriam JM, Rubenstein AB, Klymkowsky MW. Cytoplasmically anchored plakoglobin induces a WNT-like phenotype in Xenopus. Dev Biol 1997; 185(1):67-81.
157. Zecca M, Basler K, Struhl G. Direct and long-range action of a wingless morphogen gradient. Cell 1996; 87(5):833-44.
158. Tutter AV, Fryer CJ, Jones KA. Chromatin-specific regulation of LEF-1-beta-catenin transcription activation and inhibition in vitro. Genes Dev 2001; 15(24):3342-54.
159. Snider L, Thirlwell H, Miller JR et al. Inhibition of Tcf3 binding by I-mfa domain proteins. Mol Cell Biol 2001; 21(5):1866-73.
160. Kusano S, Raab-Traub N. I-mfa domain proteins interact with Axin and affect its regulation of the Wnt and c-Jun N-terminal kinase signaling pathways. Mol Cell Biol 2002; 22(18):6393-405.

161. Chan AP, Etkin LD. Patterning and lineage specification in the amphibian embryo. Curr Top Dev Biol 2001; 51:1-67.
162. Larabell CA, Torres M, Rowning BA et al. Establishment of the dorso-ventral axis in Xenopus embryos is presaged by early asymmetries in beta-catenin that are modulated by the Wnt signaling pathway. J Cell Biol 1997; 136(5):1123-36.
163. Schneider S, Steinbeisser H, Warga RM et al. b-catenin translocation into nuclei demarcates the dorsalizing centers in frog and fish embryos. Mech Devel 1996; 57:191-198.
164. Roel G, Hamilton FS, Gent Y et al. Lef-1 and Tcf-3 transcription factors mediate tissue-specific Wnt signaling during Xenopus development. Curr Biol 2002; 12(22):1941-5.
165. Zorn AM, Barish GD, Williams BO et al. Regulation of Wnt signaling by Sox proteins: XSox17 alpha/beta and XSox3 physically interact with beta-catenin. Mol Cell 1999; 4(4):487-98.
166. Kim CH, Oda T, Itoh M et al. Repressor activity of Headless/Tcf3 is essential for vertebrate head formation. Nature 2000; 407(6806):913-6.
167. Lemaire P, Garrett N, Gurdon JB. Expression cloning of Siamois, a Xenopus homeobox gene expressed in dorsal-vegetal cells of blastulae and able to induce a complete secondary axis. Cell 1995; 81(1):85-94.
168. Brannon M, Gomperts M, Sumoy L et al. A b-catenin/XTcf-3 complex binds to the siamois promoter to regulate dorsal axis specificiation in Xenopus. Genes & Devel 1997; 11:2359-2370.
169. Roel G, van den Broek O, Spieker N et al. Tcf-1 expression during Xenopus development. Gene Expr Patterns 2003; 3(2):123-6.
170. Bowles J, Schepers G, Koopman P. Phylogeny of the SOX family of developmental transcription factors based on sequence and structural indicators. Dev Biol 2000; 227(2):239-55.
171. Dailey L, Basilico C. Coevolution of HMG domains and homeodomains and the generation of transcriptional regulation by Sox/POU complexes. J Cell Physiol 2001; 186(3):315-28.
172. Laudet V, Stehelin D, Clevers H. Ancestry and diversity of the HMG box superfamily. Nucleic Acids Res 1993; 21(10):2493-501.
173. Mertin S, McDowall SG, Harley VR. The DNA-binding specificity of SOX9 and other SOX proteins. Nucleic Acids Res 1999; 27(5):1359-64.
174. Zhang C, Basta T, Jensen ED et al. The catenin/VegT-regulated early zygotic gene Xnr5 is a direct target of SOX3 regulation in Xenopus. Development 2003; 130:5609-5624.
175. Kamachi Y, Uchikawa M, Kondoh H. Pairing SOX off: With partners in the regulation of embryonic development. Trends Genet 2000; 16(4):182-7.
176. Weiss MA. Floppy SOX: Mutual induced fit in HMG (High-Mobility Group) Box-DNA recognition. Mol Endocrinol 2001; 15(3):353-362.
177. Schepers GE, Teasdale RD, Koopman P. Twenty pairs of sox: Extent, homology, and nomenclature of the mouse and human sox transcription factor gene families. Dev Cell 2002; 3(2):167-70.
178. Uchikawa M, Kamachi Y, Kondoh H. Two distinct subgroups of Group B Sox genes for transcriptional activators and repressors: Their expression during embryonic organogenesis of the chicken. Mech Dev 1999; 84(1-2):103-20.
179. Penzel R, Oschwald R, Chen Y et al. Characterization and early embryonic expression of a neural specific transcription factor xSOX3 in Xenopus laevis. Int J Dev Biol 1997; 41(5):667-77.
180. Avilion AA, Nicolis SK, Pevny LH et al. Multipotent cell lineages in early mouse development depend on SOX2 function. Genes Dev 2003; 17(1):126-40.
181. Harley VR, Lovell BR, Goodfellow PN et al. The HMG box of SRY is a calmodulin binding domain. Febs Lett 1996; 391(1-2):24-8.
182. Kamachi Y, Cheah K, Kondoh H. Mechanism of regulatory target selection by the SOX High-Mobility-Group domain proteins as revealed by comparison of SOX1/2/3 and SOX9. Mol Cell Biol 1999; 19(1):107-120.
183. Hyodo-Miura J, Urushiyama S, Nagai S et al. Involvement of NLK and Sox11 in neural induction in Xenopus development. Genes Cells 2002; 7(5):487-96.
184. Takash W, Canizares J, Bonneaud N et al. SOX7 transcription factor: Sequence, chromosomal localisation, expression, transactivation and interference with Wnt signalling. Nucleic Acids Res 2001; 29(21):4274-83.
185. Fawcett SR, Klymkowsky MW. Embryonic expression of Xenopus laevis SOX7. Gene Expression Report 2004; 4:29-33.
186. Shiozawa M, Hiraoka Y, Komatsu N et al. Cloning and characterization of Xenopus laevis xSox7 cDNA. Biochim Biophys Acta 1996; 1309(1-2):73-6.
187. Hiraoka Y, Komatsu N, Sakai Y et al. XLS13A and XLS13B: SRY-related genes of Xenopus laevis. Gene 1997; 197(1-2):65-71.
188. Kishi M, Mizuseki K, Sasai N et al. Requirement of Sox2-mediated signaling for differentiation of early Xenopus neuroectoderm. Development 2000; 127(4):791-800.

189. Mizuseki K, Kishi M, Shiota K et al. SoxD: An essential mediator of induction of anterior neural tissues in Xenopus embryos. Neuron 1998; 21(1):77-85.
190. Martin BL, Harland RM. Hypaxial muscle migration during primary myogenesis in Xenopus laevis. Dev Biol 2001; 239(2):270-80.
191. Honore SM, Aybar MJ, Mayor R. Sox10 is required for the early development of the prospective neural crest in Xenopus embryos. Dev Biol 2003; 260(1):79-96.
192. Aoki Y, Saint-Germain N, Gyda M et al. Sox10 regulates the development of neural crest-derived melanocytes in Xenopus. Dev Biol 2003; 259(1):19-33.
193. LaBonne C. Vertebrate development: Wnt signals at the crest. Curr Biol 2002; 12(21):R743-4.
194. Garcia-Castro MI, Marcelle C, Bronner-Fraser M. Ectodermal Wnt function as a neural crest inducer. Science 2002.
195. McGrew LL, Takemaru K, Bates R et al. Direct regulation of the Xenopus engrailed-2 promoter by the Wnt signaling pathway, and a molecular screen for Wnt-responsive genes, confirm a role for Wnt signaling during neural patterning in Xenopus. Mech Dev 1999; 87(1-2):21-32.
196. Kazanskaya O, Glinka A, Niehrs C. The role of Xenopus dickkopf1 in prechordal plate specification and neural patterning. Development 2000; 127(22):4981-92.
197. Kiecker C, Niehrs C. A morphogen gradient of Wnt/beta-catenin signalling regulates anteroposterior neural patterning in Xenopus. Development 2001; 128(21):4189-201.
198. Domingos PM, Itasaki N, Jones CM et al. The Wnt/beta-catenin pathway posteriorizes neural tissue in Xenopus by an indirect mechanism requiring FGF signalling. Dev Biol 2001; 239(1):148-60.
199. Maretto S, Cordenonsi M, Dupont S et al. Mapping Wnt/{beta}-catenin signaling during mouse development and in colorectal tumors. Proc Natl Acad Sci USA 2003.
200. Laurent MN, Blitz IL, Hashimoto C et al. The Xenopus homeobox gene twin mediates Wnt induction of goosecoid in establishment of Spemann's organizer. Development 1997; 124(23):4905-16.
201. McKendry R, Hsu SC, Harland RM et al. LEF-1/TCF proteins mediate wnt-inducible transcription from the Xenopus nodal-related 3 promoter. Dev Biol 1997; 192(2):420-31.
202. Vonica A, Gumbiner BM. Zygotic Wnt activity is required for Brachyury expression in the early Xenopus laevis embryo. Dev Biol 2002; 250(1):112-27.
203. Arnold SJ, Stappert J, Bauer A et al. Brachyury is a target gene of the Wnt/beta-catenin signaling pathway. Mech Dev 2000; 91(1-2):249-58.
204. Yang J, Mei W, Otto A et al. Repression through a distal TCF-3 binding site restricts Xenopus myf-5 expression in gastrula mesoderm. Mech Dev 2002; 115(1-2):79-89.
205. Gradl D, Kuhl M, Wedlich D. The Wnt/Wg signal transducer beta-catenin controls fibronectin expression [In Process Citation]. Mol Cell Biol 1999; 19(8):5576-87.
206. Vallin J, Thuret R, Giacomello E et al. Cloning and characterization of three Xenopus Slug promoters reveal direct regulation by Lef/{beta}-catenin signaling. J Biol Chem 2001; 11:11.
207. Hilton E, Rex M, Old R. VegT activation of the early zygotic gene Xnr5 requires lifting Tcf-mediated repression in the Xenopus blasula. Mech Dev 2003; 120:1127-1138.
208. Zhang C, Basta T, Hernandez-Lagunas L et al. Repression of nodal expression by maternal B1-type SOXs regulates germ layer formation in Xenopus and zebrafish. Devel Biol 2004; 273:23-37.

Chapter 19

Cadherin-Mediated Cell-Cell Adhesion and the Microtubule Network

Cécile Gauthier-Rouvière, Marie Causeret, Franck Comunale and Sophie Charrasse

Summary

Classical cadherin adhesion molecules are not only essential for the formation of cell-cell junctions but also act as adhesion-activated signaling receptors involved in a diverse range of physiological processes. Cadherins through their association with catenin proteins interact with the actin cytoskeleton and cadherin-mediated signaling pathways, acting in part through Rho GTPases, regulate cadherin anchoring to the actin cytoskeleton. The microtubule (MT) network recently emerged as having a role in cadherin-mediated cell-cell adhesion. Indeed, MT have been shown to serve as tracks for directed cadherin-containing vesicles movement toward the cell periphery and for the turnover of the junction. In addition, cadherin-based adhesion regulates MT dynamics, which become stabilized. Finally recent data have proposed that association of proteins of the catenin family to MT might be important for linking the MT ends to the F-actin-rich cortex and thus orienting mitotic spindles and the placement of the cytokinetic furrow during cell division.

Introduction

Cadherins are homophilic cell-cell adhesion molecules essential for the organization of cells into tissues during embryonic development. They are also involved in cell growth, migration and differentiation.[1,2] Cell-cell adhesion is often modified in cancer cells and during cell invasion.[3,4] We focus in this chapter on the classical cadherins, defined by their characteristic and highly conserved cytoplasmic domain. Cadherins found in adherens junctions constitute the major family of transmembrane glycoproteins that mediate cell-cell adhesion by virtue of their ability to self-associate in a Ca^{2+}-dependent manner. This homophilic binding is mediated by the N-terminal extracellular domain, which consists of five 110 amino acid repeats (EC1-EC5). Cadherins provide anchoring sites for the actin cytoskeleton through the binding of catenins.[5] β-catenin and γ-catenin (plakoglobin) bind directly to the distal region of the cadherin cytoplasmic tail and interact with α-catenin, which associates with actin filaments.[6] A fourth catenin, the phosphoprotein p120 interacts with the juxtamembrane region of cadherins thereby modulating their dimerization and adhesive function[7] (see Fig. 1).

Cadherins are not only structural components of adherens junction but also act as adhesive receptors and induce signaling programs leading for example to cell differentiation.[8] In particular, Rho family GTPases activity is changed by the formation of cadherin-dependent cell-cell contacts. Rac1 is activated after VE- or E-cadherin activation,[9-13] so is Cdc42.[14] RhoA activity is inhibited in E-cadherin-expressing MDCK cells while it is stimulated after N-cadherin-mediated adhesion[8,9] (Fig. 2). These differences are likely due to the cell types used. In fact, cadherins

Rise and Fall of Epithelial Phenotype: Concepts of Epithelial-Mesenchymal Transition, edited by Pierre Savagner.

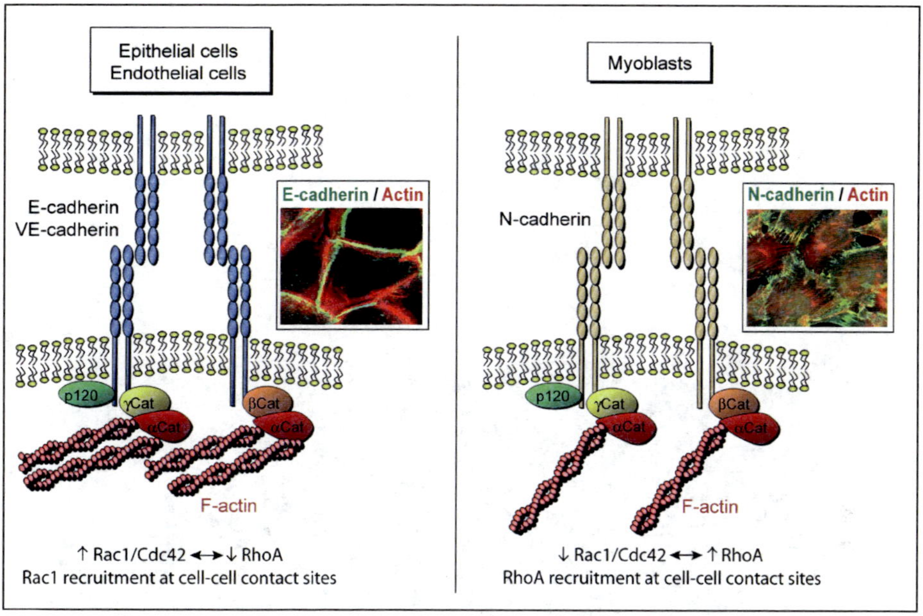

Figure 1. Signaling pathways triggered by cadherin receptors differ with the cell type. A schematic diagram of the proteins that constitute the cadherin-dependent cell-cell adhesion site. Dimers of cadherin molecules associated with cytoplasmic catenins are found at the cell surface. Whereas the overall organization of the cadherin/catenin complexes are similar between epithelial and nonepithelial cells, the downstream signaling pathways differ. The molecular mechanisms remain to be determined. Note that the organization of the junction as well as the F-actin cytoskeleton are extremely different between a keratinocyte (image obtained from V. Braga) and a myoblast.

display tissue specific expression (E- in epithelia, N- in neuronal, cardiac and skeletal muscle...) as well as spatio-temporally regulated expression profile which might generate complex downstream signaling pathways to coordinate tissue development.

In addition, adherens junctions are highly dynamic structures that turn over rapidly. During embryonic development or tumor progression, changes in cadherin function and availability at cell-cell contacts have been reported. For example, during epithelium-mesenchyme transitions, which occur during specific stages of embryonic development (see also Chapters 1-4) but also under pathological conditions (see also Chapters 8, Arnoux et al and 9, Van Marck and Bracke), intercellular adhesions are disrupted through downregulation of E-cadherin activity.[4] Similarly, during migration of neural crest cells, the localization and function of N-cadherin is regulated (see refs. 15,16 and see also Chapter 3, Newgreen). One possible mechanism for modulation of adhesive function could occur through the turn-over of cadherins at the cell surface. The availability of cadherins might be modulated through changes in either secretory or endocytotic pathways. The observation that a recycling E-cadherin pool increases in the absence of stable cell-cell contacts supports this hypothesis.[17] Rho-family GTPases RhoA, Rac1 and Cdc42 are key regulators of cadherin-mediated cell-cell adhesion (see ref. 18 and see also Chapter 17: Nakagawa et al). Again the consequence of GTPase signaling depends on the cellular context. In epithelial cells and keratinocytes, inactivation of RhoA inhibits the accumulation of E-cadherin at sites of cell-cell contact the same way Rac1 inactivation does.[19-23] Consistently, Tiam1, a guanine exchange factor (GEF) for Rac1, increases E-cadherin-mediated adhesion and inhibits hepatocyte growth factor-induced-cell scattering in MDCK cells.[24] In epithelial cells, Rac1 activation after initial contact formation seems to be important to extend

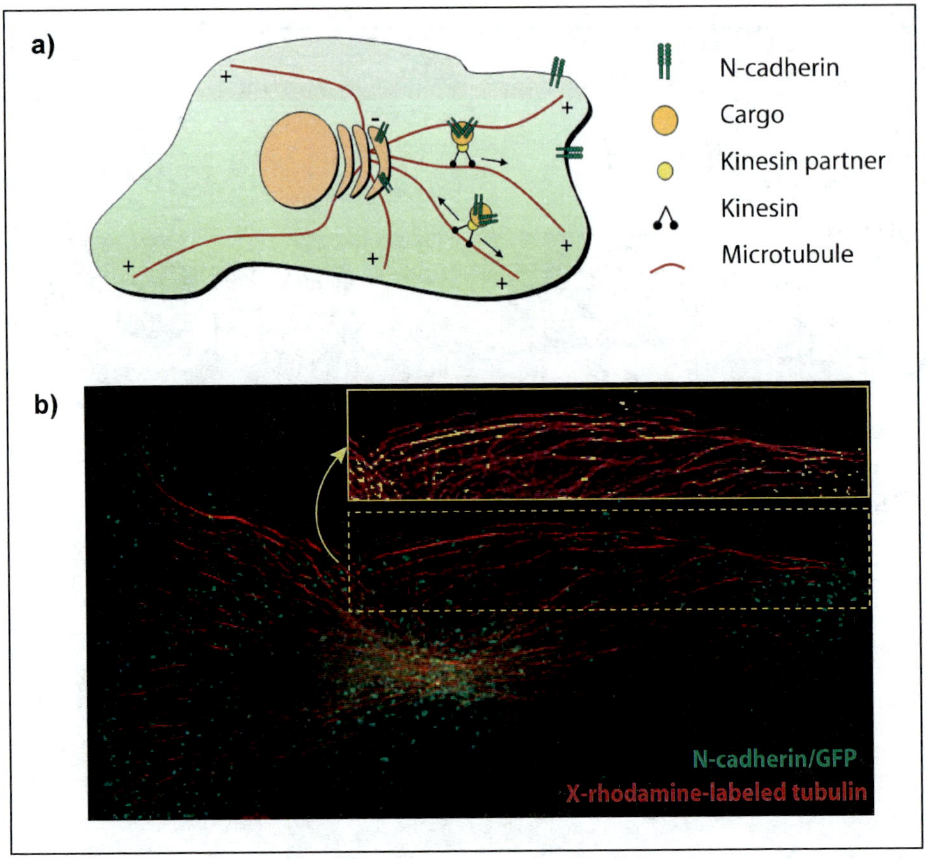

Figure 2. MT are involved in a kinesin-dependent transport of N-cadherin-containing cargo. a) Schematic distribution of N-cadherin in the golgi, in secretory and endocytotic vesicles and at the plasma membrane. The N-cadherin-containing vesicles are moved along MT by kinesin motor proteins. b) N-cadherin/GFP-expressing C2C12 myoblast was microinjected with X-rhodamine-labeled tubulin and the distribution of GFP and rhodamine-labeled probes was examined by time lapse microscopy. Images were captured every 10 s. for 10 min. and stacks of images were deconvolved using the Huygens System image restoration software. One deconvolved plan visualized using the Iview command of the Imaris software is shown. Respective voxel colocalization of N-cadherin/GFP with MT in the selected area (dashed line) obtained using the Imaris Colocalization module is shown in the yellow scare. MT (in red) and the voxel colocalization (in yellow) are shown.

the regions of contact.[11,25] Nevertheless, in other studies in the same cell type, Rac1 has opposite effects. Beyond experimental differences, this might illustrate that Rac1 is a key regulator of both adherens junction assembly and disassembly. Although Rac1 regulates adherens junction assembly through reorganization of the actin cytoskeleton, it also regulates adherens junction disassembly via endocytosis of E-cadherin.[26] Cdc42 is also required for E-cadherin-mediated cell-cell adhesion.[22,27] In myoblasts, stable localization of N-cadherin at cell-cell contacts requires RhoA activity, whereas inhibition of either Rac1 or Cdc42 has no effect[8] (MC and CGR, personal observations). Further studies are required to elucidate if these discrepancies are due to the cellular context, the cadherin molecule itself or illustrate the existence of a complex kinetics of Rho GTPases activation after cadherin engagement.

Activated GTP-bound Rho GTPases interact with specific effectors, some of which are involved in the regulation of cadherin-mediated adhesion. The Rac1 and Cdc42 effector IQGAP1 negatively regulates E-cadherin-mediated cell-cell adhesion by interacting with β-catenin.[28-30] The RhoA effector mDia1, a member of the formin-homology family of proteins, promotes the formation of α-catenin/β-catenin complexes and localizes adherens junction components to the cell periphery.[31] E-cadherin interacts with the Arp2/3 actin nucleator complex, whose activity is controlled by Rac1 and Cdc42.[32] Finally, cytosolic p120 can interact with Vav2, a guanine exchange factor for Rho GTPases, suggesting that p120 localization is associated with Rho GTPase activation.[33,34] Various mechanisms may therefore exist to ensure Rho GTPases-dependent cadherin adhesion.

While it is clear that cadherin complexes associate with the actin cytoskeleton, evidences involving the MT network in the regulation of cadherin-mediated adhesion came only very recently. MT are essential for many cellular functions including vesicle transport, cell motility, polarity and division. MT are polarized tubular structures which are produced by linear polymerization of α/β tubulin heterodimers, with a faster assembly at the plus end over the minus end. The plus ends of MT display dynamic instability modulated by a number of MT-associated proteins (MAP).[35] Post-translational modifications of tubulin have also been described.[36] For example the best studied which is tubulin detyrosination, occurs in stabilized MT. Nevertheless, it seems that detyrosination is not involved in MT stabilization per se but regulates the interaction of motor proteins and organelles with stable MT.[37] In migrating cells, in fibroblasts, and in white blood cells, MT are organized with their minus ends anchored to the centrosome adjacent to the nucleus or free and facing toward the cell center and their plus ends extending toward the cell periphery. In many differentiated cells, such as the polarized epithelial cells and neurons, MT are noncentrosomal but are nevertheless uniformly polarized with their plus ends oriented towards the basal surface, or the growth cone, respectively.[38] In these cell types, MT are less dynamic but also support vesicular trafficking and polarity. Finally, connections between MT and F-actin exist since MT interact with F-actin and MT plus ends are targeted to focal adhesion complexes and modulate their turnover.[39-41]

Delivery of Cadherin and Catenin to the Cell Periphery through Microtubules

In 1983, Geller and Lilien have proposed that gp130/4.8, now called N-cadherin, is dynamically inserted into the plasma membrane through a MT-dependent fashion.[42] To reach this conclusion, these authors have dissociated embryonic chick neural retina cells by trypsinization in the absence of divalent cations and then analyzed iodinatable polypeptides appearing at the cell surface during cellular repair by two-dimensional polyacrylamide gel electrophoresis. Recently, analyzing the behavior of a GFP-tagged N-cadherin construct by videomicroscopy in myoblasts or fibroblasts, we have demonstrated that the initial formation of N-cadherin-dependent cell-cell contacts results from the recruitment of the intracellular pool of N-cadherin to the plasma membrane.[43] Most of this intracellular pool is present in vesicular structures associated with and moving along MT in a kinesin-dependent way (Fig. 2). This MT-dependent secretory pathway might be an efficient and controlled mechanism to deliver N-cadherin to specific sites at the plasma membrane. This key role for MT in the maintenance of cell-cell adhesion was also reported in newt lung epithelial cells but not in keratinocytes.[44,45] Other studies are required to understand such discrepancies and to analyze whether other cadherin might be similarly transported along MT. N-cadherin is the only cadherin described so far as using a MT-dependent secretory pathway to reach the plasma membrane. MT might also be important for the endocytosis of cadherin occuring during normal and pathological epithelium to mesenchyme transition.[46] In addition, p120 catenin was recently found associated with MT.[46a,46b] This observation open new research avenues and it is possible that MT localization is important for the activity of p120 towards Rho family GTPases. Moreover, Rho proteins have been shown to be involved in MT-associated transport. RhoA and RhoG interact with kinectin, a membrane-anchoring protein for kinesin motor.[47] In addition, lysophosphatidic acid induces Rho-mediated stabilization of MT.[48] Once capped and stabilized, these MT might be for example detyrosinated and thus be involved in preferential interaction of vesicles and organelles leading to a cadherin-mediated polarization of MT toward the cell-cell contacts. Thus Rho GTPases could affect membrane trafficking of cadherin/catenin-containing cargo through their effects on a MT-dependent transport.

Cadherin-Dependent Cell-Cell Contact Regulates Microtubules Stability

As mentioned in the introduction, cadherin-based adhesion initiates intracellular signals allowing adapted cell responses during cell proliferation, migration and differentiation. One of these cell responses is the regulation of MT dynamic. Indeed, the dynamic behavior of MT plus ends is affected by cell-cell contacts.[44] Analysis of individual MT by time-lapse digital fluorescence microscopy reveals that plus end dynamic instability is suppressed in fully contacted cells, with individual MT exhibiting an extended state of pause, suggested that they become capped. Several proteins have been described to localize specifically to MT plus ends, including CLIP-170, EB-1, adenomatous polyposis coli (APC) and dynein. Interestingly, dynein, a MT-based motor which belongs to a multiprotein complex containing dynactin, p150Glued and dynamitin involved in both anterograde and retrograde MT transport, is localized at adherens junction in epithelial cells where it interacts by β-catenin.[49] These data suggest that cytoplasmic dynein found at adherens junctions might capture and tether MT at these sites. Whether dynein at cell-cell contact sites is involved in a retrograde transport of cargo remains to be analyzed. Another interesting candidate is the MT tip protein CLIP-170 described to associate with the Rac1 and Cdc42 effector IQGAP.[50] In fibroblasts, IQGAP is localized at the polarized leading edge and might interact with both CLIP-170 and activated Rac1 and Cdc42 acting as a linker between the plus ends of MT and cortical actin. IQGAP is also found localized at cell-cell contact sites in epithelial cells. At this place, IQGAP may also act as a linker between activated Rac1 or Cdc42 and CLIP-170 and participate in MT capture. Since Rac1 activity increases after E-cadherin-mediated adhesion, this is conceivable. Two other candidates for a function in the possible link between cadherin and MT dynamics are APC and the APC-binding protein EB1.[51,52] Moreover, cadherin-mediated MT stabilization may also act through regulating the activity of proteins such as stathmin/op18 or MAP. In addition to regulating MT plus end dynamics, cadherin-mediated adhesion stabilizes MT minus ends.[53] The importance of MT capture processes for the formation of cell-cell junction per se and /or for the cellular organization or polarization remains to be determined (Fig. 3). Most of the proteins which control MT dynamic and capture are themselves controlled by Rho GTPases, key regulators of the actin cytoskeleton indicating that these two cytoskeletal components are coordinately regulated.

M-cadherin, which is specifically expressed in skeletal muscle cells is also associated with MT.[54] This association was reported to occur at the cell boundary and might allow the organization of the MT network parallel to the longitudinal axis. Interestingly, differentiating myoblasts display stabilized microtubules.[55] M-cadherin-dependent cell signaling may have a role in both the regulation of MT stability and polarity.

Catenin-Dependent Regulation of Spindle Localization

The position of the mitotic spindle plays a key role in the spatial control of cell division. In combination to polarity cues, the position of the spindle ensures the correct segregation of cell-fate determinants during development.[56] The correct spindle position is achieved through interactions between spindle, astral MT and cortical actin. Interestingly, in the syncytial Drosophila embryo, the β-catenin homolog Armadillo is required to tether the spindle to cortical actin.[57] The entire complex contains α-catenin, β-catenin and a protein related to APC, and it

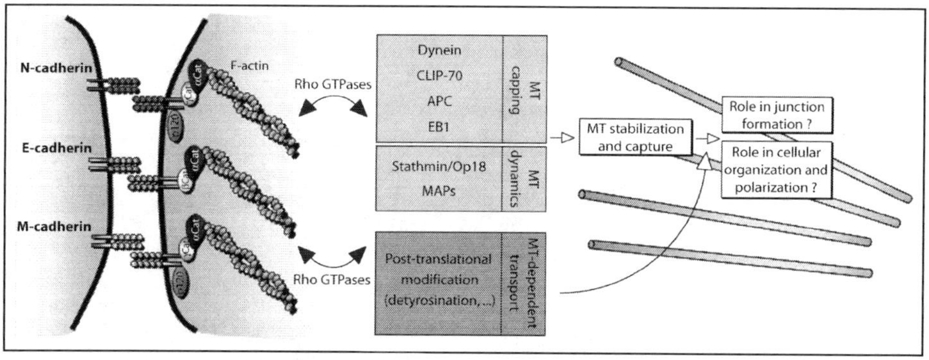

Figure 3. Potential mechanisms for how cadherin-mediated adhesion could regulate aspects of the microtubule cytoskeleton. Cadherin-mediated adhesion activates Rho GTPases which in turn might induce change in MT organization and dynamics. Rho GTPases control downstream pathways through the regulation of specific effectors such as IQGAP, Par6 and mDia which have all been described to interact with the MT tip protein CLIP-170, Dynein/dynactin and EB1. These pathways regulate MT dynamics and capture. Another pathway regulated by the Rac GTPase through the kinase Pak results in the phosphorylation and the decrease of the MT catastrophe activity of stathmin/Op18. Stabilized MT might be post-translationaly modified being then distinguishable from their dynamic counterparts. These stable modified MT might be preferentially bound by kinesin motors making them specific highways for vesicle transport.

is likely to interact with a protein of the EB1 family which localizes to the plus end of growing MT.[58] Thc same type of complex might allow cadherin/catenin to regulate MT dynamics in interphasic cells (see previous section and Fig. 4). Asymmetrical distributed DE-cadherin is essential to determine the mitotic spindle orientation.[59] The small GTPase Cdc42, acting through a Par6-atypical protein kinase C (aPKC) complex, is also required to establish asymmetric cell

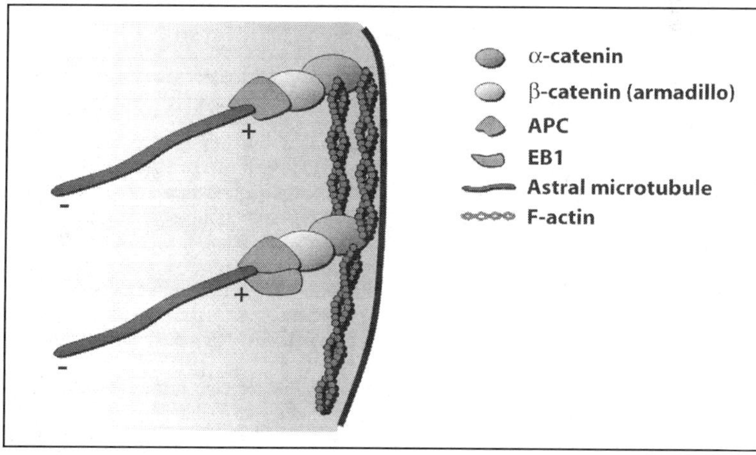

Figure 4. Hypothetical model of the interaction of the α-catenin/β-catenin/APC complex with cortical F-actin and spindle and astral MTs. α-catenin anchors the complex to cortical F-actin. The anchoring to astral MT is mediated either by APC or through the interaction between APC and EB1, two proteins located at the plus ends of MT.

division. Whether Cdc42 regulates glycogen synthase kinase-3β and APC in the context of asymmetric cell division remains to be determined.[60]

Concluding Remarks

Taken together, the observations discussed above point out a picture in which cadherin-mediated adhesion not only controls the actin cytoskeleton but also the MT dynamics and polarity. Rho GTPases are key players in these processes and are involved in the bidirectional signaling between cadherin and these two cytoskeletal elements. The formation of new cadherin cell-cell contacts may lead to the activation of small GTPases, which would then induce changes in both MT-based transport and MT organization and dynamics. Alternatively, MT reorganization induced after cadherin-dependent adhesion could somehow regulate the activity of Rho GTPases. Several RhoGEFs have been described to interact with MT, their activity might be sensitive to the MT organization and dynamics.

Acknowledgements

We are grateful to A. Blangy and P. Fort for critical reading of the manuscript and to V. Braga and P. Anastasiadis for sharing image and data. Research activity was supported by grants from Association pour la Recherche contre le Cancer, Ligue contre le Cancer and Association Française contre les Myopathies.

References

1. Gumbiner BM. Cell adhesion: The molecular basis of tissue architecture and morphogenesis. Cell 1996; 84:345-357.
2. Takeichi M. Cadherin cell adhesion receptors as a morphogenetic regulator. Science 1991; 251:1451-1455.
3. Christofori G. and Semb H. The role of the cell-adhesion molecule E-cadherin as a tumour- suppressor gene. Trends Biochem Sci 1999; 24:73-76.
4. Takeichi M. Cadherins in cancer: Implications for invasion and metastasis. Curr Opin Cell Biol 1993; 5:806-811.
5. Kemler R. From cadherins to catenins: Cytoplasmic protein interactions and regulation of cell adhesion. Trends Genet 1993; 9:317-321.
6. Jou TS, Stewart DB, Stappert J et al. Genetic and biochemical dissection of protein linkages in the cadherin-catenin complex. Proc Natl Acad Sci USA1995; 92:5067-5071.
7. Yap AS, Niessen CM, Gumbiner BM. The juxtamembrane region of the cadherin cytoplasmic tail supports lateral clustering, adhesive strengthening, and interaction with p120ctn. J Cell Biol 1998; 141:779-789.
8. Charrasse S, Meriane M, Comunale F et al. N-cadherin-dependent cell-cell contact regulates Rho GTPases and beta-catenin localization in mouse C2C12 myoblasts. J Cell Biol 2002; 158:953-965.
9. Noren NK, Niessen CM, Gumbinrt BM et al. Cadherin engagement regulates Rho family GTPases. J Biol Chem 2001; 276:33305-33308.
10. Lampugnani MG, Zanetti A, Breviario F et al. VE-cadherin regulates endothelial actin activating Rac and increasing membrane association of Tiam. Mol Biol Cell 2002; 13:1175-1189.
11. Kovacs EM, Ali RG, McCormack AJ et al. E-cadherin homophilic ligation directly signals through Rac and phosphatidylinositol 3-kinase to regulate adhesive contacts. J Biol Chem 2002; 277:6708-67182.
12. Nakagawa M, Fukata M, Yamaga M et al. Recruitment and activation of Racl by the formation of E-cadherin- mediated cell-cell adhesion sites. J Cell Sci 2001; 114:1829-1838.
13. Arthur WT, Noren NK, Burridge K. Regulation of Rho family GTPases by cell-cell and cell-matrix adhesion. Biol Res 2002; 35:239-246.
14. Kim SH, Li Z, Sacks DB. E-cadherin-mediated cell-cell attachment activates Cdc42. J Biol Chem 2000; 275:36999-370050.
15. Akitaya T, Bronner-Fraser M. Expression of cell adhesion molecules during initiation and cessation of neural crest cell migration. Dev Dyn 1992; 194:12-20.
16. Monier-Gavelle F, Duband JL. Control of N-cadherin-mediated intercellular adhesion in migrating neural crest cells in vitro. J Cell Sci 1995; 108:3839-3853.
17. Le TL, Yap AS, Stow JL. Recycling of E-cadherin: A potential mechanism for regulating cadherin dynamics. J Cell Biol 1999; 146:219-232.

18. Fukata M, Kaibuchi K. Rho-family GTPases in cadherin-mediated cell-cell adhesion. Nat Rev Mol Cell Biol 2001; 2:887-897.
19. Braga VM, Machesky LM, Hall A et al. The small GTPases Rho and Rac are required for the establishment of cadherin-dependent cell-cell contacts. J Cell Biol 1997; 137:1421-1431.
20. Braga VM. Small GTPases and regulation of cadherin dependent cell-cell adhesion. Mol Pathol 1999; 52:197-202.
21. Braga VM, Del Maschio A, Machesky L et al. Regulation of cadherin function by Rho and Rac: Modulation by junction maturation and cellular context. Mol Biol Cell 1999; 10:9-22.
22. Kuroda S, Fukata M, Fujii K et al. Regulation of cell-cell adhesion of MDCK cells by Cdc42 and Rac1 small GTPases. Biochem Biophys Res Commun 1997; 240:430-435.
23. Takaishi K, Sasaki T, Kotani H et al. Regulation of cell-cell adhesion by rac and rho small G proteins in MDCK cells. J Cell Biol 1997; 139:1047-1059.
24. Hordijk PL, ten Klooster JP, van der Kammen RA et al. Inhibition of invasion of epithelial cells by Tiam1-Rac signaling. Science 1997; 278:1464-1466.
25. Ehrlich JS, Hansen MD, Nelson WJ. Spatio-temporal regulation of Rac1 localization and lamellipodia dynamics during epithelial cell-cell adhesion. Dev Cell 2002; 3:259-270.
26. Akhtar N, Hotchin NA. RAC1 regulates adherens junctions through endocytosis of E-cadherin. Mol Biol Cell 2001; 12:847-862.
27. Kodama A, Takaishi K, Nakano K et al. Involvement of Cdc42 small G protein in cell-cell adhesion, migration and morphology of MDCK cells. Oncogene 1999; 18:3996-4006.
28. Fukata M, Kuroda S, Nakagawa M et al. Cdc42 and Rac1 regulate the interaction of IQGAP1 with beta-catenin. J Biol Chem 1999; 274:26044-26050.
29. Fukata M, Nakagawa M, Itoh N et al. Involvement of IQGAP1, an effector of Rac1 and Cdc42 GTPases, in cell-cell dissociation during cell scattering. Mol Cell Biol 2001; 21:2165-2183.
30. Kuroda S, Fukata M, Nakagawa M et al. Cdc42, Rac1, and their effector IQGAP1 as molecular switches for cadherin-mediated cell-cell adhesion. Biochem Biophys Res Commun 1999; 262:1-6.
31. Sahai E, Marshall CJ. ROCK and Dia have opposing effects on adherens junctions downstream of Rho. Nat Cell Biol 2002; 4:408-415.
32. Kovacs EM, Goodwin M, Ali RG et al. Cadherin-directed actin assembly: E-cadherin physically associates with the Arp2/3 complex to direct actin assembly in nascent adhesive contacts. Curr Biol 2002; 12:379-382.
33. Anastasiadis PZ, Moon SY, Thoreson MA et al. Inhibition of RhoA by p120 catenin. Nat Cell Biol 2000; 2:637-644.
34. Noren NK, Liu BP, Burridge K et al. 120 catenin regulates the actin cytoskeleton via Rho family GTPases. J Cell Biol 2000; 150:567-580.
35. Mitchison T, Kirschner M. Dynamic instability of microtubule growth. Nature 1984; 312:237-242.
36. Infante AS, Stein MS, Zhai Y et al. Detyrosinated (Glu) microtubules are stabilized by an ATP-sensitive plus-end cap. J Cell Sci 2000; 113(Pt 22):3907-3919.
37. Liao G, Gundersen GG. Kinesin is a candidate for cross-bridging microtubules and intermediate filaments. Selective binding of kinesin to detyrosinated tubulin and vimentin. J Biol Chem 1998; 273:9797-9803.
38. Keating TJ, Borisy GG. Centrosomal and noncentrosomal microtubules. Biol Cell 1999; 91:321-329.
39. Kaverina I, Rottner K, Small JV. Targeting, capture, and stabilization of microtubules at early focal adhesions. J Cell Biol 1998; 142:181-190.
40. Kaverina I, Krylyshkina O, Small JV. Microtubule targeting of substrate contacts promotes their relaxation and dissociation. J Cell Biol 1999; 146:1033-1044.
41. Salmon WC, Adams MC, Waterman-Storer CM. Dual-wavelength fluorescent speckle microscopy reveals coupling of microtubule and actin movements in migrating cells. J Cell Biol 2002; 158:31-37.
42. Geller RL, Lilien J. Repair of a calcium-dependent adhesive mechanism of embryonic neural retina cells: Kinetic and molecular analysis. J Cell Sci 1983; 60:29-49.
43. Mary S, Charrasse S, Meriane M et al. Biogenesis of N-cadherin-dependent cell-cell contacts in living fibroblasts is a microtubule-dependent kinesin-driven mechanism. Mol Biol Cell 2002; 13:285-301.
44. Waterman-Storer CM, Salmon WC, Salmon ED. Feedback interactions between cell-cell adherens junctions and cytoskeletal dynamics in newt lung epithelial cells. Mol Biol Cell 2000; 11:2471-2483.
45. Kee SH, Steinert PM. Microtubule disruption in keratinocytes induces cell-cell adhesion through activation of endogenous E-cadherin. Mol Biol Cell 2001; 12:1983-1993.
46. Morali OG, Delmas V, Moore R et al. IGF-II induces rapid beta-catenin relocation to the nucleus during epithelium to mesenchyme transition. Oncogene 2001; 20:4942-4950.
46a. Yanagisawa M, Kaverina IN, Wang A et al. A novel interaction between kinesin and p120 modulates p120 localization and function. J Biol Chem 2004; 273:9512-9521.

46b. Chen X, Kojima S, Borisy G et al. p120 catenin associates with kinesin and facilitates the transport of cadherin-catenin complexes to intercellular junctions. J Cell Biol 2003; 163:547-557.

47. Vignal E, Blangy A, Martin M et al. Kinectin is a key effector of RhoG microtubule-dependent cellular activity. Mol Cell Biol 2001; 21:8022-8034.

48. Cook TA, Nagasaki T, Gundersen,GG. Rho guanosine triphosphatase mediates the selective stabilization of microtubules induced by lysophosphatidic acid. J Cell Biol 1998; 141:175-185.

49. Ligon LA, Karki S, Tokito M et al. Dynein binds to beta-catenin and may tether microtubules at adherens junctions. Nat Cell Biol 2001; 3:913-917.

50. Fukata M, Watanabe T, Noritake J et al. Rac1 and Cdc42 capture microtubules through IQGAP1 and CLIP-170. Cell 2002; 109:873-885.

51. Barth AI, Siemers KA, Nelson WJ. Dissecting interactions between EB1, microtubules and APC in cortical clusters at the plasma membrane. J Cell Sci 2002; 115:1583-1590.

52. Mogensen MM, Tucker JB, Mackie JB et al. The adenomatous polyposis coli protein unambiguously localizes to microtubule plus ends and is involved in establishing parallel arrays of microtubule bundles in highly polarized epithelial cells. J Cell Biol 2002; 157:1041-1048.

53. Chausovsky A, Bershadsky AD, Borisy GG. Cadherin-mediated regulation of microtubule dynamics. Nat Cell Biol 2000; 2:797-804.

54. Kaufmann U, Kirsch J, Irintchev A et al. The M-cadherin catenin complex interacts with microtubules in skeletal muscle cells: Implications for the fusion of myoblasts. J Cell Sci 1999; 112(Pt 1):55-68.

55. Gundersen GG, Khawaja S, Bulinski JC. Generation of a stable, posttranslationally modified microtubule array is an early event in myogenic differentiation. J Cell Biol 1989; 109:2275-2288.

56. Gonczy P. Mechanisms of spindle positioning: Focus on flies and worms. Trends Cell Biol 2002; 12:332-339.

57. McCartney BM, McEwen DG, Grevengoed E et al. Drosophila APC2 and Armadillo participate in tethering mitotic spindles to cortical actin. Nat Cell Biol 2001; 3:933-938.

58. Su LK, Burrell M, Hill DE et al. APC binds to the novel protein EB1. Cancer Res, 1995; 55:2972-2977.

59. Le Borgne R, Bellaiche Y, Schweisguth F. Drosophila E-cadherin regulates the orientation of asymmetric cell division in the sensory organ lineage. Curr Biol 2002; 12:95-104.

60. Etienne-Manneville S, Hall A. Cdc42 regulates GSK-3b and adenomatous polyposis coli to control cell polarity. Nature 2003; 421:753-756.

Chapter 20

Matrix Metalloproteases and Epithelial-to-Mesenchymal Transition: Implications for Carcinoma Metastasis

Christine Gilles, Donald F. Newgreen, Hiroshi Sato and Erik W. Thompson

General Considerations of the EMT

The epithelial to mesenchymal transition (EMT) is characterized by the loss of epithelial characteristics and the gain of mesenchymal attributes in epithelial cells. It has been associated with physiological and pathological processes requiring epithelial cell migration and invasion. Initially, EMT was observed in embryological and adult development with many well characterized examples including the conversions of epiblast to primary mesenchyme (gastrulation), somite to sclerotome, somite to dermis, myotome to migratory myoblast, dorsal neural tube to neural crest, placodal ectoderm to cranial ganglion precursor, intermediate mesoderm to nephric mesenchyme, lateral mesoderm to connective/muscular tissue, endocardium to cardiac cushion mesenchyme and trophectoderm invasion.[1,2] In addition, evidence is mounting to support an important role of EMT pathways in the progression of carcinoma to metastasis providing epithelial tumour cells with the ability to migrate, invade the surrounding stroma and disseminate in secondary organs.[3-5]

Target Genes of the EMT

A variety of general hallmarks exist for the assignment of epithelial versus mesenchymal phenotype (Fig. 1). Intermediate filament proteins provide a convenient and abundant marker, with keratins indicating epithelium and vimentin indicating a mesenchymal phenotype.[6] This relationship breaks down in early development, where some cells which are clearly epithelial, as judged by junctional and basal lamina criteria, lack cytokeratins and many also possess vimentin. However, in these cases, it is the vimentin positive epithelia, such as the neuroepithelium, which often subsequently undergo dramatic reorganizations, including tissue folding and EMTs.[7,8] Another commonly employed index for the epithelial state is the presence and junctional localization of the classically epithelial homotypic cell adhesion molecule E-cadherin and/or the associated catenins, forming the adherens junctions. These cell-cell adhesion-related criteria are almost entirely absent in mesenchymal cells.[3] Even though keratin / vimentin and E-cadherin have been and are still the widest used markers of the EMT, a variety of other mechanisms and molecules have now been implicated in physiological and pathological EMT pathways, as reviewed by others (reviewed in refs. 2, 4-6, 9). They include a reorganization of other cell-cell contact complexes (tight junctions, desmosomes), a modification of cell-substrate adhesion complexes, the synthesis of extracellular matrix proteins normally expressed by mesenchymal cells such as fibronectin or collagens I/III, and the expression of several proteases including matrix metalloproteases (MMPs)

Rise and Fall of Epithelial Phenotype: Concepts of Epithelial-Mesenchymal Transition, edited by Pierre Savagner.

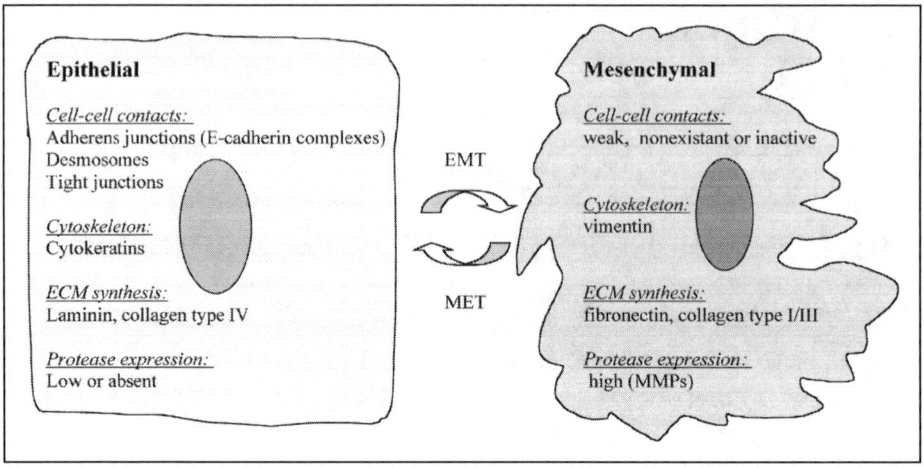

Figure 1. Molecular traits of the EMT. Epithelial and mesenchymal cells are shown schematically, and the differences commonly seen between them is grouped into four major categories: Cell-cell contacts, cytoskeleton, ECM synthesis, and protease expression. Although this is somewhat generalized, most EMT systems show these changes.

which are also predominantly expressed by stromal cells (Fig. 1). These molecular changes confer on epithelial cells the ability to scatter, migrate and degrade ECM components, all properties they do not display in a normal cohesive epithelium.

Inducers and Regulators of the EMT

Several growth factors (epidermal growth factor "EGF", basic fibroblast growth factor "FGF-2", hepatocyte growth factor "HGF", transforming growth factor β1 "TGF-β1") have been documented to trigger or at least modulate EMT phenomena.[4,5,9] The reorganization of cell adhesion molecules (E-cadherin complexes, zonula occludens) has also been shown to trigger EMT changes in different cell systems.[3,10] Also, even though they are rather considered as target genes of the EMT, several ECM components or even MMPs can in some cell systems serve as initiator of EMT changes.[11,12]

The implication of several signaling pathways in the control of the EMT is now clearly established. Thus *ras*, MAPK, PIP3K, *rho*, *rac* and *src* have been shown to control EMT events and regulate EMT target genes (reviewed in refs. 4, 5, 9).

Also, transcription factors of the Snail family (Snail, Slug) and of the ETS family, as well as the transcription factor SIP-1, have been directly implicated in the regulation of EMT target genes.[13-19] This is also well established for β-catenin, which, once relocalized from the membrane E-cadherin complexes, can translocate in the nucleus where it can act as a co-transcription factor and directly regulate gene expression through its binding to TCF/LEF transcription factors.[20]

Many cellular and molecular aspects of EMT have been characterised, and our particular interest has been in the changes associated with a class of extracellular proteases, the Matrix Metalloproteinases (MMPs). The goal of this article is to review the literature and our own data implicating MMPs and their regulation in EMT processes, both normal and pathologic.

Matrix Metalloproteases

The MMP family currently comprises 23 human homologs.[21-24] MMP protein structure is made of specific domains some of which are common to all MMPs. These conserved domains include a "pre" domain which directs the MMPs to the endoplasmic reticulum, a "pro" domain which maintains the MMP in an inactive form and a "catalytic" domain (Fig. 2). Most MMPs

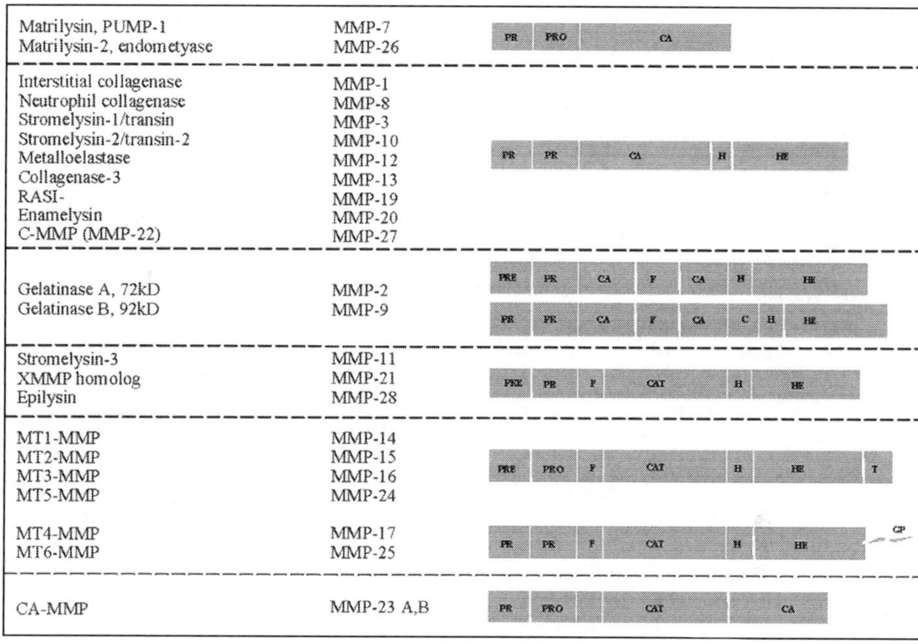

Figure 2. Schematic diagram showing the grouping of MMPs based on domain structure. Drawing courtesy of Dr. Neeracha Ruangpanit.

also contain a c-terminal hemopexin-like domain, which mediates interactions with substrates and in some cases, directs substrate specificity and participates in substrate binding. This is attached to the catalytic domain by a flexible linker termed the hinge region. A specialized, fibronectin-like gelatin binding domain is found in the two gelatinases (MMP-2, MMP-9) and facilitates binding to type IV collagen. The generation of active enzymes, known as the activation process, requires the cleavage of the "pro" domain. In some cases, this is effected constitutively by furin-like enzymes which cleave a consensus sequence near the end of the prodomain. MMP activity can be regulated by specific tissue inhibitors of MMPs (TIMP-1 to -4).

Collectively, MMPs can degrade virtually every component of the ECM. Initially, MMPs were thought to predominantly degrade specific components of the ECM thereby providing new substrates facilitating migration and invasion. Since then, it has become clear that by degrading ECM components, MMPs can also modulate signaling pathways from the ECM and modulate the bioavailability of growth factors. Furthermore, others substrates have now been identified, the cleavage of which is also involved in increased migratory and invasive properties. Thus, cell adhesion molecules (E-cadherin, CD44, αv integrin)[25-27] or growth factor receptors (FGF receptor 1, members of the EGF receptor family HER2 and HER4, c-met)[22,28-30] can be processed by MMP-dependent proteolysis. Particular attention has been paid to the gelatinases MMP-2 and MMP-9 (gelatinase A and B, respectively), previously denoted type-IV collagenases (72 kDa and 92 kDa type IV collagenases, respectively), since they specifically can degrade the type IV collagen. This forms a major component of the basement membranes normally segregating epithelial tissues from surrounding mesenchyme. Loss of the basement membrane is one of the most reliable signs of poor prognosis in most carcinoma systems.[31,32] Like most other MMPs,[23,33] MMP-2 is secreted in a latent form, requiring activation. This is effected on the cell surface by a membrane-associated subclass of MMPs called the membrane-type MMPs (MT-MMP).[34] Six MT-MMPs have been identified so far (MMP-14, 15, 16, 17, 24, 25), all of which except for MMP-17 have pro-MMP-2 activation function.[34,35] MT1-MMP was the first MT-MMP identified as an activator of pro-MMP-2.[36] Activation of MMP-2 by MT1-MMP is particularly well documented and involves the formation of a ternary complex between MT1-MMP, pro-MMP-2 and TIMP-2. Besides their implication in MMP-2 activation, MT-MMPs can also contribute directly to cell migration and invasion by degrading specific substrates. Thus MT1-MMP can cleave several specific substrates including collagens (I, II, III), laminin and fibronectin. Through their anchoring at the plasma membrane, MT-MMPs, and particularly MT1-MMP, have been shown to play a key role in the pericellular proteolysis associated with cell migration and invasion.[21,34,37,38]

MMPs in general have been implicated in many steps of malignancy, including primary tumour growth, angiogenesis, invasion of the basement membrane and stroma, and metastatic progression.[22,24,33] With the notable exception of MMP 7/matrilysin,[39] the consensus view is that MMPs are in general not produced by epithelial cells but rather by the surrounding stromal cells.[33,40] However, we will discuss here data leading to the conclusion that expression of "stromal" MMP's is one of the major attributes that epithelial cells acquire after undergoing the EMT.

One of the major implications of this review is thus to counteract the notion that MMPs are exclusively produced by the peri-tumoural stroma. In contrast, we suggest that under the appropriate stimuli, genetic or epigenetic, certain epithelial cells will undergo EMT-like changes, and exhibit MMP production. We will summarize observations regarding MMP alterations and regulation in a number of EMT systems, both normal and neoplastic. Although our review will focus primarily on the MT-MMP/ MMP-2 axis, other MMP associations from our own work and the published literature will be summarized. The relationships which exist between the EMT and MMP-2-activation, especially those which are common to different systems, may provide insights into the implication of MMPs in EMT pathways and in their ability to modulate the migratory and invasive phenotype of epithelial cells.

MMPs in Carcinoma Model Systems for the EMT

In Vitro Observations

There are many reports of the expression of MMPs in epithelial tumour cell lines. We will only discuss here studies reporting MMP expression in epithelial cell lines in relation to well defined EMT changes. Many data have been generated by comparing different cell lines of the same origin showing different invasive potentials. Also, some cell lines have been shown to be inducible for EMT changes, either by exogenous factors or by migration opportunity in in vitro wound assays. Using these in vitro models, it has become clear that the EMT is a dynamic process and that different intermediate phenotypes can be observed. Also, whether only one EMT pathway exists is still unknown, and it can thus be considered that different EMT pathways could generate different phenotypes. Accordingly, variation may also be observed in the overall migratory/invasive behavior. For instance, based on signal transduction criteria, cell scattering can be considered to be different from, or a part of, full EMT events.[41] As such, it does not necessarily correlate with increased motility and migration. Vice-versa, active cell migration does not necessary imply comprehensive cell scattering. For instance, the migration of some tumour cell types or the archetypal developmental EMT and cell migration, such as that of the neural crest (see Chapter 3), when viewed with time lapse in situ, shows that (former) epithelial cells can migrate as dynamic groups and are not always scattered individuals.[42,43] The in vitro observations discussed below are summarized in Table 1.

NBT-II Cells

Perhaps the oldest and most studied carcinoma model of the EMT is the NBT-II rat bladder carcinoma system (reviewed in ref. 9). These cells show EMT changes to a variety of specific stimuli including collagens, HGF, FGF-1 and TGFα. Responses to FGF-1 have been mapped to a splice variant of the FGFR-2,[9,44] and shown to be mediated, in part at least, by the Snail family mesenchymal-inducing transcription factor Slug.[15] Regarding MMP expression, one of the most rapidly detected changes in stimulated NBT-II cells is the secretion of MMP-9 and MMP-2, some of which appears in the media in the active form,[45,46] which further suggests MT1-MMP expression.

Table 1. MMP expression and regulation in in vitro EMT models

Origin of the Cell Line	Inducer of the EMT/ Comparison of Different Lines with Different Invasive Degrees	EMT Traits	MMP Expressed in EMT
Rat bladder carcinoma NBT- II cells	Numerous inducers (bFGF, HGF, collagens,...)	Cell scattering, reorganization of E-cadherin	MMP-2[45]
Madin Darby canine kidney (MDCK)	Numerous inducers (bFGF, HGF, oncogenes...)	Cell scattering, reorganization of E-cadherin, tube formation in collagen gels	MT1-MMP[38,105,106]
Human breast cancer cell lines	comparison	Loss of E-cadherin, vimentin expression, invasiveness	MT1-MMP MMP-2 activation[54-57]
Human breast MCF10A cells	Oncogenes (v-Ha-ras and erbB2)	Loss of E-cadherin, vimentin expression, invasiveness	MMP-2[63]
Human MCF10A breast cell line	Wound assay	E-cadherin reorganization, vimentin expression, migration	MT1-MMP[66]
Mouse mammary SpC-2 cells	MMP-3	Loss of E-cadherin, vimentin expression, invasiveness	MMP-3, MMP-9, MMP-13[12]
Mouse mammary NMuMG cells	TGF-β	Transformation to extremely spindle morphology from cobblestone appearance, branching tubules in 3D collagen gels.	MT1-MMP Sato, unpublished data
Mouse mammary tumour cell lines	Comparison	Vimentin expression, Invasiveness	MMP-2 activation[76]
Rat mammary carcinoma cell lines	Comparison	Epithelial (E) and Mesenchymal (M) variants of the same tumour based on morphology	MMP-3, MMP-9, MMP-13[78]
Human bronchial cell lines	Comparison	Lack of E-cadherin, vimentin expression, invasiveness	MMP-2, MT1-MMP[53,79]
Human bronchial epithelial cells	Wound assay	Vimentin expression, migration	MMP-9, MMP-3, MMP-11[113, 114]
Human prostate cancer cell lines	Comparison	Lack of E-cadherin, vimentin expression, invasiveness	MT1-MMP Williams & Thompson, unpublished also[81,82]
Human cervical cell lines	Human Papillomavirus 33+ Ha-ras	Lack of E-cadherin, vimentin expression, invasiveness	MT1-MMP, MMP-2 activation[84]
Squamous cell carcinoma A431	snail	Loss of E-cadherin, vimentin expression	MMP-2[85]

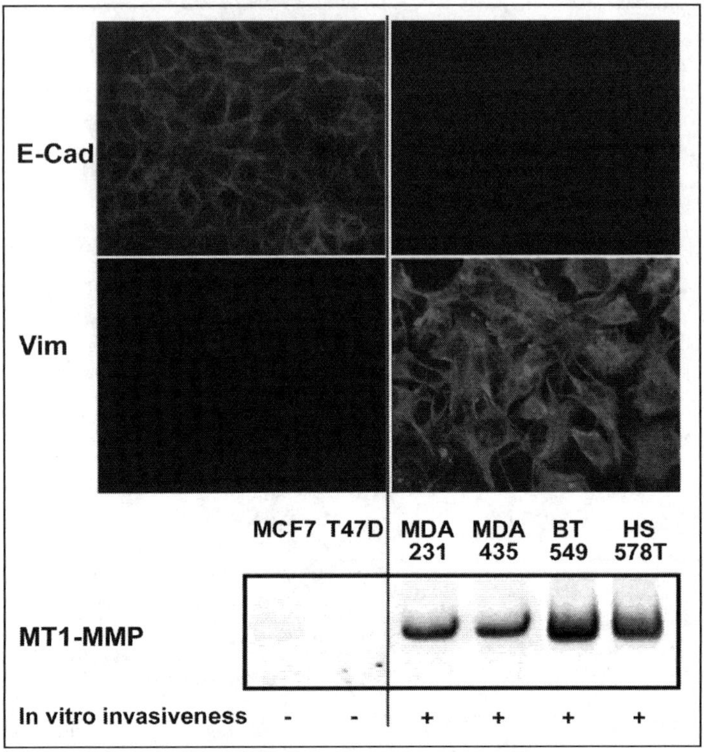

Figure 3. EMT in human breast cancer cell lines characterized by vimentin (Vim) expression, MT1-MMP expression and loss of E-cadherin (E-cad) associated with invasive properties.

Human Breast Cancer Cell Lines

Considerable circumstantial evidence exists in human breast cancer (HBC) cell lines in support of both the occurrence of EMT-like changes and their association with a more aggressive phenotype (also reviewed in refs. 47, 48). Cell lines which are invasive in vitro, many of which also metastasize in immuno-compromised hosts, show mesenchymal tendencies. They express vimentin, show reduced cytokeratins, and lack E-cadherin, in contrast to poorly invasive HBC cell lines which lack vimentin, express keratin abundantly, and in some but not all cases express functional E-cadherin.[49-53] We have also found that the vimentin-positive HBC cell lines express c-*ets*-1, a member of the ETS transcription factor family usually expressed by mesenchymal cells and largely implicated in MMP regulation.[18] When we examined proteases shown to be regulated by c-*ets*-1, we saw differential expression of MMP-1, but no expression of MMP-3, as well as differential induction of the plasmin-generating axis.[54] Also, invasive vimentin-positive HBC cell lines were found to express MT1-MMP (Fig. 3) and activate MMP-2[55] in response to ConA or collagen type I which subsequently increased MT1-MMP mRNA and protein.[56,57] Interestingly, all but one of the invasive HBC cell lines also expresses mRNA for MT3-MMP, while none expressed MT2-MMP mRNA, and both some non-invasive and some invasive lines expressed MT4-MMP.[57] A very comprehensive survey of MMP expression by numerous HBC cell lines has been published.[58,59]

Co-expression of MMPs in conjunction other mesenchymal markers in human breast cancer cell lines was also verified by independent gene array analysis by Zajchowski et al, who found that a set of 24 gene products could be used to predict the degree of invasiveness of untested cell lines.[60] Higher expression of typically epithelial gene products (e.g., keratin-18, keratin-19, plakoglobin) was seen in the less invasive cells, while high expression of mesenchymal markers (e.g., integrin α3, TIMP-2, TIMP-3, BIG-H3, PAI-1, FRA-1, vimentin, osteonectin, TSP-1, collagen α1(VI) and α2(I), and thrombospondin-1) typified the highly invasive cell lines. It was notable that MT1-MMP was co-expressed with these markers in the invasive cells. Similarly, Jechlinger et al performed microarray analysis on mammary cells undergoing an EMT following ras transfection and in the list of genes upregulated, they found MMP-2, MMP-13 and MMP-12.[61]

We have examined the potential involvement of the EMT in two other human mammary systems that were developed to model various stages of malignancy. Transformation of the A1N4 human mammary epithelial cells with SV40 middle T antigen and v-Ha-ras, or v-mos and v-Ha-ras, but not either oncogene alone, conferred dramatically increased invasiveness and stellate Matrigel outgrowth properties on the cells. Although vimentin expression was apparent in all A1N4-derived cultures, the proportion of vimentin expression increased in the invasive transformants.[62] No differences in MMP-2 or MMP-9 secretion were observed, however MT1-MMP status or collagen/ConA-induced MMP-2-activation potential was not examined. A parallel analysis was performed with various oncogene transformants derived from the MCF10A human mammary system, where again a combination of v-Ha-ras and erbB2, but not either oncogene alone, conferred a highly invasive phenotype.[63] In this case, MMP-2 expression was increased, and TIMP-2 decreased, however again, the MMP-2 activation/ MT1-MMP expression was not examined. Other studies also demonstrated an inducible EMT with a temporary expression of vimentin in MCF10A cells migrating to fill a monolayer wound, and indicated a functional role for vimentin in the migratory phenotype using antisense technology.[64] In this system, vimentin expression also associates with a reorganization of E-cadherin/ β-catenin complexes and β-catenin transactivation pathway.[65] Corresponding to these EMT changes, an induction of MT1-MMP was also found exclusively in migratory cells undergoing an EMT but not in the stationary ones.[66]

Rodent Mammary Systems

A surprising twist to the relationship between EMT and MMP came from studies in the laboratories of Bissell and Werb, with observations that stromelysin-1 (MMP-3) could initiate EMT changes in SpC-2 mouse mammary cells.[12,67] This extends considerable observations of MMP-3 expression during mammary gland morphogenesis (for review see refs. 68), and accelerated ductal development in transgenic MMTV-MMP-3 mouse mammary glands over-expressing activated MMP-3.[69,70] Such over-development can progress to carcinoma in the WAP-MMP-3-transgenic mammary gland.[70] Further studies have shown that MMP-3 is critical for mouse mammary cell invasion in vitro,[71] and that malignant mouse mammary cells, as opposed to normal counterparts, show transcriptional regulation of MMP-3 similar to that seen in fibroblasts.[72] MMP-2 and MMP-11 are also upregulated with mouse mammary gland development,[69] and although stronger in humans than mouse, so is MMP-7.[39,73] The capacity of MMP-3 to initiate the EMT in the SpC-2 mouse mammary system is unprecedented, and MMP-expression is usually placed downstream of the EMT in the same cell type. In the case of the mammary gland, however, it is predominantly stromal cells which produce the MMP-3[68,69] and this in turn could influence EMT-like changes in the neighboring mammary epithelial cells. However, MMPs, including MMP-3, also appear to be downstream targets of the EMT since endogenous MMP-3, MMP-9 and MMP-13 are upregulated following EMT induction with MMP-3.[12]

NMuMG cells, another mouse mammary epithelial cell line, undergoes a rapid and striking mesenchymal transdifferentiation in response to TGF-β, mediated by the TGF-β type I receptor/Alk-5 and SMAD proteins.[74,75] This EMT is indicated by dramatically altered cytoarchitecture, reorganization of the actin cytoskeleton to stress fibers, and down regulation of E-cadherin and β-catenin. Although MMP-2 activation analysis has not been performed on these cells after TGF-β treatment, MT1-MMP expression is seen in these cells when they form branching tubular structures in collagen gels (Sato, unpublished data).

We have recently examined another syngeneic mouse mammary cancer model system comprising a panel of cells lines clonally isolated from a naturally occurring mouse mammary tumour. These cell lines show differential metastatic potential (4T1 >> 66Cl4 >> 67NR), and further single cell cloning has yielded a variant which spontaneously metastasizes to bone from the mammary fat pad (4T1.2[76]). We have characterized whether the EMT-like patterns accompany metastatic potential in these cells, and find that while all four cell lines express vimentin and show stellate Matrigel outgrowth, the metastatic lines show much increased Boyden chamber chemoinvasion, and selectively activate MMP-2 (4T1.2>66Cl-4). This was surprising since all of the progression variants showed similar levels of MT1-MMP expression.[77] Thus, we believe that all of these cell lines have undergone some level of EMT, as indicated by their expression of vimentin and stellate outgrowth, but those which are metastatic have a more complete set of mesenchymal traits, while the non-metastatic variants either lost, or never developed, some molecular traits critical for increased MMP-2-activation and in vitro invasion. It is noteworthy that the metastatic 4T1.2 cells do not show a particularly mesenchymal appearance on plastic culture, being rather rounded.

Studying a metastatic variant of a rat carcinoma cell lines, Martorana et al also found that it has undergone an EMT and expressed higher level of MMP-3, MMP-9 and MMP-13. Furthermore, cross regulation of MMP expression was observed between the epithelial cells and the mesenchymal variants.[78]

Bronchial Cell Lines

An association between fibroblastoid features (vimentin expression, loss of E-cadherin) and increased invasiveness was also reported in human bronchial epithelial tumour cells. Indeed, the 16HBE14o-vimentin-negative, E-cadherin positive cell line formed cohesive clusters and was not invasive whereas BZR vimentin-positive cells displayed a high dispersion ability and were highly invasive in the Boyden chamber assay. Correspondingly, BZR cells synthesized MMP-2 and MT1-MMP in contrast to 16HBE14o-.[53,79] In this system, a direct link was also made between E-cadherin/β-catenin complex reorganization and MMP regulation. Treatment of non-invasive, MMP-negative 16HBE14o- cells with a soluble fragment of E-cadherin (sE-cadherin), known to be generated by protease activity and known to enhances epithelial cell migration,[25] resulted in enhanced MMP-2, MMP-9 and MT1-MMP expression.[53] This also enhanced in vitro invasiveness of the cells. On the other hand, transfection of an intact E-cadherin cDNA into invasive BZR cells clearly diminished MMP-1, MMP-3, MMP-9, and MT1-MMP. Using different in vitro (cell dispersion, modified Boyden chamber) and in vivo assays (human airway epithelial xenograft), it was also shown that E-cadherin transfectants displayed decreased invasive abilities.[80]

Prostate Cell Lines

We recently analyzed human prostate carcinoma cell lines for the same parameters as described above for the breast cancer model systems (Williams and Thompson, unpublished observations). Indeed, we found that cell lines known to be metastatic in immunocompromised mice showed increased Boyden chamber activity and a tendency towards more extensive outgrowth in matrigel, although the differential was not as clear as seen in HBC cell lines. The invasive cells however clearly express vimentin, and show high levels of MT1-MMP mRNA and protein. Consequently, they activate MMP-2 readily when stimulated with ConA. Selective expression of MT1-MMP by the more invasive prostate cancer cell lines was also reported by others.[81,82]

Squamous Carcinoma Cells

EMT events (E-cadherin loss and expression of vimentin) have also been observed in several human squamous carcinoma cells. This has been shown in a model of cervical cell lines generated by transfection of human papillomavirus type 33.[83] Only cell lines which expressed vimentin were highly invasive in the Boyden chamber assay, and these were found to synthesize MT1-MMP and were able to activate MMP-2 following ConA induction.[84]

In other established squamous cell carcinoma cell line systems, an EMT phenotype was also correlated to high levels of the transcription factor snail.[85] Furthermore, transfection of snail in vimentin-negative, E-cadherin-positive A431 cells resulted in induction of vimentin and MMP-2, and loss of E-cadherin.

From all these in vitro data, it is clear that MMP expression in epithelial cells associates with EMT processes. Among MMPs, the MT1-MMP/MMP-2 axis appear as a general pathway activated in most of the in vitro systems described above.

In Vivo Observations

Studying EMT in tumour biopsies remains a major challenge. Indeed, if one considers a "full" EMT as a dynamic phenomenon resulting in the acquisition of migratory and invasive properties by epithelial tumour cells, it is very likely to be discrete and affect only a small proportion of cells in the tumour mass. Also, if the dynamic nature of the EMT is clear, the sequence of events is still not well defined and may vary. Different phenotypes might thus be observed in a tumour mass. Accordingly, the expression of EMT genes such as E-cadherin and vimentin is very heterogenous within a given tumour.

Nevertheless, the existence of EMT phenomena in tumour biopsies is now well demonstrated. For instance, considering the two well known markers of EMT (E-cadherin reorganization and vimentin expression), numerous studies have reported the reorganization of E-cadherin complexes or vimentin expression in a variety of cancer types. Regarding MMPs, a mass of data have shown the implication of several MMPs in tumour progression (reviewed in refs. 21, 22). From these data, it is clear that the peritumoural stromal cells are the major source of MMPs in most cancers. However, even though it has rarely been studied in relationship to other EMT genes, the expression of MMPs in tumour cells has been also reported in many cancers.[86] This has particularly been shown for MT1-MMP in carcinomas including cervical,[87] lung,[88,89] prostate,[90] colorectal,[91] gastrointestinal,[92] oesophageal,[93,94] larynx,[95] breast,[96,97] oral,[98] ovarian,[99,100] head and neck,[101] hepatocellular and pancreatic[102] carcinomas.

Given the associations presented above from a number of in vitro model systems, it seems possible that the EMT in vivo may contribute to the expression of several MMPs in epithelial tumour cells and this results in tumours that do poorly. However, this is not proven, and the possibility remains that this is due to a more reactive stroma. It is interesting to note, however, the documentation in breast carcinomas of MMP-11 expression in metaplastic breast carcinoma cells which have undergone a degree of EMT-like changes.[103] The authors speculate that this may be related to the poorer prognosis associated with such carcinomas. Similarly, Trudel et al reported the expression of MMP-2 in both stromal and epithelial tumour cells of prostate cancer.[90] Importantly, MMP-2 expression by stromal cells was not associated with progression whereas MMP-2 expression by >50% of malignant epithelial cells was associated with decreased disease-free survival.[90] MMP-3 is also upregulated in a model of experimental mouse skin carcinogenesis, when squamous cell carcinomas progress to spindle cell carcinomas.[104]

While the inherent genetic deviations and implicit uniqueness of individual cancer systems make it difficult to draw general conclusions, the fact that so many different systems show EMT like capacities which are associated with increased migration, invasion, and metastasis does offer a degree of confidence in projecting an important role for this process in carcinoma progression.

MMPs in Developmental EMT Systems

Even more supportive is the parallel observations of MMP expression in EMT systems of normal development such as kidney morphogenesis modeled with MDCK cells and avian heart development, and studies on trophoblast invasion and wounded epithelia.

Kidney Morphogenesis and Disease

Madin Darby canine kidney (MDCK) epithelial cells are non tumourigenic cells and provide a well-accepted model for many aspects of epithelial cell biology. Under basal conditions in culture, they can polarize and differentiate. Following induction with different inducers including basic fibroblast growth factor (bFGF/FGF-2), hepatocyte growth factor (HGF), collagens and oncogenes (v-src and $erbB_2$), they can easily undergo dynamic EMT changes (see Fig. 4 for $erbB_2$ example). This often correlates with the loosening of cell-cell contacts, cell scattering and the capacity to invade 3D-collagen gel cultures and form branching tubules, thus mimicking a physiological renal morphogenesis rather than a pathological process. When induced by hepatocyte growth factor (HGF) to form branching tubules in collagen gels, an induction of MT1-MMP is observed[38,105] (see also Table 1). Furthermore, antisense RNA inhibition of the induced MT1-MMP abrogates the branching morphogenesis response to HGF. Also, TIMP-2 and BB-94, which inhibit both MMP-2 and MT1-MMP, block invasion or tubular formation in collagen gel. In contrast, TIMP-1, which inhibits MMP-2 but not MT1-MMP, fails to interfere with branching, implicating MT1-MMP rather than MMP-2 in the branching morphogenesis. Interestingly, we also see induced MT1-MMP after src transformation of MDCK cells,[106] and this is accompanied by overexpression of TIMP-1.[107] MT1-MMP induction is also seen after erbB2 transfection in MDCK cells (Fig. 5).[108] Although we have not examined vimentin expression in the MDCK cells undergoing branching morphogenesis, the *src* transformed cells show loss of cohesive cobblestone morphology in cell culture, and orientate as stellate single cells spread on the culture surface. They further acquire tumourigenesis in both the subcutus and kidney of athymic mice, and are metastatic to the lung when inoculated orthotopically.[106]

In vivo, and in organ culture systems, MMP-2 and MMP-9 have been demonstrated in conjunction with mouse renal tubulogenesis, however antibodies against MMP-9 only, and not MMP-2, blocked renal morphogenesis. [109] Although this study did not directly correlate this MMP expression with EMT, the EMT is known to be important in kidney morphogenesis. In contrast, MMP has been directly associated with the EMT changes seen in the tubular epithelium during kidney fibrosis associated with diabetic nephropathy.[110,111] Here, a surprising role of MMP-2 has emerged, being necessary and sufficient to induce the EMT in renal cell models.[112]

Wounded Bronchial Epithelium

Considerable attention has been paid to the MMP expression in a system where confluent human surface respiratory epithelial cells are wounded, and the cells undergo EMT-like changes (vimentin expression) to migrate into the naked culture surface. In this system, we see focal expression of MMP-9,[113,114] MMP-3 and MMP-11[115] in the cells undergoing migration at the wound edge, and this migration can be blocked with MMP-inhibitors.[114]

Avian Heart Development

EMT events have been well characterized in the transformation of endothelial cells lining the atrio-ventricular canal of avian heart to form progenitors of the septa and heart valves.[116] An induction of MT1-MMP and/orMMP-2 have been shown in EMT regions[117-120] together with an induction of Slug and c-*ets*-1, transcription factors known to be involved in EMT events and to regulate MMPs.[121,122]

Placental Implantation

Another area in which a well described EMT has been studied with respect to MMPs is the trophoblast implantation (reviewed in refs. 123, 124). Regarding MMP expression, MT1-MMP has been observed in invading trophoblasts in vivo[125] and both MMP-2 and MMP-9 in the trophoblastic columns.[126] In vitro, a clear requirement of MMP-9 for Matrigel invasion of cytotrophoblasts has also been demonstrated and correlated with the abundance of MMP-9 in the first trimester.[127]

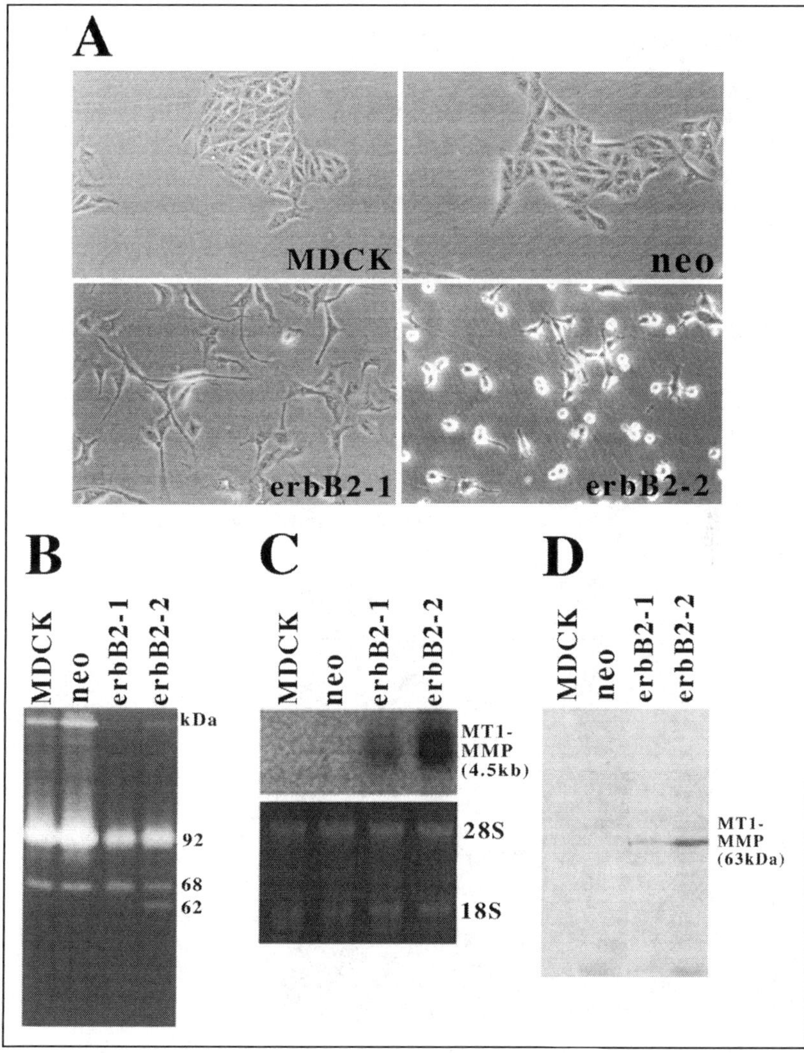

Figure 4. Induction of MT1-MMP expression by erbB2. A) Subconfluent cultures of MDCK cells (MDCK), MDCK cells transfected with control plasmid (neo), and two clones of MDCK cells transformed with erbB2 gene (erbB2-1, and erbB2-2, respectively) are shown. B) Gelatin zymography. Aliquots of culture supernatants from confluent MDCK cells (Lane MDCK), MDCK cells transfected with control plasmid (Lane neo), and two clones of erbB2-transformed cells (Lanes erbB2-1 and erbB2-2, respectively) were subjected to gelatin zymography analysis. C) Northern hybridization analysis. Total RNAs (10 μg) from MDCK cells (Lane MDCK), MDCK cells transfected with control plasmid (Lane neo) and two clones of erbB2-transformed cells (Lanes erbB2-1 and erbB2-2, respectively) were electrophoresed in a gel, stained with ethidium bromide (bottom), and then probed after Northern blotting with 32P-labeled erbB2 (top) or MT1-MMP cDNA fragments. D) Western analysis of MT1-MMP expression. MDCK cells (Lane MDCK), MDCK cells transfected with control plasmid (Lane neo), and two clones of erbB2-transformed cells (Lanes erbB2-1 and erbB2-2, respectively) were solubilized and subjected to SDS-PAGE. After blotting onto nitrocellulose filters, MT1-MMP protein was detected with a monoclonal antibody against MT1-MMP (113-5B-7).

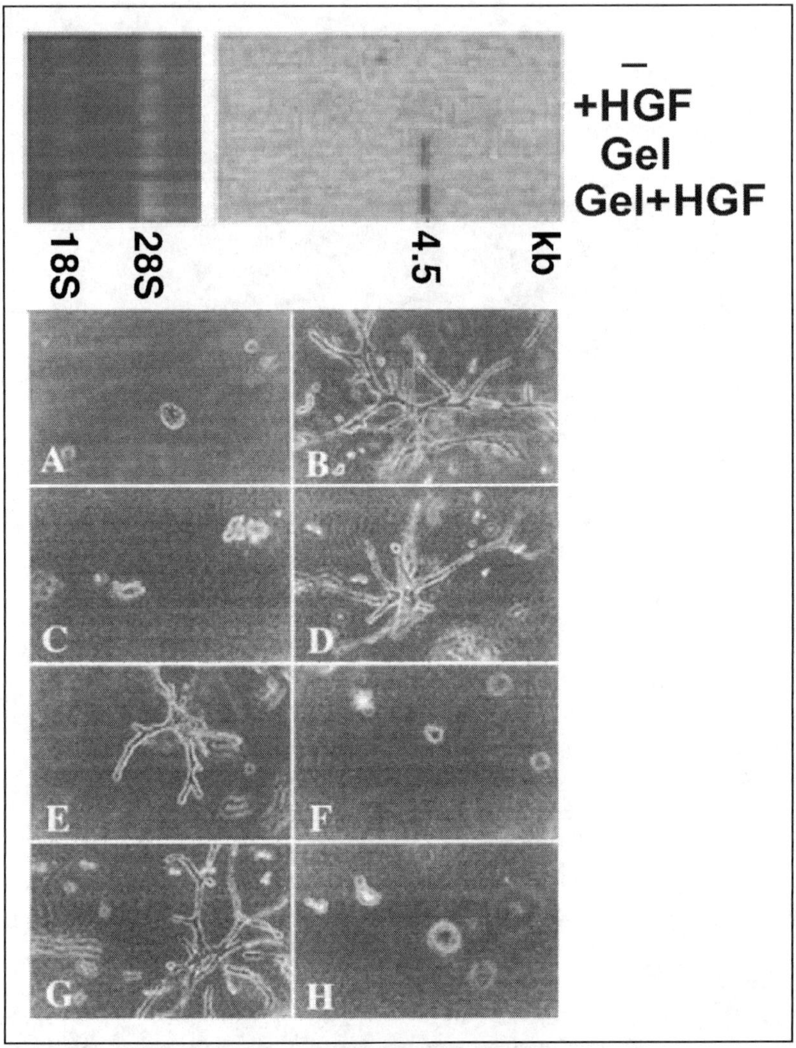

Figure 5. Upper panel) MT1-MMP induction in MDCK cells requires 3-dimensional gel. Lower Panel) Effect of MMP inhibitors on tubulogenesis. MDCK cells were cultured in collagen gel matrix for 7 days in the absence (panel A) or presence of 50 ng/ml HGF (panel B). MMP inhibitor BB-94 (1.0 x 10-7 M) was added to HGF-collagen gel culture (panel C). MDCK cells stably transfected with control plasmid, TIMP-1 or TIMP-2 gene were cultured in HGF-collagen gel as described above (panels D, E and F, respectively). Recombinant TIMP-1 (50 ng/ml) and TIMP-2 (50 ng/ml) were included in HGF-collagen gel culture of MDCK cells (panels G and H, respectively). Magnification = 100x.

Neural Crest Migration

The migration of neural crest cells in vertebrate embryogenesis is initiated by EMT of the dorsal most cells of the neural epithelium, and is perhaps the most intensively studied developmental EMT and cell migration[2] (see Chapter 3). This EMT is normally executed via stimulation by BMP-4 and crucially involves expression of the zinc-finger transcription factor genes Slug or Snail.[128] As in the heart, *c-ets-1* transcription factor is also expressed during the neural crest EMT.[129] Neural crest cells just prior to and during migration express genes for a number of extracellular proteases. MMP-8 is expressed in neural crest cells during EMT in mouse embryos.[130] Experiments in chicken embryos reveal the expression and importance of MMP-2 in these dynamic processes.[131] It is expressed when neural crest cells detach from the neural epithelium during EMT, but switched off as the cells separate. A similar scenario is seen in the sclerotome and in the dermis, where MMP-2 expression is seen when the EMT is initiated and maintained during the migration of the cells, but down-regulated once the cells cease movement. Interestingly, MMP-2-specific knockdown with morpholino antisense oligonecleotides, and global MMP- inhibition using BB-94, confirm that MMP-2 is required for neural crest EMT, but these do not inhibit migration if applied after the EMT. In contrast, MMP-2 inhibition during mesenchyme production from the somite blocks both EMT parameters and cell migration.[119] A full audit of MMP expression in relation to neural crest cell morphogenesis has not been undertaken, so the possibility of redundancy has not been addressed.

Regulation of MMPs by EMT-Associated Transcription Factors

If the data presented above clearly demonstrate that MMP expression is associated with EMT changes, several studies have also shown a direct regulation of MMPs by EMT-associated transcription factors. Thus, MMP-2 regulation by Snail has been demonstrated in squamous carcinoma cells.[85] Also, even though they do not necessarily establish a link with EMT processes, the regulation of MMPs by transcription factors of the ETS family in a variety of carcinoma cells has been clearly established (review in refs. 132, 133). Accumulating data also demonstrate the implication of the β-catenin co-transcriptional activity in the regulation of MMP. Thus MMP-7,[134-136] MT1-MMP,[137] MMP-26[138] are targets of the β-catenin/TCF pathway.

Summary and Conclusions

The EMT appears as a sequence of changes which can lead to the expression of migratory and invasive properties by epithelial cells. As shown in numerous tumoural and non tumoural in vitro systems, but also in vivo on tumour biopsies and on developmental models, the expression of MMPs (and particularly of MT1-MMP) is clearly part of EMT processes. Accordingly, the regulation of several MMPs by transcription factors (snail, ETS, β-catenin), known to regulate EMT pathways, have clearly been established. Consequently, MMPs are rather considered as target genes of EMT pathways and MMP expression as a late event of the EMT. This may be due, in part, to direct regulation of MMP gene transcription by the factors which drive EMT. Nevertheless, some MMPs (such as MMP-3) have been shown to initiate EMT changes. Also, several MMPs have been shown to be able to cleave E-cadherin, thereby inducing E-cadherin complex fragility and EMT changes. At present, the MMP field is undergoing a re-evaluation stage following the relatively poor results seen with clinical trials of MMP inhibitors.[139,140] One major factor often raised in this re-evaluation is our lack of precise knowledge regarding which specific MMPs are mediating pro-tumoural processes. Those associated with EMT are clearly contenders for this category, and may warrant specific targeting. A number of MMPs have been associated, as detailed above, with the EMT, however, MT1-MMP stands out in this regard. Although stromal cells are a major source of MMP in tumours, expression of MMPs (particularly MT1-MMP) by parenchymal cells is a clear EMT step which can ensure a pericellular proteolysis of basement membrane components but also other substrates, and thereby facilitate migration and invasion.

Acknowledgements

This work was supported by grants from the Fonds National de la Recherche Scientifique (FNRS, Belium), the Federation Belge Contre le Cancer, the Fortis Banque Assurances, the Interuniversity Attraction Poles Programme - Belgian Science Policy (Brussels, Belgium). CG is a Research Associate from the FNRS (Belgium).

Part of this work was also supported by grants from the US Department of Defence, Congressionally Directed Medical Research Program, Breast Cancer Research Section, grant number DAMD17-03-1-0416; the Victoriam Breast Cancer Research Consortium and an NIH Par 99-128: Insight Award to Stamp Out Breast Cancer.

References

1. Hay ED, Zuk A. Transformations between epithelium and mesenchyme: normal, pathological, and experimentally induced. Am J Kidney Dis 1995; 26(4):678-690.
2. Duband JL, Monier F, Delannet M et al. Epithelium-mesenchyme transition during neural crest development. Acta Anat 1995; 154(1):63-78.
3. Birchmeier C, Birchmeier W, Brand-Saberi B. Epithelial-mesenchymal transitions in cancer progression. Acta Anat 1996; 156(3):217-226.
4. Thiery JP. Epithelial-mesenchymal transitions in development and pathologies. Curr Opin Cell Biol 2003; 15(6):740-746.
5. Gotzmann J, Mikula M, Eger A et al. Molecular aspects of epithelial cell plasticity: implications for local tumour invasion and metastasis. Mutat Res 2004; 566(1):9-20.
6. Hay ED. An overview of epithelio-mesenchymal transformation. Acta Anat 1995; 154(1):8-20.
7. Viebahn C, Lane EB, Ramaekers FC. Keratin and vimentin expression in early organogenesis of the rabbit embryo. Cell Tissue Res 1988; 253(3):553-562.
8. Erickson CA, Tucker RP, Edwards BF. Changes in the distribution of intermediate-filament types in Japanese quail embryos during morphogenesis. Differentiation 1987; 34(2):88-97.
9. Boyer B, Valles AM, Edme N. Induction and regulation of epithelial-mesenchymal transitions. Biochem Pharmacol 2000; 60(8):1091-1099.
10. Reichert M, Muller T, Hunziker W. The PDZ domains of zonula occludens-1 induce an epithelial to mesenchymal transition of Madin-Darby canine kidney I cells. Evidence for a role of beta-catenin/ Tcf/Lef signaling. J Biol Chem 2000; 275(13):9492-9500.
11. Tucker GC, Boyer B, Gavrilovic J et al. Collagen-mediated dispersion of NBT-II rat bladder carcinoma cells. Cancer Res 1990; 50(1):129-137.
12. Lochter A, Galosy S, Muschler J et al. Matrix metalloproteinase stromelysin-1 triggers a cascade of molecular alterations that leads to stable epithelial-to-mesenchymal conversion and a premalignant phenotype in mammary epithelial cells. J Cell Biol 1997; 139:1861-1872.
13. Batlle E, Sancho E, Franci C et al. The transcription factor snail is a repressor of E-cadherin gene expression in epithelial tumour cells. Nat Cell Biol 2000; 2(2):84-89.
14. Cano A, Perez-Moreno MA, Rodrigo I et al. The transcription factor snail controls epithelial-mesenchymal transitions by repressing E-cadherin expression. Nat Cell Biol 2000; 2(2):76-83.
15. Savagner P, Yamada KM, Thiery JP. The zinc-finger protein slug causes desmosome dissociation, an initial and necessary step for growth factor-induced epithelial-mesenchymal transition. J Cell Biol 1997; 137(6):1403-1419.
16. Savagner P. Leaving the neighborhood: molecular mechanisms involved during epithelial-mesenchymal transition. Bioessays 2001; 23(10):912-923.
17. Nieto MA. The snail superfamily of zinc-finger transcription factors. Nat Rev Mol Cell Biol 2002; 3(3):155-166.
18. Oikawa T, Yamada T. Molecular biology of the Ets family of transcription factors. Gene 2003; 303:11-34.
19. Comijn J, Berx G, Vermassen P et al. The two-handed E box binding zinc finger protein SIP1 downregulates E-cadherin and induces invasion. Mol Cell 2001; 7(6):1267-1278.
20. Giles RH, van Es JH, Clevers H. Caught up in a Wnt storm: Wnt signaling in cancer. Biochim Biophys Acta 2003; 1653(1):1-24.
21. Sternlicht MD, Werb Z. How matrix metalloproteinases regulate cell behavior. Annu Rev Cell Dev Biol 2001; 17:463-516.
22. Egeblad M, Werb Z. New functions for the matrix metalloproteinases in cancer progression. Nat Rev Cancer 2002; 2(3):161-174.
23. Benaud C, Dickson RB, Thompson EW. Roles of the matrix metalloproteinases in mammary gland development and cancer. Breast Cancer Res Treat 1998; 50(2):97-116.
24. Lafleur MA, Handsley MM, Edwards DR. Metalloproteinases and their inhibitors in angiogenesis. Expert Rev Mol Med 2003; 5:1-39.
25. Noe V, Fingleton B, Jacobs K et al. Release of an invasion promoter E-cadherin fragment by matrilysin and stromelysin-1. J Cell Sci 2001; 114(Pt 1):111-118.
26. Kajita M, Itoh Y, Chiba T et al. Membrane-type 1 matrix metalloproteinase cleaves CD44 and promotes cell migration. J Cell Biol 2001; 153(5):893-904.
27. Deryugina EI, Ratnikov B, Monosov E et al. MT1-MMP initiates activation of pro-MMP-2 and integrin alphavbeta3 promotes maturation of MMP-2 in breast carcinoma cells. Exp Cell Res 2001; 263(2):209-223.
28. Codony-Servat J, Albanell J, Lopez-Talavera JC et al. Cleavage of the HER2 ectodomain is a pervanadate-activable process that is inhibited by the tissue inhibitor of metalloproteases-1 in breast cancer cells. Cancer Res 1999; 59(6):1196-1201.

29. Vecchi M, Rudolph-Owen LA, Brown CL et al. Tyrosine phosphorylation and proteolysis. Pervanadate-induced, metalloprotease-dependent cleavage of the ErbB-4 receptor and amphiregulin. J Biol Chem 1998; 273(32):20589-20595.
30. Nath D, Williamson NJ, Jarvis R et al. Shedding of c-Met is regulated by crosstalk between a G-protein coupled receptor and the EGF receptor and is mediated by a TIMP-3 sensitive metalloproteinase. J Cell Sci 2001; 114(Pt 6):1213-1220.
31. Barsky SH, Siegal GP, Jannotta F et al. Loss of basement membrane components by invasive tumours but not by their benign counterparts. Lab Invest 1983; 49(2):140-147.
32. Woodhouse EC, Chuaqui RF, Liotta LA. General mechanisms of metastasis. Cancer 1997; 80(8 Suppl):1529-1537.
33. Nelson AR, Fingleton B, Rothenberg ML et al. Matrix Metalloproteinases: Biologic Activity and Clinical Implications. J Clin Oncol 2000; 18(5):1135.
34. Zucker S, Pei D, Cao J et al. Membrane type-matrix metalloproteinases (MT-MMP). Curr Top Dev Biol 2003; 54:1-74.
35. English WR, Puente XS, Freije JM et al. Membrane type 4 matrix metalloproteinase (MMP17) has tumour necrosis factor-alpha convertase activity but does not activate pro-MMP2. J Biol Chem 2000; 275(19):14046-14055.
36. Sato H, Takino T, Okada Y et al. A matrix metalloproteinase expressed on the surface of invasive tumour cells. Nature 1994; 370(6484):61-65.
37. Seiki M, Mori H, Kajita M et al. Membrane-type 1 matrix metalloproteinase and cell migration. Biochem Soc Symp 2003(70):253-262.
38. Hotary K, Allen E, Punturieri A et al. Regulation of cell invasion and morphogenesis in a three-dimensional type I collagen matrix by membrane-type matrix metalloproteinases 1, 2, and 3. J Cell Biol 2000; 149(6):1309-1323.
39. Wilson CL, Matrisian LM. Matrilysin: an epithelial matrix metalloproteinase with potentially novel functions. Int J Biochem Cell Biol 1996; 28(2):123-136.
40. Almholt K, Johnsen M. Stromal cell involvement in cancer. Recent Results Cancer Res 2003; 162:31-42.
41. Janda E, Lehmann K, Killisch I et al. Ras and TGF[beta] cooperatively regulate epithelial cell plasticity and metastasis: dissection of Ras signaling pathways. J Cell Biol 2002; 156(2):299-313.
42. Nabeshima K, Inoue T, Shimao Y et al. Matrix metalloproteinases in tumour invasion: role for cell migration. Pathol Int 2002; 52(4):255-264.
43. Kulesa P, Ellies DL, Trainor PA. Comparative analysis of neural crest cell death, migration, and function during vertebrate embryogenesis. Dev Dyn 2004; 229(1):14-29.
44. Savagner P, Valles AM, Jouanneau J et al. Alternative splicing in fibroblast growth factor receptor 2 is associated with induced epithelial-mesenchymal transition in rat bladder carcinoma cells. Mol Biol Cell 1994; 5(8):851-862.
45. Gavrilovic J, Moens G, Thiery JP et al. Expression of transfected transforming growth factor alpha induces a motile fibroblast-like phenotype with extracellular matrix-degrading potential in a rat bladder carcinoma cell line. Cell Regul 1990; 1(13):1003-1014.
46. Jouanneau J, Gavrilovic J, Caruelle D et al. Secreted or nonsecreted forms of acidic fibroblast growth factor produced by transfected epithelial cells influence cell morphology, motility, and invasive potential. Proc Natl Acad Sci USA 1991; 88(7):2893-2897.
47. Bae SN, Arand G, Azzam H et al. Molecular and cellular analysis of basement membrane invasion by human breast cancer cells in Matrigel-based in vitro assays. Breast Cancer Res Treat 1993; 24(3):241-255.
48. Gilles CT. The epithelial to mesenchymal transition and metastatic progression in carcinoma. The Breast J 1996; 2(2):83-96.
49. Sommers CL, Thompson EW, Torri JA et al. Cell adhesion molecule uvomorulin expression in human breast cancer cell lines: relationship to morphology and invasive capacities. Cell Growth Differ 1991; 2(8):365-372.
50. Sommers CL, Heckford SE, Skerker JM et al. Loss of epithelial markers and acquisition of vimentin expression in adriamycin- and vinblastine-resistant human breast cancer cell lines. Cancer Res 1992; 52(19):5190-5197.
51. Thompson EW, Paik S, Brunner N et al. Association of increased basement membrane invasiveness with absence of estrogen receptor and expression of vimentin in human breast cancer cell lines. J Cell Physiol 1992; 150(3):534-544.
52. Sommers CL, Byers SW, Thompson EW et al. Differentiation state and invasiveness of human breast cancer cell lines. Breast Cancer Res Treat 1994; 31(2-3):325-335.
53. Nawrocki Raby B, Polette M, Gilles C et al. Quantitative cell dispersion analysis: new test to measure tumour cell aggressiveness. Int J Cancer 2001; 93(5):644-652.

54. Gilles C, Polette M, Birembaut P et al. Expression of c-ets-1 mRNA is associated with an invasive, EMT-derived phenotype in breast carcinoma cell lines. Clin Exp Metastasis 1997; 15(5):519-526.
55. Azzam HS, Arand G, Lippman ME et al. Association of MMP-2 activation potential with metastatic progression in human breast cancer cell lines independent of MMP-2 production. J Natl Cancer Inst 1993; 85(21):1758-1764.
56. Pulyaeva H, Bueno J, Polette M et al. MT1-MMP correlates with MMP-2 activation potential seen after epithelial to mesenchymal transition in human breast carcinoma cells [published erratum appears in Clin Exp Metastasis 1997; 15(3):338]. Clin Exp Metastasis 1997; 15(2):111-120.
57. Gilles C, Polette M, Seiki M et al. Implication of collagen type I-induced membrane-type 1-matrix metalloproteinase expression and matrix metalloproteinase-2 activation in the metastatic progression of breast carcinoma. Lab Invest 1997; 76:651-660.
58. Giambernardi TA, Grant GM, Taylor GP et al. Overview of matrix metalloproteinase expression in cultured human cells. Matrix Biol 1998; 16(8):483-496.
59. Grant GM, Giambernardi TA, Grant AM et al. Overview of expression of matrix metalloproteinases (MMP-17, MMP-18, and MMP-20) in cultured human cells. Matrix Biol 1999; 18(2):145-148.
60. Zajchowski DA, Bartholdi MF, Gong Y et al. Identification of gene expression profiles that predict the aggressive behavior of breast cancer cells. Cancer Res 2001; 61(13):5168-5178.
61. Jechlinger M, Grunert S, Tamir IH et al. Expression profiling of epithelial plasticity in tumour progression. Oncogene 2003; 22(46):7155-7169.
62. Thompson EW, Torri J, Sabol M et al. Oncogene-induced basement membrane invasiveness in human mammary epithelial cells. Clin Exp Metastasis 1994; 12(3):181-194.
63. Giunciuglio D, Culty M, Fassina G et al. Invasive phenotype of MCF10A cells overexpressing c-Ha-ras and c-erbB-2 oncogenes. Int J Cancer 1995; 63(6):815-822.
64. Gilles C, Polette M, Zahm J et al. Vimentin contributes to human mammary epithelial cell migration [In Process Citation]. J Cell Sci 1999; 112(Pt 24):4615-4625.
65. Gilles C, Polette M, Mestdagt M et al. Transactivation of vimentin by beta-catenin in human breast cancer cells. Cancer Res 2003; 63(10):2658-2664.
66. Gilles C, Polette M, Coraux C et al. Contribution of MT1-MMP and of human laminin-5 gamma2 chain degradation to mammary epithelial cell migration. J Cell Sci 2001; 114(Pt 16):2967-2976.
67. Sternlicht MD, Lochter A, Sympson CJ et al. The stromal proteinase MMP3/stromelysin-1 promotes mammary carcinogenesis. Cell 1999; 98(2):137-146.
68. Uria JA, Werb Z. Matrix metalloproteinases and their expression in mammary gland. Cell Res 1998; 8(3):187-194.
69. Witty JP, Wright JH, Matrisian LM. Matrix metalloproteinases are expressed during ductal and alveolar mammary morphogenesis, and misregulation of stromelysin-1 in transgenic mice induces unscheduled alveolar development. Mol Biol Cell 1995; 6(10):1287-1303.
70. Sternlicht MD, Bissell MJ, Werb Z. The matrix metalloproteinase stromelysin-1 acts as a natural mammary tumour promoter. Oncogene 2000; 19(8):1102-1113.
71. Lochter A, Srebrow A, Sympson CJ et al. Misregulation of stromelysin-1 expression in mouse mammary tumour cells accompanies acquisition of stromelysin-1-dependent invasive properties. J Biol Chem 1997; 272(8):5007-5015.
72. Lochter A, Werb Z, Bissell MJ. Transcriptional regulation of stromelysin-1 gene expression is altered during progression of mouse mammary epithelial cells from functionally normal to malignant. Matrix Biol 1999; 18(5):455-467.
73. Heppner KJ, Matrisian LM, Jensen RA et al. Expression of most matrix metalloproteinase family members in breast cancer represents a tumour-induced host response. Am J Pathol 1996; 149(1):273-282.
74. Miettinen PJ, Ebner R, Lopez AR et al. TGF-beta induced transdifferentiation of mammary epithelial cells to mesenchymal cells: involvement of type I receptors. J Cell Biol 1994; 127(6 Pt 2):2021-2036.
75. Piek E, Moustakas A, Kurisaki A et al. TGF-β type I receptor/ALK-5 and smad proteins mediate epithelial to mesenchymal transdifferentiation in NMuMG breast epithelial cells [In Process Citation]. J Cell Sci 1999; 112(Pt 24):4557-4568.
76. Lelekakis M, Moseley JM, Martin TJ et al. A novel orthotopic model of breast cancer metastasis to bone. Clin Exp Metastasis 1999; 17(2):163-170.
77. Tester AM, Ruangpanit N, Anderson RL et al. MMP-9 secretion and MMP-2 activation distinguish invasive and metastatic sublines of a mouse mammary carcinoma system showing epithelial-mesenchymal transition traits. Clin Exp Metastasis 2000; 18(7):553-560.
78. Martorana AM, Zheng G, Crowe TC et al. Epithelial cells up-regulate matrix metalloproteinases in cells within the same mammary carcinoma that have undergone an epithelial-mesenchymal transition. Cancer Res 1998; 58(21):4970-4979.

79. Polette M, Gilles C, de Bentzmann S et al. Association of fibroblastoid features with the invasive phenotype in human bronchial cancer cell lines. Clin Exp Metastasis 1998; 16(2):105-112.
80. Nawrocki-Raby B, Gilles C, Polette M et al. E-Cadherin mediates MMP down-regulation in highly invasive bronchial tumour cells. Am J Pathol 2003; 163(2):653-661.
81. Jung M, Romer A, Keyszer G et al. mRNA expression of the five membrane-type matrix metalloproteinases MT1-MT5 in human prostatic cell lines and their down-regulation in human malignant prostatic tissue. Prostate 2003; 55(2):89-98.
82. Daja MM, Niu X, Zhao Z et al. Characterization of expression of matrix metalloproteinases and tissue inhibitors of metalloproteinases in prostate cancer cell lines. Prostate Cancer Prostatic Dis 2003; 6(1):15-26.
83. Gilles C, Piette J, Peter W et al. Differentiation ability and oncogenic potential of HPV-33- and HPV-33 + ras-transfected keratinocytes. Int J Cancer 1994; 58(6):847-854.
84. Gilles C, Polette M, Piette J et al. Epithelial-to-mesenchymal transition in HPV-33-transfected cervical keratinocytes is associated with increased invasiveness and expression of gelatinase A. Int J Cancer 1994; 59(5):661-666.
85. Yokoyama K, Kamata N, Fujimoto R et al. Increased invasion and matrix metalloproteinase-2 expression by Snail-induced mesenchymal transition in squamous cell carcinomas. Int J Oncol Apr 2003; 22(4):891-898.
86. Polette M, Nawrocki-Raby B, Gilles C et al.Tumour Invasion and Matrix Metalloproteases. 2004; In press.
87. Gilles C, Polette M, Piette J et al. High level of MT-MMP expression is associated with invasiveness of cervical cancer cells. Int J Cancer 1996; 65(2):209-213.
88. Polette M, Nawrocki B, Gilles C et al. MT-MMP expression and localisation in human lung and breast cancers. Virchows Arch 1996; 428(1):29-35.
89. Yamamura T, Nakanishi K, Hiroi S et al. Expression of membrane-type-1-matrix metalloproteinase and metalloproteinase-2 in nonsmall cell lung carcinomas. Lung Cancer 2002; 35(3):249-255.
90. Trudel D, Fradet Y, Meyer F et al. Significance of MMP-2 expression in prostate cancer: an immunohistochemical study. Cancer Res 2003; 63(23):8511-8515.
91. Kikuchi R, Noguchi T, Takeno S et al. Immunohistochemical detection of membrane-type-1-matrix metalloproteinase in colorectal carcinoma. Br J Cancer 2000; 83(2):215-218.
92. Ohtani H, Motohashi H, Sato H et al. Dual over-expression pattern of membrane-type metalloproteinase-1 in cancer and stromal cells in human gastrointestinal carcinoma revealed by in situ hybridization and immunoelectron microscopy. Int J Cancer 1996; 68(5):565-570.
93. Etoh T, Inoue H, Yoshikawa Y et al. Increased expression of collagenase-3 (MMP-13) and MT1-MMP in oesophageal cancer is related to cancer aggressiveness. Gut 2000; 47(1):50-56.
94. Ohashi K, Nemoto T, Nakamura K et al. Increased expression of matrix metalloproteinase 7 and 9 and membrane type 1-matrix metalloproteinase in esophageal squamous cell carcinomas. Cancer 2000; 88(10):2201-2209.
95. Cazorla M, Hernandez L, Nadal A et al. Collagenase-3 expression is associated with advanced local invasion in human squamous cell carcinomas of the larynx. J Pathol 1998; 186(2):144-150.
96. Ueno H, Nakamura H, Inoue M et al. Expression and tissue localization of membrane-types 1, 2, and 3 matrix metalloproteinases in human invasive breast carcinomas. Cancer Res 1997; 57(10):2055-2060.
97. Ishigaki S, Toi M, Ueno T et al. Significance of membrane type 1 matrix metalloproteinase expression in breast cancer. Jpn J Cancer Res 1999; 90(5):516-522.
98. Nakamura H, Ueno H, Yamashita K et al. Enhanced production and activation of progelatinase A mediated by membrane-type 1 matrix metalloproteinase in human papillary thyroid carcinomas. Cancer Res 1999; 59(2):467-473.
99. Davidson B, Goldberg I, Gotlieb WH et al. High levels of MMP-2, MMP-9, MT1-MMP and TIMP-2 mRNA correlate with poor survival in ovarian carcinoma. Clin Exp Metastasis 1999; 17(10):799-808.
100. Davidson B, Goldberg I, Berner A et al. Expression of membrane-type 1, 2, and 3 matrix metalloproteinases messenger RNA in ovarian carcinoma cells in serous effusions. Am J Clin Pathol 2001; 115(4):517-524.
101. Imanishi Y, Fujii M, Tokumaru Y et al. Clinical significance of expression of membrane type 1 matrix metalloproteinase and matrix metalloproteinase-2 in human head and neck squamous cell carcinoma. Hum Pathol 2000; 31(8):895-904.
102. Maatta M, Soini Y, Liakka A et al. Differential expression of matrix metalloproteinase (MMP)-2, MMP-9, and membrane type 1-MMP in hepatocellular and pancreatic adenocarcinoma: implications for tumour progression and clinical prognosis. Clin Cancer Res 2000; 6(7):2726-2734.

103. Ahmad A, Hanby A, Dublin E et al. Stromelysin 3: an independent prognostic factor for relapse-free survival in node-positive breast cancer and demonstration of novel breast carcinoma cell expression. Am J Pathol 1998; 152(3):721-728.
104. Wright JH, McDonnell S, Portella G, et al. A switch from stromal to tumour cell expression of stromelysin-1 mrna associated with the conversion of squamous to spindle carcinomas during mouse skin tumour progression. Mol Carcinog 1994; 10(4):207-215.
105. Kadono Y, Shibahara K, Namiki M et al. Membrane type 1-matrix metalloproteinase is involved in the formation of hepatocyte growth factor/scatter factor-induced branching tubules in madin-darby canine kidney epithelial cells. Biochem Biophys Res Commun 1998; 251(3):681-687.
106. Kadono Y, Okada Y, Namiki M et al. Transformation of epithelial Madin-Darby canine kidney cells with p60(v-src) induces expression of membrane-type 1 matrix metalloproteinase and invasiveness. Cancer Res 1998; 58(10):2240-2244.
107. Noritake H, Miyamori H, Goto C et al. Overexpression of tissue inhibitor of matrix metalloproteinases-1 (TIMP-1) in metastatic MDCK cells transformed by v-src. Clin Exp Metastasis. 1999; 17(2):105-110.
108. Miyamori H, Hasegawa K, Kim KR et al. Expression of metastasis-associated mts1 gene is co-induced with membrane type-1 matrix metalloproteinase (MT1-MMP) during oncogenic transformation and tubular formation of Madin Darby canine kidney (MDCK) epithelial cells. Clin Exp Metastasis 2000; 18(1):51-56.
109. Lelongt B, Trugnan G, Murphy G et al. Matrix metalloproteinases MMP2 and MMP9 are produced in early stages of kidney morphogenesis but only MMP9 is required for renal organogenesis in vitro. J Cell Biol 1997; 136(6):1363-1373.
110. Li Y, Yang J, Dai C et al. Role for integrin-linked kinase in mediating tubular epithelial to mesenchymal transition and renal interstitial fibrogenesis. J Clin Invest 2003; 112(4):503-516.
111. Kalluri R, Neilson EG. Epithelial-mesenchymal transition and its implications for fibrosis. J Clin Invest 2003; 112(12):1776-1784.
112. Cheng S, Lovett DH. Gelatinase A (MMP-2) is necessary and sufficient for renal tubular cell epithelial-mesenchymal transformation. Am J Pathol 2003; 162(6):1937-1949.
113. Buisson AC, Zahm JM, Polette M et al. Gelatinase B is involved in the in vitro wound repair of human respiratory epithelium. J Cell Physiol 1996; 166(2):413-426.
114. Legrand C, Gilles C, Zahm JM et al. Airway epithelial cell migration dynamics. MMP-9 role in cell-extracellular matrix remodeling. J Cell Biol 1999; 146(2):517-529.
115. Buisson AC, Gilles C, Polette M et al. Wound repair-induced expression of a stromelysins is associated with the acquisition of a mesenchymal phenotype in human respiratory epithelial cells. Lab Invest 1996; 74(3):658-669.
116. Markwald R, Eisenberg C, Eisenberg L et al. Epithelial-mesenchymal transformations in early avian heart development. Acta Anat 1996; 156(3):173-186.
117. Alexander SM, Jackson KJ, Bushnell KM et al. Spatial and temporal expression of the 72-kDa type IV collagenase (MMP-2) correlates with development and differentiation of valves in the embryonic avian heart. Dev Dyn 1997; 209(3):261-268.
118. Cai DH, Vollberg TM Sr., Hahn-Dantona E et al. MMP-2 expression during early avian cardiac and neural crest morphogenesis. Anat Rec 2000; 259(2):168-179.
119. Duong TD, Erickson CA. MMP-2 plays an essential role in producing epithelial-mesenchymal transformations in the avian embryo. Dev Dyn 2004; 229(1):42-53.
120. Song W, Jackson K, McGuire PG. Degradation of type IV collagen by matrix metalloproteinases is an important step in the epithelial-mesenchymal transformation of the endocardial cushions. Dev Biol 2000; 227(2):606-617.
121. Macias D, Perez-Pomares JM, Garcia-Garrido L et al. Immunoreactivity of the ets-1 transcription factor correlates with areas of epithelial-mesenchymal transition in the developing avian heart. Anat Embryol (Berl) 1998; 198(4):307-315.
122. Carmona R, Gonzalez-Iriarte M, Macias D et al. Immunolocalization of the transcription factor Slug in the developing avian heart. Anat Embryol (Berl) 2000; 201(2):103-109.
123. Vicovac L, Aplin JD. Epithelial-mesenchymal transition during trophoblast differentiation. Acta Anat 1996; 156(3):202-216.
124. Nawrocki B PM, Maquoi E, Birembaut P. Expression of matrix metalloproteinases and their inhibitors during human placental development. Trophoblast Research 1997; 10:97-113.
125. Nawrocki B, Polette M, Marchand V et al. Membrane-type matrix metalloproteinase-1 expression at the site of human placentation. Placenta 1996; 17(8):565-572.
126. Polette M, Nawrocki B, Pintiaux A et al. Expression of gelatinases A and B and their tissue inhibitors by cells of early and term human placenta and gestational endometrium. Lab Invest 1994; 71(6):838-846.

127. Librach CL, Werb Z, Fitzgerald ML et al. 92-kD type IV collagenase mediates invasion of human cytotrophoblasts. J Cell Biol 1991; 113(2):437-449.
128. del Barrio MG, Nieto MA. Overexpression of Snail family members highlights their ability to promote chick neural crest formation. Development 2002; 129(7):1583-1593.
129. Fafeur V, Tulasne D, Queva C et al. The ETS1 transcription factor is expressed during epithelial-mesenchymal transitions in the chick embryo and is activated in scatter factor-stimulated MDCK epithelial cells. Cell Growth Differ 1997; 8(6):655-665.
130. Giambernardi TA, Sakaguchi AY, Gluhak J et al. Neutrophil collagenase (MMP-8) is expressed during early development in neural crest cells as well as in adult melanoma cells. Matrix Biol 2001; 20(8):577-587.
131. Cai DH, Brauer PR. Synthetic matrix metalloproteinase inhibitor decreases early cardiac neural crest migration in chicken embryos. Dev Dyn 2002; 224(4):441-449.
132. Westermarck J, Kahari VM. Regulation of matrix metalloproteinase expression in tumour invasion. Faseb J May 1999; 13(8):781-792.
133. Singh S, Barrett J, Sakata K et al. ETS proteins and MMPs: partners in invasion and metastasis. Curr Drug Targets 2002; 3(5):359-367.
134. Brabletz T, Jung A, Dag S et al. beta-catenin regulates the expression of the matrix metalloproteinase-7 in human colorectal cancer. Am J Pathol 1999; 155(4):1033-1038.
135. Crawford HC, Fingleton BM, Rudolph-Owen LA et al. The metalloproteinase matrilysin is a target of beta-catenin transactivation in intestinal tumours. Oncogene 1999; 18(18):2883-2891.
136. Crawford HC, Fingleton B, Gustavson MD et al. The PEA3 subfamily of Ets transcription factors synergizes with beta-catenin-LEF-1 to activate matrilysin transcription in intestinal tumours. Mol Cell Biol 2001; 21(4):1370-1383.
137. Takahashi M, Tsunoda T, Seiki M et al. Identification of membrane-type matrix metalloproteinase-1 as a target of the beta-catenin/Tcf4 complex in human colorectal cancers. Oncogene 2002; 21(38):5861-5867.
138. Marchenko GN, Marchenko ND, Leng J et al. Promoter characterization of the novel human matrix metalloproteinase-26 gene: regulation by the T-cell factor-4 implies specific expression of the gene in cancer cells of epithelial origin. Biochem J 2002; 363(Pt 2):253-262.
139. Coussens LM, Fingleton B, Matrisian LM. Matrix metalloproteinase inhibitors and cancer: trials and tribulations. Science 2002; 295(5564):2387-2392.
140. Overall CM, Lopez-Otin C. Strategies for MMP inhibition in cancer: innovations for the post-trial era. Nat Rev Cancer 2002; 2(9):657-672.

Index

D

E

F

G

H

I

J

K

L

M

N

P

R

S